THE ROUTLEDGE HANDBOOK OF LOGICAL EMPIRICISM

Logical empiricism is a philosophical movement that flourished in the 1920s and 30s in Central Europe and in the 1940s and 50s in the United States. With its stated ambition to comprehend the revolutionary advances in the empirical and formal sciences of their day and to confront anti-modernist challenges to scientific reason itself, logical empiricism was never uncontroversial. Uniting key thinkers who often disagreed with one another but shared the aim to conceive of philosophy as part of the scientific enterprise, it left a rich and varied legacy that has only begun to be explored relatively recently.

The Routledge Handbook of Logical Empiricism is an outstanding reference source to this challenging subject area, and the first collection of its kind. Comprising 41 chapters written by an international and interdisciplinary team of contributors, the *Handbook* is organized into four clear parts:

- The Cultural, Scientific and Philosophical Context and the Development of Logical Empiricism
- Characteristic Theses of and Specific Issues in Logical Empiricism
- Relations to Philosophical Contemporaries
- Leading Post-Positivist Criticisms and Legacy

Essential reading for students and researchers in the history of twentieth-century philosophy, especially the history of analytical philosophy and the history of philosophy of science, the *Handbook* will also be of interest to those working in related areas of philosophy influenced by this important movement, including metaphysics and epistemology, philosophy of mind and philosophy of language.

Thomas Uebel is Professor Emeritus of Philosophy at the University of Manchester, UK. His books include *Empiricism at the Crossroads: The Vienna Circle's Protocol-Sentence Debate* (2007) and an edition of Neurath's *Economic Writings* (2004).

Christoph Limbeck-Lilienau is a Research Fellow at the University of Toronto, Canada, and principal investigator in the project *Naturalizing Meaning* (2021–24, FWF-grant J-4502-G). He is the author of *Der Wiener Kreis* (with F. Stadler, 2015) and editor of *The Philosophy of Perception* (2019).

ROUTLEDGE HANDBOOKS IN PHILOSOPHY

Routledge Handbooks in Philosophy are state-of-the-art surveys of emerging, newly refreshed, and important fields in philosophy, providing accessible yet thorough assessments of key problems, themes, thinkers, and recent developments in research.

All chapters for each volume are specially commissioned, and written by leading scholars in the field. Carefully edited and organized, *Routledge Handbooks in Philosophy* provide indispensable reference tools for students and researchers seeking a comprehensive overview of new and exciting topics in philosophy. They are also valuable teaching resources as accompaniments to textbooks, anthologies, and research-orientated publications.

Also available:

THE ROUTLEDGE HANDBOOK OF PHILOSOPHY AND IMPROVISATION IN THE ARTS
Edited by Alessandro Bertinetto and Marcello Ruta

THE ROUTLEDGE HANDBOOK OF IDEALISM AND IMMATERIALISM
Edited by Joshua Farris and Benedikt Paul Göcke

THE ROUTLEDGE HANDBOOK OF PHILOSOPHY OF ECONOMICS
Edited by Conrad Heilmann and Julian Reiss

THE ROUTLEDGE HANDBOOK OF LOGICAL EMPIRICISM
Edited by Thomas Uebel and Christoph Limbeck-Lilienau

THE ROUTLEDGE HANDBOOK OF PHILOSOPHY OF AGENCY
Edited by Luca Ferrero

THE ROUTLEDGE HANDBOOK OF PROPOSITIONS
Edited by Adam Russell Murray and Chris Tillman

For more information about this series, please visit: www.routledge.com/Routledge-Handbooks-in-Philosophy/book-series/RHP

THE ROUTLEDGE HANDBOOK OF LOGICAL EMPIRICISM

Edited by Thomas Uebel and Christoph Limbeck-Lilienau

Routledge
Taylor & Francis Group
LONDON AND NEW YORK

Cover credit: © Getty Images

First published 2022
by Routledge
2 Park Square, Milton Park, Abingdon, Oxon OX14 4RN

and by Routledge
605 Third Avenue, New York, NY 10158

Routledge is an imprint of the Taylor & Francis Group, an informa business

British Library Cataloguing-in-Publication Data
A catalogue record for this book is available from the British Library

Library of Congress Cataloging-in-Publication Data
Names: Uebel, Thomas, editor. | Limbeck-Lilienau, Christoph, editor.
Title: The Routledge handbook of logical empiricism / edited by Thomas Uebel and Christoph Limbeck-Lilienau.
Description: Abingdon, Oxon ; New York, NY : Routledge, 2022. | Series: Routledge handbooks in philosophy | Includes bibliographical references and index.
Identifiers: LCCN 2021034655 (print) | LCCN 2021034656 (ebook) | ISBN 9781138122000 (hbk.) | ISBN 9780367610036 (pbk.) | ISBN 9781315650647 (ebk.)
Subjects: LCSH: Logical positivism.
Classification: LCC B824.6 .R68 2022 (print) | LCC B824.6 (ebook) | DDC 146/.42—dc23
LC record available at https://lccn.loc.gov/2021034655
LC ebook record available at https://lccn.loc.gov/2021034656

ISBN: 978-1-138-12200-0 (hbk)
ISBN: 978-0-367-61003-6 (pbk)
ISBN: 978-1-315-65064-7 (ebk)

DOI: 10.4324/9781315650647

Typeset in Bembo
by Apex CoVantage, LLC

CONTENTS

NOTES ON CONTRIBUTORS

Anastasios Brenner is Professor of Philosophy at Université Paul-Valéry Montpellier 3, France. He is a specialist of French philosophy of science and has explored the origins and development of this tradition as well as its interaction with logical empiricism, namely in *Les origines françaises de la philosophie des sciences* (PUF, 2003). He has also examined the issue of rational values: *Raison scientifique et valeurs humaines* (PUF, 2011), and edited *French Studies in the Philosophy of Science* (with J. Gayon, Springer, 2009).

A.W. Carus gained his PhD in Philosophy at the University of Chicago under the supervision of Howard Stein and currently teaches at the Munich Center for Mathematical Philosophy (Ludwig-Maximilians-Universität Munich). He is the author of papers on Carnap, including a series of papers with Steve Awodey on Carnap and Gödel, of *Carnap and Twentieth-Century Thought: Explication as Enlightenment*, and co-editor of Carnap's *Collected Works, Vol. 1* (*Early Writings*).

Jordi Cat is Associate Professor at the Department of History and Philosophy of Science and Medicine, Indiana University, Bloomington. His research interests include philosophy, history, and history of philosophy of science. He is editor of *Neurath Reconsidered* (with A. Tuboly, Springer, 2019), and author of *Fuzzy Pictures as Philosophical Problem and Scientific Practice* (Springer, 2016), *Maxwell, Sutton and the Birth of Color Photography* (Palgrave 2013), and *Otto Neurath: Philosophy between Science and Politics* (with N. Cartwright, L. Fleck and T. Uebel, CUP 1996).

Sean Crawford is Lecturer in Philosophy at the University of Manchester. His research concerns topics in the philosophy of mind and language in both historical and contemporary manifestations. His publications include essays on the explanation of action, the nature of perceptual and object-dependent thought, propositions and propositional attitudes, as well as discussions of the philosophy of psychology of the logical positivists.

Richard Creath is President's Professor of Life Sciences and of Philosophy and Director of the Program in History and Philosophy of Science at Arizona State University. He is the editor of several books and the author of many papers in philosophy of science and on Rudolf Carnap and W.V. Quine. He is also General Editor of *The Collected Works of Rudolf Carnap* (OUP).

Hans-Joachim Dahms is a Research Fellow at the Institute Vienna Circle, now retired. His numerous publications on the philosophy and history of the sciences and institutional university history include *Positivismusstreit* (Suhrkamp, 1994) and co-editorship of *Die Universität Göttingen unter dem Nationalsozialismus* (1987, 2nd enl. ed, 1998). He is currently finishing a book on the movement of *Neue Sachlichkeit* in Germany in the 1920s.

Christian Damböck is Privatdozent, Department of Philosophy, and Research Fellow, Institute Vienna Circle, University of Vienna. He is author and editor of papers and books on the philosophy of science and the humanities in the nineteenth and twentieth centuries; moral noncognitivism and theories of democracy; philosophy of logic; the philosophies of Wilhelm Dilthey, Hermann Cohen, Richard Avenarius, and Ernst Mach. Currently, he is working on an edition of Rudolf Carnap's diaries and scientific correspondence.

William Demopoulos was Professor of Philosophy at the University of Western Ontario and held numerous visiting professorships and fellowships. He was the editor of the *Western Ontario Series in Philosophy of Science* (Springer) and *Frege's Philosophy of Mathematics* (Harvard, 1995), and the author of a great number of papers, some collected in *Logicism and its Philosophical Legacy* (Cambridge, 2013). He is also the author of the posthumous work *On Theories. Logical Empiricism and the Methodology of Modern Physics* (Harvard, 2022). He died in 2017 while this volume was in preparation: his was the first contribution received.

Fynn Ole Engler is Research Associate at the University of Rostock and Visiting Scholar at the Max Planck Institute for the History of Science in Berlin. He is an editor of volumes of the Moritz Schlick Edition and author of *Gespaltene Vernunft* (with Jürgen Renn, Matthes & Seitz, 2018). His research focuses on the history and philosophy of science in the twentieth century, in particular on the philosophy of Albert Einstein.

Massimo Ferrari is Professor of History of Philosophy at the University of Turin. His research is devoted to neo-Kantianism, logical empiricism, pragmatism, and the history of philosophy of science; a central concern is Moritz Schlick's intellectual biography. Among others works, he is author of *Ernst Cassirer: Stationen einer philosophischen Biographie* (Meiner, 2003).

Johannes Friedl is Assistant Professor at the University of Graz, with previous appointments at the University of Vienna and at the *Forschungsstelle und Dokumentationszentrum für österreichische Philosophie* at Graz. Co-editor of two volumes of the *Moritz-Schlick-Gesamtausgabe* and author of *Konsequenter Empirismus: Die Entwicklung von Moritz Schlicks Erkenntnistheorie im Wiener Kreis* (Springer, 2013) and papers on the Vienna Circle, he is leading a research project on the correspondence between Carnap and Neurath.

Michael Heidelberger is Professor Emeritus at the University of Tübingen, where he held the chair for Logic and Philosophy of Science. He specializes in the history of the philosophy of science, mainly of the late nineteenth and early twentieth centuries, and focuses on philosophy and history of physics and psychology. He worked on Gustav Theodor Fechner, Hermann von Helmholtz, and Ernst Mach. He is co-editor of the philosophical works of Helmholtz (3 vols., Meiner, 2017).

Gürol Irzik is Professor of Philosophy at Sabanci University. Previously he taught at Bogazici, Duke, and Auckland Universities. He was also a Visiting Fellow at the Center for Philosophy of Science, University of Pittsburgh. Having published widely on the philosophy of science, he is currently working on justice in the production of knowledge and public trust in science.

Frederique Janssen-Lauret is Lecturer in Philosophy at the University of Manchester. She specializes in philosophical logic and history of analytic philosophy, especially female logicians. She is co-translator of Quine's *The Significance of the New Logic* (CUP, 2018) and editor of *Quine: Structure and Ontology* (OUP, 2020). She has published papers in *Synthese*, *The Monist*, *Australasian Journal of Philosophy*, *British Journal for the History of Philosophy*, and in volumes published by OUP, CUP, Springer, and Palgrave.

James "Jack" Justus is Professor of Philosophy at Florida State University. His research interests include, besides history of analytic philosophy (especially Carnap and logical empiricism), philosophy of science (especially biology), environmental philosophy, formal epistemology, metaphilosophy, and philosophy of mathematics. He has published in numerous philosophical and scientific journals and, most recently, *Philosophy of Ecology: An Introduction* (CUP, 2021).

Artur Koterski is Associate Professor at Maria Curie-Sklodowska University in Lublin, Poland. His areas of research are history of philosophy of science and methodology, particularly focused on the inter-war period. He authored two books on the history of the problem of demarcation: *Verificational Criteria of Demarcation in the Vienna Circle's Philosophy of Science* and *Falsificationist Criteria of Demarcation in the 20th-Century Philosophy of Science* (both in Polish); he also edited and translated various works by logical empiricists.

Gregory Lavers is Associate Professor in the Department of Philosophy at Concordia University, where he has taught since 2005. His research and publications focus on issues at the intersection of philosophy of mathematics, philosophy of logic and the history of analytic philosophy.

Christoph Limbeck-Lilienau is a Research Fellow at the University of Toronto and principal investigator in the project *Naturalizing Meaning* (2021–24, FWF-grant J 4502-G). Previously, he was Research Fellow at the Institute Vienna Circle and Lecturer at the Department of Philosophy of the University of Vienna. He published on Carnap, the Vienna Circle, pragmatism and the philosophy of perception. He is the author of *Der Wiener Kreis* (with F. Stadler, LIT, 2015) and co-editor of *The Philosophy of Perception* (De Gruyter, 2019).

Nikolay Milkov is Professor at the University of Paderborn. He had visiting positions in Oxford, Pittsburgh, McMaster University and St. Petersburg. Milkov authored *A Hundred Years of English Philosophy* (Springer, 2003) and *Early Analytic Philosophy and the German Philosophical Traditions* (Bloomsbury, 2020). His edited works include *The Berlin Group and the Philosophy of Logical Empiricism* (with V. Peckhaus, Springer, 2012), Hans Reichenbach, *Ziele und Wege der heutigen Naturphilosophie* (Meiner, 2011), and *Die Berliner Gruppe: Texte zum Logischen Empirismus* (Meiner, 2015).

Elisabeth Nemeth is Professor of Philosophy, now retired, at the University of Vienna. Currently she is President of the Austrian Ludwig Wittgenstein Society and co-editor of the journal *HOPOS*. Recent publications include "Kants Erkenntnisvermögen als historische und soziologische Kategorie. Ein Versuch zu Edgar Zilsel und Michael Friedman," in *Natur und Freiheit*

(M. L. Waibel, M. Ruffing and D. Wagner, eds, de Gruyter, 2018), and *Edgar Zilsel: Philosopher, Historian, Sociologist* (D. Romizi, M. Wulz, E. Nemeth, eds., Springer, 2021).

Matthias Neuber is Privatdozent in Philosophy at the University of Mainz. Previously Academic Visitor at the University of Helsinki and the Center for Philosophy of Science at the University of Pittsburgh, he works on Kant, neo-Kantianism, logical empiricism, and early twentieth-century American realism. His publications include *Die Grenzen des Revisionismus: Schlick, Cassirer und das 'Raumproblem'* (Springer, 2012) and *Der Realismus im logischen Empirismus: Eine Studie zur Geschichte der Wissenschaftsphilosophie* (Springer, 2018) and edited volumes.

Ilkka Niiniluoto is Professor Emeritus of Theoretical Philosophy at the University of Helsinki. His academic career in this university includes: Master of Science (1968), Doctor of Philosophy (1974), Associate Professor of Mathematics (1973–1977), Professor of Theoretical Philosophy (1977–2014), Rector (2003–2008), and Chancellor (2008–2013). He was President of the Philosophical Society of Finland in 1975–2015. Niiniluoto has worked in philosophical logic, philosophy of science, epistemology, and philosophy of technology. His main works include *Is Science Progressive?* (1984), *Truthlikeness* (Kluwer, 1987), *Critical Scientific Realism* (OUP, 1999), and *Truth-Seeking by Abduction* (Springer, 2017).

John O'Neill is Professor of Political Economy at the University of Manchester. He has written widely on philosophy, political economy and the environment, including a number of papers and chapters on Neurath's philosophy and political economy. His books include *Markets, Deliberation and Environment* (Routledge, 2007), *The Market: Ethics, Knowledge and Politics* (Routledge, 1998), and *Ecology, Policy and Politics: Human Well-Being and the Natural World* (Routledge, 1993). He is co-author of *Environmental Values* (with A. Holland and A. Light, Routledge, 2008).

Flavia Padovani is Associate Professor of Philosophy at Drexel University. Her research addresses issues in both history and philosophy of science and general philosophy of science. She published in *Synthese*, *Studies in History and Philosophy of Modern Physics*, *Studies in History and Philosophy of Science* and *European Studies in Philosophy of Science*, and co-edited various volumes such as *Objectivity in Science: New Perspectives from Science and Technology Studies* (Springer, 2015) and a forthcoming *Synthese* special issue, *All Things Reichenbach*.

Christopher Pincock is Professor of Philosophy, Ohio State University. His research focuses on the philosophy of science, the philosophy of mathematics, and the history of analytic philosophy. He is the author of the book *Mathematics and Scientific Representation* (OUP, 2012) and co-editor of *Philosophy of Science: The Central Issues*, second edition (Norton, 2012) and *Innovations in the History of Analytical Philosophy* (Palgrave, 2017).

Stathis Psillos is Professor of Philosophy of Science and Metaphysics at the University of Athens, Greece and a member of the Rotman Institute of Philosophy at the University of Western Ontario. He is the author or editor of eight books and over 160 papers on scientific realism, causation, explanation, and the history of philosophy of science. He is member of the Academy of Europe, the International Academy of Philosophy of Science, a former president of the European Philosophy of Science Association (EPSA), and former editor of *Metascience* (2009–2014).

Erich H. Reck is Professor of Philosophy at the University of California, Riverside. He is the author of a series of articles on early analytic philosophy, the philosophy of mathematics, and logic. He is also the editor of several books, including *From Frege to Wittgenstein* (OUP, 2002), *Frege's Lectures on Logic: Carnap's Student Notes, 1910–1914* (with S. Awodey, Open Court, 2004), *Gottlob Frege: Critical Assessments of Leading Philosophers*, 4 vols. (with M. Beaney, Routledge, 2006), and *The Historical Turn in Analytic Philosophy* (Palgrave, 2016).

Alan Richardson is Professor of Philosophy at the University of British Columbia. He is the author of many works on logical empiricism, especially Rudolf Carnap and Hans Reichenbach, including *Carnap's Construction of the World* (CUP, 1997). His editorial works include *The Cambridge Companion to Logical Empiricism* (with T. Uebel, CUP, 2007). He is on the editorial board of *The Collected Works of Rudolf Carnap*.

Sahotra Sarkar is a Professor in the Departments of Philosophy and of Integrative Biology at the University of Texas at Austin. He specializes in the history and philosophy of science, environmental philosophy, conservation biology, and disease ecology. He is the author of six books and more than 250 articles in philosophy and biology. He is the editor of 24 works including the six-volume *Science and Philosophy in the Twentieth Century: Basic Works of Logical Empiricism* (Garland, 1996).

Georg Schiemer is Associate Professor at the Department of Philosophy, University of Vienna and External Fellow at the Munich Center for Mathematical Philosophy at LMU Munich. He is also principal investigator of the project *The Roots of Mathematical Structuralism* funded by an ERC Starting Grant (2017–2022, project number 715222). His research focuses on the history and philosophy of mathematics and early analytic philosophy. He is also interested in logic, the philosophy of logic, and formal philosophy of science.

Anne Siegetsleitner is Professor of Practical Philosophy at the University of Innsbruck. Her work focuses on ethics, social and political philosophy with special interests in bioethics, information ethics and the history of humanist philosophy. In her contributions to ethics and morality in the Vienna Circle (e.g., *Ethik und Moral im Wiener Kreis*, Böhlau, 2014), she undertakes a long-overdue revision of the prevailing view of the role and conception of ethics and morality in this influential group.

Jan Sprenger is Professor of Logic and Philosophy of Science at the University of Turin in Italy, having previously taught at Tilburg University in the Netherlands. His main research and publication topics are statistical inference, causality, probability and scientific objectivity. Together with Stephan Hartmann, he recently published *Bayesian Philosophy of Science* (OUP, 2019).

Friedrich Stadler is Professor Emeritus for History and Philosophy of Science, University of Vienna, Permanent Fellow of the Institute Vienna Circle, and Director of the Vienna Circle Society. Author of several books and editor or co-editor of four book series, he is former President of the European Philosophy of Science Association and the Austrian Ludwig Wittgenstein Society, and Member of the Commission for History and Philosophy of Science, Austrian Academy of Sciences. Awards he has received include the Jan Patočka Medal of the Czech Academy of Sciences (2016) and the George Sarton Medal of the University of Ghent (2017).

Michael Stöltzner is Professor of Philosophy at the University of South Carolina. He often conducts historical investigations to support philosophical claims. His most significant scholarly contributions pertain to the history of logical empiricism and its embedding in late nineteenth- and twentieth-century physics, the philosophy of physics and applied mathematics, and the philosophy of elementary particle physics. He has also worked on issues in general philosophy of science, among them models, causation, explanation, experiment, and applied science.

David J. Stump is Emeritus Professor of Philosophy at the University of San Francisco. He is the author of *Conceptual Change and the Philosophy of Science: Alternative Conceptions of the A Priori* (Routledge, 2015), co-translator of a new version of Henri Poincaré's *Science and Hypothesis* (Bloomsbury, 2018), and co-editor of *The Disunity of Science* (with P. Galison, Stanford, 1996). He is author of numerous journal articles on Poincaré, Duhem and history and philosophy of mathematics. He is currently President of HOPOS, the International Society for the History of the Philosophy of Science.

Marta Sznajder is a Postdoctoral Researcher at the Faculty of Philosophy at the University of Groningen, specializing in the history of inductive logic and philosophy of probability. She leads the research program "In Inductive Logic, There Are No Morals: Carnap's Philosophy of Scientific Reasoning," which grew out of her doctoral work on Carnap's "Basic System," and runs a biographical project on the inductive logician, Janina Hosiasson.

Mark Textor is Professor of Philosophy at King's College London. He has worked on the history of analytic philosophy (especially Frege), Austrian philosophy and philosophy of mind and language. His most recent books are *Brentano's Mind* (OUP, 2017) and *The Disappearance of the Soul and the Turn against Metaphysics: Austrian Philosophy 1874–1918* (OUP, 2021).

Thomas Uebel is Professor Emeritus of Philosophy at the University of Manchester. His research concerns the histories of analytical philosophy and philosophy of science and contemporary epistemology and philosophy of social science. Having published widely in journals and collections of essays, his books include *Empiricism at the Crossroads. The Vienna Circle's Protocol-Sentence Debate* (Open Court, 2007) and an edition of Neurath's *Economic Writings* (with R.S. Cohen, Kluwer, 2004).

Pierre Wagner is Professor in the Philosophy Department, University Paris 1 Panthéon-Sorbonne and Director of the IHPST (Institut d'histoire et de philosophie des sciences et des techniques), Paris. His research interests include philosophy of logic and the history of the philosophy of science. He has edited *Carnap's Logical Syntax of Language* (Palgrave Macmillan, 2009) and *Carnap's Ideal of Explication and Naturalism* (Palgrave Macmillan, 2012), and co-edited *Précis de philosophie de la logique et des mathématiques* (Éditions de la Sorbonne, 2021).

Jan Wolenski is Professor at the University of Information, Technology and Management, Rzeszów, Poland. Until his retirement in 2010 he was Professor of Philosophy at Jagiellonian University in Cracow. He is a Member of Polish Academy of Sciences, Polish Academy of Arts and Sciences, Academia Europea, Brazilian Academy of Philosophy, and International Academy of Philosophy; and author of 30 books, including *Logic and Philosophy in the Lvov-Warsaw School* (Kluwer, 1989) and *Semantics and Truth* (Springer, 2019), and 900 papers. His areas of interests include logic and its history, epistemology, philosophy of law, and history of philosophy.

INTRODUCTION

Christoph Limbeck-Lilienau and Thomas Uebel

The last forty years or so have seen exciting developments in analytic philosophy. One of the most exciting is its discovery of its own history. Logical empiricism forms a very significant part of it, and important facts about it have been learnt in this time. It turns out that much that is written about it in standard histories of philosophy—and taught in introductory philosophy courses—is at least seriously misleading. The present *Handbook* has been put together to give a convenient entry into what may easily appear a forbidding thicket of specialist literature. While we cannot claim completeness for our coverage of the wide-ranging phenomenon that is logical empiricism, we hope that readers will find that both its historical development and its doctrinal contents are treated in sufficient breadth and detail to bring them up to date on what is their topic of initial interest or need to know, and to encourage further study. In this introduction we give a very brief description of logical empiricism and its reception to provide a framework for the reader with which to approach the subsequent overview descriptions of the different essays in this volume.

A very brief outline of logical empiricism

The philosophical movement called "logical empiricism"—sometimes also called "logical positivism" or "neopositivism"—is best conceived as characterized by a set of philosophical positions developed in the 1920s and 1930s by two groups, the Vienna Circle and the Berlin Group, importantly along with certain metaphilosophical ambitions. The two did not always match. In consequence, their positions were amended, elaborated, and refined by members, students, and sympathizers until the 1960s. By then logical empiricism was increasingly criticized and rejected by a younger generation of analytic philosophers. Once out of favor, an oversimplified picture of it gained hold that tended to stress only its earliest doctrines.

Despite differences in style and content, the initial two groups had this common ambition: to create a philosophy which Reichenbach, the leader of the Berlin Group, called "scientific." Their efforts were motivated by the belief that the revolutionary changes in physics, the foundational crisis in mathematics, and the emergence of the social sciences demanded a new beginning for philosophy. Comprehending the modern world meant comprehending modern science, and this required that philosophy itself become scientific. It had to stop the pursuit of

 DOI: 10.4324/9781315650647-1

individual system-building and become a collective endeavor, and it had to replace intuitions of deeper meanings with logical arguments open to inspection and informed criticism.

Early logical empiricism was characterized by a set of claims which were formulated in programmatic writings such as the "manifesto" composed by members of the Vienna Circle to celebrate the decision by Schlick, their titular leader, to resist a call back to Germany and stay in Vienna (Verein Ernst Mach 1929). A central claim concerned the nature of philosophy. It was viewed primarily as logical analysis, especially the logical analysis of the language of scientific theories. Connected to this logical analysis was a conception of the bounds of meaning, verificationism. According to it, all meaningful statements must be analyzed in terms of empirically testable statements. A consequence of logical analysis and this strict empiricist outlook was the rejection of metaphysics and the rejection of the view that moral and normative statements report something about the world entirely independent of the reporter. Such statements were seen as having no cognitive content.

Logical analysis with its verificationist bent and its anti-metaphysical consequences came to be regarded as the defining characteristic of logical empiricism. While this methodological outlook was widely shared among its practitioners, individual doctrines were soon criticized, amended, and interpreted in different ways. (Just how cognitive content was to be characterized, for instance, remained an issue.) In the 1930s, its anti-metaphysical methodology was complemented by the emphasis on physicalism and on the unity of science. Physicalism as initially developed by Carnap and Neurath claimed that all scientific statements can be translated into a physicalist language, that is, into a language speaking only about spatiotemporal objects and properties ultimately regarded as physical. The unity of science was seen as the unity of the scientific method, the unity of the language of science and the unity of the objects of science conceived as spatiotemporal physical objects. Both physicalism and the unity of science opposed the then popular categorical division of *Natur-* from *Geisteswissenschaften*, the natural from the human sciences.

To complement this brief picture, it must be emphasized that, besides their concern with empirical science generally, reflections about the nature of logic and of the foundations of mathematics played an essential role, especially within the Vienna Circle. Here, an initial tentative embrace of logicism was progressively abandoned for a position which saw in the foundations of mathematics a matter of choice between several possible mathematical languages. In the Berlin Group, the rational reconstruction of specific theories of the special sciences played an essential role, especially of physics, but also of psychology.

Although they do not do justice to the complexity of logical empiricism, with its often divergent views and personal differences, the central claims mentioned in this section can be said to have set the parameters for the debates within the movement. Certainly, the earlier bold and polemical claims against "metaphysics" were softened somewhat in the later phase of logical empiricism when its practitioners, then in exile, were no longer confronted by its excesses on a daily basis, as they had been. It may even seem that a more tolerant attitude was adopted towards certain aspects of metaphysics, when, as in the case of Carnap, ontological concerns were refashioned as a choice between linguistic frameworks. But traditional metaphysics continued to be condemned as fruitless speculation.

The historical development of logical empiricism

The development of logical empiricism can be divided roughly into three phases: (1) the initial articulation of logical empiricist positions within the Vienna Circle and the Berlin Group; (2) the phase of internationalization beginning in the mid-1930s, a phase which corresponds to the emigration of most of the logical empiricists to the Anglo-American world; (3) logical empiricism in the postwar period, then mainly limited to the USA and Great Britain.

(1) The Vienna Circle formed in 1924 around Schlick, who had been appointed professor of philosophy at the University of Vienna in 1922. Since its beginning, Neurath and Hans Hahn were active members of the Circle (with Philipp Frank a correspondent in Prague), along with Viktor Kraft and Schlick's students Herbert Feigl and Friedrich Waismann. They were joined during the 1920s by Carnap and Hahn's students Karl Menger and Kurt Gödel. Other members soon included Gustav Bergmann and Rose Rand. (Others, like Edgar Zilsel, preferred to remain "associates.") The Berlin Group formed around Reichenbach in 1928, independently of the much larger Berlin "Society for Empirical Philosophy," which was also led by him since 1929, and had Kurt Grelling, Walter Dubislav, Carl Gustav Hempel, and Olaf Helmer as its members. (Richard von Mises was prominent in the Society but not a member of Reichenbach's Group.)

The initial phase of the Vienna Circle was marked by a close reading of Ludwig Wittgenstein's *Tractatus* (1922) whose conception of logical propositions as tautologies and of logical analysis as the decomposition of all propositions into elementary propositions made a deep impression upon the Circle. Yet despite their initial appreciation of Wittgenstein, the members of the Circle developed alternative conceptions. Carnap, who had finished most of the writing of *The Logical Structure of the World* (1928) just before joining the Circle, became one of the driving forces in its discussions. While the *Tractatus* had left unspecified the nature of the elementary propositions, the terminus of logical analysis, Carnap and the Circle gave an empiricist twist to this picture in interpreting them as propositions about experience. Concerning the foundations of logic and mathematics, prompted by Frank Ramsey, they extended Wittgenstein's conception of logical propositions as tautologies to the propositions of mathematics.

While at the end of the 1920s, these conceptions, along with the doctrine of the unity of science, were firmly in place, new developments emerged at the beginning of the 1930s, with physicalism championed by Neurath and Carnap and Carnap's adoption of Neurath's thoroughgoing fallibilism. Though the view continued to be that the claims of the empirical sciences were justified through their link to empirical statements, these statements were now viewed as statements about physical objects and their physical properties, not as statements about subjective experience (as in the *Aufbau*). As all empirical claims were held to be translatable into such physicalist statements, all the claims of the empirical sciences were understood as statements about spatiotemporal entities. As such, they were revisable like any other statement about the world. Schlick objected to the elimination of subjective experience from the justification of empirical claims as well as to their wholesale fallibilization. These were the themes of the so-called protocol-sentence debate of the first half of the 1930s. Another major change came with Carnap's *Logical Syntax of Language* (1934) which introduced logical pluralism (tolerance) and finally put the notion of logical analysis on a firm footing. Now the logical form of statements became expressible in a metalanguage, a position strongly opposed by Schlick and Waismann, who followed Wittgenstein in his Tractarian rejection of the legitimacy of metalinguistic discourse. Instead of an "elucidatory" activity of clarifying meaning, as pursued by Schlick, philosophy now became, with Carnap, the "logic of science." After the forced emigration of Neurath and the death of Hahn in 1934, these debates came to an abrupt end through the assassination of Schlick in 1936.

The Berlin Group was forced to disband already earlier under the pressure of the Nazi race laws and their immediate implementation in 1933. Less concerned with overarching philosophical doctrines, its members advanced their work in their chosen specialties. Grelling continued his work in logic (a paradox had been named after him in 1908), Dubislav refined his theory of definitions, and Reichenbach continued his long-standing interest in the axiomatization of physical theories, probabilistic accounts of causality, and the development of his version of a frequentist theory of probability. After Reichenbach's dismissal from the university, his students Hempel and Helmer were forced to complete their PhDs with sympathetic colleagues, like the psychologist Wolfgang Köhler.

(2) Parallel to the eradication of logical empiricism in Central Europe from 1933 onwards, an increasing internationalization of the movement took place that built in part on earlier visits to Vienna (and Berlin) by philosophers from the Nordic countries (Eino Kaila, Jørgen Jørgensen, Arne Naess), Poland (Alfred Tarski), England (A. J. Ayer), the USA (W. V. O. Quine, Ernest Nagel, Charles Morris) and China (Tscha Hung). This internationalization was driven by the organization of a series of International Congresses for the Unity of Science (Prague, 1934; Paris, 1935; Copenhagen, 1936; Paris, 1937; Cambridge, England, 1938; Cambridge, Massachusetts, 1939; Chicago, 1941) and by the emigration of the logical empiricists (1931: Feigl to USA; 1933: Reichenbach and Richard von Mises to Turkey, 1938 to USA; 1934: Neurath to Holland, 1940 to UK; 1935: Carnap to USA, Hempel to Belgium, 1937 to USA; 1936: Dubislav to Czechoslovakia, Menger to USA; 1937: Waismann to UK, Grelling to Belgium; 1938: Frank, Kaufmann, Bergmann, Else Frenkel and Egon Brunswik to USA; Zilsel, Rand to UK, later to USA; 1940: Gödel to USA). A central aim of the Congresses was to establish connections of logical empiricism with philosophical tendencies in other countries which shared a comparable project of a "scientific philosophy" (the title of the first big Congress in Paris, 1935). Strong alliances were formed particularly with members of the Polish school (Tarski, Ajdukiewicz) as well as with American philosophers close to pragmatism (Morris, Nagel). Some of these were fortified by a large-scale publication project initiated by Neurath and stretching into the third phase, originally entitled *International Encyclopedia of Unified Science*: begun in 1938, this series of mostly small monographs was concluded (except for three late replacement additions including Kuhn [1962] and the index) by the mid-1950s under the title *Foundations of Unified Science*.

Although topics pertaining to the logic of science and the connection between the sciences were strongly emphasized in the second phase, the new alliances also shaped the further development of logical empiricism. This is especially so with Carnap's adoption of Tarski's semantic theory of truth in 1935, marking a major step away from the exclusive emphasis on syntax in the analysis of language. Less important initially but increasingly significant in later years was the growing awareness of the pragmatic aspects of language, given the promptings of Morris. In addition, Carnap's principle of logical tolerance now extended to the view that the criteria of empiricist probity—the criteria of confirmation of empirical statements—were themselves conventionally chosen, to be discussed as revisable proposals. Overall in this process of internationalization, logical empiricism was seen as part of a bigger movement sometimes called "scientific empiricism" (Morris, Carnap).

(3) The third phase of logical empiricism, its post-World War II development, was largely limited to the Anglo-American world, with Carnap, Reichenbach, Feigl, and Hempel as the main driving forces in the USA, and Ayer and Waismann the remaining representatives in England, albeit very much after their own fashion. In its North American environment, logical empiricism continued to thrive at first. Reichenbach had gathered some talented and later very influential students around himself at the University of California in Los Angeles (UCLA) before his early death in 1953. Carnap continued to work on semantics and modal logic before turning to inductive logic, first at the University of Chicago and from 1954 at UCLA. And while funds for Frank's Institute for the Unity of Science in Cambridge, Massachusetts, dried up, Feigl was able to found the first Center for the Philosophy of Science in 1953 at the

University of Minnesota in Minneapolis, and Hempel established himself as a major force in the philosophy of science, first at Yale and from 1955 at Princeton. Over time, new focal points for logical empiricist investigations emerged, especially with regard to the analysis of scientific theories and the concepts of confirmation and explanation (Carnap, Hempel), but also with the development of new theories in the philosophy of mind (Feigl).

One important doctrinal change that marked the entire third period (but was first pioneered by Carnap in 1939) was this. While the early logical empiricists tried to define all empirical concepts in terms of their observational basis and later specified a looser connection through so-called reduction sentences, now a categorical distinction was accepted between theoretical terms, which referred to theoretical entities, and observational terms. The new task was to formulate a precise theory of the meaning of theoretical terms, and different proposals were made. This shift adapted to the newest trends in scientific theories, not only in physics but also in cognitive psychology, where theoretical terms and entities became more and more important. In the philosophy of mind, this also led to the view that mental terms referred to unobservable physical states and to an explicit rejection of a too-limited behaviorist approach. As regards the theory of scientific theories, by the late 1950s Carnap was led to adopt and develop Ramsey's method of regimenting scientific theories by so-called Ramsey sentences.

In Great Britain, where Neurath had died unexpectedly in 1945, Ayer continued to work on empiricist epistemology by developing a more precise formulation of the empirical meaning criterion and, opposing physicalism, by working on new arguments for epistemological phenomenalism. Thereby he continued a project he saw Carnap to have started in the *Aufbau* but abandoned afterwards. Waismann developed a new and highly original approach to the philosophy of language. Natural languages, he held, contain different sub-languages adapted to different spheres of experience, different "language strata": like different scientific languages they are not reducible one to another. His notion of "open texture" initiated a new approach to the indeterminacy of empirical concepts which was influential in the philosophy of law.

Logical empiricism was always contested and probably never, even in the USA after World War II, was the dominant position in philosophy, although it certainly attracted much attention. Having been challenged by W. V. O. Quine's criticism of the "dogmas of empiricism" from 1951 onwards, logical empiricism was under siege by the 1960s. Perhaps most consequential was that logical empiricism was seriously contested in its main field, the philosophy of science. Scientific realism rejected its conception of scientific theories, and the new history of science rejected what it took to be its view of scientific change and progress. Although the differences between these new views and those of the logical empiricists were often greatly exaggerated, this created a climate where logical empiricism appeared "old fashioned." At the same time, the new left mischaracterized the political allegiances of logical empiricism (whether so intended or not, the unspecified broadsides against "positivism" and "operationalism" in Marcuse's countercultural bestseller *One-Dimensional Man* [1964] were read in this fashion), so that the strong ties of the logical empiricists to progressive, and often socialist, causes were forgotten. Although the theoretical output of its survivors was still impressive, logical empiricism was widely proclaimed "dead," and a new generation of analytic philosophers turned away from it.

Logical empiricism reconsidered

It is probably no coincidence that it was not until the late 1970s, when logical empiricism was no longer a "live option," that a historical re-evaluation began. Philosophers raised under "post-positivism" began to look into the history of analytic philosophy, and those interested in the history of philosophy of science turned to the origins of logical empiricism and its often not easily accessible German-language sources.

It would lead too far to list all important way-stations of this rediscovery, but a few early ones in the English-language literature must be mentioned. Indispensable, to begin with, was the work by Henk L. Mulder, Robert S. Cohen and Brian McGuinness as general editors of the *Vienna Circle Collection* (1973–2004) which made many primary sources available to an English-language readership for the first time (e.g., essays and/or monographs by Schlick, Reichenbach, Neurath, Frank, Menger, Kraft, and others). Of great importance were also the volumes deriving from conferences marking the centenaries of Schlick and Neurath in 1982 (Haller 1982) and Carnap and Reichenbach in 1991 (Spohn 1991; Sarkar 1992; Salmon and Wolters 1993), Michael Friedman's first essay on Carnap's *Aufbau* (1987), the first large-scale monograph on logical empiricism by Albert Coffa (1991), and the edition of the Carnap-Quine correspondence (Creath 1991).

Overall, two trends can be distinguished in this historical reassessment: first, an emphasis on the European context of the emergence of logical empiricism, especially the influences of neo-Kantianism, French conventionalism and of the earlier scientific philosophy of scientists such as Helmholtz, Mach, and Boltzmann; second, an emphasis on the internal diversity and pluralism within logical empiricism which had been forgotten. In consequence a number of lines of inquiry emerged. The anti-foundationalist epistemology of Neurath, Carnap, and Frank was rediscovered and emphasized, as well as their much less naïve view about the history of science and scientific change. Schlick's pre-Vienna Circle work was taken note of and his distinction as the first philosopher of note to have understood Einstein's theories of relativity duly appreciated. The received view of the Carnap-Quine debate about the analytic/synthetic distinction and ontology was questioned, and Carnap's tolerant view about philosophy as "explication" was underlined. Further instances could be cited, not of the least importance is the recovery of the political dimension possessed by the work of the logical empiricists in Central Europe in the 1920s and 1930s.

The present volume presents many of the results of this historical reassessment of logical empiricism. While a new attention to some of its ideas is detectable in several areas of contemporary analytic philosophy, from meta-ontology to the thriving area of conceptual engineering, a wholesale revival is certainly not on the cards. A more accurate view of the history of analytic philosophy can only be salutary, however, and we hope that the present volume will contribute to this.

The chapters in this volume

The present volume covers logical empiricism from its origins and early development to its late phase and to the criticism it received in the 1960s, concluding with chapters on its legacy. We organized the volume into four main parts. Part I looks at the historical context of the emergence of logical empiricism and retraces its historical development. Part II is dedicated to specific theses and positions within logical empiricism, from logicism and verificationism to Neurath's views about economics. Part III analyzes the interaction of logical empiricism with other philosophers and philosophical movements. Part IV is dedicated to post-positivist criticism from Quine to Kuhn and to the legacy of logical empiricism. Hereafter, we briefly offer an overview of the topics and chapters in each section.

Part I on the historical context and the development of logical empiricism begins with a series of chapters on the early influences on logical empiricism in its formative years. Michael Stöltzner analyzes how the crisis in physics and mathematics at the turn of the twentieth century shaped the view of science of the young logical empiricists. The foundational crisis in mathematics and the crisis in the transition to relativity theory were essential in the logical empiricist's redefinition of scientific rationality. The cultural and socio-political context at the turn of the century is described in a chapter by Hans-Joachim Dahms on the German Youth Movement, a social movement in pre-World War I Germany which greatly marked the reformist social and ethical views of later logical empiricists such as Carnap and Reichenbach. Matthias Neuber looks at the ramified neo-Kantian influence in the formative years of many logical empiricists, from the influence of the Marburg School on Carnap to that of Alois Riehl's critical realism on Schlick. Christian Damböck analyzes the historicist background of Carnap's early intellectual milieu and argues that Carnap was influenced by the early, more empiricist historicism of Dilthey, while both he and Neurath rejected the later historicism with its strong contrast between the humanities and the natural sciences. Michael Heidelberger looks specifically at Helmholtz's empiricist approach to geometry and at the way this shaped Schlick, Carnap, and Reichenbach in their understanding of the relation of geometry to experience.

While the previous chapters looked at the German influences on logical empiricism, specifically on its German members, the next series of chapters highlight the Austrian and French influences which strongly shaped the formation of the Austrian members. The intellectual impact of Mach on a younger generation in Vienna is analyzed by Elisabeth Nemeth. She emphasizes especially how Mach's historico-critical method marked Frank's approach to the history of sciences and Neurath's approach to economics. Mark Textor analyzes the Austrian tradition represented by Bolzano and Brentano, retracing how the philosophy of Bolzano marked Hahn's conception of analytic propositions and of infinity, while Schlick responded to Brentano's understanding of judgments. French conventionalism was another essential influence on early logical empiricism. Anastasios Brenner describes the origins of conventionalism and the transmission of its ideas to Neurath, Hahn, Frank, and Schlick. The tremendous importance of Einstein in the formation of logical empiricism is analyzed in detail by Fynn Ole Engler, who emphasizes not only that logical empiricists defended the theory of relativity from early on, but also shows the way Schlick influenced Einstein's understanding of relativity.

The next three chapters describe the main groups which constituted logical empiricism thereby retracing its development. Christoph Limbeck-Lilienau looks at the early discussions, around 1907–10, of Neurath, Hahn, and Frank on philosophy of science and their early reception of the French conventionalists and Russell's logic. He then highlights an early manifestation of logical empiricism at the Erlangen Conference of 1923. Friedrich Stadler discusses the evolution of the Vienna Circle and its public wing, the "Society Ernst Mach," accentuating especially the way these two main institutions of the logical empiricists in Vienna were embedded in a network of socially progressive groups. Nikolay Milkov describes the formation and evolution of the Berlin Group around Reichenbach with Grelling, Dubislav and Hempel. He analyzes their cooperation with the Vienna Circle as well as the philosophical differences between the two groups. Frederique Janssen-Lauret shows that the often-underestimated role of women in or in close collaborations with logical empiricism needs a substantial correction. She highlights the contributions of Rose Rand and Else Frenkel within logical empiricism, as well as the contributions Susan Stebbing and members of the Polish school made to the debates of logical empiricism.

Part II on the main theses and specific issues within logical empiricism begins with chapters on logic and semantics. Erich H. Reck looks at the debates on logic and the foundations of mathematics in the early formative years, highlighting the impact of Hilbert, Russell, and Wittgenstein. Pierre Wagner analyzes the early conceptions of truth of Schlick and Carnap, as well as Neurath's rejection of the notion of truth. He describes how the correspondence theory met with skepticism and how, later on, Carnap adopted Tarski's conception of truth. James Justus discusses the motivation for and different formulations of verificationism, from early attempts to the later versions by Ayer and Carnap. He argues, against a very widespread presumption to the contrary, that they can be defended against criticism by careful amendments. The central importance of noncognitivism in the logical empiricist's view about ethics is analyzed and its force clarified by Anne Siegetsleitner. She underlines that some of the logical empiricists such as Schlick and Kraft diverged from the noncognitivist position. Jordi Cat looks at the unity of science thesis and argues that the logical empiricists did not equate this thesis with a form of foundationalism or reductionism.

The next set of chapters looks at different aspects of the conception of scientific theories and their function within logical empiricism. Stathis Psillos gives a detailed analysis of Hempel's model of scientific explanation, explaining the reasoning behind the deductive-nomological view and assessing the criticisms addressed to it. William Demopoulos discusses the partial interpretation view of scientific theories of later logical empiricism. He reconstructs Carnap's adoption of the method of regimenting theories by means of the Ramsey sentence and critically analyzes Carnap's distinction between the empirical and the analytic part of a theory via the Carnap sentence. David J. Stump investigates the role played by the relativized a priori in the logical empiricist's reconstruction of scientific theories and looks at the tension between this view and the general rejection of a synthetic a priori within logical empiricism. Finally, Georg Schiemer investigates the way logicism was reconstructed by the logical empiricists and adapted to the new conception of logic as tautological. He shows that, prior to Carnap's *Logical Syntax*, the only way for the logical empiricists to save certain fundamental axioms of mathematics within logicism was the adoption of a form of if-thenism.

Three chapters are dedicated to the central role of induction and probability. Marta Sznajder compares the frequentist and the logical interpretation of probability within early logical empiricism and then analyzes Carnap's inductive logic and its later reception in the context of Bayesianism. Reichenbach's lifelong preoccupation with the problem of induction is retraced by Flavia Padovani from Reichenbach's early neo-Kantian roots to his later more pragmatic approach to induction. Jan Sprenger's chapter analyzes in detail Hempel's theory of confirmation and the problems it encounters. He highlights especially the raven paradox and the different solutions to it.

A final segment of Part II addresses topics in the philosophy of mind, ontology and the philosophy of the social sciences. The philosophy of mind in the Vienna Circle as well as in the period after 1945 is retraced by Sean Crawford, who describes the early views of Schlick, Carnap, and Feigl before looking at Feigl's later identity theory. He shows that all of them are committed to a form of physicalism and that it is false to describe their early physicalist position as a form of logical behaviorism. Although metaphysics is generally believed to be antithetic to logical empiricism, Gregory Lavers argues that from early on Carnap had a specific view of ontology as nontheoretical and relative to the choice of a language. Despite this continuity in Carnap's thought, he also emphasizes the transformations in his position on ontology due to the adoption of Tarski's semantic approach. Neurath's lifelong reflections about economics and its place in a philosophy of the social sciences are analyzed by John O'Neill. He shows that Neurath's defense of economic planning involved a multidimensional conception of economic rationality and the rejection of "pseudo-rational" decision making based on a single measure of economic success.

The third part of the volume analyzes the interaction between logical empiricists and other philosophers and representatives of connected philosophical movements of the time. A first series of chapters focuses on the relation to Wittgenstein, Ernst Cassirer, and Karl Popper.

Johannes Friedl gives an overview of Wittgenstein's influence which accompanies the Vienna Circle during its whole existence, from the early reception of the *Tractatus* to the deepening internal split dividing the Circle into pro- and anti-Wittgenstein wings. Matthias Neuber analyzes the commonalities between Cassirer and Carnap, Schlick, Reichenbach and Frank. He highlights especially their common rejection of metaphysics and their common interest in a logic of relations. The conflicted relation with Popper and the critical rationalists is the topic of Artur Koterski's chapter. He emphasizes how the problem of the empirical base of a scientific theory generated a debate among the critical rationalists which was quite similar to the protocol-sentence debate in the Vienna Circle.

The next four chapters look at the connection of logical empiricism to allied philosophical movements in Poland, the Nordic countries, Great Britain, and the USA. Jan Wolenski gives an overview of the Lvov-Warsaw School and analyzes the strong ties it had to the Vienna Circle, from the common interest in logic to the impact of Tarski's semantics and the common work on the analysis of scientific theories, e.g., on confirmation. Ilkka Niniiluoto traces the evolution of a scientifically oriented philosophy in Denmark, Finland, Sweden, and Norway and investigates how philosophy in these countries was shaped by the impact of logical empiricism. Christopher Pincock analyzes the early interaction of logical empiricists with philosophers in England and the USA, focusing on Stebbing's and Ayer's discussion of the nature of logical analysis and on how Morris and Nagel looked for an ally in their defense of a consistent form of naturalism. Massimo Ferrari looks specifically at the connections to pragmatism, from the early reception by logical empiricists, especially by Neurath and Frank, to the later interaction with pragmatists in the 1930s.

Concluding the volume, Part IV is dedicated to the post-positivist critics of logical empiricism, Quine and Thomas Kuhn, and to the general legacy of logical empiricism in today's analytic philosophy and philosophy of science. Quine's confrontation with Carnap is analyzed by Richard Creath, who reassesses their opposition with regard to the position of foundationalism and the analytic/synthetic distinction, emphasizing that the differences between the logical empiricists and post-positivism are far less pronounced than generally thought. Gürol Irzik highlights the commonalities between Kuhn and Carnap and challenges the narrative that Kuhn gave a direction to the philosophy and history of science which was opposed to the logical empiricists.

The last four chapters analyze the legacy of logical empiricism. Thomas Uebel highlights that the logical empiricism of Carnap, Neurath, and Frank allowed for a bipartite metatheory conception of the nature of philosophy which gives room not only to the logical analysis of science but also to the pragmatic dimension, that is, to the social and historical aspects of science. This second part was often forgotten in later assessments of logical empiricism. Sahotra Sarkar analyzes the continuing positive impact of the logical empiricists' approach to formal epistemology, emphasizing their role as pioneers in this field. A. W. Carus challenges the frequently asserted claim that Carnap had a limited and one-sided conception of rationality which would exclude the normative side of reason. He shows that Carnap's position can be reconstructed as a dialectical combination of reason and understanding in the Kantian sense, i.e., as an integration of scientific with normative rationality. Alan Richardson looks at the American legacy of logical empiricism and the endeavor to establish a "scientific philosophy." He analyzes the complex destiny of this project, the resistances to and the distortions of it, taking into account the institutional and social setting of the reception of this émigré movement.

Readers will find, we hope, that this *Handbook* pursues no party line or expresses a philosophical consensus. To be sure, all contributors to this volume are convinced of the philosophical importance of the works they discuss, but this appreciation need not extend to agreement with the doctrines at issue; in fact, it often does not (examples are too numerous to be cited). Nor is there agreement among the contributors to this volume about various issues (for instance, the interpretation of Mach or the value of noncognitivism). The shared belief that logical empiricism is a worthwhile object of study—if only to understand the history of analytical philosophy better than we do so far—entails no doctrinal or interpretational allegiances.

Other secondary sources

At present, the secondary literature on logical empiricism is burgeoning (as on history of analytic philosophy and history of philosophy of science generally). As we do not provide further overviews of the whole movement, the life work of individual members, or of logical empiricist work in individual subdisciplines of philosophy, we refer readers to the free-to-access and ever-growing *Stanford Encyclopedia of Philosophy* (Zalta 1995–) with its regularly updated detailed entries and bibliographies (so far on Rudolf Carnap, Herbert Feigl, C. G. Hempel, Otto Neurath, Hans Reichenbach, Moritz Schlick, and on logical empiricism and the Vienna Circle). Incisive discussions of different aspects of Carnap's work in particular are also provided in Friedman and Creath (2007) and of discussions of topics in the philosophy of science developed by logical empiricism under much broader headings than used here in Richardson and Uebel (2007).

Acknowledgment

Christoph Limbeck-Lilienau's work on this "Introduction" is funded by the Austrian Science Fund (FWF), project number J 4502-G.

References

Carnap, R. (1928) *Der logische Aufbau der Welt*, Berlin: Weltkreis-Verlag. Trans. *The Logical Structure of the World*, Berkeley: University of California Press, 1967. Repr. Chicago: Open Court, 2003.

——— (1934) *Logische Syntax der Sprache*, Vienna: Springer. Rev. ed. trans. *The Logical Syntax of Language*, London: Kegan Paul, Trench, Trubner & Cie, 1937, repr. Chicago: Open Court, 2002.

Coffa, A. (1991) *The Semantic Tradition from Kant to Carnap: To the Vienna Station* (ed. by L. Wessels), Cambridge: Cambridge University Press.

Creath, R. (ed.) (1991) *Dear Carnap, Dear Van: The Quine-Carnap Correspondence and Related Work*, Berkeley: University of California Press.

Friedman, M. (1987) "Carnap's *Aufbau* Reconsidered," *Nous* 21: 521–45. Repr. in M. Friedman, *Reconsidering Logical Positivism*, Cambridge: Cambridge University Press, 1999, pp. 89–113.

Friedman, M. and Creath, R. (eds.) (2007) *The Cambridge Companion to Carnap*, Cambridge: Cambridge University Press.

Haller, R. (ed.) (1982) *Schlick und Neurath—ein Symposion, Grazer Philosophische Studien* 16/17, Amsterdam: Rodopi.

Kuhn, T. S. (1962) *The Structure of Scientific Revolutions*, Chicago: University of Chicago Press.

Marcuse, H. (1964) *One Dimensional Man. Studies in the Ideology of Advanced Industrial Society*, Boston: Beacon Press, 2nd ed., 1991, repr., London: Routledge, 2002.

Mulder, H. L., Cohen, R. S. and McGuinness, B. (1973–2004) *The Vienna Circle Collection*, Dordrecht: Reidel/Kluwer, vol. 22.

Richardson, A. and Uebel, T. (eds.) (2007) *The Cambridge Companion to Logical Empiricism*, Cambridge: Cambridge University Press.

Salmon, W. and Wolters, G. (eds.) (1993) *Logic, Language and the Structure of Scientific Theories*, Pittsburgh: University of Pittsburgh Press.

Sarkar, S. (ed.) (1992) *Carnap: A Centenary Reappraisal, Synthese* 93 (1–2).

Spohn, W. (ed.) (1991) "Hans Reichenbach, Rudolf Carnap: A Centenary," *Erkenntnis* 35.

Verein Ernst Mach (1929) *Wissenschaftliche Weltauffassung. Der Wiener Kreis*, Vienna: Wolf. Trans. "The Scientific Conception of the World. The Vienna Circle," in O. Neurath, *Empiricism and Sociology* (ed. by R. S. Cohen and M. Neurath), Dordrecht: Reidel, 1973, pp. 299–318; rev. trans. (with orig. annotated bibliography) "The Scientific World-Conception. The Vienna Circle," in F. Stadler and T. Uebel (eds.), *Wissenschaftliche Weltauffassung. Der Wiener Kreis. Hrsg. vom Verein Ernst Mach (1929)*, Vienna: Springer, 2012, pp. 75–116.

Wittgenstein, L. (1922) "Logisch-Philosophische Abhandlung," *Annalen der Naturphilosophie* 14: 185–262. Bilingual ed. trans. by F. Ramsey and C. K. Ogden, *Tractatus Logico-Philosophicus*, London: Kegan Paul, Trench Trubner & Co., 1922, rev ed. 1933. Repr. London: Routledge, Kegan Paul, 1983; trans. by D. F. Pears and B. F. McGuinness, London: Routledge, Keagan Paul, 1961. Repr. 1974.

Zalta, E. N. (1995–) *Stanford Encyclopedia of Philosophy*, https://plato.stanford.edu/.

PART I

The cultural, scientific, and philosophical context and the development of logical empiricism

1

THE FOUNDATIONAL CRISIS OF MODERN PHYSICS AND ITS CULTURAL SIGNIFICANCE

Michael Stöltzner

The foundational crises in logic and physics that occurred during the first decades of the twentieth century had a strong influence on the formation of logical empiricism. Some of the movement's key members were involved in efforts to overcome them by providing new foundations to logic, mathematics, and physics. This includes Philipp Frank, Hand Hahn, and Richard von Mises, who had assumed professorships (in physics, mathematics, and applied mathematics, respectively) already before 1914, and those who pioneered the philosophical debates centering on these crises during the 1920s, among them Moritz Schlick, Hans Reichenbach, and Rudolf Carnap. The aim of this chapter is to put these crises into a broader historical context and to investigate: (1) how logical empiricists combined the diagnosis of crisis with the demand for a new construction (*Neuaufbau*); (2) how their historiography of this process oscillated between close-ups and long-shots (compare Kracauer 1969); (3) how combining both perspectives enabled a scientific philosophy flexible enough to address distinct scientific, philosophical, and social challenges.

From the perspective of logical empiricism, it seems natural to identify the crisis in physics with the theories of relativity developed by Einstein in 1905 and 1915. After all, both were inspired by Mach's criticism of Newtonian mechanics and aimed at the core of Kant's a priori justification of the Newtonian framework, the pure intuitions of Euclidean space and absolute time. Einstein was the towering physicist of a generation, an icon of modernism, and a primary target of the political enemies of the young Weimar Republic. Modern theoretical physics of Einstein's ilk, so the propagandists of the *Deutsche Physik* (German Physics) to come, was mathematically abstract and unintuitive (*unanschaulich*). The defensive belt around relativity theory figured key logical empiricists: Schlick had written a much-praised philosophical study and took part, at Planck's behest, in a plenary session at the 1922 *Naturforscherversammlung* - the centennial meeting of the largest German society of natural scientists and physicians - arranged to win German scientists and physicians for the theory so vigorously attacked in the press (Schlick 1917, 1922). Reichenbach's involvement ranged from combative popular articles to a new axiomatics (Gimbel 2006). Frank had earned his Prague professorship, succeeding Einstein, with early work on the special theory; he criticized the philosophical roots of the demand for *Anschaulichkeit* (visualizability) (1928), and wrote one of Einstein's first biographies (1947). There he divided anti-relativists into three camps: right-wing propagandists and anti-Semites, unimaginative experimental physicists, and traditional scholastic philosophers.

 DOI: 10.4324/9781315650647-3

Yet, Frank and other logical empiricists often adopted a broader perspective and, especially during the 1920s and 1930s, involved themselves in fights about the interpretation of physical theories, primarily to combat what they viewed as a dangerous return to metaphysical scholasticism. Moreover, the fights about relativity theory cannot be seen in isolation from other physical and social developments that occurred on different time scales. The demise of the old order and the political crises after 1918 coincided with ruptures in multiple scientific fields; some had already been settled internally but now entered the philosophical or cultural discourse, while others were felt within the scientific community but would shape the broader perception of physics only at the end of the decade.

After characterizing the subtle dialectic between close-ups and long shots in the historiography of logical empiricists and their allies, I will discuss their role in the fight for scientific modernism in the 1920s with some examples. To disentangle the various physical fields and discursive formations, I will give a brief sketch of the demise of classical physics, analyzing the debates about atomism and statistical physics, the relativity theories, and quantum mechanics.

Crises and new constructions: close-ups and long shots

In 1929, Frank was president of the biennial congress of German mathematicians and physicists that convened at Prague. Presented to a satellite meeting organized by the Vienna Circle and its Berlin allies, the Circle's famous manifesto called for a consistent scientific world-conception. Frank's opening speech (1929) did not outline a historical narrative for a movement but tried to interest the many scientists assembled at Prague in a new scientific epistemology replacing the traditional philosophical views still common within the community. Frank mainly targeted Emil Du Bois-Reymond's (1872) distinction between scientific questions still unresolved (*Ignoramus*) within the then prevailing mechanical worldview, and philosophical riddles forever unsolvable (*Ignorabimus*) by science. This distinction blatantly contradicted the optimistic thrust of the manifesto: "The scientific world conception knows *no unsolvable riddle*" (Verein Ernst Mach 1929/2012: 82, orig. emphasis). While there was some sympathy for Frank's call for a fresh start, his critics, among them Arnold Sommerfeld, were not swayed by the combination of Machian empiricism, James's pragmatism, and French conventionalism.

In later debates about the interpretation of quantum mechanics, Frank zoomed from the same long shot to a different close-up. It would be a fatal mistake to believe that the fall of the mechanical worldview and the abundance of abstract mathematical entities implied that "the universe is now more like a great [organic] idea than a great machine" (1935/1987: 112). In actual fact, only "the equations have changed, the [measured] quantities are different. . .; but the general scheme according to which a physical theory is constructed still has the same fundamental character today as it had in Newton's time" (ibid.: 128–9). Neither in Du Bois-Reymond's days nor in the age of quantum mechanical indeterminism was there any space for metaphysical inquiry dealing with problems unresolvable with scientific means.

It would be a mistake to take Frank's rhetorical strategy simply as another instance of the flexibility that distinguished the scientist-philosophers of the German-speaking world from traditional system philosophers, including those who fully integrated the findings of modern science, such as the neo-Kantian Cassirer. By the mid-1930s, logical empiricists had settled on a spectrum of methods they considered as key to a scientific philosophy of science.

In 1933, Karl Menger organized a lecture series in Vienna whose introduction acknowledged that "in recent decades, . . . [the natural sciences] have been shaken by severe crises." But a "*Neuaufbau* has begun in each and every field" (Mark et al. 1933: iii, trans. MS). The chemist Hermann Mark showed that the demise of the mechanical worldview began with experimental discoveries in the late nineteenth century. The physicist Hans Thirring described three fields in which increasingly drastic changes of the basic concepts of substance, space, time, and causality were required: (1) atomism and statistical physics; (2) relativity theory; (3) quantum theory. These three developments occurred on different time scales: The atomism controversy was resolved when relativity entered the scene. Quantum theory developed in different stages from Planck's radiation formula of 1900 and the Bohr–Sommerfeld theory of 1913 to the quantum mechanics of 1926. Despite their long historical arc, these foundational changes came together after 1918. Eddington's 1919 solar eclipse expedition brought experimental evidence for Einstein's largely theoretical breakthrough of 1915. The older Bohr–Sommerfeld quantum theory increasingly became untenable in the calculations of the early 1920s, giving way to quantum mechanics that had its own its interpretational challenges.

Menger's collection featured an essay by Hahn on the crisis of *Anschauung* that painstakingly explained why modern mathematics had moved away from the intuitive foundations it had received in Kant and much of nineteenth-century mathematical practice. To keep geometry and (differential and variational) calculus consistent, and able to deal with new physical theories, new conceptual foundations and techniques had become necessary. Hahn himself was convinced that these foundations were closely connected to a new understanding of logic, but both crises were not identical. The specific foundational crisis in mathematics arose from set-theoretic paradoxes discovered at the turn of the century; it prompted three different foundational programs championed by eminent mathematicians. The new understanding of logic promised a consistent empiricism in assessing physical theories and fundamentally changed the nature of philosophical activity. In the early 1930s, Gödel's incompleteness theorem showed that at least the philosophical hopes for logical foundations were unwarranted and that the crisis had not been resolved completely.

Physics and mathematics were not the only fields that underwent crises in the early 1920s. In biology, the traditional developmental mechanics had found viable competitors due to the availability of new experimental strategies that allowed searching for epigenetic and vitalist factors (see Hofer 2013). The crisis of psychology (Bühler 1926) had resulted from the pluralism of approaches that had not yielded a broadly shared methodological standard. Even among logical empiricists, there was disagreement on whether to favor behaviorism or gestalt psychology.

Amid the turmoil: Spengler's challenge to scientific modernism

There were compelling external reasons for the sentiment of crisis. The demise of the established order after 1918, the political fights in the young Austrian and German republics, and the experiences of a war based on twentieth-century technology provided the backdrop for broad anti-scientific and irrationalist tendencies that are best expressed in Oswald Spengler's monumental *The Decline of the West*. Spengler's picture of Western science was shaped by the mechanical worldview; he took the statistical laws long successfully applied throughout physics as a surrender to fate, a deep crisis after which the age of science would eventually give way to an epoch of belief. No wonder that also relativity theory was often seen as a proof of an all-encompassing relativism threatening the cultural organisms of Spengler's historical morphology.

In a controversial study of the popular and philosophical writings of German physicists, Paul Forman (1971) has taken Spengler's crusade against modern science as emblematic of the demands of the cultural milieu of the early Weimar Republic that was characterized by the rise of *Lebensphilosophie*, existentialism, and vitalism, and that caused physicists to abandon prematurely the requirement of a causal description of atomic phenomena. But Forman not only overlooks the long arc and specificity of the causality debate among German physicists that reached back to the debates about statistical mechanics (see Stöltzner 2011), but he also isolates atomic physics from the other physical disciplines, especially the fight about relativity theory.

While I will address the physical debates in the following sections, let me point out that even the issue of vitalism is not that simple. Driesch and his allies could point to newly discovered scientific phenomena, the self-restoration and self-organization in lower organisms that could not be explained by the traditional physiological and physico-chemical approaches. But the second law of thermodynamics seemed to pose a nomological limit to the formation of complex organisms unless one could identify some additional force that Driesch called entelechy. This alternative harked back to debates about statistical physics waged more than a decade before, but it was nurtured by the fact that several issues about entropy and the second law had remained unclear since then. Only when the philosophical ambitions of vitalists became more outspoken, in the cultural climate after 1918 and during the discussion about the interpretation of quantum mechanics in the 1930s, did Frank and other logical empiricists feel compelled to attack them as part of their broader anti-metaphysical agenda.

All this is not to deny that German-speaking physicists reacted to the cultural and social challenges. Most of them were traditional *Bildungsbürger* (educated bourgeois), conservative or liberal, some had anti-Semitic tendencies, few were politically progressive. But the combination of their political, cultural, and scientific identities took different forms. Here are two examples. The Viennese experimental physicist Franz Serafin Exner was thrilled by Spengler's cultural morphology; he lamented the decline of culture into civilization and opposed the artists of the Vienna Secession. Exner was a typical advocate of nineteenth-century academic liberalism (Hofer and Stöltzner 2012) and an academic mandarin (Ringer 1969). Early on, Exner declared that Boltzmann's indeterminist foundations of physical science represented progress. For him, statistical laws kept modern society predictable and allowed the operation of the rational state that guaranteed the mandarin's role. Exner even extended the probabilistic interpretation of the second law of thermodynamics to a theory of culture (see Stöltzner 2002).

Max Planck, in contrast, plead for a determinist basis of the phenomena cited by Exner. He rejected Machian empiricism in favor of a neo-Kantianism with a realist twist and an emphasis on general principles. Planck secured the acceptance of special relativity in the physics community even though he gave it an anti-Machian twist. Planck was among the first physicists who worked exclusively on theory, and whose chair was not ranked lower than that of the local experimentalist. He signed the infamous "Manifesto of the Ninety-Three," giving unconditional support to the German military in 1914, but stopped short of endorsing Wilhelm Wien's manifesto against international collaborations and the use of the English language, which Exner signed, if only reluctantly. Both manifestos burnt bridges for many years after 1918. Fighting for the revival of German physics, Planck became a devout institutionalist and steered financial support to those young theoretical physicists who would bring about a quantum revolution that he himself met with skepticism. He also supported the academic careers of Schlick and Reichenbach.

A key forum for the broader debates among German scientists was the weekly *Die Naturwissenschaften*, especially after the *Physikalische Zeitschrift* and Ostwald's *Annalen der Naturphilosophie* ceased publication. Its founding editor, Arnold Berliner, was an industrial physicists and man of culture who provided ample space not only to broad expositions of scientific discoveries and novel theories, but also to their philosophical and social implications. This included rebuttals of the critics of relativity theory and Spengler's *Decline*. Especially during the 1920s, Berliner's journal provided the main venue of publication and reviews for the future logical empiricists with a scientific background and a forum presenting their views alongside those of Heisenberg, Born, Bohr.

Fin-de-siècle physics and the cracks in the mechanical worldview

At the time of Du Bois-Reymond's *Ignorabimus* speech, many still cherished hopes to obtain a mechanical explanation of all physical phenomena. Yet electromagnetism would defy any reduction even in the sense of Maxwell's mechanical analogies. Many thus abandoned the ideal of explanation altogether in favor of an exhaustive description of physical phenomena. Instead of mechanical reduction, a different promise of unity emerged: general principles, such as the conservation of energy and the principle of least action, extended their sway across several domains. Planck went with the abstract principle, but its seemingly tangible nature and technological significance favored energy. Were all natural phenomena just transformations from one form of energy into another? Led by the physical chemist Ostwald, energeticism became a powerful if short-lived, and primarily German, movement that extended its sway far beyond the domains of exact science into the broader culture (Ostwald 1912).

Fin-de-siècle physics brought many experimental inspirations for the twentieth century. Spectroscopists collected large amounts of data and condensed them into semi-empirical formulas that would inform Bohr's quantum theory. The many technical studies on incandescent light bulbs allowed for a quick test of Planck's 1900 semi-empirical formula describing black-body radiation. Michelson–Morley's attempt to measure the properties of the luminiferous ether—a strange interstellar medium of very low density that seemed essential for Maxwell's electrodynamics—came out negative. Lorentz's problem-fix put the equations of the special theory of relativity already on the table, but without the universal generalization they would obtain from Einstein. The investigations of electric discharges in vacuum tubes brought three discoveries in a row: X-rays by Röntgen in 1895, radioactivity by Bequerel in 1896, the electron by Thomson in 1897. The first caused an enormous stir because it brought pictures of a new world for immediate use by medicine and technology. The list of alleged and real discoveries of new rays grew rapidly (Kragh 1999: 37). Kragh concludes that to laymen used to mechanical thinking, electrodynamic phenomena had a taste of immateriality "sometimes characterized as neoromantic" (1999: 10). It would still take a while until radioactivity changed from a miraculous property of certain substances, over whose acquisition physicists and spa doctors fought bitterly, into a systematically studied phenomenon that eventually challenged the determinist worldview.

Statistical physics and the fight about atomism

While Boltzmann saw himself still as a loyal scion of the mechanical worldview, members of the younger generation, among them Sommerfeld, were rather swayed by atomism's unifying mathematical power, in contrast to the often handwaving reasoning of energeticists. In their eyes, Boltzmann had emerged victoriously from the debate with Helm at the 1895 *Naturforscherversammlung*. To Ostwald and Mach, instead, mathematization and unification under a few principles provided only a practical advantage. Their goal was a descriptivist physics that could eventually be freed of theoretical hypotheses, most prominently atoms, and whose basic quantities could be traced back to direct measurements, to sense physiology. Planck agreed with Boltzmann that the second law could not be reduced to energy conservation but was initially loath to accept his statistical methods before being forced to do so gradually in black-body radiation.

Statistical methods had been well entrenched in the social sciences and technology. When they extended their sway into foundational physics, the problem arose as to how to interpret foundational probabilities. In 1919, Richard von Mises succeeded in putting Fechner's relative frequency interpretation on a sound mathematical basis, but problems remained until deep into the 1930s. The interpretation of probability would remain controversial also among logical empiricists until the 1940s and influence their understanding of causality.

Boltzmann and other advocates of statistical mechanics looked at the second law for a proof of atomism, but the transition between the microscopic level of random atomic collisions and the macroscopic level of observable classical physics required assumptions, most prominently the *Stosszahlansatz* (molecular chaos hypothesis) and ergodicity, that remained murky deep into the twentieth century. Boltzmann faced two notable and pretty intuitive objections, the reversibility paradox and the recurrence paradox, that he could answer only in a highly abstract manner by referring to the typicality of initial conditions and estimating the time scale of recurrence. Another question was whether the second law prevented the emergence of local order necessary for life. If such a local violation of a statistical law was compensated by far-away regions of the universe, would their clocks tick in the opposite direction? The irreversibility of the statistical law was considered the only way to introduce a direction of time into a world of mechanics and energy conservation whose basic laws were invariant under time-reversal. Hence, the question about the relationship between the microcosm and the macrocosm was passed on to cosmology. But a physical cosmology without large-scale data and without even understanding of what made the sun shine was speculative at best. At any rate, there was enough material for academic polemics about the eternal question of atomism and fodder for the popular and philosophical literature contemplating the heat death of the universe.

While the precise foundations of the second law remained a matter of debate—one that grew increasingly mathematical—the question of atomism was resolved in a surprisingly *anschaulich* way that swayed theoreticians and experimentalists alike. The irregular motions of small particles, e.g., gamboge or smoke, suspended in a liquid or in air had been known for decades since the botanist Robert Brown first saw them under the microscope. They had resisted diverse attempts at explanation, including atomistic ones, which all suffered from misunderstanding the randomness of the motions. The decisive step, accomplished simultaneously by Einstein and Exner's former assistant Marian von Smoluchowski, consisted in interpreting Brownian motions as fluctuation phenomena, that is, as witnesses to a domain of transition in which the thermodynamic limit was not yet reached and fluctuations in the summary effect of very many impacts of molecules of the liquid or air on the Brownian particle were measurable. Fluctuation physics would also provide a handle on the indeterministic nature of radioactive decay. Thus, atomism, a problem that had tormented philosophers for millennia, was confirmed in a tangible way, far from what was supposed to be a foundational domain. To be sure, there remained rather abstract problems and paradoxes that would figure prominently in the debates about the proper foundations of probability. The story of the empirical conformation of relativity theory, on the other hand, was markedly different.

Relativity theory: a crisis overcome meets a political crisis

Special relativity elevated the Lorentz transformation from electrodynamics to a universal property of all physical processes. In doing so, Einstein critically analyzed and reinterpreted concepts as basic as simultaneity and causal influence. While he could cite the negative outcome of the Michelson–Morley experiment, there were no measurements precise enough to supply decisive experimental evidence for relativistic effects in the mechanical realm. Yet Einstein's 1905 paper convinced Planck, who used his clout to secure the new theory a surprisingly quick acceptance in the theoretical physics community, purging it from Machan positivism. Einstein himself presented relativity theory in seemingly *anschaulich* thought experiments about trains and embankments, and a young Max Laue (1911) penned a popular description of the principle of relativity to provide a basis for the philosophical debates where the theory met resistance from neo-Kantians—with the notable exception of Ernst Cassirer.

Surprisingly, also Machians, among them Friedrich Adler, rejected relativity theory as grounded in abstract hypotheses, although Einstein had adopted Mach's critique of Newton's three laws and absolute space and time as a motivation for the theory. When the young relativist Frank came to Prague in 1912, he faced the combined criticism of Machians and metaphysicians. This influenced his later presentation of Mach as a critic of the mechanical worldview and advocate for a renewed enlightenment.

The special theory of relativity had internal shortcomings. Clarifying them in general relativity, Einstein carried the revolution even further by geometrizing space and time and equating them to matter-energy. The general theory of relativity was highly abstract and based on novel mathematics that could no longer be pictured in Minkowski diagrams. Once again, Einstein devised thought experiments; Laue, Max Born, and Einstein himself penned popular presentations. Philosophers like Schlick and Reichenbach held that relativity required a renewal of philosophy (see CH. 9). There was enough material for scholarly polemics and public controversy. But the fights about relativity theory after 1918 obtained a decidedly political dimension.

There were only two routes to test the predictions of general relativity. The anomalous perihelion motion of Mercury was known for a long time, but physicists had learned to live with it in various ways, including effective theories of gravitation that made minor modifications to Newton's laws. To measure the deflection of star light by the sun was difficult and required an expedition to observe a solar eclipse. When Eddington reported on the 1919 British expedition in *Nature*, he announced the confirmation of Einstein's theory. While the British physics community initially reacted with skepticism, abandoning Newton's worldview soon became one option in a scientific and comparably tempered public debate. When the news made it into the German press, the reaction was markedly different (see Rowe 2006). A British expedition had confirmed the theory of a German physicist whose pacifism and progressive attitude were well-known. Einstein's theory was met with fierce criticism that no longer could be alleviated by several popularization attempts that also involved Berliner's *Naturwissenschaften*. The experimental physicist Ernst Gehrcke penned a series of increasingly polemical papers that targeted, in a mostly superficial way, the weak spots of Eddington's reasoning and advocated alternatives; Einstein's thought experiments provided ample ammunition for popular critics who dragged them through a classical world of philosophical prejudice and common sense. The polemics merged into the existing political struggles of the early Weimar Republic. Did the internationalist Einstein conspire with British physicists who had shut out Germans from the international community? Was Einstein advocating an abstract mathematics foreign to German science and technology? Of course, anti-Semitism had its part. The public fight also split the broader community at the 1920 *Naturforscherversammlung*, the first after the war, that featured a short debate between Einstein and Philipp Lenard. The frontlines for the fights splitting the German physics community were drawn, and by and large they followed the existing political front lines. Frank reports an anecdote about Einstein being told by a French historian that the supporters (and opponents) of his theory were the supporters (and opponents) of Dreyfus. Relativity would become a comparable wedge issue in the Germany of the 1920s and 1930s.

Quantum mechanics: a crisis ended giving way to a crisis of interpretation

In 1911 Rutherford had shown that atoms were largely empty, which he explained by negative charges orbiting a positive nucleus. But this violated the basic laws of electrodynamics. The

Bohr–Sommerfeld model suspended its validity for stable orbits and allowed a mathematically sophisticated characterization of the transitions between different energy levels described by three quantum numbers. In 1922, the Stern–Gerlach experiment demonstrated that there was indeed a third quantum number, delivering a final blow to classical electrodynamics. But the Bohr–Sommerfeld model did not enjoy this confirmation for long. It ran into a crisis itself: Born's 1923 attempt to obtain a description of multi-electron atoms, such as helium, by means of a perturbative expansion failed miserably.

The resolution of the crisis in atomic theory came in two versions that appeared starkly different at first but quickly turned out to be formally equivalent, Schrödinger's wave mechanics and Heisenberg's matrix mechanics. While Heisenberg's theory was highly abstract and less suitable for practical calculations, Schrödinger's theory appeared more *anschaulich*, dealing with some non-classical waves that lived in a higher-dimensional space. Both brands of quantum mechanics allowed theoreticians quickly to come up with many results concerning atomic phenomena; Dirac's equation generalized quantum mechanics to relativistic phenomena, and the 1930s saw further attempts to arrive at a quantum theory of the electromagnetic field. The new construction of atomic and subatomic physics had quickly come into gears. But a new crisis had emerged.

The foundational crisis brought about by quantum mechanics was neither a failure of theory nor a lack of empirical confirmation. It was a struggle about interpretation that left most practicing physicists uninterested until the rise of quantum optics in the 1980s. In what was later dubbed the Copenhagen interpretation developed by Bohr and Heisenberg, there was, for one, a deterministic evolution of a wave packet that consisted of a superposition of many quantum states. But the interaction with a macroscopic measurement device trigged a collapse of this wave packet into a single value. The wave function itself had no physical meaning; only its absolute square described the probability to find a particle in a given state.

Thought experiments, such as the Heisenberg microscope and the double-slit experiment, devised to justify the Copenhagen interpretation, showed how much the new theory violated our traditional intuitions and contradicted the concept of a particle trajectory that was at the roots of Newtonian mechanics. Heisenberg read the Copenhagen interpretation as a positivist restriction to observable quantities. Einstein and Schrödinger, two advocates of statistical laws, disagreed with Heisenberg's brand of positivism and devised famous through experiments to argue that quantum mechanical descriptions were incomplete: the cat paradox and the EPR problem. They became emblematic in the interpretational debates to come.

While the Copenhagen interpretation was largely accepted in the physics community, it attracted a plethora of traditional metaphysical questions. If indeterminism was at the roots of physics, was there a scientific basis for free will? Since atomic transitions occurred between two states, was there any return for teleological factors? Heisenberg's uncertainty relation posed a limit on the simultaneous measurability of certain physical quantities. Bohr developed this fact into a theory of complementary descriptions that he extended into the relation between animate and inanimate matter. Pascual Jordan, one of the leading quantum physicists, developed a sympathy for vitalism. The growth of the related philosophical and popular literature in the 1930s prompted logical empiricists to try to purge the Copenhagen interpretation of metaphysics and integrate its philosophical consequences into their philosophical outlook. By that time, however, they had not been involved in the logical analysis of the theory in a way comparable to that of relativity theory. This only changed in the 1940s.

References

Bühler, K. (1926) "Die Krise der Psychologie," *Kant-Studien* 31: 455–526.

Du Bois-Reymond, E. (1872) *Über die Grenzen des Naturerkennens*, Leipzig: Veit.

Forman, P. (1971) "Weimar Culture, Causality, and Quantum Theory, 1918–1927: Adaption by German Physicists and Mathematicians to a Hostile Intellectual Environment," *Historical Studies in the Physical Sciences* 3: 1–114.

Frank, P. (1928) "Über die Anschaulichkeit physikalischer Theorien," *Die Naturwissenschaften* 16: 121–8.

——— (1929) "Was bedeuten die gegenwärtigen physikalischen Theorien für die allgemeine Erkenntnislehre," *Die Naturwissenschaften* 17: 971–7 and 987–94, Trans. "Physical Theories of the Twentieth Century and School Philosophy," in Frank, *Modern Science and Its Philosophy*, Cambridge, MA: Harvard University Press, 1956, pp. 96–125.

——— (1935) *Das Ende der mechanistischen Physik*, Wien: Gerold. Trans. "The Fall of Mechanistic Physics," in B. McGuinness (ed.), *Unified Science. The Vienna Circle Monograph Series*, Dordrecht: Reidel, 1987, pp. 110–29.

——— (1947) *Einstein—His Life and Times*, New York: Knopf.

Gimbel, S. (ed.) (2006) *Defending Einstein. Hans Reichenbach's Writings on Space, Time and Motion*, Cambridge: Cambridge University Press.

Hofer, V. (2013) "Philosophy of Biology in Early Logical Empiricism," in H. Andersen et al. (eds.), *New Challenges to Philosophy of Science*, Dordrecht: Springer, pp. 351–63.

Hofer, V. and Stöltzner, M. (2012) "What Is the Legacy of Austrian Academic Liberalism?" *NTM* 20: 31–42.

Kracauer, S. (1969) *History: The Last Things Before the Last*, New York: Oxford University Press.

Kragh, H. (1999) *Quantum Generations: A History of Physics in the Twentieth Century*, Princeton: Princeton University Press.

Laue, M. (1911) *Das Relativitätsprinzip*, Braunschweig: Vieweg.

Mark, H., Thirring, H., Hahn, H., Nöbeling, G. and Menger, K. (1933) *Krise und Neuaufbau in den exakten Wissenschaften. Fünf Wiener Vorträge*, Wien: Deuticke.

Ostwald, W. (1912). *Der energetische Imperativ*, Leipzig: Akademische Verlagsgesellschaft.

Ringer, F. K. (1969) *The Decline of the German Mandarins: The German Academic Community 1890–1933*, Cambridge, MA: Harvard University Press.

Rowe, D. (2006) "Einstein's Allies and Enemies: Debating Relativity in Germany, 1916–1920," in V. F. Hendricks et al. (eds.), *Interactions: Mathematics, Physics and Philosophy, 1860–1930*, Dordrecht: Springer, pp. 231–80.

Schlick, M. (1917) "Raum und Zeit in der gegenwärtigen Physik: Zur Einführung in das Verständnis der allgemeinen Relativitätstheorie," *Die Naturwissenschaften* 5: 161–7 and 177–86. 3rd enlarged ed. (1920) trans. *Space and Time in Contemporary Physics*, Oxford: Oxford University Press, repr. with trans. of rev. 4th ed. (1922) in Schlick (1979), pp. 207–69.

——— (1922) "Die Relativitätstheorie in der Philosophie," *Verhandlungen der Gesellschaft Deutscher Naturforscher und Ärzte* 87: 58–69. Trans. "The Theory of Relativity in Philosophy," in Schlick (1979), pp. 343–53.

——— (1979) *Philosophical Papers, vol. 1 (1909–1922)* (ed. by H. L. Mulder and B. van de Velde-Schlick), Dordrecht: Reidel.

Stöltzner, M. (2002) "Franz Serafin Exner's Indeterminist Theory of Culture," *Physics in Perspective* 4: 267–319.

——— (2011) "The Causality Debates and Their Preconditions. Revisiting the Forman Thesis from a Broader Perspective," in C. Carson, A. Kojevnikov and H. Trischler (eds.), *Quantum Mechanics and Weimar Culture. Selected Papers by Paul Forman and Contemporary Perspectives on the Forman Thesis*, Singapore: World Scientific, pp. 505–22.

Verein Ernst Mach (1929) *Wissenschaftliche Weltauffassung. Der Wiener Kreis*, Vienna: Wolf. Trans. "The Scientific Conception of the World. The Vienna Circle," in O. Neurath, *Empiricism and Sociology* (ed. by R. S. Cohen and M. Neurath), Dordrecht: Reidel, 1973, pp. 299–318; rev. trans. (with orig. annotated bibliography) "The Scientific World-Conception. The Vienna Circle," in F. Stadler and T. Uebel (eds.), *Wissenschaftliche Weltauffassung. Der Wiener Kreis. Hrsg. vom Verein Ernst Mach (1929)*, Vienna: Springer, 2012, pp. 75–116.

2

THE GERMAN YOUTH MOVEMENT AT THE START OF THE TWENTIETH CENTURY AND LOGICAL EMPIRICISM

Hans-Joachim Dahms

One can talk about generations in two different meanings of the word. In a weak sense, it covers cohorts of people of similar age; in a strong sense, it includes awareness of common goals and attitudes, and also common organisations and media. In twentieth-century Germany there were at least three different "youth generations" in the strong sense of a "youth movement" (*Jugendbewegung*): first, the one active before, during, and right after the First World War (whose members were born around 1890); as the third, the "generation of 1968" (born around the end of the Second World War); and in the middle, alas, the young Nazi students (born shortly before the First World War).

This chapter concerns the first of these youth generations. There exists a large literature about it consisting, first, of the memoirs of participants (see Kindt and Vogt 1964) and, since their passing, of a variety of reassessments of important events like the march to the Hohe Meissner in 1913 in protest against German militarism, which deliver a very divided picture. There are the very critical judgments by Laqueur (1962), Pross (1964), and, most recently, Niemeyer (2013), who all emphasize a dangerous romanticism and right-wing, even racist tendencies; but there are also positive assessments by Reulecke (2009) or Wipf (2004), who underline its positive aspects as a protest movement against the authoritarian Wilhelminian Empire. These divergent assessments are due in large part to different strata of the youth movement being under investigation: whereas the critical ones treat more the non-academic chapters of it, like the Wandervogel (Wandering Bird) movement at the high school level, the more positive ones focus on the university sections of the youth movement like the Freistudentenschaft (Free Students), the association of non-incorporated students (students not belonging to the traditional fraternities). Here, we deal exclusively with the latter. As their cultural background, the youth movement gives an important insight into the early intellectual development of Rudolf Carnap, Hans Reichenbach, and several other members of the Berlin Group. It also represents an important historical context which can explain some of their later social and political engagements.

The Free Student movement

A suitable point of departure is this paragraph from Carnap's "Intellectual Autobiography":

> Before the war, I, like most of my friends, had been uninterested and ignorant in political matters. We had some general ideals, including a just, harmonious and

DOI: 10.4324/9781315650647-4

> rational organization within the nation and among the nations. We realized that the existing political and economic order was not in accord with these ideals, and still less the customary method of settling conflicts of interests among nations by war. *Thus the general trend of our political thinking was pacifist, anti-militarist, anti-monarchist, perhaps also socialist.* But we did not think much about the problem of how to implement these ideals by practical action. The war suddenly destroyed our illusion that everything was already on the right path of continuous progress.
>
> (Carnap 1963: 9; emphasis added)

The Free Student movement started in the late 1890s in several German university towns, before it became a nationwide organization in 1910. It aimed to organize all the non-incorporated students and to offer help with housing, extra-mural work-offers, cheap access to books, etc. (Jarausch 1984: 97–98). These efforts were welcomed also outside the movement itself and helped to spread its more controversial ideas.

Two convictions characterized the academic youth movement: first, its anti-bourgeois rejection of the traditional fraternities with their rituals of heavy drinking and sable-fighting and their political conservatism and chauvinism; second, their resolve to lead a self-determined, liberated way of life: no alcohol, no tobacco, sometimes also vegetarianism. Women were admitted as members to the organization (unlike in the fraternities). In addition, a common life closer to nature was celebrated (long hikes, open-air parties, and dancing). Since some of their principles (like the admission of women) were controversial, they embraced a "principle of tolerance" in order to maintain the unity of their organization. One of them, Karl Korsch, even spoke of "a standpoint of absolute tolerance" (1909). That said, there was no tolerance of anti-Semitism in the Free Student movement (unlike in most student fraternities and in many chapters of the Wandervogel groups). Some sought to realize what they spoke of as their "self-education" in collective activities to overcome the constraints of traditional *Bildung* and gain a better understanding of the driving forces of modern life. These included contemporary art, like expressionist painting, and modern architecture as well as modern science, like contemporary physical theories and Freudian psychoanalysis, even "economy, geography, statistical foundations, contemporary history" (Jarausch 1984: 98).

The historian Hans Ulrich Wipf (2004: 14; for more, see Stambolis 2013) counts Walter Benjamin, Siegfried Bernfeld, Alfred Döblin, Kurt Lewin, Helmuth Plessner, and Arnold Zweig, besides Carnap, Korsch, and Reichenbach, among the best-remembered members of the Free Student movement, and adds: "Common to all of them is the fact that they had to flee national-socialist terror." Notably, of these Benjamin, Carnap, Plessner, and Reichenbach became philosophers, while the psychologist Lewin and the law professor and political activist Korsch developed strong interests in philosophy and logical empiricism. Lewin attended the famous Erlangen Conference of 1923 (see CH. 10) and Korsch frequently visited, sometimes with Bertolt Brecht, events organized by the Berlin Group of logical empiricists (Buckmiller 1994: 114–15, 126–7).

The free-student groups in Jena, Berlin and Göttingen

Of the numerous academic groups belonging to the youth movement, only the Jena and Berlin chapters can be considered here, though the Göttingen group must be mentioned too.

Carnap was a member of the romantic Sera Circle (see Werner 2003, 2014) and also belonged to the Free Student movement. In Jena, both groups merged in 1911. Carnap specified what he found attractive in an unpublished portion of his autobiography:

> For those whose work is of a purely theoretical nature, there is the danger of a too narrow concentration on the intellectual side of life, so that the properly human side

> may be neglected. I think it was very fortunate for my personal development during these decisive years (i.e.: student years) that I could participate both in Freiburg and in Jena in the common life in such fine and inspired groups of the Youth Movement.
> (Carnap papers, folder EM 03; see also Carus 2007: 50–56)

Carnap averred that "the aim was to find a way of life which was genuine, sincere, and honest, in contrast to the fakes and frauds of traditional bourgeois life, a life, guided by the own conscience and the own standards of responsibility and not by the obsolete norms of tradition" (ibid.) Carnap never felt as much at home as in his Jena circles. About his time in Vienna he wrote: "None of the members of the Vienna Circle had taken part in the youth movement, and I did not feel myself strong and productive enough to transform singlehandedly the group of friends into a living community, sharing the style of life which I wanted." His feelings of alienation grew still further after his emigration to the USA (Carnap papers, folder CM 03, B 36).

The Sera Circle appears to have had around 100 members before the First World War. Important for Carnap were Wilhelm Flitner, later on a well-known professor of pedagogy in the tradition of Wilhelm Dilthey's and Herman Nohl's *Lebensphilosophie* (philosophy of life) in Hamburg and Tübingen; Hans Freyer, the first designated professor of sociology in Germany in Leipzig from 1925; and Franz Roh, the art historian and art critic and the mastermind behind the famous Mannheim 1925 exhibition *Neue Sachlichkeit* (New Objectivity), which started another cultural movement.

The Göttingen chapter of the Free Student movement (see Dahms 1996) was stronger than elsewhere. It was led by the charismatic philosopher Leonard Nelson, who gathered around him young students from different disciplines who later became leading academics (Peckhaus 1990). Nelson tried to revive a minor Kantian tradition, embodied by Fries and Apelt in the nineteenth century, and to make these ideas fruitful for a proper understanding in mathematics and physics, later on also in political philosophy. His group belonged for a while to the founders of a small nationwide sub-group of the movement, the Akademischer Freibund, who were more outspokenly left-wing liberal. Nelson published in 1910 together with members from other local groups a program-brochure, which leans heavily towards social liberalism (Nelson et al. 1910). So, it comes as no surprise that the group struggled against the conservative fraternities who tended to disturb their public events and discussions and against the authorities of the university as well.

Hans Reichenbach and his brother Bernhard, later a journalist and politician, belonged to the Berlin chapter of the Free Student movement, as did the psychoanalyst Bernfeld and Walter Benjamin. Notably, two of the most important scientific philosophers besides Reichenbach in the later Berlin Group of logical empiricism (see CH. 13) had come from Göttingen: Kurt Grelling, after constant quarrels with the dominant Nelson, and later also Walter Dubislav. Both are remembered mainly for their achievements in mathematical logic, Grelling for the paradox named after him, Dubislav for a book on definition (1926) which went through several editions. But both their lives ended in tragedy: Dubislav committed suicide in his exile from Nazi Germany in Prague in 1937, Grelling was captured during the German occupation of France in 1941, deported to Auschwitz concentration camp, and killed there on 18 September 1942.

Cultural and political profile of the free students

Two large-scale events shortly before the First World War characterize the profile of the youth movement. The first was the Werkbund festival in summer 1913 near Jena, the second the march to the Hohe Meissner mountain in Northern Hesse in October of that year.

The first of those gatherings was the artists festival on 7 June 1913, organized jointly by the Sera Circle and the German Werkbund (see Dahms 2021), which accompanied the congress of the Werkbund that year in Leipzig and took place some 30 km north of Jena on the meadows of the Saale River opposite the Rudelsburg castle. The idea for this appears to have come from the Jena-based publisher Eugen Diederichs who belonged to the founders of the German Werkbund, an association of architects, designers, industrialists, and publishers. This group aimed at a thorough modernization of architecture, design, and applied arts, and already then favored a certain "objectivity" (*Sachlichkeit*) of form and functionalism. Not only Sera Circle students from Jena, but also students from the school of applied arts (*Kunstgewerbeschule*) in Weimar, the predecessor of the Bauhaus, took part in the event. After song contests, performances of expressionist dance, and other cultural activities, the participants traveled on the river on rafts at dawn, and some, including Carnap, spend the night in the ruins of the castle. It became clear to the participants that they shared not only a common lifestyle but also awareness of parallel tendencies in philosophy, science, modern architecture, and art.

The remarkable convergence between this particular strand of the youth movement and modernism found a late echo in the invitation to the Dessau Bauhaus in 1929, which the "middle" Bauhaus director Hannes Meyer (between the better-known Walter Gropius and Ludwig Mies van der Rohe) extended to Otto Neurath, who together with Josef Frank had taken over the Austrian Werkbund in November 1928 with the aim to strengthen its international ties. Neurath as well as Carnap and Herbert Feigl gave talks (see Dahms 2004). Feigl described the similarities between modern architecture and the Vienna Circle's scientific world-conception in a long letter to Schlick afterwards, but Carnap's surviving notes for his five lectures in late October 1929 are most noteworthy for our purposes. In one of them he revived his early ideas about the relationship between science and life. Science, he argued, cannot function as a leader in life, but it can show which goals can be envisaged at all, which aims are compatible with each other and which not, and which means would help to attain those goals. The German sociologist Max Weber had taken this position in his famous talks "Science as a Profession" and "Politics as a Profession and Vocation," given twelve years earlier before an audience of free students (see Weber 1919). The other important talk by Carnap at the Bauhaus was about the misuse of language. It contained *in nuce* the guiding idea of his famous article "The Elimination of Metaphysics" (1932), but the examples given were different. In the published article, Carnap added an analysis of Heidegger's famous sentence "*Das Nichts selbst nichtet*" (taken from his *Was ist Metaphysik?*) as an outstanding instance of syntactical nonsense.

The second major event in 1913 was the march to the Hohe Meissner mountain in northern Hesse. It was organized mainly by free students from Jena and Göttingen in protest against the inauguration of the Völkerschlachtdenkmal (the monument to the battle of the nations) in Leipzig in October 1913 (see Reulecke 2009; Stambolis and Reulecke 2015). This massive monument, almost 100 meters tall, was built in commemoration of the battle of Leipzig in the war of liberation against Napoleon a hundred years earlier. Its inauguration was attended by the emperor William II himself (who in that year celebrated the 25th anniversary of his reign), all the German kings and noblemen, higher officers and representatives from politics and society. The Austrian heir to the throne Franz Ferdinand was also present, whose murder in Sarajevo was a decisive factor in the start of First World War less than a year later. In addition, several thousand students from the fraternities in their uniforms and with their traditional flags and 40,000 athletes attended the ceremony, all in all around 100,000 people.

Every German and also many Austrian newspaper dedicated multiple front pages to the event. But there was a notable protest against this militarist and chauvinist pomp. It was the march from the castle of Ludwigstein in Northern Hesse to the highest mountain in the area, the Hohe Meissner, twenty kilometers away, in which around 3000 members of the German youth movement participated on 10 October 1913. Carnap and Reichenbach were among them as was Benjamin (Germer 1997: 12; Brodersen 1990: 71). The Jena and the Göttingen chapters of the youth movement played a leading role in the preparation of the event. Both groups were also active in the program of the day and the Jena Sera Circle even contributed a performance of Goethe's *Iphigenie*.

The high point of the entire event was the collective pledge by those present:

> The free German youth wants to form its life after its own determination and in inner truthfulness. It defends this inner freedom under all circumstances in a unified way. In order to further the mutual understanding free German youth days will be held. All common gatherings of the free German youth take place without alcohol and nicotine.
> (Quoted in Germer 1997: 12)

This formula was a compromise between different currents among the organizers, who had to rely here as elsewhere on their principle of tolerance in order to keep the movement together. The pledge was already criticized by some at the event itself, and it can of course be criticized today as a slogan that meant something different for everyone who used it (e.g., Laqueur 1962: 51).

At the end of the meeting one of the leaders of the movement, the (then 38-year-old) Gustav Wyneken, gave a speech in which he underlined that the memory (not the Leipzig monument) of the war of liberation against Napoleon would remain the "eternal symbol of patriotism." He also insisted that never again should foreign armies march through Germany. But, he added, German troops also should abstain from invading foreign countries.

The Free Student movement in the First World War

The pacifist attitude did not remain intact when the world war began. As is well known, only very few people in Germany and Austria remained internationalists and pacifists. Among intellectuals there was dispute about the violation of Belgium's neutrality by the German high command. When the German army invaded Belgium (and destroyed the famous library at Louvain), only very few voiced criticism, whereas very many professors signed an appeal "An die Kulturwelt" (To the world of culture) in support of the aggression (Flasch 2000).

The youth movement also was divided. Some critical voices were heard. When Wyneken changed his mind and, in a speech to the Munich chapter of the free students in November 1914 (Wyneken 1915), voiced support for the German invasion of Belgium, Reichenbach wrote an open letter to him, exclaiming:

> You old ones, who inflicted on us this miserable catastrophe, you dare at all to speak to us of ethics and to set aims for our lives? You, who not even gave assurance that every living being in your cultural community had the right to personal security against the predatory tendencies of their fellow humans, you have forgone the right to be our leaders. We look with contempt on you and your glorious time.
> (Quoted in Brodersen [1990, 84]; for the ensuing correspondence between Reichenbach and Wyneken, see Germer [1997: 20–25])

Likewise, Benjamin broke every tie to Wyneken, his former teacher in Berlin, because of the latter's "treason" (Brodersen 1990: 85).

In contrast to his report in his autobiography, at the beginning of the war Carnap was not a pacifist but joined the army as a volunteer. When one of the leading members of the Sera Circle died already in November 1914, his enthusiasm received a first blow. Many more were to follow. In the autumn of 1917, Carnap was removed from the front to begin a course as an aircraft pilot, but he could not complete this training as he was sent as an officer to a military institution in Berlin which was in charge of developing wireless radiotelegraphy. By February 1918 he started to send circular letters to his Free Student friends. Having access, as a lieutenant in the capital, to the international press which described the war situation differently from the censured German press, he was well informed and developed a view strongly in favor of a *Verständigungsfrieden* (peace through negotiation), not the *Siegfrieden* (peace by victory) propagated by political nationalists. His second letter concluded: "The sacrifices of this war were not made in vain, and our dead comrades did not die in vain, if this war leads to a situation, which makes a new war—at least among Europeans and their descendants, and at least for the whole forseeable future—impossible" (quoted in Werner 2015: 479). The circular letters were stopped by order of a higher-ranking officer on 11 September 1918, only two months before the revolution and the downfall of the Wilhelminian Empire.

After the First World War: "perhaps also socialist"?

An important question concerns the political position of the youth movement and the Free Students among them *after* the First World War. After the defeat of the Central European powers in November 1918, Carnap's earlier ambivalent attitude ("perhaps also socialist") no longer held true, and he became a socialist party member. In spring 1919 in many university towns, socialist student groups emerged (Mohn 1977: 76–78). In Göttingen Nelson founded the Internationale Sozialistische Jugendbund (International Socialist Youth Association) a sort of pressure group intent on infiltrating the leftist Weimar Republic parties, first the Social Democrats and then the Communists. When they were thrown out, Nelson founded a political party of his own, the Internationale Sozialistische Kampfbund (International Socialist Fighting Front) in 1925. Immediately after the German Revolution in Berlin, the Sozialistische Studentenpartei (Socialist Student Party) was founded, with Reichenbach and his brother Bernhard as leading members. Carnap belonged to the founders of a similar group, the Sozialistische Studentengruppe (Socialist Student Group), in Jena in March 1919. It comprised only twelve members, with no female students and no other former Sera Circle members among them. The statute of the group describes well the mixture of some romantic but mostly leftist ideas:

1 The socialist students group is the association of students, who want the reign of the spirit (*Geist*) and therefore says yes to the aim of the proletarian revolution, the classless society, and put their whole personality in the service of the proletarian movement. It is created and maintained by common spirit and common direction of will and so has no fixed form.

2 . . .

3 The immediate purpose of the socialist student group is the preparation for the work towards the socialist society and in it.

4 In order to achieve this the following seems necessary:

a) a common scientific treatment of economics and social problems,
b) an active contact with the proletarian, especially the youth movement,
c) in deed, word and writings support the idea of the classless society—especially among students.

(Quoted in Steinmetz et al. 1958: 534–5)

The first big activity of the group consisted in hosting a German and Austrian congress of all those socialist groups already on the Easter weekend from 20 to 22 April 1919 in Jena (ibid.; see also Carnap's diary). Around 100 students from 22 universities in Germany and the University of Vienna attended. Visitors from the Soviet Union and Italy also gave talks. After a welcoming speech of the local Jena workers and soldiers council, an (unnamed) socialist student from Vienna gave the first speech, in which he criticized the prevailing reactionary tendencies at the universities and demanded easier access of proletarian pupils to them. Hans Reichenbach was scheduled as speaker for the Berlin Group. The participation of the Munich group was also on the program: perhaps they reported about the ideas and achievements of the Bavarian revolution (and even Otto Neurath's participation in them). That experiment collapsed only a few days later in Munich (Dahms and Neumann 1994).

The enthusiasm of the socialist student groups faded away soon with the revival of conservative and reactionary groups, which could rely on the full backing of the university authorities. After receiving their doctorates early on during the Weimar Republic, Carnap and his friends from the Jena chapter of the Free Student movement pursued different directions politically.

Importance of the youth movement for logical empiricism

Their differentiation and in some cases confrontation reflected their philosophical outlooks: Carnap, Reichenbach, Dubislav, and Grelling contributed to the development of what later on became logical empiricism. But not all of the former Free Students took that course. Flitner remained faithful to *Lebensphilosophie* in the sense of Dilthey and Nohl. Freyer even became an "*Edelfaschist*" ("noble" fascist), as Edgar Zilsel (1932) described him after the publication of his book *Revolution von Rechts* (Freyer 1931, Revolution from the Right).

Concerning the later development of logical empiricism in the USA and Great Britain, it is useful to distinguish between Carnap and Reichenbach as persons on the one hand and their philosophical teachings on the other, although it is not easy sometimes to disentangle them. The *attitudes* they developed as former members of the youth movement and adhered to in later life are the following:

- a feeling to have a mission,
- anti-authoritarianism,
- anti-traditionalism,
- internationalism,
- a predilection for controversial discussions and collective work.

These are traits they shared in part with the student movement of the late 1960s. Their predilection for life in nature and large hiking tours in the wilderness and especially their strict anti-alcoholism and anti-tobacco stance marks a striking difference though.

Concerning *philosophical and political ideas* matters were more complicated. The following principles seem to be relevant:

- a thorough scientism,
- along with it a noncognitivist position towards ethical questions, perhaps also
- the principle of tolerance.

The first two of these principles need to be discussed together. All logical empiricists believed that scientific rationality should be the role model of every rational discourse and behavior, also outside science in everyday life and politics. They also thought that there did not exist rationality outside science, with the consequence that ethics and aesthetics was relegated to the realm of feeling and emotion.

Reichenbach expressed support for ethical noncognitivism already in his dispute with Wyneken. Later on, he repeatedly warned against the hope to get advice from books of professional philosophers on matters moral and political. Instead, one should engage practically in social movements to learn about adequate conduct:

> Whoever wants to study ethics, therefore, should not go to the philosopher; he should live in the community of a group where life is made vivid by competing volitions, be it the group of a political party, or of a trade union, or of a professional organization, or of a ski club, or a group formed by common study in a classroom. There he will experience what it means to set his volitions against that of other persons and what it means to adjust oneself to group will. If ethics is the pursuit of volitions, it is also the conditioning of volitions through group environment.
>
> (Reichenbach 1951: 297)

In addition he delivered a sketch of a noncognitivist conception of ethics, according to which value judgments are hidden imperatives. In my view, this does not hold, since many value judgments cannot be translated into imperatives. (The moral evaluations of persons, groups, states, etc., and of actions in the past provides counterexamples; see Dahms [1994: 342–5].)

Carnap's development in this regard appears to be more complicated. According to Mormann (2007), towards the end of the First World War, Carnap still had strong neo-Kantian leanings with an emphasis on a cognitivist metaethics and belief in objective values and switched only later on—possibly under the influence of his Viennese friends—to a noncognitivist metaethics. By contrast, other researchers locate Carnap's noncognitivism much earlier (see Damböck forthcoming; Carus forthcoming). Whether his (to my mind untenable) view of value statements as cognitively meaningless was saved by later emendations must be left open here. We also must leave it open whether the youth movement's principle of tolerance has anything to do, apart from its name, with Carnap's famous principle of tolerance in logic and the foundations of mathematics (1934: §17). In any case, it is plain that the youth movement worked as a source of inspiration for its former members in their later lives, be it for them as individual persons or as professional philosophers, and be it positively so or negatively. The leading thinkers of logical empiricism are no exception.

References

Brodersen, M. (1990) *Spinne im eigenen Netz. Walter Benjamin: Leben und Werk*, Bühl-Moos: Elster. Trans. *Walter Benjamin: A Biography*, London: Verso, 1996.

Buckmiller, M. (1994) "Karl Korsch: Zwischen materialistischer Dialektik und positiver Wissenschaft," in Danneberg, Kamlah, and Schäfer (1994), pp. 113–30.

Carnap, R. (1932) "Überwindung der Metaphysik durch logische Analyse der Sprache," *Erkenntnis* 2: 219–41. Trans. "The Elimination of Metaphysics Through Logical Analysis of Language," in A. J. Ayer (ed.), *Logical Positivism*, New York: Free Press, 1959, pp. 60–81.

——— (1934) *Logische Syntax der Sprache*, Vienna: Springer. Rev. ed. trans. *The Logical Syntax of Language*, London: Kegan, Paul, Trench Teubner & Cie, 1937.

——— (1963) "Intellectual Autobiography," in P. A. Schilpp (ed.), *The Philosophy of Rudolf Carnap*, La Salle, IL: Open Court, pp. 3–85.

Carus, A. W. (2007) *Carnap and Twentieth-Century Thought: Explication as Enlightenment*, Cambridge: Cambridge University Press.

——— (forthcoming) "Carnap's Fundamental Philosophical Commitment: From Religious Origins to Kantian Non-Cognitivism, 1911–21," in E. Ramharter (ed.), *The Vienna Circle and Religion*, Cham: Springer.

Dahms, H.-J. (1994) "Hans Reichenbachs Beziehungen zur Frankfurter Schule—nebst Bemerkungen zum Wahren, Schönen und Guten," in Danneberg, Kamlah, and Schäfer (1994), pp. 333–49.

——— (1996) "Politischer und religiöser Liberalismus. Bemerkungen zuihrem Verhältnis im wilhelminischen Kaiserreich am Beispiel der 'Religionsgeschichtlichen Schule'," in Lüdemann, G. (ed.) *Die "Religionsgeschichtliche Schule". Facetten eines theologischen Umbruchs*, Frankfurt am Main: Lang, 225–242.

——— (2004) "Neue Sachlichkeit in the Architecture and Philosophy of the 1920s," in S. Awodey and C. Klein (eds.), *Carnap Brought Home. The View from Jena*, Chicago: Open Court, pp. 357–77.

——— (2021) "Rudolf Carnap: Philosoph der Neuen Sachlichkeit," in C. Damböck and G. Wolters (eds.), *Der junge Carnap in historischem Kontext: 1918–1935*, Cham: Springer, pp. 75–106.

Dahms, H.-J. and Neumann, M. (1994) "Sozialwissenschaftler in der Münchner Räterepublik," in C. Klingemann (ed.), *Jahrbuch für Soziologiegeschichte 1992*, Opladen: Leske + Budrich, pp. 115–46.

Damböck, C. (forthcoming) "Carnap's Non-Cognitivism and His Views on Religion, Against the Background of the Herbartian Philosophy of His Grandfather Friedrich Wilhelm Dörpfeld," in E. Ramharter (ed.), *The Vienna Circle and Religion*, Cham: Springer.

Danneberg, L., Kamlah, A. and Schäfer, L. (eds.) (1994) *Hans Reichenbach und die Berliner Gruppe*, Braunschweig: Vieweg.

Dubislav, W. (1926) *Die Definition*, Berlin: Weiss, 4th ed., Hamburg: Meiner.

Flasch, K. (2000) *Die geistige Mobilmachung. Die deutschen Intellektuellen und der erste Weltkrieg*, Berlin: Alexander Fest.

Freyer, H. (1931) *Revolution von Rechts*, Jena: Diederichs.

Germer, K. (1997) *Hans Reichenbach. Sein Leben und Wirken. Eine wissenschaftliche Biographie*, Osnabrück: Phoebe Autorenpress.

Jarausch, K. (1984) *Deutsche Studenten 1800–1970*, Frankfurt: Suhrkamp.

Kindt, W. and Vogt, K. (eds.) (1964) *Der Meißnertag 1963. Reden und Geleitworte*, Düsseldorf: Diederichs.

Korsch, K. (1909) "Monismus, Reinkevortrag, Toleranz und Freistudenten," *Jenaer Hochschulzeitung* 7.5. 1909. Repr. in Korsch, *Recht, Geist und Kultur. Schriften 1908–1918* (ed. by M. Buckmiller), Hanover: Offizin, 1990.

Laqueur, W. (1962) *Die deutsche Jugendbewegung. Eine historische Studie*, Köln: Wissenschaft und Politik.

Mohn, E. (1977) *Der logische Positivismus. Theorien und politische Praxis seiner Vertreter*, Frankfurt am Main: Campus.

Mormann, T. (2007) "Carnap's Logical Empiricism, Values and American Pragmatism," *Journal for General Philosophy of Science* 38: 127–46.

Nelson, L., Bousset, W., Cahn, E. and Ohr, W. (1910) *Was ist Liberal?*, Munich: Buchhandlung Nationalverein.

Niemeyer, C. (2013) *Die dunklen Seiten der Jugendbewegung: Vom Wandervogel zur Hitlerjugend*, Tübingen: Francke.

Peckhaus, V. (1990) *Hilbert-Programm und Kritische Philosophie. Das Göttinger Modell interdisziplinärer Zusammenarbeit zwischen Mathematik und Philosophie*, Göttingen: Vandenhoeck & Ruprecht.

Pross, H. (1964) *Jugend, Eros, Politik. Die Geschichte der deutschen Jugendverbände*, Wien: Scherz.

Reichenbach, H. (1951) *The Rise of Scientific Philosophy*, Berkeley: University of California Press.

Reulecke, J. (2009) "Hoher Meißner 1913–2013: Zum Umgang mit einem Jubiläum: Ein essayistischer Annäherungsversuch," *Jahrbuch Archiv der deutschen Jugendbewegung*, Neue Folge 4: 51–66.

Stambolis, B. (ed.) (2013) *Jugendbewegt geprägt. Essays zu autobiographischen Texten von Werner Heisenberg, Robert Jungk und vielen anderen*, Göttingen: Vandenhoek.

Stambolis, B. and Reulecke, J. (2015) *100 Jahre Hoher Meißner (1913–2013)—Quellen zur Geschichte der Jugendbewegung*, Göttingen: Vandenhoek.

Steinmetz, M. et al. (1958) *Geschichte der Universität Jena 1548/58–1958*, Jena: Fischer.

Weber, M. (1919) *Wissenschaft als Beruf*, Leizig: Duncker und Humblot. Trans. "Science as a Profession and Vocation," in Weber, *Collected Methdological Writings* (ed. by H. H. Brun and S. Whimster), London Routledge, 2014, pp. 335–53.

Werner, M. (2003) *Moderne in der Provinz, Kulturelle Experimente im Fin de Siecle Jena*, Göttingen: Wallstein.

——— (2014) "Jugend im Feuer. August 1914 im Serakreis," *Zeitschrift für Ideengeschichte* 7 (2): 19–34.

——— (2015) "Freideutsche Jugend und Politik. Rudolf Carnaps *Politische Rundbriefe*," in F. W. Graf, E. Hanke and B. Picht (eds.), *Geschichte intellektuell. Theoriegeschichtliche Perspektiven*, Tübingen: Mohr Siebeck, pp. 465–86.

Wipf, H. U. (2004) *Studentische Politik und Kulturreform. Geschichte der Freistudenten-Bewegung 1896–1918*, Schwalbach: Wochenschau.

Wyneken, G. (1915) *Der Krieg und die Jugend. Öffentlicher Vortrag gehalten am 25. November 1914 in der Münchner Freien Studentenschaft*, Munich: Steinicke.

Zilsel, E. (1932) "Die geistige Situation der Zeit?" *Der Kampf* 25: 168–76. Repr. in Gerald Mozetic (ed.), *Edgar Zilsel: Wissenschaft und Weltanschauung. Aufsätze 1929–1933*, Wien: Böhlau, 1992, pp. 45–57.

3

DILTHEY, HISTORICISM, AND LOGICAL EMPIRICISM

Christian Damböck

The topic of this chapter is complex and problematic. Historicism is a current of nineteenth-century philosophy (I will call it "objectivist historicism" or "hermeneutics") which was heavily distorted and finally extinguished in the context of major developments of twentieth-century philosophy. There is one nonstandard reading of this history, however, according to which a certain aspect of the nineteenth-century original conception—which in (2017) I called "German empiricism"—survived in the context of early logical empiricism. Obviously, this is a complicated story because those early logical empiricists who fit our narrative best (Rudolf Carnap and Otto Neurath) sometimes explicitly and polemically rejected hermeneutics and *Einfühlung* (empathy). It turns out, however, that what Carnap and (more importantly) Neurath criticized at the beginning of the 1930s where the then-distorted varieties of the old nineteenth-century conception, while they intended to uphold the original conception. The reason for this was simple. Objectivist nineteenth-century hermeneutics, as developed by scholars such as August Boeckh, Moses (Moritz) Lazarus, Chaijm (Heyman) Steinthal, and Wilhelm Dilthey, was perfectly compatible with an empiricist world-conception, whereas the then-contemporary conceptions of Oswald Spengler, Martin Heidegger, Eduard Spranger, and Otto Friedrich Bollnow, tended to be radically anti-empiricist, and even anti-scientific. In order to unravel these complexities I first sketch the long-term development of historicism and hermeneutics during the last two centuries. Then I focus on the cases of Carnap and Neurath. For reasons of space, I cannot take into account other logical empiricists such as Philipp Frank, Victor Kraft, Edgar Zilsel, and Moritz Schlick or Hans Reichenbach, although they all also might have *some* historicist background: the case studies of Carnap and Neurath illustrate a significant pattern in any case.

Historicism and hermeneutics

Nineteenth-century German historicism is characterized, first of all, by a double rejection: the rejection of crude historical speculations such as Georg Wilhelm Friedrich Hegel's idealistic reconstruction of world history and world spirit, and the rejection of Henry Thomas Buckle's positivistic attempts to explain historical developments by means of a small number of "natural laws" of history. The key aspect of what we today call "historicism" in a somewhat sympathetic spirit—in the nineteenth century, the term was mostly used in a pejorative way: there was

DOI: 10.4324/9781315650647-5

hardly any "historicist" who identified herself in this fashion—was that the task of history was shifted from the mere description of *spatiotemporal* facts (as it was dominant in all varieties of historiography until the eighteenth century) toward the level of *mental* facts. Wilhelm Dilthey identified the task of the historian to develop a "critique of historical reason" where spatiotemporal facts serve only as the historian's sources for a reconstruction of the historical changes in abstract categories and transcendental ideas. This new form of historical experience was also provocatively called "transcendental experience" by Dilthey (Damböck 2017: ch. 3, 2012). The abstract entities historicized here also comprised sociological structures, or what Lazarus and Steinthal called "*Volksgeist*." Dilthey, Lazarus, and Steinthal developed their competing empirical approaches toward the mental by means either of a certain psychological method, what Dilthey called "descriptive psychology," or of what Lazarus and Steinthal called "*Völkerpsychologie*" (see Damböck, Feest, and Kusch 2020). Dilthey's method remained important in later developments: on the one hand, as an empirical method with the inclusion of experimental approaches—here the thought psychology of the Würzburg school around Oswald Külpe and Karl Bühler as well as the Berlin school of Gestalt psychology around Max Wertheimer, Wolfgang Köhler, and Kurt Lewin could be mentioned; on the other hand, at the extreme opposite side of the epistemic spectrum, as an entirely non-empirical and non-experimental method of "*geisteswissenschaftliche Psychologie*" in the sense of the Dilthey student Spranger. The empirically minded branch of these developments was also important for Carnap's early intellectual development and served as a conceptual background to his *Aufbau* (Feest 2007).

Historicism was partly an elaboration and partly a criticism of the Age of Enlightenment. It adopted the idea that what we call "reality" is partly composed of, if not entirely constructed by means of, abstract notions in the human mind, i.e., it called on what Immanuel Kant called the "Copernican turn" of philosophy. But historicism also criticized the Kantian idea that this human reality of mental facts might be something eternally fixed and unshakable: it also meant the historicization of the mental world. The mental world, for a historicist, is *a product* of history, rather than its immutable background. In this form, historicism did not emerge earlier than 1860; previously, in the first half of the nineteenth century, historians did not question that history reflects either merely external spatiotemporal events or a reservoir of eternal ideas or an eternally fixed logic of the spirit, in the sense of Hegel. But scholars such as Dilthey, Steinthal, and Lazarus rejected these ideas around 1860.

We can distinguish a number of different approaches that frame the history of historicism from the middle of the nineteenth century up to the present. The main distinction is that between *objective* approaches, for whom the historian merely reconstructs a certain historical context, and all those approaches for whom the historian has to do significantly more. For the objectivist, different historical contexts c and $c^\star$ require different representations y and $y^\star$. Objectivists divided over how y and $y^\star$ are differentiated. On the one hand, there were those who merely replaced the idea of an eternally fixed representation of the world with the idea of representations that change through history. Given a certain x, we obtain a certain y, where the law of transition is fixed. The role of the historian was limited to getting y right, The context $c^\star$ of the historical reconstruction remains irrelevant; only the historical context c itself matters. In sharp contrast to this *aprioricist* conception (as defended by Hermann Lotze and the Southwest-German school of Wilhelm Windelband and Heinrich Rickert), *empirically minded* thinkers like Dilthey or Hermann Cohen were convinced that any representation y, being an abstract conceptual structure, is *a construction* that relies heavily on the respective context in which it is made. For them, $y^\star$ still can be *objective* because it is not *the preference* of the constructor or her *emotional state* that furnishes the context $c^\star$, but *the scientific perspective that is taken up* (the state of scientific knowledge plus the general scientific viewpoint as determined by the latter): $y^\star$ is *a* *result* of both c (the investigated context) *and* c^* (the context of investigation). What empirically minded thinkers like Dilthey and Cohen rejected was the reliance on all kinds of emotional and moral skills: the historian was neither a politician nor the founder of a *Weltanschauung*. This latter conception, rather, was invented by *subjectivist* historicists such as Gustav Droysen and Heinrich Treitschke, who took the historian to be a scientific moralist who instructed politicians. Droysen and Treitschke rejected the "eunuchoid objectivity" (Droysen) of the objectivists and (though still sharing some ideas of objectivist historicism which was simply *not enough* for them) developed a decidedly *subjective* approach toward history. To overcome the idea of objectivism also was the ambition of large parts of twentieth-century continental philosophers (then in a setting that was no longer committed to objectivism at all), who took the task of the historian (or the interpreter, respectively) to be the "destruction" (Martin Heidegger) or "deconstruction" (Jacques Derrida) of history, rather than its *re*construction.

Besides these programmatic aspects, historicism also possessed an important *methodological* perspective. Each variety of historicism was closely tied to a certain variety of *hermeneutics*, the method of *understanding*. The formula is very simple: early objectivist historicists were strongly committed to objectivist hermeneutics, whereas later subjectivists developed more subjectivist conceptions (some of them being utterly at odds with the earlier varieties). Hermeneutics emerged as *the* methodology of the humanities in German-speaking Europe already in the first half of the nineteenth century when historicism was not yet in play. Classical philologists like Boeckh and "positivist" theologians like Friedrich Schleiermacher developed hermeneutics as a method to reconstruct the very meaning of texts from historically distant sources, like the poetry of ancient Greek or the Bible. This method was based on the idea of *analogical reasoning*. We learn understanding in the course of personal communication with others. These "elementary forms of understanding" (Dilthey) provide the main principles to decipher the utterances of others. We assume that people in different cultures behave similarly to some extent and try to figure out the differences, on a strictly empirical basis. Thus, to understand the poem of an ancient poet, we first have to study all kinds of sources that allow us to reconstruct ancient language, art, politics, economy, warfare, as well as all kinds of aspects of the ancients' everyday life. On the basis of these comprehensive studies—indeed, the development of a universal *encyclopedia of the ancient world*—we reread the poem and only then try to grasp the author's intention intuitively. Intuition (*Gefühl*) does play a role in the hermeneutics of Boeckh and Schleiermacher, but in a very restricted sense. It constitutes only the final step of the process of understanding, where the hermeneuticist has to trust her "intuition which bases itself on the [perceived] similarity with the explanandum" (Boeckh 1886: 86). This final step can be successful only because it is conditioned by the accumulated empirical knowledge of the subject matter. This process of accumulation is somewhat circular, to be sure, because most of the empirical sources in question are *historical texts* and therefore have to be rendered accessible by means of hermeneutics in themselves. This "hermeneutic circle" (the term was invented by Boeckh), however, does not hamper the possibility of understanding, but rather is said to lead to a gradual sharpening of hermeneutic skills and a more holistic picture of historical knowledge (as in Otto Neurath's boat or W. V. O. Quine's web of belief). A consequence of this cumulative conception is that historians often know significantly more about the historical context than the historical actors and can exceed and correct their self-understanding, for these actors are *limited* by the inaccessibility of numerous economical, political, etc., aspects of the historical context that become visible only afterwards. This is the core feature of the objectivist and "positivist" conception of hermeneutics. Not only is it possible to grasp the meaning of a historical utterance in the way it was intended by the utterer, but, under certain conditions, it is possible to grasp that meaning even more comprehensively than the utterer herself. The role of intuition is strictly limited insofar as empirical knowledge (1) allows distinguishing between good and bad intuition (the latter being based on insufficient knowledge); and (2) restricts intuition to those intuitive scenarios of "elementary forms of understanding" familiar to the hermeneuticist from her everyday communication.

The limitation of Boeckh's objectivist conception of hermeneutics was, according to Dilthey and other historicists, that the process of accumulation of historical knowledge was thought to terminate at some point (see (Dilthey 1900/1996: 256). According to Boeckh, our knowledge would be complete and perfect, and final understanding of historical sources would have been attained. Although the apriorіcists (Lotze, Windelband, Rickert) held similar views, the empirically minded historicists thought that different contexts of investigation may lead to different outcomes and therefore rejected the idea of complete knowledge. However, even Dilthey and his allies shared the hypotheses that objective understanding is possible and that a hermeneuticist is able to understand a historical figure even better then she understood herself. Understanding can never be *final* for Dilthey, but it can be objective, and so objectively true or false (viz. relative to a certain context of investigation *c**). Dilthey's hermeneutics reduce extremely complex and remote historical scenarios to situations accessible to simple intuitions about cases of everyday communication. The "eunochoid objectivity" mocked by Droysen is exactly what "positivist" hermeneuticists like Dilthey had in mind.

The later development of (post-objectivist) hermeneutics in the twentieth century must be divided into at least two different currents. There is hermeneutics as developed by students and followers of Dilthey, such as Spranger and Bollnow who rejected the strong connection Boeckh and Dilthey saw between the natural sciences and the human sciences (historical encyclopediae consist of empirical source studies and the collection of hard facts). For Spranger, the human sciences had to be developed in complete isolation from the natural sciences, and the role of intuition was much enhanced. (Incidentally, the term "*Einfühlung*," crucial for Spranger, did not play any significant role in Boeckh or Dilthey, the latter preferring objectivist "*Nachfühlung*" [1900/1996: 235].) Bollnow supplemented his radically subjective conception of understanding with a notion of "objectivity" that no longer meant intersubjectivity and general validity but "appropriateness": there is true and objective knowledge which is not generally valid but restricted to a single knower (say, an epistemic *Führer*). Spranger and Bollnow took hermeneutics to form an exclusive alternative to the natural sciences and their epistemology of general validity. The second important current of twentieth-century hermeneutics is more explicitly directed against Dilthey. Philosophers like Heidegger, Hans-Georg Gadamer, and Jürgen Habermas rejected Dilthey's "positivist self-misunderstanding" (Habermas) and replaced Dilthey's "sovereign" notion of understanding (Gadamer) with the romanticism of "encountering the incomprehensible." The most radical forms of anti-objectivist hermeneutics are represented by Heidegger and Derrida, for whom the subject, and with it objectivity (in the sense of general validity of analogical conclusions between different minds), disappears: *re*construction becomes replaced with (arbitrary) *de*construction.

This historical development is, I believe, very important for gaining a proper assessment of the logical empiricists' stance toward hermeneutics and historicism, for some early logical empiricists (Carnap and Neurath) were deeply influenced by (and for some time even committed to) *objectivist* varieties of historicism and hermeneutics, whereas all logical empiricists always rejected all non-objectivist versions of historicism and hermeneutics. The logical empiricist reception of historicism and hermeneutics belongs to the rare but important cases of receptions of objectivism unshaken by twentieth-century trends of subjectivism, deconstructivism, postmodernism. Logical empiricism appears to be compatible with only one of the various different appearances of hermeneutics and historicism, namely, empirically minded objectivism.

Table 3.1 Overview of nineteenth- and twentieth-century hermeneutics

Role of analogical reasoning	*Role of intuition*	*Objectivism*	*Non-Objectivism*
Analogical reasoning possible and crucial	*Cumulative objective knowledge plus a limited amount of "intuition"*	Aprioricism: Lotze, Windelband, Rickert	
		Empirically minded: Steinthal, Lazarus, Cohen, Dilthey	Emotivism: Droysen, Treitschke
Analogical reasoning possible though not crucial	*Overwhelming importance of "empathy," empirical knowledge relatively irrelevant*		Genius- or *Führer*-Objectivity: Spranger, Bollnow
Analogical reasoning impossible or irrelevant			Anti-positivist Hermeneutics: Gadamer, Habermas
	Entire irrelevance of the "subject"		Deconstructivism: Heidegger, Derrida

Carnap: hermeneutics in the *Aufbau*

Carnap's *Der Logische Aufbau der Welt* (Carnap 1928, henceforth: *Aufbau*) has a complex history (Damböck 2021, 2022). The first version was written before Carnap came to Vienna in 1926: the bulk of it was submitted for his habilitation in December 1925. The published version of 1928 contains a foreword that was written under the direct influence of the Viennese discussions (and Neurath, in particular) and also the final five paragraphs of the book seem to be heavily influenced by the Viennese discussions (and by Carnap's reading of Wittgenstein). The major parts of the book, however, to the best of our knowledge, were hardly revised after 1925: it seems that Carnap made radical cuts (the first version consisted of 226 sections, the published version has only 183) but did not alter the existing text in any substantial way. That the *Aufbau* was written mostly independently of the Viennese context is underscored by the fact that Carnap worked out important aspects of it already before 1925—and even before 1922/23 when he first got in touch with Moritz Schlick, Reichenbach, and Neurath. The first sketches for the *Aufbau* date from 1920, written in the course of discussions Carnap had with his friends from the Jena Sera Circle (a group associated with the German Youth Movement; see CH. 2, Damböck et al. 2022) at his house in Buchenbach near Freiburg (Dahms 2016). All of these early discussion partners—the sociologist Hans Freyer, the art historian Franz Roh, the pedagogue Wilhelm Flitner—belonged to the Dilthey School. It is not surprising, then, that Diltheyian notions dominate Carnap's early sketch of the "skeleton of the theory of knowledge" (ASP RC 081-05-04).

The "primary given," according to Carnap's 1920 sketch, is the "experiences" (*Erlebnisse*) or "facts of consciousness" (*Bewusstseinstatsachen*). Like "elementary experiences" in the *Aufbau*, these elementary phenomenal facts comprise all mental phenomena we could ever be party to. Some of these experiences are intentional: they have "objects" and are called "representations" (*Vorstellungen*); some representations, in turn, are sensory experiences (*Empfindungen*). The objects of representations are either (other) representations or physical objects. One distinguished physical object is "my body." "Others" have similarities with "my body": "Some sequences of physical events of my body are often simultaneous with certain sequences of experiences; if I find sequences of events in other bodies, then I produce the fiction that even their 'experiences of another self' occur" (ASP RC 081-05-04).

Two aspects of this treatment are retained in the published version of *Aufbau*. First, the comprehensiveness of "experiences": the task is to distinguish and reconstruct different modes of experience. Second, one crucial step in the course of these reconstructions is the step from one's own to other minds. What allows this step is the "fiction" of analogical inference. Thus, the *Aufbau* can be regarded as a hermeneutic conception in two ways. The entire group of so-called *higher spheres*—other minds, intersubjective and scientific objects—are constituted by means of analogical inference. In other words, there is no behaviorism in the *Aufbau* in any reductive sense: behavior is only *the bridge* that allows us to draw analogical conclusions *between different minds* (another option would be physiology §140). The other hermeneutic aspect in the *Aufbau* is represented by the fundamental notion of *elementary experiences*. Crucially, the employment of elementary experiences led to a more comprehensive notion of the empirical than is typically found in classical British and French empiricism and includes not only sense-data but all kinds of mental phenomena. In this Carnap follows the empirically minded objectivist hermeneuticists.

Neurath's hermeneutic conception of history

Unlike Carnap, who implemented hermeneutics on a rather ahistorical level—the constitution of other minds via analogical reasoning—the hermeneutic aspects in Neurath's philosophy also comprise the historical perspective (Uebel 2019). Neurath studied in Berlin in the early 1900s, among others, with Georg Simmel and Friedrich Paulsen. His dissertation from 1906 was supervised by the historian of antiquity Eduard Meyer and the economist Gustav Schmoller (Sandner 2014: 34–48), both key representatives of hermeneutics and historicism of an objectivist fashion. In his dissertation (1906–07), Neurath interprets a passage from Cicero in a typically historicist way, uncovering the various different historical perspectives on Cicero's text, being inevitably based on analogies between the recent world of the historian and the historical period in question. Neurath defends objectivist hermeneutics, but one that covers the different perspectives that emerge from different historical contexts; he is by no means an aprioricist, in the sense of Rickert or Windelband but is empirically minded. Neurath also repeatedly refers to Boeckh, highlighting the importance of his encyclopedic treatment of ancient history and to other important representatives of an objectivist and encyclopedic treatment of philology and universal history such as Wolf, Niebuhr, and Droysen (e.g., in 1909: 143).

Neurath's famous *Anti-Spengler* from 1921 is a powerful defense of an objectivist conception of hermeneutics, against Spengler's subjectivist approach. Neurath does not criticize the idea of universal history as adopted by Spengler, but his relativistic method. Whereas the *good* relativism of historicism assumes that historical understanding must be based on analogical reasoning and therefore is always relative to some degree to the context where understanding takes place, the *bad* relativism (viz. Spengler's) rejects the idea of analogical reasoning as a whole and replaces it with an appeal to the power of the interpretive genius.

Subjectivist hermeneutics, for Neurath, not only leads to fictitious results, it is self-contradictory. Neurath illustrates the latter by means of a brief dialogue from Zhuang Zhou (Chuang Tze) that Neurath chose as a motto of *Anti-Spengler*. The key idea of this motto is iterated in Neurath's work over and over again: understanding *presupposes* analogical reasoning. Both understanding *and* the claim of the impossibility of understanding are based on analogical conclusions: they both prove the possibility of understanding. "If Spengler knows that we always misunderstand, he must at least think himself free from such misunderstanding, since he does not merely remain silent about other cultures, but also talks [about them]" (Neurath 1921/1973: 204).

Besides its subjective nature, Spengler's account, according to Neurath, suffers from another typical failure of historical accounts devoid of analogical reasoning, namely "pseudorationalism." Spengler does not adopt intuition in the good sense which Neurath recognized as "something quite sober, a clear and comprehensive view which lacks proofs or even provability" (ibid.: 207). Rather, Spengler's "reckless way" (ibid.: 204) renders intuition entirely nebulous, "exercising a spell on many when it is surrounded by all kinds of mystery" (ibid.: 207).

Most importantly, Neurath's rejection of Spengler's subjectivist hermeneutics and historicism also converges with his famous "holism." The metaphor of Neurath's boat, which became famous as a key element of his later protocol-sentence theory, already shows up in the just-cited passage of *Anti-Spengler*. "We are like sailors who on the open sea must reconstruct their ship but are never able to start afresh from the bottom. Where a beam is taken away a new one must at once be put there, and for this the rest of the ship is used as support" (ibid.: 199). Remarkably, the context of this early appearance of the boat metaphor is Spengler's "reckless" exaggeration of relativism. Our actions suffer from overestimations of commonalities between people who communicate but are no less impaired by an *underestimation* of the possibility of understanding. Thus, the boat metaphor illustrates nothing else than the only possible encyclopedic (or holistic) solution to the hermeneutic circle: each single act of understanding has to be based on all of our knowledge and at the same time must be allowed to modify and increase it (when a plank is replaced). Neurath mentions Duhem as his witness here, but there can be no doubt that his own practice of objectivist hermeneutics forms the background of this key passage of *Anti-Spengler*. The subject is not the natural sciences but history, the human sciences, and the possibility of objective understanding; he is following Boeckh and his Berlin teachers here as much as Duhem.

Logical empiricism from 1929 onwards: empathy skepticism

From the late 1920s onward, Neurath, Carnap, and their followers (Hempel, in particular) became increasingly critical of hermeneutics and the possibility of an empathy-driven branch of science. Spengler was a main target here, but possibly even Neurath's teacher Eduard Meyer, who became a strong advocate and defender of Spengler in the early 1920s (Meyer 1925). The discussions with Karl Bühler might also have played an important role here. Bühler was a strong defender of objectivist historicism and hermeneutics, and he rejected Spranger's tendencies to divide the human and the natural sciences (see Bühler 2000: §8). Bühler, Carnap, and Neurath had various discussions on the topic of understanding and construction of other minds. Carnap mentions these discussions several times in his diary, between 1925 and 1934. For example, on 4 June 1930, he discussed with Bühler and Lazarsfeld the "possibility of a behavioristic and nevertheless intersubjectively verifiable psychology of experience" (ASP RC 025-73-04). One could also rephrase this topic as "possibility of an objectivist hermeneutics." Bühler, Carnap, and Neurath certainly disagreed over key theses and concepts in the Vienna Circle's discussions concerning physicalism. But there is no indication that they differed with regard to their shared objectivist attitude towards hermeneutics. Finally, Neurath's criticism of Max Weber was also an important motive for his rejection of aspects of historicism. Here, however, the point of criticism was not subjectivism, but rather the opposite: Neurath accused Weber of being a representative of Rickert-style aprioricism (Platonism).

What Carnap and Neurath actually rejected when they rejected empathy after 1929 were autonomous subjective forms of understanding that were no longer based on analogical reasoning at all, i.e., those varieties of "anti-positivist" hermeneutics that became dominant in the twentieth century, from Spengler, Spranger, Heidegger, and Bollnow, until Gadamer, Habermas, and Derrida. Logical empiricism is clearly incompatible with any understanding of understanding that does not employ analogical reasoning. Because understanding and the humanities as a whole shifted away from the idea of analogical conclusion during the 1920s, and developed a "pseudorationalist" conception of empathy as a method of grasping the "unintelligible" (Gadamer) and the "mysterious" (Spengler), logical empiricism could not but demur. Their critical rejection of (pseudorationalist) empathy was formulated for the first time in the Vienna Circle's manifesto (Verein Ernst Mach 1929), and then in various writings of Neurath and Carnap, from 1928 onward. However, close readings allow us to distinguish in all these cases between a negative attitude towards empathy (*Einfühlung*) that does not rely on analogical conclusion and a positive reception of understanding and intuition as being based on analogical conclusion. After 1945, one can find two diverging developments. The orthodoxy of logical empiricism (Hempel, until the late 1950s, and younger authors such as Theodore Abel and Richard Rudner) takes the hypothetico-deductive conception to be entirely incompatible with the method of understanding (in all its varieties). From the late 1950s onward, however, Hempel developed more careful analyses that pick up the notion of analogical reasoning in a positive way again and try to show that the latter could be very well compatible with the hypothetico-deductive conception (Uebel 2009).

The argument from analogy revisited

Reasoning by analogy allows us to talk about the mental states of others and to set up a framework that consistently allows us to distinguish between understanding and misunderstanding. Carnap's and Neurath's use of this form of reasoning makes clear that they were not possessed of eliminativist ambitions. But if, as has recently been argued, reading Carnap as a logical behaviorist who reductively defines mental states in terms of behavioral manifestations is mistaken (Crawford 2013), then the question arises as to how he and Neurath could respond to criticism common since Ryle (1949) that analogical inference to other minds is fallacious. Would they have been happy to follow Ayer's defense of it (1956: 219–22) and be able to improve upon it? Probably not, because neither Carnap and Neurath nor Dilthey used the argument from analogy for a (logical or ontological) *proof* of the existence of other minds. Rather, they used arguments from analogy because everything in the empirical world is *as if* other minds existed. Whether they *really* exist or not is something that these philosophers were not concerned about. They required only a sound background for their framework that involves talk about other minds, understanding, and misunderstanding. Here, arguments from analogy proved to be the appropriate tools.

References

ASP = *Archives of Scientific Philosophy. Hillman Library. University of Pittsburgh, Carnap Papers.*

Ayer, A. J. (1956) *The Problem of Knowledge*, Harmondsworth: Penguin Books.

Boeckh, A. (1886) *Encyklopädie und Methodologie der philologischen Wissenschaften* (ed. by E. Bratuschek and R. Klußmann), Leipzig: Teubner.

Bühler, K. (1928) *Die Krise der Psychologie*, Jena: Gustav Fischer. Repr. Weilerwist: Velbrück, 2000.

Carnap, R. (1928) *Der Logische Aufbau der Welt*, Berlin: Weltkreis Verlag. Trans. *The Logical Structure of the World*, London: Routledge & Kegan Paul, 1967.

Crawford, S. (2013) "The Myth of Logical Behaviourism and the Origins of Identity Theory," in M. Beaney (ed.), *The Oxford Handbook of the History of Analytic Philosophy*, Oxford: Oxford University Press, pp. 621–55.

Dahms, H.-J. (2016) "Carnap's Early Conception of a 'System of the Sciences': The Importance of Wilhelm Ostwald," in C. Damböck (ed.), *Influences on the Aufbau*, Dordrecht: Springer, pp. 163–86.

Damböck, C. (2012) "Rudolf Carnap and Wilhelm Dilthey: 'German' Empiricism in the *Aufbau*," in R. Creath (ed.), *Rudolf Carnap and the Legacy of Logical Empiricism*. Dordrecht: Springer, pp. 67–88.

——— (2017) *⟨Deutscher Empirismus⟩. Studien zur Philosophie im deutschsprachigen Raum 1830–1930*, Cham: Springer.

——— (2021) "Die Entwicklung von Carnaps *Aufbau* 1920–1928," in C. Damböck and G. Wolters (eds.), *Young Carnap in an Historical Context: 1918–1935*, Cham: Springer, 19–53.

——— (2022) "Is there a Hermeneutic Aspect in Carnap's *Aufbau*?," in A. T. Tuboly and A. Sivadó (eds.), *The History of Understanding in Analytic Philosophy. Before and After Logical Empiricism*, London: Bloomsbury.

Damböck, C., Feest, U. and Kusch, M. (eds.) (2020) "Descriptive Psychology and Völkerpsychologie—in the Contexts of Historicism, Relativism, and Naturalism," *HOPOS* 10 (1).

Damböck, C., G. Sandner and M. Werner: *Logischer Empirismus, Lebensreform und die Deutsche Jugendbewegung*, Cham: Springer, 2022.

Dilthey, W. (1900) *Die Entstehung der Hermeneutik*, Tübingen: J.C.B. Mohr (Paul Siebeck). Trans. "The Rise of Hermeneutics," in W. Dilthey, *Hermeneutics and the Study of History. Selected Works, Volume VI* (ed. by R. A. Makkreel and F. Rodi), Princeton: Princeton University Press, pp. 235–59.

Feest, U. (2007) "Science and Experience/Science of Experience: Gestalt Psychology and the Anti-Metaphysical Project of the Aufbau," *Perspectives on Science* 15: 1–25.

Meyer, E. (1925) *Spenglers Untergang des Abendlandes*, Berlin: Karl Curtius.

Neurath, O. (1906–07) "Zur Anschauung der Antike über Handel, Gewerbe und Landwirtschaft," *Jahrbücher für Nationalökonomie und Statistik* 32: 577–606 and 34: 145–205. Repr. in O. Neurath, *Gesammelte ökonomische, soziologische und sozialpolitische Schriften Teil 1* (ed. by R. Haller and U. Höfer). Vienna: Hölder-Pichler-Tempsky, 1997, pp. 25–110.

——— (1909) *Antike Wirtschafsgeschichte*, Leipzig: Teubner. Excerpts [but not chapter quoted above] trans. "Economic History of Antiquity," in Neurath, *Economic Writings. Selections 1904–1945* (ed. by T. Uebel and R. S. Cohen), Dordrecht: Kluwer, pp. 120–53.

——— (1921) *Anti-Spengler*, Munich: Callwey. Excerpts trans. "Anti-Spengler," in Neurath, *Empiricism and Sociology* (ed. by M. Neurath and R. S. Cohen), Dordrecht: Reidel 1973, pp. 158–213.

Ryle, G. (1949) *The Concept of Mind*, London: Hutchinson's University Library.

Sandner, G. (2014) *Otto Neurath. Eine politische Biografie*, Vienna: Zsolnay.

Uebel, T. (2009) "Opposition to *Verstehen* in Orthodox Logical Empiricism," in U. Feest (ed.), *Historical Perspectives on Erklären and Verstehen*, Dordrecht: Springer, pp. 291–309.

——— (2019) "Neurath on Verstehen," *European Journal of Philosophy* 27: 912–38.

Verein Ernst Mach (1929) *Wissenschaftliche Weltauffassung. Der Wiener Kreis*, Vienna: Wolf. Trans. "The Scientific Conception of the World. The Vienna Circle," in O. Neurath, *Empiricism and Sociology* (ed. by R. S. Cohen and M. Neurath), Dordrecht: Reidel, 1973, pp. 299–318; rev. trans. (with orig. annotated bibliography) "The Scientific World-Conception. The Vienna Circle," in F. Stadler and T. Uebel (eds.), *Wissenschaftliche Weltauffassung. Der Wiener Kreis. Hrsg. vom Verein Ernst Mach (1929)*, Vienna: Springer, 2012, pp. 75–116.

4

VARIETIES OF NEO-KANTIAN INFLUENCES

Matthias Neuber

It is well known and widely documented by now that logical empiricism was, to some degree, inspired and conditioned in its early development by certain neo-Kantian influences. In this chapter, I consider how these influences were distributed among the diverse logical empiricist approaches. After a brief overview of the three most dominant types of neo-Kantianism, I discuss the influence of the Marburg School on Rudolf Carnap, then the influence of the Baden School on Carnap and Moritz Schlick, and finally the influence of Alois Riehl on Schlick.

Three types of Neo-Kantianism

Neo-Kantianism had its origin in work by thinkers such as Kuno Fischer, Eduard Zeller, Otto Liebmann, Friedrich Albert Lange, and Hermann von Helmholtz. All of them were opposed both to speculative metaphysics in the vein of German idealism and to materialism as it had developed in the first half of the nineteenth century (see Beiser 2014). Neo-Kantianism became the prevailing current in German academic philosophy between approximately 1870 and the outbreak of World War I (see Köhnke 1986; Heis 2018). Importantly, neo-Kantianism comes in diverse versions, the most significant of which were the Marburg School, the Baden (or Southwest) School, and the so-called critical realism of Alois Riehl.

The Marburg School originated with the publication in 1871 of Hermann Cohen's *Kants Theorie der Erfahrung* (*Kant's Theory of Experience*). Cohen finished his habilitation under Lange at the University of Marburg in 1873. In 1876, Cohen became full professor and successor of Lange in Marburg. His *Logik der reinen Erkenntnis* (Logic of Pure Knowledge), published in 1902, might be considered his principal contribution to theoretical philosophy. In 1881, Paul Natorp finished his habilitation under Cohen at Marburg, and in 1893, he became full professor there. Natorp's *Die logischen Grundlagen der exakten Wissenschaften* (*The Logical Foundations of the Exact Sciences*), published in 1910, was his main work in theoretical philosophy, especially philosophy of science. In 1899, Ernst Cassirer finished his dissertation under Cohen and Natorp in Marburg. His influence on and interaction with logical empiricists is discussed separately (see CH. 30).

The Marburg School's distinction can be characterized by the combination of the following three features: the commitment to Kant's transcendental method in combination with a predominant focus on the exact sciences, particularly physics; the rejection of the

 DOI: 10.4324/9781315650647-6

Kantian assumption of things-in-themselves; and the rejection of the Kantian dualism of sensibility and understanding. Clearly, the Marburg School was all but dogmatic. As Natorp put it, "it was never anyone's wish nor intent to cling to Kant's doctrines in an absolute way" (1912: 180 in Heis 2018: §1). Marburgian neo-Kantianism is most adequately described as an autonomous current in *transcendental revisionism*, which attempts to reconcile the original Kantian doctrine with the developments of modern mathematics (the advent of non-Euclidean geometries in the first place) and modern physics (the advent of relativity theory in the first place). The Marburg neo-Kantians combined this approach with an explicit commitment to idealism. Dispensing with the assumption of ontologically independent things-in-themselves, the object of knowledge rather is an ideal limit point towards which the conceptual development of science is converging. Moreover, sensibility is no longer regarded as an independent "faculty" of cognition. "In the end, 'intuition' no longer remains a cognitive factor which stands across from and is opposed to thinking. It *is* thinking" (ibid.: 186 in Heis 2018: §2.2, emphasis added). Consequently, for the Marburg School, *objectivity* is not achieved by copying sensual reality. Rather, objectivity is generated within the system of scientific concepts itself. Correspondingly, scientifically described reality is incorporated within the realm of pure thought. Cohen fittingly coined the label "*logical* idealism" for their epistemological conception (see Munk 2005).

The Baden School likewise was revisionist. Its founder, Wilhelm Windelband, declared that "to understand Kant means to go beyond him" (1884: vi; trans. MN). He taught first at Freiburg, then Strasbourg, and from 1903 at Heidelberg. Heinrich Rickert, the most prominent representative, also taught at Freiburg until he became Windelband's successor in Heidelberg in 1915. A further representative is Emil Lask, who studied and taught there; aged only 40, he was killed in the Great War in 1915.

The Baden School also rejected the Kantian dualism of sensibility and understanding and focused on science. It focused not on the natural sciences but the *Geisteswissenschaften* (the human sciences), especially history, where they sought to identify a priori elements. For Rickert it is a priori that every historical individual is located in a context of values. Consequently, values figure prominently in his work (see Zijdervelt 2006; Staiti 2017). Likewise, Windelband argued that history has its own "logic" for *nomothetic* sciences such as physics must be distinguished from *ideographic* ones like history. The former seek generalizations about what happens invariably, the latter seek to elucidate the particular, what happens only once. For Windelband, this was

> a purely methodological classification of the empirical sciences that is grounded upon sound logical concepts. The principle of classification is the formal property of the theoretical or cognitive objectives of the science in question. One kind of science is an inquiry into general laws. The other kind of science is an inquiry into specific historical facts. In the language of formal logic, the objective of the first kind of science is the general apodictic judgment; the objective of the other kind of science is the singular, assertoric proposition.
>
> (1894: 291 in Heis 2018: §3.3)

Rickert elaborated on this principal methodological distinction in two influential monographs (1896–1902, 1899).

With Alois Riehl, we encounter a third independent type of neo-Kantianism. Riehl ultimately succeeded Wilhelm Dilthey at the University of Berlin in 1905. In his three-volume *Der philosophische Kriticismus und seine Bedeutung für die positive Wissenschaft* (*Philosophical*

Criticism and its Significance for Positive Science, 1876–1887), Riehl attempted to combine Kantian criticism with certain elements of British (especially Lockean) empiricism and developed his ideas in direct confrontation with the "positive" (experiential) sciences. For him, Kant's *Critique of Pure Reason* was intimately correlated with a "theory of reality" (1876/1908: 562). In a section titled "*Erscheinung, Ding an sich, Noumenon*," Riehl argued for the reality of things-in-themselves, thereby refusing to interpret them as mere limiting concepts. Correspondingly, he argued that we acquire knowledge of things-in-themselves *via* our knowledge of appearances (1887/1926: 164–5). This conception he called "critical realism" (ibid: 163; for details, see Neuber 2014). Despite this radical difference, Riehl agreed with the members of the two "official" neo-Kantian schools in downgrading the scientific relevance of intuition. For critical realists, knowledge of things-in-themselves is possible since intuition plays only the role of an "epistemic starting point" that we eventually abandon for the sake of purely conceptual knowledge.

> We see, in fact, how science reduces the content of experience to its lawlike elements, to what recurs in similar form, to what is accessible to quantitative determination and is thus expressible in numerical operations, in short, to the *conceivable*. Everything else is an object, not of *conceiving*, but of immediate *acquaintance*, and hence of feeling, sensation, and perception (1879/1925: 221; trans. MN).

Riehl's distinction between perceptual—i.e., everyday—"acquaintance" (*Wissen*) and quantitatively determined—i.e., scientific—"conceiving" (*Begreifen*) can hardly be overestimated in its historical significance: a very similar distinction can be found in the early work of Bertrand Russell (see Russell 1910–11) and, as we shall see, in Schlick's *Allgemeine Erkenntnislehre*.

The influence of Marburg neo-Kantianism

Turning to the influence of these neo-Kantian schools on logical empiricism, we note that Carnap was subject to double exposure. Carnap's dissertation advisor at Jena was Bruno Bauch, who completed his dissertation under Rickert at Freiburg in 1902 and considered himself a member of the Baden School (see Schlotter 2004). Carnap himself reports:

> I studied Kant's philosophy with Bruno Bauch in Jena. In his seminar, the *Critique of Pure Reason* was discussed in detail for an entire year. I was strongly impressed by Kant's conception that the geometrical structure of space is determined by the form of our intuition. The after-effects of this influence were still noticeable in the chapter on the space of intuition in my dissertation, *Der Raum*.
>
> (1963: 4)

Carnap adds: "Knowledge of intuitive space I regarded at that time [1922], under the influence of Kant and the neo-Kantians, especially Natorp and Cassirer, as based on 'pure intuition' and independent of contingent experience" (ibid.: 12).

Let's begin with Carnap's Marburg connection. According to Michael Friedman, "Carnap's project [in the *Aufbau*] has less affinity with traditional empiricism and more with Kantian and neo-Kantian conceptions of knowledge" (1987, 1999b: 98). Alan Richardson even goes as far as to claim that "Carnap's principal goals in the *Aufbau* were not empiricist" (2003: 64). Rather, Richardson echoes Friedman in that there was "deep agreement with neo-Kantian thinking about knowledge and the theory of knowledge" (ibid.: 64–5). Carnap explicitly remarks upon the kinship between his own "construction theory" (orig. "theory of constitution") and the "genetic" view of knowledge of the Marburg School.

> Construction theory agrees with *transcendental idealism* in the conception that all objects of cognition are constructed (in idealistic language: 'are created in thought'); in fact, the constructed objects are objects of conceptual knowledge only *qua* logical forms which are generated in a certain way. Ultimately, this holds also for the basic elements of the constructional system.
>
> (1928/1967: §177)

Natorp in particular set the stage for this sort of 'constructivism' (see Edgar 2015: 143–6; Luft 2016: 332–4). Natorp's principal claim was that *objectivity* can only be achieved by a constructive move. Thus, he declared: "To authenticate the laws of knowledge we must move in the *constructive* direction of knowledge, in the direction of objectification, for it is the ultimate *objective* unities we seek" (1887/1981: 264). Carnap clearly echoed this view by claiming that "the necessity of constructing the world of physics rests on the circumstance that only this world, but not the perceptual world. . ., can be made intersubjective in an unequivocal, consistent manner" (1928/1967: § 136).

In §12 of the *Aufbau*, Carnap further points out that it is particularly Cassirer who has recognized that it is *relational concepts* that need to be applied in science (see Cassirer 1910/1923, esp. ch. 2). This is highly significant, since the "fundamental thesis of construction theory" is that "*each scientific statement can in principle be so transformed that it is nothing but a structure statement*" (1928: §16; orig. emphasis). Interestingly enough, a very similar view can be found in the early Schlick (see Neuber 2013).

However, Carnap's agreement with the Marburg School was limited. He saw no necessity for synthetic a priori judgments which, for the Marburgians, still played an essential role (see Friedman 1999a: 159–60). Moreover, he rejected their conception of the object of knowledge as infinite task.

> In the opinion of the Marburg Neo-Kantian School . . . the object is the eternal X, and its determination an aim that can never be accomplished. Against this, it must be pointed out that a finite number of characteristics suffices for the construction of the object, thus for its definite description within the field of objects in general. If such a description is given, then the object is no longer an X, but something that is uniquely determined, whose complete description, however, still remains a task that cannot be completed.
>
> (1928/1967: § 179, trans. amended)

For Carnap, objects in general—be they mathematical or physical—are *definitely determined* by being "constituted" or "constructed" on the basis of relational concepts. Friedman concluded correctly that "Carnap is much more radical than Cassirer" (1999a: 159 n. 78) who still recognized a fundamental difference between the constructability of mathematical objects and physical (empirical) objects. Furthermore, the Marburg neo-Kantians (even less the Badensians) were not yet in possession of a "sufficiently rich and determinate conception of logic itself" (ibid.: 141).

The influence of Baden neo-Kantianism

The second line of neo-Kantian influence in Carnap's *Aufbau*, exerted by the Badensians, pertained to the realm of values. To be sure, Carnap retrospectively conceded that he "had written almost nothing about values" (1963: 1000). However, as Thomas Mormann observed, "in the *Aufbau* values were considered as philosophically respectable entities that could be constituted from value experiences" (2007: 127). There, it appears, Carnap was inspired by Rickert's views. (In 1911 and 1912, Carnap had attended Rickert's lectures at Freiburg and, after the war, finished his dissertation in Jena under Rickert's student Bauch.) As Rickert himself clearly indicated, his understanding of history and the cultural sciences is deeply entrenched in a fundamental system of values.

> Fundamental progress in the cultural sciences with respect to their objectivity, their universality, and their arrangement in a coherent system, is dependent on progress in the development of an objective and systematically articulated concept of culture, i.e., on the approach to a knowledge of value based on a system of valid values.
>
> (1899/1926: 139–40, trans. MN)

Furthermore, Rickert, like Carnap, employs in his 1921 *System der Philosophie* the interplay of "experiences" (*Erlebnisse*) and conceptual "construction" (*Aufbau*), in order to implement his particular variant of "scientific philosophy." Rickert's writings clearly were instrumental for the constitution of values in Carnap's *Aufbau* (Mormann 2006: 176; Leinonen 2016).

As Carnap summarizes (1928/1967: 241–2), the *Aufbau* world is four-layered consisting of: (1) autopsychological objects; (2) physical objects; (3) heteropsychological objects; (4) cultural objects, specifically values. Decisive for Carnap's approach to values is that they are, as in Rickert's *System der Philosophie* (1921), constructed from a special type of experiences.

> The construction of values from certain experiences, namely value experiences, is in many ways analogous to the construction of physical things from "perceptual experiences". . . . For the construction of ethical values, for example, we must consider (among others) experiences of conscience, experiences of duty and of responsibility, etc. For aesthetic values, we take into account experiences of (aesthetic) pleasure or other attitudes in the appreciation of art, experiences of artistic creation, etc.
>
> (1928/1967: §152)

As a viable alternative to the construction of values from an autopsychological basis, Carnap mentions Wilhelm Ostwald's construction from a physical (energetic) basis in the latter's *Die Philosophie der Werte* (1913).

The question arises: why do values disappear from Carnap's list of respectable, scientifically constructable objects at the beginning of the 1930s? As Mormann puts it starkly:

> While the world of the *Aufbau* had been a world with history, values, and value judgments, for which it made sense to speak of human practice, from ["Elimination of Metaphysics" (1931)] onwards these features disappeared from the agenda of his scientific philosophy. The world of science became a world of facts (or their linguistic counterparts) without values.
>
> (2007: 134)

But did Carnap really switch from a commitment to the objectivity of value judgments to a radically noncognitivist (emotivist) point of view, which entails that value judgments are cognitively meaningless and therefore metaphysical? Discussion about this continues (see Damböck 2021; Carus 2021). Thomas Uebel (2010) has provided one key to understanding either Carnap's apparent change to noncognitivism or its more forceful expression. To begin with, empirically tangible value pluralism—endorsed in Carnap's 1929 Bauhaus lectures (see da Cunha 2017)—stood against both Rickert's idealism and Ostwald's energeticism, the consequence being that values themselves were excluded by Carnap from the ongoing project of an *Einheitswissenschaft*. Moreover, there were genuinely political reasons that reinforced Carnap's decision: both Rickert and Bauch tended to the extreme right and thus stood in clearest opposition to Carnap's and the Vienna Circle's progressive humanism.

Ever conscious of these matters, Neurath not only objected to Carnap's treatment of values in the *Aufbau* but also criticized Schlick for the latter's charitable approach toward Windelband and Rickert's attempt at a demarcation of the cultural from the natural sciences (see Uebel 2010: 119–20). Schlick agreed that the cultural sciences are primarily concerned with particular events and persons and even endorsed their conception of their mission:

> Behind what seems to be the ultimate purpose of research in the humanities, the description of the particular, there lies, in truth, a more ultimate purpose still: the enrichment of the mind [*Seele*] through the immediate reliving of great thoughts and emotions of the past. In other words, the humanities, in the final analysis, are not purely theoretical at all, are not directed to pure knowledge, but serve in the end as means to experience [*Erleben*].
>
> (1934/1979: 150)

For Neurath, this approximated the Baden School's "official" doctrine. However, as Anne Siegetsleitner (2014: 246–7) noted, Schlick's appreciation of the Badensians' methodological dualism was a restricted one. Having criticized Rickert's attempt at a demarcation between the sciences in one of his very first papers (Schlick 1910), he objected again that the cultural sciences do *not* possess different basic concepts. They employ concepts in use in everyday life which are reducible to natural scientific, specifically psychological concepts.

> The humane or cultural sciences, in other words, have absolutely no essentially basic concepts of their own, but borrow them from other levels of knowledge; their own concern is solely with complex structures derived therefrom (just as meteorology, say, has no specifically fundamental concepts of its own, but borrows them all from physics).
>
> (1934/ 1979: 149)

Schlick restricted the cultural sciences' significance to the realm of *Erleben* and denied that they contribute on their own to the gain of *knowledge* (ibid.: 150).

Both Carnap and Schlick, then, were in certain respects inspired by conceptions of the Baden School, but in both cases the influence was very limited.

The influence of Riehl

Schlick's article from 1934 was by no means his first articulated encounter with neo-Kantian doctrines. Already his early *Allgemeine Erkenntnislehre* was deeply influenced by critical realism in the vein of Riehl (see Neuber 2012: ch. 2; Heidelberger 2006). Riehl was a rather influential figure in his time: Richard Hönigswald and Oswald Spengler wrote their dissertations under his supervision and Rickert his well-known habilitation *Der Gegenstand der Erkenntnis* (Rickert

1904). Moreover, he also served for several years as co-editor of the influential *Vierteljahrsschrift für wissenschaftliche Philosophie und Soziologie*, a journal that "sought a kind of harmonization of positivism and neo-Kantianism" (Ferrari 2003: 64; see also Köhnke 1986: 388–404). Schlick, who had corresponded with Riehl since March 1910, published some articles and about 30 reviews in the *Vierteljahrsschrift* between 1911 and 1916 (Ferrari 2003). Schlick quotes Riehl extensively and primarily positively in his *Erkenntnislehre*. It was especially his combination of criticism and realism that attracted the early Schlick. Friedman correctly claimed that "Schlick was not a positivist or strict empiricist in 1918, but a neo-Kantian or 'critical' realist" (1983/1999: 20).

In a paper from 1919, Schlick expresses the opinion that "the only natural continuation of Kant's theory of knowledge, to which his system points from various angles, lies not in the idealist but the realist direction, and we arrive at it by a revision of Kant's utterances about the so-called thing-in-itself and its knowability" (1919/1979: 282). This could have been written by Riehl, in whose view things-in-themselves must be taken ontologically seriously. Like him, the early Schlick promoted a program of Kant-inspired scientific philosophy. Thus, in his first article on Einstein's theory of relativity, Schlick points out:

> We have known since the days of Kant that the only fruitful method of all theoretical philosophy consists in critical inquiry into the ultimate principles of the special sciences. . . . It is primarily, or even exclusively, the principles of the exact sciences that are of major philosophical importance, for the simple reason that in these disciplines alone we do find foundations so firm and sharply defined, that a change in them produces a notable upheaval, which can then also acquire an influence on our world-view.
>
> (1915/1979: 153)

The *critical component* of critical realism is accounted for here: "Schlick aimed to do for Einstein's physics what Kant had done for Newton's, namely, to explain and exhibit the special features of this physics that make it a model or paradigm of coherent rational knowledge of nature" (Friedman 2001: 14).

The most obvious line of Riehl's influence is detectable in the distinction between acquaintance and knowledge. For example, Schlick states:

> Kant has uncritically presupposed that in order to know an object, an *intuition* of the object is ultimately in some way necessary. In the very first sentence of the transcendental aesthetics he says this with complete clarity. But in truth intuition gives us no knowledge whatever; it is wholly inessential for this purpose. It provides, to be sure, an *acquaintance* with objects, but never knowledge of them.
>
> (1919/1979: 282)

In consequence, Schlick arrives at a conception of purely conceptual scientific knowledge that bears obvious similarities with the one propounded by Riehl (see Neuber 2012: 68–82). Of course, in Vienna, Schlick's position changed significantly: things-in-themselves were supplanted by (Russellian) "logical constructions" and critical realism by "coherent empiricism" (see Friedl 2013; Neuber 2016). Yet, the Riehlian distinction between acquaintance and knowledge continued to form an integral part of Schlick's philosophy. Accordingly, it can be concluded that Riehl, at least in this respect, anticipated logical empiricism.

Conclusion

This chapter has shown that, early in their careers, logical empiricists were influenced by various types of neo-Kantianism. The early Carnap's *Aufbau* project integrated elements both from the Marburg and the Baden Schools of neo-Kantianism, whereas Schlick's account was partially inspired by Badensian and Riehlian points of view. To be sure, neither Schlick nor Carnap took over neo-Kantianism unreservedly. Rather, they appropriated and significantly transformed certain of its conceptions, so the neo-Kantian influence was somewhat diffuse.

Another logical empiricist that needs to be mentioned in this connection is Hans Reichenbach, whose early philosophical approach had a definitely Kantian ring to it. The idea of the relativized a priori in his *Relativitätstheorie und Erkenntnis apriori* (1920) comes to mind. However, it is hard to determine whether he was, like Schlick and Carnap, inspired by contemporary neo-Kantian approaches or whether he just elaborated on the original Kantian doctrine. At any rate, these debts of his are discussed elsewhere in this volume (see CHS. 21, 24, and 30). And what about Neurath, who categorically *rejected* neo-Kantianism? Here it must be noted that logical empiricism generally, even the Vienna Circle on its own, was a pluralistic movement: Neurath and Schlick are particularly good examples. Neo-Kantian influences were salient rather for its German than for its Austrian representatives.

All in all, then, the diverse neo-Kantian influences on logical empiricism are important to understand how the logical empiricist movement emerged. Yet, these influences should not be overestimated, as sometimes happens in recent research.

References

Beiser, F. (2014) *The Genesis of Neo-Kantianism, 1796–1880*, Oxford: Oxford University Press.

Carnap, R. (1922) *Der Raum. Ein Beitrag zur Wissenschaftslehre*, Kant Studien Ergänzungshefte 56. Trans. "Space. A Contribution to the Theory of Science," in Carnap, *Collected Works* (ed. by A. W. Carus et al.), Oxford: Oxford University Press, 2019, pp. 22–208.

——— (1928) *Der logische Aufbau der Welt*, Berlin: Weltkreis-Verlag. Trans. *The Logical Structure of the World*, Berkeley: University of California Press, 1967. Repr. Chicago: Open Court, 2003.

——— (1931) "Überwindung der Metaphysik durch logische Analyse der Sprache," *Erkenntnis* 2: 219–41. Trans. "The Elimination of Metaphysics through Logical Analysis of Language," in A. J. Ayer (ed.), *Logical Positivism*, New York: Free Press, 1959, pp. 60–81.

——— (1963) "Intellectual Autobiography" and "Comments and Replies," in P. A. Schilpp (ed.), *The Philosophy of Rudolf Carnap*, La Salle: Open Court, pp. 3–84 and 859–1016.

Carus, A. W. (2021) "Carnap's Fundamental Philosophical Commitment: From Religious Origins to Kantian Non-Cognitivism, 1911–21," in Ramharter (2021).

Cassirer, E. (1910) *Substanzbegriff und Funktionsbegriff*, Berlin: Bruno Cassirer. Trans. "Substance and Function." in Cassirer, *Substance and Function and Einstein's Theory of Relativity*, 1923, repr. New York: Dover, 1953, pp. 3–346.

Cohen, H. (1871) *Kants Theorie der Erfahrung*, Berlin: Dümmler. Excerpts trans. "*Kant's Theory of Experience*," in S. Luft 2015, pp. 107–16.

——— (1902) *Logik der reinen Erkenntnis*, Berlin: Bruno Cassirer.

da Cunha, I. F. (2017) "Utopias and Forms of Life," *Revista de filosofia* 24: 121–48.

Damböck, C. (2021) "Carnap's Non-Cognitivism and his Views on Religion, Against the Background of the Herbartian Philosophy of His Grandfather Friedrich Wilhelm Dörpfeld," in Ramharter (2021).

Der Gegenstand der Erkenntnis. Einführung in die Transzendentalphilosophie. Tübingen: Mohr Siebeck 1904.

Edgar, S. (2015) "Intersubjectivity and Physical Laws in Post-Kantian Theory of Knowledge: Natorp and Cassirer," in S. Luft (ed.), *The Philosophy of Ernst Cassirer: A Novel Assessment*, Berlin: de Gruyter, pp. 141–62.

Ferrari, M. (2003) "An Unknown Side of Moritz Schlick's Intellectual Biography: The Reviews for the 'Vierteljahrsschrift für wissenschaftliche Philosophie und Soziologie (1911–1916)," in F. Stadler (ed.), *The Vienna Circle and Logical Empiricism: Re-evaluation and Future Perspectives*, Dordrecht: Kluwer, pp. 63–77.

Friedl, J. (2013) *Konsequenter Empirismus. Die Entwicklung von Moritz Schlicks Erkenntnistheorie im Wiener Kreis*, Vienna: Springer.

Friedman, M. (1983) "Critical Notice: Moritz Schlick, *Philosophical Papers*," *Philosophy of Science* 50: 498–514. Repr. with postscript in Friedman (1999b), pp. 17–43.

——— (1987) "Carnap's *Aufbau* Reconsidered," *Nous* 21: 521–45. Repr. in Friedman (1999b), pp. 89–113.

——— (1999a) "Postscript: Carnap and the Neo-Kantians," in Friedman (1999b), pp. 152–63.

——— (1999b) *Reconsidering Logical Positivism*, Cambridge: Cambridge University Press.

——— (2001) *Dynamics of Reason*, Stanford: CSLI Publications.

Heidelberger, M. (2006) "Kantianism and Realism: Alois Riehl (and Moritz Schlick)," in M. Friedman and A. Nordmann (eds.), *The Kantian Legacy in Nineteenth Century Science*, Cambridge, MA: MIT-Press, pp. 227–47.

Heis, J. (2018) "Neo-Kantianism," in E. N. Zalta (ed.), *Stanford Encyclopedia of Philosophy*, https://plato.stanford.edu/entries/neo-kantianism/.

Köhnke, K. (1986) *Entstehung und Aufstieg des Neukantianismus. Die deutsche Universitätsphilosophie zwischen Idealismus und Positivismus*, Frankfurt a. M.: Suhrkamp. Trans. *The Rise of Neo-Kantianism: German Academic Philosophy between Idealism and Positivism*, Cambridge: Cambridge University Press.

Leinonen, M. (2016) "Assessing Rickert's Influences on Carnap," in C. Damböck (ed.), *Influences on Carnap's Aufbau*, Cham: Springer, pp. 213–32.

Luft, S. (ed.) (2015) *The Neo-Kantian Reader*, New York: Routledge.

——— (2016) "Reconstruction and Reduction: Natorp and Husserl on Method and the Question of Subjectivity," *Meta* VIII: 326–70.

Mormann, T. (2006) "Werte bei Carnap," *Zeitschrift für philosophische Forschung* 60: 169–89.

——— (2007) "Carnap's Logical Empiricism, Values, and American Pragmatism," *Journal for General Philosophy Science* 38: 127–46.

Munk, R. (2005) *Hermann Cohen's Critical Idealism*, Dordrecht: Springer.

Natorp, P. (1887) "Ueber objektive und subjektive Begründung der Erkenntnis," *Philosophische Monatshefte* 23: 257–86. Trans. "On the Objective and Subjective Grounding of Knowledge," *Journal of the Society of Phenomenology* 12 (1981): 245–66. Repr. in Luft 2015, pp. 164–79.

——— (1910) *Die logischen Grundlagen der exakten Wissenschaften*, Leipzig & Berlin: Teubner. Excerpts trans. "The Logical Foundations of Exact Science," in Luft (2015), pp. 198–213.

——— (1912) "Kant und die Marburger Schule," *Kant-Studien* 17: 193–221. Trans. "Kant and the Marburg School," in Luft (2015), pp. 180–97.

Neuber, M. (2012) *Die Grenzen des Revisionismus. Schlick, Cassirer und das "Raumproblem"*, Vienna: Springer.

——— (2013) "Treffpunkt Struktur—Cassirer, Schlick und Carnap," *Archiv für Geschichte der Philosophie* 95: 206–33.

——— (2014) "Critical Realism in Perspective: Remarks on a Neglected Current in Neo-Kantian Epistemology," in M. C. Galavotti et al. (eds.), *New Directions in the Philosophy of Science*, Cham: Springer, pp. 657–73.

——— (2016) "Schlick und die 'Wende der Philosophie'—Vom kritischen Realismus zum logischen Empirismus (und wieder zurück?)," in M. Neuber (ed.), *Husserl, Cassirer, Schlick. 'Wissenschaftliche Philosophie' im Spannungsfeld von Phänomenologie, Neukantianismus und logischem Empirismus*, Cham: Springer, pp. 207–35.

Ostwald, W. (1913) *Die Philosophie der Werte*, Leipzig: Kröner.

Ramharter, E. (ed.) (2021) *The Vienna Circle and Religion*, Cham: Springer.

Reichenbach, H. (1920) *Relativitätstheorie und Erkenntnis apriori*, Berlin: Springer. Trans. *The Theory of Relativity and A Priori Knowledge*, Los Angeles: University of California Press, 1965.

Richardson, A. (2003) "Conceiving, Experiencing, and Conceiving Experiencing: Neo-Kantianism and the History of the Concept of Experience," *Topoi* 22: 55–67.

Rickert, H. (1896–1902) *Die Grenzen der naturwissenschaftlichen Begriffsbildung*, Freiburg: Mohr, 6th rev. ed. Tübingen: Mohr Siebeck, 1929. Part. trans. *The Limits of Concept Formation in Natural Science*, Cambridge: Cambridge University Press, 1986.

——— (1899) *Kulturwissenschaft und Naturwissenschaft*, Freiburg: Mohr, 7th rev. ed. Tübingen: Mohr Siebeck, 1926.

——— (1904) *Der Gegenstand der Erkenntnis. Eine Einführung in die Transzendentalphilosopie*, Tübingen: Mohr Siebeck.

——— (1921) *System der Philosophie. Erster Teil*, Freiburg: Mohr.

Riehl, A. (1876) *Der philosophische Kriticismus: Geschichte und System, Bd. 1*, Leipzig: Engelmann, 2nd ed., 1908.

——— (1879) *Der philosophische Kriticismus: Geschichte und System, Bd. 2*, Leipzig: Engelmann, 2nd ed., 1925.

——— (1887) *Der philosophische Kriticismus: Geschichte und System, Bd. 3*, Leipzig: Engelmann, 2nd ed., 1926.

Russell, B. (1910–11) "Knowledge by Acquaintance and Knowledge by Description," *Proceedings of the Aristotelian Society (New Series)* XI: 108–28.

Schlick, M. (1910) "Die Grenze der naturwissenschaftlichen und philosophischen Begriffsbildung," *Vierteljahrsschrift für wissenschaftliche Philosophie und Soziologie* 34: 121–42. Trans. "The Boundaries of Scientific and Philosophical Concept Formation," in Schlick (1979), vol. 1, pp. 25–40.

——— (1915) "Die philosophische Bedeutung des Relativitätsprinzips," *Zeitschrift für Philosophie und philosophische Kritik* 159: 129–75. Trans. "The Philosophical Significance of the Principle of Relativity," in Schlick (1979), vol. 1, pp. 141–52.

——— (1918) *Allgemeine Erkenntnislehre*, Berlin: Springer, 2nd rev. ed. 1925. Trans. *General Theory of Knowledge*, Lasalle: Open Court, 1974.

——— (1919) "Erscheinung und Wesen," *Kant-Studien* 23: 188–203. Trans. "Appearance and Essence," in Schlick (1979), vol. 2, pp. 270–87.

——— (1934) "Philosophie und Naturwissenschaft," *Erkenntnis* 4: 370–96. Trans. "Philosophy and Natural Science," in Schlick (1979), vol. 2, pp. 139–52.

——— (1979) *Philosophical Papers* (ed. by H. Mulder and B. van de Velde-Schlick), Dordrecht: Reidel, vol. 2.

Schlotter, S. (2004) *Die Totalität der Kultur. Philosophisches Denken und politisches Handeln bei Bruno Bauch*, Würzburg: Königshausen & Neumann.

Siegetsleitner, A. (2014) *Ethik und Moral im Wiener Kreis. Zur Geschichte eines engagierten Humanismus*, Wien: Böhlau.

Staiti, A. (2017) "Ethical Validity and Its Ontological Bearer in Heinrich Rickert's Metaethics," *Quaestio* 17: 623–35.

Uebel, T. (2010) "'BLUBO-Metaphysik': Die Verwerfung der Werttheorie des Südwestdeutschen Neukantianismus durch Carnap und Neurath," in A. Siegetsleitner (ed.): *Logischer Empirismus, Werte und Moral. Eine Neubewertung*, Vienna: Springer, pp. 103–29.

Windelband, W. (1884) *Präludien. Aufsätze und Reden zur Philosophie und ihrer Geschichte*, Tübingen: Mohr Siebeck, 4th enlarged ed. in 2 vols., 1911.

——— (1894) *Geschichte und Naturwissenschaft. Rede zum Antritt des Rectorats der Kaiser-Wilhelm-Universität Strassburg*, Strassburg: J. H. Ed. Heitz. Trans. "History and Natural Science," in Luft (2015), pp. 287–98.

Zijdervelt, A. C. (2006) *Rickert's Relevance: The Ontological Nature and Epistemological Functions of Values*, Leiden: Brill.

5
HERMANN VON HELMHOLTZ AND LOGICAL EMPIRICISM

Michael Heidelberger

Prominent scientists in the German-speaking world of the late nineteenth and early twentieth centuries showed great willingness to reflect philosophically on their work. Such reflection was increasingly taken up by philosophers and, in a more modest way, even by the public at large. The resulting attitude shaped several generations in their outlook on science and affected all the currents that came together in logical empiricism. This chapter concentrates on Hermann von Helmholtz (1821–1894), who played an outstanding role for the philosophy of science of logical empiricism.

The emergence of a problem

Perhaps the most important German author to have raised the level of reflection on the sciences during the nineteenth century was the philosopher Friedrich Albert Lange (1828–1875). His *History of Materialism*, especially the second volume of its greatly expanded second edition of 1875, provides an in-depth survey and philosophical evaluation of the most significant scientific developments of the nineteenth century. In the last quarter of the century, this role was increasingly taken over by scientists themselves (Du Bois-Reymond, Ostwald, Kirchhoff, Hertz, Helmholtz, Mach, Boltzmann). The physiologist and physicist Helmholtz became the model philosopher-scientist of his time. He transcended the trend of the time to specialization and often touched upon foundational issues in the different scientific fields he discussed, allowing him to lay claim to a unified view of science. His "vision of the wholeness and unity of knowledge" (Cahan 2018: 2) was unequaled among his peers.

The importance of Helmholtz for logical empiricism turns on his work on the relation of mathematics or mathematized empirical theory to experience. Helmholtz was one of very few scientists who dealt with the *mathematical nature* of science, especially of theoretical physics and geometry, in a philosophical way and supplied new and promising perspectives.

Philosophy of applied mathematics became an urgent concern around the turn of the century for several reasons.

(1) The relation between geometry and experience formed the subject of Helmholtz's systematic investigations since about 1868. Nicolai Lobachevsky and János Bolyai had recently discovered non-Euclidean geometry. Both expressed doubts that Euclidean geometry really described the space we live in: the parallel axiom of Euclidean geometry can be replaced by one

 DOI: 10.4324/9781315650647-7

that contradicts it without creating an inconsistency with the remaining axioms. With Euclid's axioms under threat, Helmholtz's empirical view of geometry provoked debate in the whole of Europe. Previously, non-Euclidean geometry either was unknown or raised only minimal interest, but now discussions extended beyond mathematics into intellectual culture in general. Helmholtz's claims suffered a wave of "refutations," but they tapered off rapidly in the early 1880s, and in spite of these criticisms, the interest in his arguments survived well into a later period when interest in the mathematical nature of advanced science was revitalized.

(2) Contemporary philosophy of science concentrated more and more on theoretical physics, especially electromagnetic theory. The theoretical physicist Paul Volkmann (1856–1938) noted that "the unexpected [recent] upswing of epistemological research [in physics] seems to be essentially a result of the Faraday-Maxwell direction" (1896: iii). Whereas in traditional mechanistic physics the relation of the formal-mathematical apparatus to experience seemed relatively unproblematic, Maxwell's equations posed many questions about the exact nature of the physical reality underlying them.

(3) There was an increasing interest in the work of foreign philosopher-scientists who focused on the role of mathematics in science. Henri Poincaré's (1854–1912) *La Science et l'Hypothèse* of 1902 appeared in an authorized German translation already in 1904 as the first foreign translation, with editorial notes of almost 100 pages; a second edition followed in 1906 and a third in 1914. For Poincaré, the fundamental principles of a physical theory, especially of geometry and mechanics, are nothing but conventions. The geometry of our world is underdetermined by experience and leaves room for conventional choice. Pierre Duhem's (1861–1916) *La Théorie physique, son objet et sa structure* of 1906, translated into German in 1908, offered a different form of conventionalism, arguing for an entirely instrumentalist role of mathematics in physics.

While there was a limited reception in Germany of James Clerk Maxwell's (1831–1879) philosophy as well as that of William Kingdon Clifford (1845–1879), the Italian mathematician Federigo Enriques's (1871–1946) book *Problems of Science* (1906) fit well for the times. Translated into German in 1910, it contains valuable material on geometry, mechanics, and theoretical physics, discussing many of the works and topics mentioned in this chapter.

(4) With the discovery of non-Euclidean geometries, views of the ontology of mathematics changed profoundly and with it the meaning of the concept of "pure" and "applied" mathematics. At the beginning of the nineteenth century, the Aristotelian view still prevailed that mathematical objects result from abstraction from the sensible properties of ordinary objects and idealizing them. Every mathematical object can reversely be "enriched" with empirical properties and gradually be more and more "applied" to the phenomenal world. By contrast, the modern view maintains that mathematics is a free creation of formal structures, of "empty conceptual schemata" (Einstein 1921/2007: 148). Accordingly, an application of a mathematical schema is the result of an empirical interpretation of its basic concepts. The resultant distinction between pure and applied geometry or mathematical and physical geometry made successful applications in empirical explanation a mystery.

(5) Finally, there are Albert Einstein's (1879–1955) theories of special and general relativity theory of 1905 and 1915. Immanuel Kant (1724–1804) was widely thought to have solved the problem of the relation of mathematical theory and reality as regards Newtonian mechanics and many who adhered to his doctrine of the synthetic a priori character of mathematics were convinced that, correctly understood, his framework could withstand the new challenges. Yet with the advent of Einstein, the defense of the orthodox Kantian standpoint became downright irrational, and so-called neo-Kantians introduced ingenious substitutes for certain "Kantian" elements. (Helmholtz himself is a case in point, notwithstanding his empiricist criticism of Kant. Many founding members of the Vienna Circle came from a similar background.)

Einstein's general theory of relativity of 1915 was the first to apply non-Euclidean geometry successfully to nature. With its subsequent confirmation by Arthur Eddington in 1919 it became clear that a profound change was taking place in the conception of the relation between mathematics and experience. Since the new theory establishes a direct connection between a non-Euclidean structure and spatiotemporal curvature, its application of geometry is inextricably intertwined with the conception of space-time as introduced by Einstein in 1905. One can say with Hilary Putnam: "[T]he overthrow of Euclidean geometry is the most important event in the history of science for the epistemologist" (1979: x).

Helmholtz on geometry

Helmholtz's starting point (1868) was that elementary geometry cannot be completely characterized by the usual axioms of Euclid. We need additional assumptions that are tacitly presupposed but easily overlooked. (The visual evidence of geometry derives from *contingent* characteristics; it is not self-evident.) To free ourselves from the vexations of intuition, we must formulate the presuppositions explicitly in terms of analytical geometry. Helmholtz had almost finished his investigation when he learned of Bernhard Riemann's (1826–1866) famous *Habilitationsvortrag* of 1854, published only in 1868, in which similar ideas were developed.

Riemann drew inspiration from the philosopher Johann Friedrich Herbart (1776–1841) for his concept of the manifold. Herbart had observed that sensations given in perception form "qualitative continua." Starting from a given pitch of a tone, we can reach any other by continuously increasing or decreasing the starting pitch: pitch is ordered in a "series-form." In this series, each pitch is uniquely determined in relation to another pitch by exactly one quantity—the quantity of increase or decrease necessary to reach this pitch in the series. Each pitch thus has a certain distance from the starting pitch. Comprehending a tone as the combination of pitch and loudness (intensity), we have to consider in addition a *second* quantity, in order to reach any other tone from a given one: a tone can change in two directions or dimensions. And three quantities are needed to reach, for example, a certain color from a given one. Our sensations come thus in different forms of "series," and each sensation in the series is determined by a certain quantity. Expressed in Riemann's terminology, pitch forms a (continuous) manifold of one dimension, tone of two and color of three, and so on. Generalizing this, we can talk of an "*n* times extended manifold" or a "manifold of *n* dimensions." This is easily applied to the objects of geometry: a line is a manifold of one dimension, a plane or surface of two and space of three dimensions, and so on.

The task now was to identify the quantity (or quantities) that determine(s) a point in the manifold. This is tantamount to finding a general expression for the (shortest) *distance* between two points on a line, on a plane, in space, in an *n*-dimensional manifold, or, the "determination of the measure-relations" in a manifold, as Riemann said. (Actually, he looked for an expression of the *infinitesimal* distance.) Riemann argued for the hypothesis that the geometry of space can be founded upon the generalized Pythagorean, inaugurating a completely new way to conceive geometry as the embodiment of distances or "a metric."

The next property of manifolds introduced by Riemann was *curvature*. Here, he relied on the work of his teacher Carl Friedrich Gauss (1777–1855), who had developed a rigorous definition of curvature some years before. Riemann showed that in a manifold of constant curvature, any part of the manifold (e.g., a triangle on a sphere) can be moved without distortion and abolition of congruence. The congruence is independent of place, direction, and path of transportation. Spaces with constant curvature can be divided in spaces with positive, negative, or zero curvature. Applied to the two-dimensional case, the surface of a sphere has positive curvature (angular sum > 180°, elliptic geometry of Riemann). A saddle-shaped surface has negative curvature (angular sum < 180°, hyperbolic geometry of Lobachevsky and Bolyai). Euclidean geometry, finally, has zero curvature (angular sum = 180°) and is called "flat." To determine the nature of the space we live in, we have to choose among the many spatial structures that are possible according to Riemann's scheme. An appropriate choice has to be based on the empirical hypothesis of the infinitesimal Pythagorean. (Riemann also noted that the Euclidean axioms are only approximately true: even if they have not been falsified so far, their truth depends on the degree of precision we have reached in applying them.)

Now, Helmholtz thought that Riemann had put the cart before the horse. Instead of developing possible spaces and asking which one fits best our real one, we should first ask under what conditions measurement is at all possible. This is tantamount to asking under what conditions rigid bodies (or one-dimensional lines) can be brought into congruence with each other. The most important condition Helmholtz identified is their free mobility, i.e., that they do not change their form if moved in space. This conception is perceiver-centered, unlike Riemann's, because it starts from the problem of how a perceiving subject can develop the idea of space. Bodily movement is decisive: if you are unable to move your body or your eyes, you cannot learn what an object is and how it relates to your body and to other objects, and then you have not acquired a sense of space. Experience from motion allows you to judge the distances to objects in your environment. In measurement, you move a yardstick and bring it into agreement with the measured body—a case of congruence. Helmholtz wrote: "The fact that congruence can be observed is the original fact, upon which all our notions of space are based" (1870: 645). Space is based on fact, and not on a hypothesis as Riemann had it.

Originally Helmholtz argued that the only infinite three-dimensional space in which congruence holds is the Euclidean one, but the mathematician Eugenio Beltrami (1835–1900) soon informed him that it holds in hyperbolic space as well. Beltrami also gave a conditional proof of the consistency of hyperbolic space (or of "pseudospherical" space, as he called it); he specified a model of it in Euclidean space, thus showing that hyperbolic geometry is consistent if Euclidean geometry is. (Helmholtz used this model to show that one can develop an intuition of hyperbolic space.) Free mobility is also possible in elliptic space, as Riemann has shown, and as Helmholtz came to realize. From 1870 onward, Helmholtz employed his and Riemann's results to criticize the Kantian doctrine of the a priori and necessary character of Euclidean geometry. To show that we can develop an intuition of a non-Euclidean world, Helmholtz considered the sense-impressions that a convex mirror causes as an analogy to the case of hyperbolic space. A person living in the mirror-world would reach exactly the same results of measurement as in the world that is not mirrored. The fact that it is much more difficult to imagine what it means to live in a hyperbolic than in a Euclidean world cannot, according to Helmholtz, serve as a counterexample to the possibility of our space being negatively curved. There are, however, experiences with which the two non-Euclidean worlds (with constant curvature) can be distinguished from the Euclidean case as long as the curvatures are large enough to create effects that are perceivable with the senses or with suitable instruments. Inhabitants of a constantly curved world can find out about their kind of space by measuring the angular sum of triangles. The criterion of rigidity, Helmholtz argued, must therefore depend on some physical feature of the world.

In 1879, in an appendix to his seminal "The Facts in Perception," Helmholtz introduced the expression "physical geometry" for a geometry that is exclusively based on measurement using physical means and not on any transcendental intuition of space. It is very tempting to contrast this with the phrase "(system of) geometrical axioms" that he also used and interpret it in the sense of Einstein's "empty conceptual schemata." To declare Helmholtz the founder of the distinction between formal and physical geometry would, however, be completely anachronistic. Seeing mathematical theories as empty structures to be physically interpreted was possible only after Hilbert's work ca. 1900 (prepared by Moritz Pasch's work of 1882). In addition, Helmholtz made it clear that he used the term "physical geometry" only for the purpose of freeing the *axioms* of geometry from all reliance on pure intuition in the sense of Kant. But there is no indication in the text that he thought he had succeeded. Helmholtz rather assumed that a general form of spatial intuition existed *besides* or *before* the empirically acquired axioms: "space can be transcendental without the axioms being so" (1879/1977: 149; see Biagioli 2014). The transcendental part of geometry relates to the spatial order in which our sensations offer themselves to us (1879/1977: 126–31).

The reception of Helmholtz

As already noted, the advent of Einstein's relativity theories created a situation that demanded an in-depth reconsideration of the relation between mathematics (especially geometry) and physical theory, taking into account also earlier considerations like Poincaré's and Helmholtz's. All the leading members of logical empiricism took on this task, and all of them had a more or less articulate view of Helmholtz's pioneering work.

Helmholtz's conception of geometrical space was founded on the intimate relation of congruence with motion (free mobility). The Norwegian mathematician Sophus Lie (1842–1899), assisted by Friedrich Engel, succeeded in the 1880s in expressing this close relationship with the help of the concept of a mathematical group, more exactly, a continuous transformation group operating on a three-dimensional manifold (Lie 1893: sect. V). Independently, transformation groups also became the starting point for Poincaré's philosophy of geometry. Poincaré, who had appreciated group-theoretic ideas before, wholeheartedly embraced the group approach to geometry and related it to the question of the transcendental in geometry: "What we call geometry is nothing but the study of formal properties of a certain continuous group; so that we may say space is a group. The notion of this continuous group exists in our mind prior to all experience; but the assertion is no less true of the notion of many other continuous groups; for example, that which corresponds to the geometry of Lobatchévski." We choose among these groups of displacements that which comes closest to experience. Yet this choice is "not imposed by experience. It is simply guided by experience. But it remains free; we choose this geometry rather than that geometry, not because it is more true, but because it is the more convenient" (Poincaré 1898: 41–42).

One could argue that it was the step to *variable* curvature in general relativity (and its faint foreshadowing at the end of Riemann's habilitation lecture) that actually revolutionized our view of geometry and not so much the Helmholtz-Lie view of the constancy of spatial curvature. For Einstein, however, it was crucial to assume "that the 'body' of geometry is realized in principle by rigid bodies in nature. . . [and that] the 'distance' of geometry agrees [thereby] with a natural object." He continued: "This point of view was especially clearly advocated by Helmholtz, and we can add that without him the formulation of relativity theory would have been practically impossible" (Einstein 1925/2007: 160–1). He contrasted Helmholtz's outlook strongly with Poincaré's conventionalism, conceding that Poincaré's view could in the end be more adequate for the systematic representation of a finished physics. For the present unfinished and provisional situation, however, he believed Helmholtz's view to be better suited (see Einstein 1921/2007: 148–9; Friedman 2014: 204–7; Ryckman 2017: ch. 7).

Moritz Schlick (1882–1936) wholeheartedly praised Helmholtz's treatment of geometrical knowledge: "Helmholtz's greatest epistemological achievement, his theory of space, is quite certainly true" (1921a/1979: 340). Yet, he criticized Helmholtz's notion of a general form of spatial intuition. This criticism was based on the view of "intuition" that Schlick had analyzed and rejected earlier. An old view of geometry has it that we need some kind of direct intuition in order to grasp the truth of its axioms (e.g., Blaise Pascal's *raisons du cœur*). Helmholtz rejected this view as far as the axioms were concerned, but he nevertheless wanted to retain some kind of intuition in order to ground geometry in it. Schlick argued not against this idea as such, but against Helmholtz's version of it. Still, in 1915 he held that the Kantian forms of intuition must be retained once they are freed "from all quantitative, all mathematical, all metrical attributes" so that "only qualitative attributes of space" or "in short the . . . specifically spatial of space" remain—whatever this may be (1915/1979: 163). Schlick championed the "method of coincidence" as a purely conceptual substitute for intuition only in later work, when he criticized Helmholtz's retention of the transcendental as still too entangled in traditional notions of intuition. There is a (partially unsolved) debate in the literature on whether Schlick *misunderstood* Helmholtz's sign theory of perception, according to which our sensations must be viewed as signs rather than pictures or images of external objects, or whether his epistemology is a systematic *extension* of Helmholtz's theory (the details of this need not concern us here; see Friedman 1997; Oberdan 2015).

Since Helmholtz thought that the space in which we live can be apprehended in a number of entirely different ways as long as we change our system of mechanical principles accordingly, Schlick read him as a forerunner of Poincaré, who wrote that space is "in reality amorphous and the things which are therein alone give it a form" (1908/1914: 417). Schlick (1922) adopted this idea but criticized Poincaré for prematurely insisting that a physicist would actually always retain Euclidean geometry for simplicity's sake, if a deviation from the Euclidean case occurred, and declare light-rays (or whatever plays the role of straight lines) as curved. With Einstein, the physics community chose the alternative possibility.

Hans Reichenbach (1891–1953), who more than any other logical empiricist immersed himself in the philosophy of space and time, also valued the work of Helmholtz greatly and saw in him the founder of the modern philosophy of space:

> Helmholtz laid the philosophical foundations. In particular, he recognized the connection of the problem of geometry with that of rigid bodies and interpreted correctly the possibility of a visual representation of non-Euclidean spaces. It is his merit, furthermore, to have clearly stated that Kant's theory of space is untenable in view of recent mathematical developments. Helmholtz' epistemological lectures must therefore be regarded as the source of modern philosophical knowledge of space.
>
> (1928/1958: 36)

The penultimate sentence is true if one benevolently understands it in the sense of "as far as the mathematics of the problem is concerned." Helmholtz thought, as we have seen, that beyond the mathematical (axiomatic) side of the space problem, there are indeed good reasons speaking in favor of Kant's view. It is clear, however, that Reichenbach did not concede such a possibility.

Reichenbach formulated a version of conventionalism that was standard until recently. It rests on the difference between differential and universal forces. A universal force, as for example, gravity, acts on all objects alike and cannot be shielded against. In contrast, a differential force, as for example heat, influences bodies differently, depending on some physical property of the bodies. What can we say in the light of this distinction about the conditions a solid object has to fulfill in order to meaningfully serve as a measuring rod? It should be corrected against all the differential forces acting on its length. What about the universal forces? If we assume that there is no universal force acting on it, we must take the result of our measurements as indication that we live in such and such a geometrical space. (If our rigid rods, for example, measure angular sums of triangles less than 180°, we have to conclude that we live in a hyperbolic space.) If we assume, however, that there is a universal force acting on the measuring device, we can blame on it any deviation from the Euclidean case and avoid the conclusion that space is non-Euclidean. So, the decision to correct or not to correct a measuring rod for a universal force amounts to a conventional choice of geometry. The reason is that a constantly curved space can also be described as a space without curvature but with universal forces. Reichenbach concluded: "The solution to the problem of space is therefore found only in this conception we call conventionalism and which goes back to Helmholtz and Poincaré" (Reichenbach 1922/2006: 135).

Reichenbach and his fellow logical empiricists demanded that infinitesimally small (Minkowskian) rigid rods have to be conventionally stipulated in order to fix the metric of general relativity. However, the German mathematician and physicist Hermann Weyl (1885–1955) objected that these rods cannot be rigid because of the inhomogeneity of the gravitational field. It should be decided on the basis of the theory itself which entities can suitably be used as measuring instruments. Einstein came to accept this view, but he nevertheless insisted that one should use the prima facie empirical validity of rigid rods for weak gravitational fields as a regulative empirical principle in the working out of the new theory. Otherwise, the theory would methodologically be too far removed from the empirical realm. As a result, Einstein and the logical empiricists talked at cross purposes for the rest of their lives. The controversy is perhaps partly due to the logical empiricists' neglect of the Riemannian over the Helmholtzian tradition (see Giovanelli 2013).

In his dissertation, Rudolf Carnap (1891–1970) distinguished between formal, intuitive, and physical space in order to clarify the situation in which philosophers, mathematicians and physicists came (and still come) to different conclusions. According to him, both philosophers defending Kant and mathematicians disputing him "were right and could easily have reached agreement if clarity had prevailed concerning the three different meanings of space" (1922/2019: 121). At the time, Carnap took knowledge of physical space to be "entirely empirical, in agreement with empiricists like Helmholtz and Schlick" (1963: 11, see also 957). Yet, he thought that we have to acknowledge an intuitive space that is based on "pure intuition," is independent of contingent experience, and can be represented entirely by topological axioms. Carnap acknowledged he had been influenced in his view of intuitive space by the neo-Kantians Paul Natorp and Ernst Cassirer, but above all by Edmund Husserl's "immediate grasp of essences." Both, neo-Kantians and Husserl, rejected Kant's original sharp distinction between intuition and thought, which Carnap did not discuss. Although the mathematician Weyl was himself fond of Husserl, he dryly wrote in a review of Carnap's dissertation: "The work seems to have originated from a logical tendency to order. A deeper epistemological analysis, especially of the relation between intuitive and physical space, is lacking" (Weyl 1922: 632).

Carnap returned to the problem of space only much later and devoted a large part of an influential introduction to the philosophy of science to it. No trace was left of the earlier idea of intuitive space. Carnap stressed the importance of thoroughly discussing the nature of geometry because it leads to an analysis of the basic structure of modern physics. And he noted that mathematical and physical geometry "are excellent paradigms of two fundamentally different ways of gaining knowledge: the aprioristic and the empirical" (1966: 125). A subsection entitled "Kant's Synthetic A Priori" is exclusively devoted to the distinction between mathematical and physical geometry and its conflict with Kant's view.

Conclusion

The reception of Helmholtz's, Riemann's, and Hilbert's foundational work on geometry led all logical empiricists to the division of geometry into a mathematical (or pure) and physical (or applied) part, reflecting a dichotomy extending to all knowledge: it is either analytical a priori or synthetic a posteriori. The classical formulation of this view was given by Einstein in his "Geometry and Experience" of 1921. Schlick commented that Einstein there "formulated in wonderful conciseness [what] arguably must be regarded as a cornerstone of modern theory of exact scientific knowledge" (1921b: 435) and quoted the famous words: "As far as the laws of mathematics refer to reality, they are not certain, and as far as they are certain, they do not refer to reality" (Einstein 1921/2007: 147; compare Poincaré 1908/1914: 435). In consequence, the last remnants of "intuition" disappeared completely. Philipp Frank wrote:

> [W]e must consider that there is an ambiguity in the use of the term "intuitive." It can mean "perceivable by sense-observation," but it can also mean "perceivable by the eyes of the mind or by inner intuition." In the first sense, our knowledge of the table on which we write is "intuitive"; in the second sense, the axioms of geometry are "intuitive" to those who believe that their validity is self-evident. . . . If we keep to the first meaning of the word "intuitive," which is the only one admissible in science, we shall be adhering to the way in which the great German physiologist, mathematician, physicist, and philosopher, Hermann Helmholtz, defined an "intuitive" presentation of geometry".
>
> (1957: 89)

Here, Helmholtz's arguments that non-Euclidean geometry can be imagined and visualized are taken as a refutation that there is a "necessary form of outer intuition"—despite what Helmholtz maintained together with Kant.

References

Biagioli, F. (2014) "What Does It Mean That 'Space Can Be Transcendental Without the Axioms Being So'?: Helmholtz's Claim in Context," *Journal for General Philosophy of Science* 45: 1–21.

Cahan, D. (2018) *Helmholtz: A Life in Science*, Chicago and London: University of Chicago Press.

Carnap, R. (1922) *Der Raum: Ein Beitrag zur Wissenschaftslehre*, Berlin: Reuther & Reichard. Trans. "Space: A Contribution to the Theory of Science," in R. Carnap, *Collected Works, vol. 1. Early Writings* (ed. by A. W. Carus et al.), Oxford: Oxford University Press, 2019, pp. 21–208.

——— (1963) "Intellectual Autobiography" and "Replies and Systematic Expositions," in P. A. Schilpp (ed.), *The Philosophy of Rudolf Carnap*, La Salle, IL: Open Court, 1963, pp. 1–84 and 859–1013.

——— (1966) *Philosophical Foundations of Physics: An Introduction to the Philosophy of Science* (ed. by Martin Gardner), New York: Basic Books.

Duhem, P. (1906) *La Théorie physique. Son objet et sa structure*, Paris: Chevalier & Rivière 1906. German trans. *Ziel und Struktur der physikalischen Theorien*, Leipzig: Barth, 1908. Engl. trans. *The Aim and Structure of Physical Theory*, Princeton: Princeton University Press, 1954.

Einstein, A. (1921) *Geometrie und Erfahrung*, Berlin: Julius Springer. Trans. "Geometry and Experience," in Pesic (2007), pp. 147–57.

——— (1925) "Nichteuklidische Geometrie und Physik," *Die Neue Rundschau* 36 (1): 16–20. Trans. "Non-Euclidean Geometry and Physics," in Pesic (2007), 159–62.

Enriques, F. (1906) *Problemi della scienza*, Bologna: Zanichelli. German trans. *Probleme der Wissenschaft*, Leipzig und Berlin: Teubner 1910. Engl. trans. *Problems of Science*, Chicago: Open Court, 1914.

Frank, P. (1957) *Philosophy of Science: The Link Between Science and Philosophy*, Englewood Cliffs, NJ: Prentice-Hall.

Friedman, M. (1997) "Helmholtz's *Zeichentheorie* and Schlick's *Allgemeine Erkenntnislehre*: Early Logical Empiricism and Its Nineteenth-Century Background," *Philosophical Topics* 25: 19–50.
——— (2014) "Space, Time, and Geometry: Einstein and Logical Empiricism," in M. Janssen and C. Lehner (eds.), *The Cambridge Companion to Einstein*, Cambridge: Cambridge University Press, pp. 398–420.
Giovanelli, M. (2013) "Talking at Cross-Purposes: How Einstein and the Logical Empiricists Never Agreed on What They Were Disagreeing About," *Synthese* 190: 3819–63.
Helmholtz, H. (1868) "Ueber die thatsächlichen Grundlagen der Geometrie," *Verhandlungen des naturhistorisch-medicinischen Vereins zu Heidelberg* 4: 197–202.
——— (1870) "The Axioms of Geometry," *The Academy* 1: 128–31.
——— (1879) *Die Thatsachen in der Wahrnehmung*, Berlin: Hirschwald. Trans. "The Facts in Perception," in Helmholtz (1977), pp. 115–85.
——— (1977) *Epistemological Writings. The Paul Hertz/Moritz Schlick Centenary Edition of 1921* (ed. by R. S. Cohen and Y. Elkana), Dordrecht: Reidel.
Lie, S. (1893) *Theorie der Transformationsgruppen. Dritter Abschnitt*, Leipzig: Teubner. Repr. New York: Chelsea Publishing Co., 1970.
Oberdan, T. (2015) "From Helmholtz to Schlick: The Evolution of the Sign-Theory of Perception," *Studies in History and Philosophy of Science* 52: 35–43.
Pasch, M. (1882) *Vorlesungen über neuere Geometrie*, Leipzig: Teubner.
Pesic, P. (ed.) (2007) *Beyond Geometry: Classic Papers from Riemann to Einstein*, Mineola, NY: Dover.
Poincaré, H. (1898) "On the Foundations of Geometry," *The Monist* 9: 1–43.
——— (1902) *La Science et l'Hypothèse*, Paris: Flammarion. German trans. *Wissenschaft und Hypothese*, Leipzig: Teubner, 1904. Engl. trans. "Science and Hypothesis," in Poincaré (1913), pp. 3–197.
——— (1908) *Science et Méthode*. Paris: Flammarion. German trans. *Wissenschaft und Methode*, Leipzig: Teubner, 1914. Engl. trans. "Science and Method," in Poincaré (1913), pp. 359–546.
——— (1913) *The Foundations of Science: Science and Hypothesis, The Value of Science, Science and Method*, New York: The Science Press.
Putnam, H. (1979) *Mathematics, Matter and Method*, Cambridge: Cambridge University Press.
Reichenbach, H. (1922) "La signification philosophique de la théorie de la relativité," *Revue Philosophique de la France et de l'Étranger* 94: 5–61. Trans. "The Philosophical Significance of the Theory of Relativity," in S. Gimbel and A. Walz (eds.), *Defending Einstein: Hans Reichenbach's Writings on Space, Time, and Motion*, Cambridge: Cambridge University Press, 2006, pp. 95–160.
——— (1928) *Philosophie der Raum-Zeit-Lehre*, Berlin: de Gruyter. Trans. *The Philosophy of Space and Time*, New York: Dover, 1958.
Riemann, B. (1868) "Ueber die Hypothesen, welche der Geometrie zu Grunde liegen," *Abhandlungen der Königlichen Gesellschaft der Wissenschaften zu Göttingen. Dreizehnter Band von den Jahren 1866 und 1867*, Mathematische Classe, pp. 133–50. Trans. "On the Hypotheses That Lie at the Foundations of Geometry," in Pesic (2007), pp. 23–41.
Ryckman, T. (2017) *Einstein*, London: Routledge.
Schlick, M. (1915) "Die philosophische Bedeutung des Relativitätsprinzips," *Zeitschrift für Philosophie und philosophische Kritik* 159: 129–75. Trans. "The Philosophical Significance of the Principle of Relativity," in Schlick (1979), pp. 153–89.
——— (1921a) "Helmholtz als Erkenntnistheoretiker," in E. Warburg (ed.), *Helmholtz als Physiker, Physiologe und Philosoph. Drei Vorträge*, Karlsruhe: Müllersche Hofbuchhandlung, pp. 29–39. Trans. "Helmholtz the Epistemologist," in Schlick (1979), pp. 335–42.
——— (1921b) "[Review of Einstein 1921]," *Die Naturwissenschaften* 22: 435–6.
——— (1922) *Raum und Zeit in der gegenwärtigen Physik*, 4th ed., Berlin: Springer 1922. Trans. "Space and Time in Contemporary Physics," in Schlick (1979), pp. 207–69.
——— (1979) *Philosophical Papers, vol. I: 1909–1922* (ed. by H. L. Mulder and B. van de Velde-Schlick), Dordrecht: Reidel.
Volkmann, P. (1896) *Erkenntnistheoretische Grundzüge der Naturwissenschaften*, Leipzig: Teubner.
Weyl, H. (1922) "[Review of Carnap 1922]," *Jahrbuch über die Fortschritte der Mathematik* 48: 631–2.

6

ERNST MACH AND EARLY LOGICAL EMPIRICISM

Elisabeth Nemeth

It is hardly news that the early logical empiricists owe characteristic aspects of their philosophical views to the physicist Ernst Mach. Mach's radical empiricism, his militant and strict rejection of metaphysics, as well as his conception of the ultimate unity of the object domains of the natural sciences and the human sciences (*Geisteswissenschaften*)—the unity of science—constitute important elements particularly of the Vienna Circle's program for epistemology and the philosophy of science. The society for the promotion of the scientific world-conception co-founded by the Circle bears his name. A look beyond their programmatic pronouncements, however, shows very pronounced differences in the reception of his work. While Moritz Schlick, Hans Hahn, and Rudolf Carnap paid their respects to the great physicist and philosopher in more general terms, Philipp Frank, Otto Neurath, and Richard von Mises took account of Mach's works more widely and intensively. (For an overview of the varied Mach reception in the Circle, see Stadler [2021]; Mach was of lesser importance for Reichenbach's Berlin Group.)

It was in particular the historical-critical writings, above all *The Science of Mechanics*, that left visible traces in Frank's, Neurath's, and Mises's way of thinking. But Mach's early *Die Geschichte und die Wurzel des Satzes von der Erhaltung der Arbeit* (*History and Root of the Principle of the Conservation of Energy*) was of particular importance for them. There, Mach made fully explicit the close connection that existed between his criticism of metaphysics and his historical investigations of physical hypotheses. "We are accustomed to call concepts metaphysical, if we have forgotten how we reached them. One can never lose one's footing or come into collision with facts, if one always keeps in view the path by which one has come" (1872/1911: 17).

Historical research demonstrated for Mach that coincidences and social preconditions contribute to the development of science. (On Mach's pragmatism, see Patton [2021] and Uebel [2021].) This insight has a liberating effect: it allows us to conceive that new ways of investigations are possible.

> Historical investigation not only promotes the understanding of that which now is, but also brings new possibilities before us, by showing that which exists to be in greater measure *conventional* and *accidental*. From the higher point of view at which different paths of thought converge we may look about us with freer vision and discover routes before unknown.
>
> (Mach 1883/1960: 316)

DOI: 10.4324/9781315650647-8

It was this program of reaching a "clearer view" by means of historical-critical investigations that made Mach's work so very attractive to Frank, Neurath, and Mises—and what distinguishes them among and from other members of the Vienna Circle. It may well be that they owe their awareness of the liberating effect of Mach's teachings to the discussions they led as young scientists around 1910 in Vienna about the foundational issues of modern science in the so-called First Vienna Circle (see CH. 10). The sources documenting this group are meager, but the exceptional importance of Mach's thought for Frank, Mises, and Neurath may be regarded as an indirect pointer to the common intellectual socialization of these young scientists before World War I. In any case, Frank and Mises felt liberated by Mach's demonstration of how historical-critical studies can free one's thinking both in and outside of science from misleading concepts and ideas. For Neurath in particular, Mach's critical discussion of the law of inertia provided an important stimulus for his ambitious project to reconceive the foundations of economic science.

The philosopher of enlightenment: against the misuse of "auxiliary concepts" (Philipp Frank)

One year after Mach's death, Frank published a paper in *Die Naturwissenschaften*, a very widely read scientific weekly bridging all specialties, that spelled out the importance of Mach's epistemology as representing "the philosophy of enlightenment appropriate to our time" (1917/1949: 72). Mach's proximity to the enlightenment showed itself, so Frank, not only in those paragraphs of *The Science of Mechanics* which describe the eighteenth century in glowing terms, but predominantly in that Mach, like the early modern scientists, criticized the misuse of auxiliary concepts. "[W]hat was enlightening in Galileo's writings was his setting a limit to the misuse of auxiliary concepts [of Aristotelian philosophy, EN]. And it is this protest against the misuse of merely auxiliary concepts in general philosophical proofs that I consider to be an essential characteristic of enlightenment" (ibid.: 73). Since all thinking—even enlightenment thinking—requires the use of auxiliary concepts, to Frank, it is necessary for criticism to start anew with every age. "Every period of physics has its auxiliary concepts, and every succeeding period misuses them. Hence in every period a new enlightenment is required in order to abolish this misuse" (ibid.).

In this interminable, ongoing dialectical movement (see Uebel 2021: 89–92), Frank detected "that the philosophy of the enlightenment possesses a tragic feature," a feature which had also been recognized by Nietzsche (as, he noted, recently argued by Kleinpeter [1913]). This enlightenment "destroys the old system of concepts, but while it is constructing a new system, it is also already laying the foundation for new misuse. For there is no theory without auxiliary concepts, and every such concept is necessarily misused in the course of time" (Frank 1917/1949: 78). Here, Frank made visible certain connections between Mach's thought and twentieth-century philosophy that are anything but commonplace: especially the connection to Nietzsche and to the "dialectics" of enlightenment cast a surprising light on Mach's empiricism.

Mach's critical attitude towards the concept of the atom also becomes much more intelligible, Frank argued, once it is taken into account that criticism of the misuse of concepts played the central role for Mach. "It is true that atomistics, when applied to physiologic and psychologic problems, easily leads into a blind alley. Such questions arise as: 'How can a brain atom think?', 'How can an atom perceive green, since, after all, it is itself only a miniature picture of a macroscopic body composed of perceptions?'" (ibid.: 70). Frank conceded that Mach overextended his dislike of atomic theory, failed to recognize its use in the limited domain of physics, and that this brought his epistemology into disrepute. Especially the contrast between Mach and Ludwig Boltzmann, to Frank, was unduly sharpened for polemical purposes, but more detailed investigation shows that Mach's epistemological views are often much closer to those of his critics than one was led to expect.

> The historical error is often made of connecting the struggle of Mach and Duhem for the positivistic physics with their aversion for atomism, so that a victory for atomism was considered a defeat for positivism. In reality the champions of atomism, Maxwell and Boltzmann, were exactly of the same opinion concerning the general nature of a physical theory as Hertz and Mach. The difference in their views about the value of atomistic theories arose only because they differed in their estimates of the convenience with which actually known physical phenomena could be derived from these theories.
>
> (1937/1949: 140–1)

In Frank's view, the dividing line between Boltzmann's conception of the nature of scientific theories and the descriptivism of Kirchhoff and Mach was a much more subtle affair than commonly suggested. To show that Boltzmann himself agreed, Frank quoted from his *Lectures on the Theory of Gases*:

> The question as to the fitness of the atomistic philosophy [*Zweckmässigkeit der atomistischen Anschauungen*] is naturally wholly untouched . . . by the question whether it may not be expedient to consider our theories as pure descriptions so that we may always recall their relation to nature. Therefore the question is whether the pure differential equations or atomism will one day turn out the more complete descriptions of phenomena.
>
> (Boltzmann, quoted in Frank 1937/1949: 142–3)

The commonalities between Boltzmann and Mach are indeed remarkable. First, Boltzmann thought it a matter of course that the legitimacy of the concept "atom" depends on whether its use proves "expedient" for the future development of science. This he stated himself in his first lecture filling in for Mach after his retirement (see Boltzmann 1903/1990: 152). Second, an even more far-reaching commonality becomes evident in the paragraph Frank quoted from Boltzmann. It concerns the question in what the aim of science consists of such that it can provide a measure for the utility of scientific hypotheses. Whether the concept of pure differential equations or atomism proves more useful will be shown by which of them "will one day turn out the more complete descriptions of phenomena." Here, Boltzmann clearly refers to Kirchhoff's formulation of science as aiming for the "complete and simplest description" of phenomena, a conception wholly shared by Mach (e.g., Mach 1905/1976: 212).

In one respect Frank's 1917 paper could mislead. The notion of an "auxiliary concept" appears as a term of Mach's; however, Mach used it only to designate concepts that are of secondary importance relative to others (*sekundäre Hilfsbegriffe*) (see Mach 1883/1960: 320). Now, Frank can be read as bringing Mach's thought to bear on a type of philosophical discussion that Mach was rightly suspicious of and where concepts originally defined in physics are used in providing "general philosophical proofs" (Frank 1917/1949: 73). Designating physical concepts in these contexts as "merely auxiliary concepts" draws attention to the fact that their meaning is bound to certain procedures and usages in physics, and that in those "general philosophical proofs" they are treated as if they possessed a wholly independent meaning. Yet Frank also discussed concept formation in physics and noted ongoing obscurities. "Into these nebulosities persistent doubt can penetrate and shake the whole system of physics as the foundation of our scientific world picture. Here Mach steps in and says: All these concepts are only auxiliary concepts" (ibid.: 67, trans. amended; notably, the phrase after the colon is not a quotation). So Frank's use of "auxiliary concept" reflects a polemical context, in particular the defense of physics itself against "attacks from outside" (ibid.). Frank's rhetorical strategy is justified within this polemic, but it is problematic in suggesting an instrumentalist reading of Mach's epistemology. Common for the longest time, this interpretation has recently faced strong opposition: according to Banks (2014), Mach professed a "realistic empiricism" that places him in line with James and Russell.

It must be added that Frank did not obliterate the differences between Mach and Boltzmann, for instance, that the latter considered theories more important than the former who called them "withered leaves, which drop off after having enabled the organism of science to breathe for a time" (Mach 1872 in Frank 1917/1949: 62). Against this he stressed that the logical empiricists already early on regarded it as a weakness of Mach's epistemology that underemphasized the theoretical-logical side of science, in particular the role of mathematics. Particularly in this respect, Poincaré's conventionalism served as an important corrective (Frank 1949: 7–9).

Freedom of thought and radical pluralism (Richard von Mises)

In an essay to celebrate Mach's centenary Richard von Mises also stressed his importance as a philosopher of enlightenment. In doing so Mises concentrated upon a different aspect, namely how Mach reacted to the epistemological problem that preoccupied the scientists and philosophers of the nineteenth century: "the dilemma of the illusory world of our senses and the 'true world' of science" (Mises 1938/1987: 177).

> Anyone who thinks like a normal person . . . makes it the foundation of his view of life that there are two kinds of worlds: one which is perceived by our senses, but is more or less illusionary and perceptive, and another behind it, which is the true, the real world, but can be glimpsed only with the acute vision of the scientist.
>
> (ibid.: 166)

According to Mises, since the rise of modern science, philosophers had been disturbed by this gap but, with the exception of Hume, offered only inadequate responses to the challenge. Kant fought metaphysical dogma, but the attempt to replace such justifications of claims to knowledge with his rigidly constructed system did not amount to much more than an attempt "to prove in a grand style and with a show of greatest ingenuity that it is after all of the *schoolmaster* who is right" (ibid.: 168, orig. emphasis). Kant thought about science as if it were a Platonic idea, whereas for Mach it was an earthly affair, "a real living manifestation of the human race" (ibid.: 171).

It was in virtue of his this-worldly conception of science that Mach was able, so said Mises, to make a substantial contribution to the overcoming of the modern problem of knowledge, especially by way of historical studies. Mach dared to undertake a critical investigation of the fundamental principles of Newton at a time when for most of his colleagues those principles counted as established once and for all. It was this independence of thought that Mach defended in the dispute about atomism by fighting the fossilization of physical theories into dogmas. Subsequent developments in physics then made plain "how much these advances have owed to the freedom of thought Mach created" (ibid.).

It seems to me that Mises chose the seemingly awkward formulation that Mach "created" freedom of thought—rather than used the freedom inherent in scientific thinking—intentionally. For Mises, Mach had achieved a new freedom of thought for himself and his successors, first, by always connecting physical research with investigations of the history of science, and, second, by developing an analysis of knowledge that opened up a new way to approach the old dilemma of the illusory world of experience and the "true" world of science. "[B]y postulating the possibility of a regress from an expression of any degree of complexity to its observational or experiential elements, Mach's analysis of knowledge provides science with the foundations which secure it against contradictions and other intellectual dangers" (ibid.: 179).

Mises captured the spirit of Mach very well by adding that science needs to be freed not only from contradictions but also from "other intellectual dangers." Both the supposed essential difference between the natural sciences and the humanities and the hierarchical ordering of types of investigations hinder this freedom. This also holds of the sharp separation of everyday and scientific knowledge. Mises discerns the outstanding potential of Mach's ideas precisely in that they make plain and obvious that different perspectives can only ever reveal a certain set of relations between appearances, but never their totality. None of them, moreover, is "more true" than another. Any increase in our knowledge of nature and of ourselves comes about only by virtue of establishing an increasing number of connections between these perspectives (*Betrachtungsweisen*).

> The world we live in is too complicated for it to be either possible or useful for us to consider everything we encounter at every moment from *all* sides. To the person who surveys a property with a view to building a house, the surface of the earth is a plane; the person who plots the course of a river must attend to the unevenness of the surface, to its deviations from a plane; to a world traveller, the earth has certain properties of the surface of a sphere; and the geodesist, who carries out measurements on a very large scale and with great precision, will notice that the earth is not a sphere at all, but a spheroid. Each of these points of view is justified within its context, and *none is truer or more real* or closer to reality than any other; for the earth shares certain properties with a plane and certain others with a sphere, and we are sometimes interested in the ones and sometimes in the others; and that is all. The question whether the earth is at rest or turns about its axis resolves itself in the same way: it is at rest if in using the word "turn" we are thinking of subjective phenomena like the ones known to us from our experience with merry-go-rounds; but the hypothesis of rotation, severed from all subjective accompanying phenomena, enables us to give a simpler, more economical description of the observable cosmic processes—if this is what we are interested in.
>
> (ibid.: 179–80, orig. emphasis)

It is important to realize that granting equal justification to the standpoints of different types of investigations was, for Mach, not a philosophical demand but an eminently practical scientific one. Recent investigations of Mach's work show this *in concreto*. His famous criticism of Newton's "absolute space" and "absolute time" relied not only on his analysis of Newton's bucket experiment, but also on his studies of the physiological and psychological effects of turning around one's own body (Staley 2019: 349).

Here we find the origin of Neurath's anti-hierarchical idea of an "encyclopedic unity of science" in which the connections between distant domains of research should become visible—and his strict refusal to recognize physics as *the* fundamental science. Neurath's rejection of personal protocol languages too is related to his refusal to grant such a privilege to any one type of investigation.

The historical-critical method and the theory of elements

As noted, Mach always stressed the value of historical inquiries. One passage, first added in the fifth edition of 1904 of his *The Science of Mechanics* to his chapter on Newton, is particularly interesting in this respect. Prompted by the physicist MacGregor, Mach there reflects on his "own historical-critical method" and stated that by then—contrary to when his book was first published in 1883—an increasing number of physicists shared his view that "'absolute motion' is a conception which is devoid of content and cannot be used in science" (1883/1960: 293). But this is only the precondition for the next step that really matters to him: how now to give an intelligible meaning to the law of inertia? MacGregor (1893), Mach noted, showed in an "excellent and clearly written essay" that there are two ways of doing so:

> 1. The historical-critical way, which involves taking a fresh look at the facts upon which the law of inertia rests and considering the limits of its validity and perhaps also a new formulation of it. 2. To assume that the law of inertia in its old form acquaints us sufficiently with motions and to derive the right coordinate system *from* these motions. The presentation I give here is an example of the first method. It also contains an indication of modifications of expression which it now becomes necessary to introduce by expanding experience. Surely, the second way is psychologically more appealing in view of the trust that mechanics enjoys as the most exact natural science. In fact this way has often been pursued with more or less success, and I myself tried to do so before I came to believe it necessary to prefer the other way.
>
> (Mach 1883/1908: 257, trans. EN. In the next [7th] edition of 1912, Mach abbreviated this passage and named Newton himself as an example of the "first way": 1883/1960: 293)

Mach here referred to MacGregor's contraposition of two legitimate procedures which both aim to give "an intelligible meaning to the law of inertia" (see Nemeth 2019). This gives the historical-critical method a specific status, namely that of a distinctive method to reach a goal which other methods aim for as well. Mach even emphasizes the connection between them by pointing out that he himself pursued the second one for a long time. This in effect makes the historical-critical method potentially into an alternative which scientists could, perhaps even should, pursue in cases where they lost confidence in being able to achieve a substantial extension of knowledge on the basis of laws that have been discovered so far.

Even though this is another occasion where Mach appears to have anticipated the historiography of science we associate with Thomas Kuhn in significant respects (Banks 2014: 39), we must not neglect the differences. The first is that for Mach, the history of science is continuous with the evolution of the human species and therefore part of biological evolution. By contrast, history of science in the second half of the twentieth century employs a more sociological perspective and has no need to discuss the relation between historiography and evolutionary theory. For Mach as a natural scientist of the nineteenth century, however, this is an unavoidable topic. In his writings, therefore, we find components derived from evolutionary theory mixed with historical and epistemological investigations. (A similar tendency can also be detected in the writings of Boltzmann.) The second difference in methodology is related to the first. For Mach, historical investigations of scientific hypotheses aim to extend and improve our knowledge of nature and thus are part of natural science. Here, historical criticism is reflexive—to be sure, this is not a term Mach used, but it fits nevertheless—in that natural scientists investigate the origin of hypotheses which they themselves are working on in order to test and improve them.

Importantly, Mach's quasi-reflexive turn does not lead investigators into another domain, neither onto a philosophical meta-level nor into the writing of history. In other words, historical-critical research must not be understood as one of the many possible types of investigating the world scientifically, because the historical-critical perspective presupposes one of these types of investigation with its respective concepts and propositions. Mach's historical-critical method aims for the renewal of concepts, but for Mach this was a part of first-order research. It leads back to "the facts" in order to "take a fresh look" at them, it seeks to determine the "limits of the validity" of the hypotheses under investigation and to propose "a new formulation."

Mach spelled out the nature and relevance of his historical-critical method also in his writings on epistemological matters. Here the way leads "back" to the " '*elements*' which we cannot *currently* go beyond" (1905/1976: 8, trans. amended, orig. emphasis) and to what is the aim of all science: the investigation of "*the complex interdependence of the elements*" (ibid., orig. emphasis).

> The finding out of the direct connections between the elements is so complex a task that it cannot be solved all at once but only step by step. It was much easier to ascertain a rough and ready outline of the way in which whole *complexes* of elements or bodies depend on each one another, and it was rather a matter of chance and practical need *which* elements seemed more important, which were focussed on and *which* remained unnoticed. The individual enquirer is in the midst of developing science and must start with his predecessors' incomplete findings which at best he can correct and perfect according to his ideal. In gratefully adopting for his own work the help and hints contained in these preliminaries, he often adds the errors of predecessors and contemporaries to his own. A return to a completely naive point of view, even if it were possible, would afford to one who shed all the views of his contemporaries not only the advantage of freedom from prejudice, but also the drawback of confusion arising from the complexity of the task and the impossibility of even *starting* any enquiry. If therefore we seem here to be returning to a *primitive* standpoint, in order to conduct the enquiry along new and better paths, this is an *artificial* naivity that does not give up the advantages gained through long periods of growing civilization, but on the contrary uses insights presupposing a fairly high level of thought as to physics, physiology and psychology. Only at such a level is a resolution into 'elements' conceivable. We are thus returning to the *starting points* of enquiry with the deeper and richer insight produced by previous enquiry.
>
> (Mach 1905/1976: 9–10, orig. emphasis, trans. amended)

In this paragraph we find condensed several themes that were important to Neurath throughout his life: the dependence of concepts on practical needs, the influence on scientific research possessed by coincidence and historical factors, the insight that researchers always stand midstream, as it were, and can never begin anew his thinking ("Neurath's Boat"; see Uebel 1996). And, not to forget, the "artificial naivity" with which he set about trying to lead the concepts "along new and better paths."

Ernst Mach and Neurath's economics

From about 1910 onwards, in a number of papers, Neurath proposed a thorough renewal of economic science (see Uebel 2004; Nemeth 2013). In a letter to Mach from 1915, he himself pointed out that he drew the inspiration for his new theory of value from Mach's attempt to rethink the concept of gravitation in his *Science of Mechanics* (Nemeth 2007: 21). Indeed, it is remarkable how what Mach said about his historical-critical way of investigating the law of gravitation (see the previous section) applies to Neurath's efforts: to "take a fresh look" at "the facts" on which the propositions of contemporary economics rested in order to determine their "limits of the validity" and propose "a new formulation."

Neurath pointed to Aristotle when he defined wealth to be the object of investigation for economics (Neurath 1911/1998: 471). This perspective, he added, was largely lost in the nineteenth century when economics limited itself to investigating price formation in markets. This in turn led to the view that the theory of exchange relations in money came to be regarded as the theory of economic activity as such, which Neurath regarded as a fundamental misconception for the laws of price formation are dependent on a particular form of economic activity which did not always exist as such and need not continue either. Viewed historically, market economies of a pure form are found even in modern times only very rarely, but mostly mixed with regulations of very varied sorts (e.g., prices fixed by the state), and sometimes they were suspended altogether (e.g., in times of war). This shows that concepts legitimately formed for representing price formation under conditions of a (free) market are utterly irrelevant and useless for answering the question of what allows people to be or become rich and poor in general. A system of economic concepts that claims scientific universality must be formulated in a way that not only allows it to theorize about exchange relations in markets but also to indicate the limits of market economics. This requires a conceptual framework to be particularly comprehensive. The concept of economic efficiency (*Wirtschaftlichkeit*) must be formulated so that it becomes possible to investigate and compare the effects which different economic orders have on the wealth or poverty of a population.

Prior to World War I, Neurath developed the outlines of a theory of economics whose substance and radically rethought nature is best explained in terms of the effect on his thinking of Mach's radical innovations in physics. By defining wealth as "the epitome of pleasure (*Lust*) and displeasure (*Unlust*) found in individuals and groups of individuals" (1911/1998: 471, trans. EN) Neurath returned to a "more primitive standpoint in order to redirect the investigation to new and better ways" (see Mach from earlier). In the years up to 1917 the terminology of pleasure and displeasure receded into the background and was replaced by "happiness" (*Glück*) and "unhappiness" (*Unglück*) (1917/2004: 313), but what Mach called "artificial naivity" continues to inform Neurath's methodology and even becomes more prominent. His "The Conceptual Structure of Economic Theory and its Foundations" of 1917 dissolved the object of economics into components of well-being beyond which it was not yet possible to go. The wealth of individuals and groups is analyzed in terms of "conditions of life" (*Lebenslage*) and "qualities of life" (*Lebensstimmung*) that are as heterogeneous as the famous "elements" in Mach's *Analysis of Sensations*: "colors, sounds, temperatures, pressures, spaces, times" (Mach 1886/2000: 2). "[T]he quality of life is connected with all types of experiences, with eating, drinking, reading, artistic sensibility, religious contemplation, moral speculation, loving, hating, heroic and cowardly behavior" (1917/2004: 313). "Inventories of the standard of living" (as it was put twenty years later [1937/2004: 513]) were to provide the basis for comparative assessments of the effect of different economic measures or institutions (from free markets to centrally planned economies) on the conditions of life of populations (see CH. 28). Central roles are played then by methodical comparisons, by the method of variation, and by thought experiments—all of them core themes of Mach's epistemology and theory of science.

Neurath's later contributions to logical empiricism are firmly rooted in his early economic thought. If it is correct (which I think it is) "that we should regard Mach's element-and-function methodology as an attempt to actually *change* physical science rather than rationally reconstruct it as it exists" (Banks 2014: 38, orig. emphasis), then we can regard Neurath's economic theory as an attempt to do in social science what Mach did in physics.

References

Banks, E. (2014) *The Realistic Empiricism of Mach, James, and Russell: Neutral Monism Reconceived*, Cambridge: Cambridge University Press.

Boltzmann, L. (1903) "Ein Antrittsvortrag zur Naturphilosophie," *Die Zeit*, 11 December, Beilage, pp. 1–2. Repr. in *Principien der Naturfilosofi. Lectures on Natural Philosophy 1903–1906* (ed. by I. M. Fasol-Boltzmann), Berlin: Springer, 1990, pp. 152–6.

Frank, P. (1917) "Die Bedeutung der physikalischen Erkenntnistheorie Ernst Machs für das Geisteslebens unserer Zeit," *Die Naturwissenschaften* 5: 65–80. Trans. "The Importance for Our Times of Ernst Mach's Philosophy of Science," in Frank (1949), pp. 61–79.

——— (1937) "The Mechanical versus the Mathematical Conception of Nature," *Philosophy of Science* 4: 41ff. Repr. in Frank (1949) as "Mechanical Explanation or Mathematical Description?" pp. 138–43, and "How Idealists and Materialists View Modern Physics," pp. 186–97.

——— (1949) Modern Science and its Philosophy, Cambridge, MA: Harvard University Press.

Kleinpeter, H. (1913) *Der Phänomenalismus*, Leipzig: Barth.

MacGregor, J. G. (1893) "On the Hypotheses of Dynamics," *The London, Edinburgh and Dublin Philosophical Magazine and Journal of Science* 36: 233–64.

Mach, E. (1872) *Die Geschichte und die Wurzel des Satzes von der Erhaltung der Arbeit.* Repr. in Leipzig: Barth. Trans. *History and Root of the Principle of the Conservation of Energy*, LaSalle: Open Court, 1911.

——— (1883) *Die Mechanik in ihrer Entwicklung historisch-kritisch dargestellt*, Leipzig: Brockhaus. 6th ed. 1908, 7th ed. 1912. Trans. *The Science of Mechanics: A Critical and Historical Account of its Development*, LaSalle: Open Court, 1919, repr. 1960.

——— (1886) *Beiträge zur Analyse der Empfindungen*, Jena: Fischer. Trans. *The Analysis of Sensation and the Relation of the Physical to the Psychical*, LaSalle: Open Court, 1897, 2nd ed. 1914.

——— (1905) *Erkenntnis und Irrtum*, Leizig: Barth, 2nd ed. 1906. Trans. *Knowledge and Error*, Dordrecht: D. Reidel, 1976.

Mises, R. V. (1938) "Ernst Mach und die empiristische Wissenschaftsauffassung. Zu Ernst Machs hundertstem Geburtstag am 18. Februar 1938," S'Gravenhage: van Stockum and Zoon. Trans. "Ernst Mach and the Scientific Conception of the World," in B. McGuinness (ed.), *Unified Science*, Dordrecht: Reidel, 1987, pp. 166–90.

Nemeth, E. (2007) "Freeing up One's Point of View: Neurath's Machian Heritage Compared with Schumpeter's," in E. Nemeth, S. W. Schmitz and T. Uebel (eds.), *Otto Neurath's Economics in Context*, Dordrecht: Springer, pp. 13–36.

——— (2013) "The Philosophy of the 'Other Austrian Economics'" in H. Andersen et al. (eds.), *New Challenges to Philosophy of Science. The Philosophy of Science in a European Perspective*, Dordrecht: Springer, vol. 4, pp. 339–50.

——— (2019) "Zur historisch-kritischen Methode bei Ernst Mach," in F. Stadler (ed.), *Ernst Mach—Zu Leben, Werk und Wirkung*, Cham: Springer, pp. 21–43.

Neurath, O. (1911) "Nationalökonomie und Wertlehre," *Zeitschrift für Volkswirtschaft, Sozialpolitik und Verwaltung* 20: 52–114. Repr. in Neurath, *Gesammelte ökonomische, soziologische und sozialpolitische Schriften* (ed. by R. Haller and U. Höfer), Vienna: Hölder-Pichler-Tempsky 1998, vol. 1, pp. 470–519.

——— (1917) "Das Begriffsgebäude der Wirtschaftslehre und seine Grundlagen," *Zeitschrift für die gesammte Staatswissenschaft* 73: 484–520. Trans. "The Conceptual Structure of Economic Theory and Its Foundations," in Neurath (2004), pp. 312–42.

——— (1937) "Inventory of the Standard of Living," *Zeitschrift für Sozialforschung* 6: 140–51. Repr. in Neurath (2004), pp. 513–26.

——— (2004) *Economic Writings. Selections 1904–1945* (ed. by T. Uebel and R. S. Cohen), Dordrecht: Kluwer.

Patton, L. (2021) "Abstraction, Pragmatism, and History in Mach's Economy of Science," in Preston (2021), pp. 164–83.

Preston, J. (ed.) (2021) *Interpreting Mach: Critical Essays*, Cambridge: Cambridge University Press.

Stadler, F. (2021) "Ernst Mach and the Vienna Circle: A Re-Evaluation of the Reception and Influence of his Work," in Preston (2021), pp. 184–207.

Staley, R. (2019) "Revisiting Einstein's Happiest Thought. From the Physiology of Perception to Experimental Propositions and Principles in the History of Relativity," in F. Stadler (ed.), *Ernst Mach—Life, Work, Influence*, Cham: Springer, pp. 349–66.

Uebel, T. (1996) "On Neurath's Boat," in N. Cartwright et al. (eds.), *Otto Neurath: Philosophy Between Science and Politics*, Cambridge: Cambridge University Press, pp. 89–166.

——— (2004) "Introduction: Neurath's Economics in Critical Context," in Neurath (2004), pp. 1–108.

——— (2021) "Ernst Mach's Enlightenment Pragmatism: History and Economy in Science," in Preston (2021), pp. 84–102.

7

BOLZANO AND BRENTANO AND LOGICAL EMPIRICISM

Mark Textor

Bernard Bolzano (1781–1848), a Bohemian philosopher and mathematician best known for his work on (logical) consequence and real analysis, and Franz Brentano (1838–1917), the "grandfather" of phenomenology, are not the first names that come to mind when thinking about logical empiricism. Yet the manifesto of the Ernst Mach Society listed them as philosophers who paved the way for the scientific world-conception (1929/2012: 79). Otto Neurath, one of the co-authors of the manifesto, went on to argue that the logic in logical empiricism was influenced by Bolzano, while the empiricism owes much to Brentano (1936/1981: 688–91). Later Rudolf Haller (1968/1988: 2–3) strengthened Neurath's thesis. He saw in Brentano and the early positivist Ernst Mach (1838–1916) the main influences on the development of scientific philosophy. Brentano's main work *Psychology from an Empirical Standpoint* (1874) provides the initial impulse for the development of a specific Austrian philosophy that was opposed to the philosophies in other German-speaking countries. The Vienna Circle is a part of this development (see Smith 1997).

This chapter gives an overview over Bolzano's and Brentano's influences on the development of logical empiricism. The next section outlines how Bolzano's philosophy of logic impacted on the work of early logical empiricists. A later section relates Brentano's empiricism and his methodology to the economy philosophy of Ernst Mach and Moritz Schlick.

Bolzano's philosophy of logic

Bolzano published on all core areas of practical and theoretical philosophy. He was trained as a mathematician and in his work mathematical and philosophical methodology inform each other. He defined concepts, assessed proposed definitions in the light of counterexamples, and organized his views around fundamental principles. This makes him a role model for philosophers opposed to the speculative philosophy associated with German idealism (see Künne 1997). It is therefore unsurprising that the Philosophical Society of the University of Vienna sought to promote Bolzano's philosophy by reprinting the first volume of Bolzano's *Theory of Science* in 1914 and a new edition of his *Paradoxes of the Infinite* in 1920 (see Uebel 1999: 260–1; Fisette 2014: 361). But it is not only Bolzano's methodology that proved influential; first-generation logical empiricists also mined his philosophy of logic for new ideas.

 DOI: 10.4324/9781315650647-9

The aim of Bolzano's monumental *Theory of Science*, published in four volumes in 1837, is to develop a manual for presenting sciences in book form. In pursuing this rather practical project, Bolzano tackled a number of important philosophical questions:

- What is a science? Not a practice of inquiry into a subject matter, but the results of such a practice: a body of truths.
- What are truths? Neither sentences nor judgments, but true propositions. Propositions are abstract complexes in which parts, objective ideas, can be distinguished.
- How are propositions, whether true or not, related?

In answering the last question, Bolzano made use of a concept that he discovered when thinking about the fact that the same sentence seems to be true at one time (place) and false at another. For example, when I utter "It is raining today" on Monday and it is raining, the sentence token produced is true; if I utter "It is raining today" on Tuesday and it is not raining, the sentence token produced is false. Hence, the same sentence type can have tokens with different truth-values. Now, propositions are either true or false; they are not supposed to be true or false at times, etc. The first sentence token, Bolzano says, expresses the absolutely true proposition (P1) [It is raining today_1]. The second sentence token expresses a different absolutely false proposition (P2) [It is raining today_2]. Now (P2) shares with (P1) all constituents apart from [today_2]. Bolzano described such examples as cases of *variation* of an objective idea in a proposition, or alternatively, of *generating a proposition* from another by means of variation. "Variation" is suggestive, but also misleading. Bolzano proposed, less misleadingly put, to move from considering a proposition "on its own" to considering a totality of propositions which only differ in specific constituents. If we do so, we can discover "notable" properties of the members of the totality:

> Given a mere proposition, we could merely inquire whether it is true or false. But some very remarkable properties of propositions can be discovered if, in addition, we consider the truth values of all those propositions which can be generated from them, if we take some of their constituent ideas as variable and replace them by any other ideas whatsoever.
>
> (1837: §147; Bolzano did not use the truth-value terminology: in this respect, the translation is anachronistic)

For our purposes, the following properties of proposition discovered by "variation" are important. (The following definitions are simplifications: the ideas must meet certain constraints to be "appropriate", i.e., allow the substitution to be truth-valuable.)

> Degree of validity of a proposition: if we consider in a proposition *P* the objective idea *I* as variable, the ratio of all true *I* variants of *P* to its false *I* variants is *P*'s degree of validity with respect to *I*.
>
> (ibid.: §147)

> Analyticity: a proposition *P* is analytic if it has a constituent *I* such that all propositions that differ from *P* only in containing in the position where *I* occurs in *P* a different idea have the same truth-value as *P*.
>
> (ibid.: §148)

A special subclass of the propositions Bolzano called "analytic" are what he called "logically analytic propositions":

> A proposition is logically analytic if, and only if, all propositions that differ from *P* only in their non-logical objective ideas have the same truth-value as *P*.

These can be logically analytic true or false; the former truths are what today are called "logical truths." Bolzano assumed that in order to recognize whether a logically analytic proposition is true or false, "nothing but logical knowledge is needed, since the concepts which form the invariable part of these propositions all belong to logic" (ibid.: §148.3).

The intuitive notion of one proposition being the consequence of another is also captured in the variation framework:

> Consequence: A proposition *Q* is a consequence of a proposition *P* with respect to an idea *I* if, and only if, all propositions that differ from *Q* only in containing an idea *J* in the position where *I* occurs in *Q* have the same truth-value as all propositions that differ from *P* only in containing an idea *J* in the position where *I* occurs in *P*.
>
> (ibid.: §164.2)

Bolzano's explication of the notion of consequence is similar in spirit to Alfred Tarski's famous explication of the narrower notion of logical consequence. Importantly, it seems to suffer from similar shortcomings (see Kneale 1961: 94–95; Etchemendy 1990: 28–30; Rusnock and Burke 2010), but here I consider only the use logical empiricists made of the variation framework.

Putting variation to work

The idea to consider totalities of propositions that differ in specific constituents proved fruitful and was taken up by logical empiricists for their purposes.

Walter Dubislav, the joint founder of the Berlin Gesellschaft für empirische Philosophie, championed Bolzano's *Theory of Science* and used Bolzano's theories in his philosophical work (Kasabova 2013). He praised Bolzano for his "truly revolutionary insights into the importance of variables in scientific arguments and, based on this, the discovery of what is today called a 'sentence function'" (1930: 409, trans. MT). This sounds as if for Bolzano a language contains variables as a distinct kind of symbol with meaning and reference. (For Carnap [1934: 190], variables are symbols for which substitution is permissible.) But when Bolzano writes "***A*** is green," he designates a plurality of propositions that differ from say "Grass is green" only in containing a different subject-idea. Putting "***A***" in boldface is a way to make clear that the idea expressed can be "varied" while the rest remains fixed. Propositions are composed of objective ideas only, and these do not come in two kinds: variable and invariable. But we can consider what will happen to the truth-value of a proposition if we vary one or more of its ideas and replace it by others.

Why does Dubislav draw on Bolzano? Any empiricist needs to have an explanation of how it is possible to recognize the truth of analytic judgments *without* recourse to experience that is in harmony with the general doctrine that *all* knowledge has its source in experience. Dubislav criticized Kant's attempt to draw the analytic/synthetic distinction and turned to Bolzano for a better alternative, relying on what he called "assertion schemata" to draw the distinction (1926: 19–20, 1929). But since this notion gives rise to a number of new questions, I will set it aside and consider the question whether Bolzano's distinction between analytic and synthetic propositions coincides with the distinction between a priori and a posteriori propositions.

According to Bolzano, the proposition expressed by an utterance of "The man Caius is mortal" is analytically true. For every variation of the "Caius" idea that will make it referential will also make it true (e.g., 1837: §197). But the truth of the proposition cannot be known independently of experience. Bolzano himself did therefore not intend his analytic/synthetic distinction to answer epistemological questions. Only the sub-distinction between logically and non-logically analytic truths was to have epistemological import. Quine unwittingly rediscovered the linguistic version of Bolzano's definition of logical analyticity when he defined a logical truth as a true sentence that contains only logical words *essentially*; that is, whose non-logical words can be *varied* without changing the truth-value (1936/1976: 80–81). In a later paper, Quine himself comments that "substantially this formulation is traced back a century and a quarter, by Bar-Hillel, to Bolzano" (1963: 386).

Like Dubislav and Quine, Schlick took Bolzano's variation method to hold the key to the clarification of important philosophical concepts. Schlick followed in broad outline Bolzano's definition of validity when investigating probability. While Bolzano considered in a proposition *P* one or several ideas to be variable, and then determined the ratio of the number of true variants of *P* to the number of false variants, Schlick considered the circumstances that make a statement true. We want to say that the probability of the statement (S2) "I throw a six" relative to the statement (S1) "I throw one normal dice" is 1/6. For there are six distinct states of affairs whose obtaining make (S1) true, but only one that makes both (S1) and (S2) true. It is difficult to assess whether Schlick's proposal is superior to Bolzano's because Schlick said little about truth-making and states of affairs. Schlick remarked that for each statement, there are infinitely many states of affairs that make it true (1936b/1979: 452). My statement "I throw a six" is made true by the obtaining of the fact that I throw a six slowly, that I throw a six slowly with my left hand, etc. So, either we find a suitable and independently plausible principle to reduce the number of truth-making states of affairs, or Schlick has not improved Bolzano's idea.

These three examples show how Bolzano's idea of variation informed the work of logical empiricists on logical and epistemological concepts.

Hahn on Bolzano's formalization of mathematics

Bolzano proved (very) early on in his career (1817) what is now known as the Bolzano-Weierstrass theorem, an important result in real analysis. He called his proof of the theorem "purely analytic": the proof proceeds only from conceptual truths and definitions; it does not draw on intuitions of mathematical objects that Kantian philosophers took to be an indispensable source of mathematical knowledge. Hans Hahn, the mathematician and philosopher of mathematics of the Vienna Circle, therefore listed Bolzano as one of the pioneers of the program to completely formalize mathematics: "every new mathematical concept was to be introduced through purely logical definition, every mathematical proof was to be carried through by strictly logical means" (Hahn 1933/1980: 104; see Coffa 1991: ch. 2).

Hahn also wrote a commentary to the 1920 edition of Bolzano's *Paradoxes of the Infinite*. The title *Paradoxes of the Infinite* invites the thought that infinity gives rise to paradoxes. Bolzano's book argues for the exact opposite conclusion: The alleged paradoxes are not real paradoxes. Bolzano is generally credited with discovering and discussing a now well-known prima facie paradox about infinite sets (see his 1851: §20–22). An infinite set *M* may be a subset of an infinite set *N*, yet there is a one-to-one mapping between the elements of *M* and *N*. This seems paradoxical. Intuitively, how can one set include another, and yet a one-to-one correspondence relation obtain between them? The number of elements of a set M is the same as the number of elements of a set N if, and only if, there is a one-to-one correspondence. Bolzano shows that this does not hold for infinite sets and tries to show that this is not counterintuitive. In his commentary, Hahn updated and tightened some of Bolzano's definitions. He suggested that Bolzano's "*Inbegriffe*" are sets with at least two members. He also connected Bolzano's treatment of the cardinality of infinite sets with Cantor's later development.

Brentano's empiricism

As noted, Brentano's *Psychology from an Empirical Standpoint* (1874) is considered to be one of the main sources of Austrian philosophy. In its year of publication Brentano was appointed to a chair at the University of Vienna, which he had to give up in 1880. He continued to teach in Vienna until 1895 as *Privatdocent*. During this time he propagated an empiricist and scientific approach to philosophy.

Already in his *Habilitationsthesen* (1866/1968: 137 and 139) Brentano exhorted philosophers to protest against the distinction between exact and speculative sciences (Thesis 1), claimed that the true method of philosophy is the same as the method of the natural sciences (Thesis 4) and that the understanding contains nothing—apart from itself—that had not previously been in the senses (Thesis 13). In subsequent work. Brentano aimed to show that Thesis 13 is true: every concept we possess is either acquired by abstraction from perception or composed of concepts that are acquired in this way (I return to this later). Brentano and his students set out to show against both Kant and Hume that concepts like causation, good and true are abstracted from perceptions. When implementing this program, Brentanians rely on the methods of psychology (Thesis 4).

Brentano's main work *Psychology from an Empirical Standpoint*—the clue is in the name—applies the outlined methodology to the concept of the mental in general and to kinds of mental phenomena in particular. When one takes the empirical standpoint in psychology, one relies on consciousness or inner perception as an empirical source of knowledge of the mental events and processes. Inner perception combined with attention to what is given in episodic memory allows us to arrive at the following three empirical claims about mental phenomena:

(B1) There are mental events and processes, but no bearer of mental phenomena (see 1874/2015: 11).

(B2) Mental phenomena are directed upon an object, physical phenomena are not (see ibid.: 92–3).

(B3) Judgment is a *sui generis* mental act and distinct from predication (see ibid.: 210–11).

We are aware of mental events and processes, but not of something—a soul—that undergoes them. Brentano concluded from this, following Hume, that the soul is a fiction (ibid.: 11). We will see in the next section that Mach agrees with Brentano's classification of the soul as a fiction. Schlick will later say that the "selfless" view of experience strikes him as "one of the most important steps which philosophy must take towards the clarification of its deepest problems" (1936a/1979: 473). However, after *Psychology*, Brentano would change his mind about (B1). There is a mental substance that is perceived together with its mental activities.

(B2) is the basis for Brentano's conception of the mental. Contemporary philosophers of mind aim to capture it by the suggestive slogan "Intentionality is the mark of the mental." According to Brentano, we are aware of the distinction between the mental and the physical. If we systematize our knowledge, we can easily recognize that no mental phenomenon is a physical one, and vice versa (cf. Textor 2017: Part I).

(B3) motivated Brentano's reform of (syllogistic) logic in which propositions are replaced with terms (see Hillebrand 1891). Brentano rejected the idea that every judgment is a predication of a property to something. A judgment can be the acknowledgement of an object and its content is exhausted by one term (*eingliedrig*).

While Brentano is an empiricist about the mind, his empirical claims (B2) to (B3) as well as his methodology were criticized and finally rejected by philosophers who influenced the development of logical empiricism. Next, I will outline how and why.

Mach's (Neutral) Monism versus Brentano's Dualism

Brentano's contemporary Mach is the other main influence on the development of logical empiricism. His influence is due to the fact that he led by example: he did not philosophize about science, but practiced it (Musil 1908: 10–11).

Mach takes science to grow out of an instinct for economy. Because we have limited resources, we try to expend as little effort in thinking as possible. Science allows us to predict experiences and thereby saves us effort. In order to maximize economy, we need to unify special sciences such as biology and geology into one general science. The unification of special sciences in general and the unification of psychology and physics in particular require a new ontology that overcomes the deeply entrenched dualisms between self and world, permanent substance and changing properties, and the mental and the physical:

> Anyone who has in mind the gathering up of the sciences into a single whole, has to look for a conception to which he can hold in every department of science. Now if we resolve the whole material world into elements which are at the same time also elements of the psychical world and, as such, are also commonly called sensation, if, further, we regard it as the sole task of science to inquire into the connexion and combination of these elements, which are of the same nature in all departments, and into their mutual dependence on each other; we may then reasonably expect to build a unified monistic structure upon this conception, and thus to get rid of the distressing confusions of dualism.
>
> (Mach 1886/1914: 312)

The elements are neutral, they are neither mental nor physical, hence the title "neutral monism" for Mach's view. On a charitable reading, Mach takes the difference between the mental and the physical to be merely conceptual: we apply different concepts to the same elements when we investigate on which factors they depend (ibid: 17–18) Permanent substances and selves are useful fictions: we make believe that a relatively stable complex of elements is permanent thing (ibid.; 10–11, 1905/1976: 15).

Mach's neutral monism is in direct conflict with Brentano's (B2). It is therefore surprising that Mach never engaged with Brentano's philosophy (see Blackmore 1972: 61). Only Mach's associate Wilhelm Jerusalem engages with Brentano and, like Mach, wants "to abandon the attempt to demarcate the mental from the physical" (1895: 6, trans. MT). By contrast, Brentano read Mach's work, corresponded with him and responded to his claims. In 1893–4, just prior to Mach's appointment to Vienna, Brentano gave a seminar called "Positivism" in which he devoted a substantial part of a seminar on positivism to Mach (Brentano n.d.). Later, he wrote a chapter-by-chapter commentary to Mach's *Knowledge and Error* (Brentano 1988).

In all these writings Brentano comes back to the distinction between the mental and the physical (e.g., n.d.: 67ff.). As we have seen, Brentano argues that it is an empirical finding that mental phenomena are distinct from physical phenomena. The distinction is not merely conceptual: All mental, but no physical phenomena, are directed on an object. Describing a physical phenomenon as so directed is a mistake, not a useful fiction. Hence, Mach's *modus ponens* is Brentano's *modus tollens*: Because mental and physical phenomena are different kinds of things, we need both physics and psychology. This is fully compatible with methodological monism: both sciences operate according to the same methodology.

Schlick and Brentano's theory of judgment

Schlick is a central figure in the development of the Vienna Circle. His book *General Theory of Knowledge* (1st edition 1918, 2nd edition 1925) worked out the central arguments of his theoretical philosophy. He is an adherent of Mach's economy philosophy, but an opponent of Mach's brand of phenomenalism. However, the economy under consideration, Schlick proposes, is not one of minimizing cognitive effort, but of explaining the maximum number of phenomena with a minimum of concepts (he shares the criticisms of Brentano [1988: 75] and Planck [1910]). In the spirit of the economy view, Schlick reduces knowledge acquisition, whether scientific or common sense knowledge, to recognition: "To cognize is to re-cognize (*Wiederkennen*) or rediscover (*Wiederfinden*). And to rediscover is to equate what is known with that as which it is known" (Schlick 1918/1974: 15). So every judgment that can constitute knowledge must relate one thing or term to one thing which is already known.

If the basic idea of Schlick's reformed economy theory is right, there cannot be judgments that constitute knowledge that have only one term. Schlick's *modus ponens* is Brentano's *modus tollens*. There are judgments that have only one term: when one is aware of hearing a note, one acknowledges a mental process but does not predicate anything of it. According to Brentano's (B3), experiences teaches us that there are judgments that acknowledge an object without predicating something of it (1874/2015: 217). A fortiori, these judgments are not acts of recognition.

Schlick agreed with Brentano that, *psychologically speaking*, judging cannot be analytically defined. But

> it is obvious that acknowledgement and rejection can at most characterize judging as a psychological act. They do not reach its epistemological and logical significance and this is what is at issue in our inquiry into the nature of judgement. The logical side of Brentano's theory—the claim that basically every judgement is one-termed—is an error that can lead to serious mistakes, in particular to the attempt to detach "things" from the relations existing among them (of which we will speak at another place).
>
> (1918/1974: 41)

Schlick proposed to explain prima facie counterexamples to the two-term view such as existential and impersonal sentences ("It is raining") away. Whether he succeeded is a matter of controversy. What is not controversial is that the viability of Schlick's reformed economy philosophy depends on the success of the proposed analysis.

Take the proposition that 2 = 2. If you master all concepts it is composed of, judging that 2 = 2 will seem right. Your compulsion to judge is not a mere psychological urge but is based on the fact that the judgment "declares" or "announces" its correctness: "There are judgements *that declare their correctness*; that have the character of insight. One calls them self-evident" (1956: 141, trans. and emphasis MT). Consciousness, for instance, is a source of evident judgments about present mental acts. Brentano argued against Mach that evident judgments, and not the fitting of new observations into what is already known, is important for knowledge (1988: 70).

The question of whether evidence can be a basic concept of epistemology is one of the main points of dispute between Brentanians and proponents of the economy philosophy. Schlick is perceived by Brentanians as a particularly formidable opponent:

> In recent times, because of the weight that Brentano rightly put on the notion of evidence, there have been many attacks, the most vigorous by Schlick, against the introduction of this notion. But it is completely indispensable and of fundamental importance for epistemology.
>
> (Stumpf 1939: 56–57)

Schlick argued against Brentano's claim that inner perception consists in evident judgments (1918/1974: §§12 and 19). We have experiences and are aware of them, but this awareness neither consists in evident judgments nor can it be the source of our basic concepts. What is given in experience is imprecise, and we cannot abstract the *absolutely precise* concepts we need for science from this basis (ibid.: 29). If Schlick's arguments are convincing, Brentano's empiricism in which evidence and inner perception are key elements is not a viable philosophical program for *scientifically minded* philosophers. In his early work Schlick went on to replace Brentano's foundational picture of concept acquisition and knowledge with a holistic one in which the nature of a concept is determined by its relations to others (see Friedmann 1983). In sum, the criticism of Brentano's empiricism was a starting point for Schlick's new philosophical outlook.

Conclusion

Bolzano was a source of inspiration to logical empiricists: he offered them a new way of thinking about logical notions. While Brentano shared the empiricist leanings of the key figures of logical empiricism, the spirit of his philosophy is incompatible with the principle of economy that guided the work of these philosophers.

References

Blackmore, J. T. (1972) *Ernst Mach: His Work, Life and Influence*, Berkeley: University of California Press.

Bolzano, B. (1817) *Rein analytischer Beweis des Lehrsatzes, dass zwischen je zwey Werthen, die ein entgegengesetztes Resultat gewähren, wenigstens eine reelle Wurzel der Gleichung liege*, Prag: Gottlieb Haase.

——— (1837) *Wissenschaftslehre*, Sulzbach: Seidel. Trans. *Theory of Science*, Oxford: Oxford University Press, 2014, vol. 4.

——— (1851) *Paradoxien des Unendlichen* (ed. by Fr. Přihonsky), Leipzig: Reclam. Trans. "Paradoxes of the Infinite," in Russ (2004), pp. 591–679.

Brentano, F. (1866) "Die 25 Habilitationsthesen," in F. Brentano, *Über die Zukunft der Philosophie* (ed. by O. Kraus), Hamburg: Meiner, 2nd ed., 1968, pp. 133–43.

——— (1874) *Psychologie vom Empirischen Standpunkt*, Leipzig: Duncker & Humblot, 2nd ed. (ed. by O. Kraus) Leipzig: Meiner, 1924. Trans. *Psychology from an Empirical Standpoint*, 3rd ed., London: Routledge, 2015.

——— (1956) *Die Lehre vom Richtigen Urteil* (ed. by F. Mayer-Hillebrand), Bern: Franke.

——— (1988) *Über Ernst Machs "Erkenntnis und Irrtum"* (ed. by R. M. Chisholm and J. C. Marek), Amsterdam: Rodopi.

——— (n.d.) "Positivismus/Zeitbewegende philosophische Fragen," unpublished manuscript (1893/4), (signum LS 20), Harvard Houghton Library.

Carnap, R. (1934) *Logische Syntax der Sprache*, Vienna: Springer. Rev. ed. trans. *Logical Syntax of Language*, London: Kegan, Paul, Trench Teubner & Cie, 1937, repr. Chicago: Open Court, 2002.

Coffa, A. (1991) *The Semantic Tradition from Kant to Carnap To the Vienna Station*, Cambridge: Cambridge University Press.

Dubislav, W. (1926) *Über die sogenannten Analytischen und Synthetischen Urteile*, Berlin-Schöneberg: Herrmann Weiss.

——— (1929) "Über Bolzano als Kritiker Kants," *Philosophisches Jahrbuch* 42: 357–68.

——— (1930) "[Review of Bolzano, Bernard: *Wissenschaftslehre*. Neudruck in vier Bänden]," *Erkenntnis* 1: 408–9.

Etchemendy, J. (1990) *The Concept of Logical Consequence*. Cambridge, MA: Harvard University Press.

Fisette, D. (2014) "Austrian Philosophy and Its Institutions. Remarks on the Philosophical Society of the University of Vienna (1888–1938)," in A. Reboul (ed.), *Mind, Meaning and Metaphysics*, Dordrecht: Springer, pp. 349–74.

Friedman, M. 1983. "[Critical Notice M. Schlick *Collected Papers*]," *Philosophy of Science* 50: 498–513.

Hahn, H. (1933) "Die Krise der Anschauung," in H. Mark et al. (eds.), *Krise und Neuaufbau in den exakten Wissenschaften. Fünf Wiener Vorträge*, Vienna: Deuticke. Trans. "The Crisis in Intuition," in Hahn *Empiricism, Logic and Mathematics. Philosophical Papers* (ed. by B. McGuiness), Dordrecht: Reidel, 1980, pp. 73–103.

Haller, R. (1968) "Wittgenstein und die österreichische Philosophie," *Wissenschaft und Weltbild* 21: 78–87. Trans. "Wittgenstein and Austrian Philosophy," in Haller, *Questions on Wittgenstein*, London: Routledge, 1988, pp. 1–26.

Hillebrand, F. (1891) *Die neuen Theorien der kategorischen Schlüsse*, Vienna: Hölder.

Jerusalem, W. (1895) *Die Urteilsfunktion*, Vienna: W. Braunmüller.

Kasabova, A. (2013) "Dubislav and Bolzano," in N. Milkov and V. Peckhaus (eds.), *The Berlin Group and Logical Empiricism*, Dordrecht: Springer, pp. 205–28.

Kneale, W. (1961) "Universality and Necessity," *The British Journal for the Philosophy of Science* 12: 89–102.

Künne, W. (1997) "'Die Ernte wird erscheinen. . .' Die Geschichte der Bolzano-Rezeption (1839–1939)," expanded and corrected version in Künne, *Versuche über Bolzano*, St. Agustin: Akademia, 2008, pp. 305–405.

Mach, E. (1886) *Beiträge zur Analyse der Empfindungen und das Verhältnis des Physischen zum Psychischen*, Jena: Fischer. Trans. of enlarged ed. *The Analysis of Sensations and the Relation of the Physical to the Psychical*, Chicago: Open Court, 1914.

——— (1905) *Erkenntnis und Irrtum*, Leipzig: Barth. Trans. of 2nd ed., *Knowledge and Error*, Dordrecht: Reidel, 1976.

Musil, R. (1908) *Beitrag zur Beurteilung der Lehren Machs*, Berlin: Karl Arnold.

Neurath, O. (1936) *Le développement du Cercle de Vienne et l'avenir de l'empirisme logique*, Paris: Hermann & Cie. Trans. "Die Entwicklung des Wiener Kreises und die Zukunft des Logischen Empirismus," in Neurath, *Gesammelte philosophische und methodologische Schriften* (ed. by R. Haller and H. Rutte), Vienna: Hölder-Pichler-Tempsky, 1981, pp. 673–703.

Planck, M. (1910) "Zur Machschen Theorie der physikalischen Erkenntnis," *Vierteljahresschrift für wissenschaftliche Philosophie* 9: 497–507.

Quine, W. V. O. (1936) "Truth by Convention," in O. H. Lee (ed.), *Philosophical Essays for A. N. Whitehead*, New York: Longmans. Repr. in Quine, *The Ways of Paradox and Other Essays*, Cambridge, MA: Harvard University Press, 1976, pp. 70–106.

——— (1963) "Carnap on Logical Truth," in P. A. Schilpp (ed.), *The Philosophy of Rudolf Carnap*, La Salle, IL: Open Court, pp. 385–406.

Rusnock, P. and Burke, M. (2010) "Etchemendy and Bolzano on Logical Consequence," *History and Philosophy of Logic* 31: 3–29.

Russ, S. (ed.) (2004) *The Mathematical Works of Bernard Bolzano*, Oxford: Oxford University Press.

Smith, B. (1997) "The Neurath-Haller Thesis: Austria and the Rise of Scientific Philosophy," in K. Lehrer and J. C. Marek (eds.), *Austrian Philosophy Past and Present. Essays in Honor of Rudolf Haller*, Dordrecht: Kluwer, pp. 1–20.

Schlick, M. (1918) *Allgemeine Erkenntnislehre*, Berlin: Springer, 2nd rev. ed. 1925. Trans. *General Theory of Knowledge*, Lasalle: Open Court, 1974.

——— (1936a) "Meaning and Verification," *The Philosophical Review* 45: 339–69. Repr. in Schlick (1979), pp. 456–81.

——— (1936b) "Gesetz und Wahrscheinlichkeit," in Schlick, *Gesammelte Aufsätze 1929–36*, Vienna: Gerold, pp. 323–37. Trans. "Law and Probability," in Schlick (1979), pp. 446–55.

——— (1979) *Philosophical Papers* (ed. by B. van de Velde-Schlick and H. Mulder), Dordrecht: Reidel, vol. 2.

Stumpf, C. 1939. *Erkenntnislehre*, Leipzig: Barth. Repr. in Lengerich: Pabst Science Publishers, 2011.

Textor, M. (2017) *Brentano's Mind*, Oxford: Oxford University Press.

Uebel, T. (1999) "Otto Neurath, the Vienna Circle, and the Austrian Tradition," in A. O'Hear (ed.), *German Philosophy Since Kant*, Cambridge: Cambridge University Press, 249–69.

Verein Ernst Mach (1929) *Wissenschaftliche Weltauffassung. Der Wiener Kreis*, Vienna: Wolf. Trans. "The Scientific Conception of the World. The Vienna Circle," in O. Neurath, *Empiricism and Sociology* (ed. by R. S. Cohen and M. Neurath), Dordrecht: Reidel, 1973, pp. 299–318; rev. trans. (with orig. annotated bibliography) "The Scientific World-Conception. The Vienna Circle," in F. Stadler and T. Uebel (eds.), *Wissenschaftliche Weltauffassung. Der Wiener Kreis. Hrsg. vom Verein Ernst Mach (1929)*, Vienna: Springer, 2012, pp. 75–116.

8

FRENCH CONVENTIONALISM AND THE VIENNA CIRCLE

Anastasios Brenner

Conventionalism is a source of influence explicitly mentioned in the Vienna Circle's manifesto (Verein Ernst Mach 1929/2012: 80). Yet its meaning is often misunderstood or its scope misconstrued. One may distinguish two lines of interpretation: an analytic and a historical one. The first line more or less ignores the historical development and tends to focus on late forms of conventionalism: a Wittgensteinian "iconoclastic conventionalism" or a Quinian "global conventionalism" (Ben-Menahem 2006; Horwich 1998). Such categories leave us with several perplexing questions. Whence comes conventionalism? Although Poincaré is generally mentioned, little attention is given to the origins of this doctrine. How did it evolve during a period marked by profound scientific revolutions, from the crisis of classical physics to the general theory of relativity? Or again, what should be made of the developments in philosophy after Wittgenstein and Quine? What is at stake in the doctrine we are examining is a whole array of problems—the given and the constructed, the artificial and the natural, the contingent and the necessary—and numerous debates touching on one or another of these dichotomies have arisen in the meantime. It is not without reason that Poincaré and other conventionalist thinkers have been subject to greater scrutiny during the past twenty years.

The historical line of interpretation takes us back to the discussions at the turn of the twentieth century between Poincaré, Duhem, Le Roy, and others (Frey 1976; Brenner 2015) and scrutinizes the various channels of transmission of these discussions to the Vienna Circle. Such recourse to history reveals the shortcomings of an approach strictly based on rational reconstruction.

The context of emergence

It was Poincaré who initiated the debate on conventions. As early as 1891 in an article on "Les géométries non euclidiennes," he writes: "*The axioms of geometry are not synthetic a priori judgments nor experimental facts*. They are *conventions*; our choice, among all possible conventions, is *guided* by experimental facts; but remains *free*" (Poincaré 1891: 773, orig. emphasis, trans. AB). Poincaré is dismissing the conceptions of geometry formulated by Kant and Mill. The former held that the propositions of geometry—as opposed to those of logic—were synthetic a priori; in other words, they extend our knowledge while being valid independently of experience. Mill, in contrast, claimed that geometry is founded on facts, albeit of an abstract

 DOI: 10.4324/9781315650647-10

nature. Poincaré rejected these two traditional conceptions by calling on the discovery of non-Euclidean geometries and the numerous consequences mathematicians had drawn from these since the 1830s.

One can indeed construct different geometrical systems. Being free from contradiction, all these systems are equally legitimate; they are also fruitful from a mathematical point of view. To resort to Euclidean geometry to describe the physical world, as was the rule at the time, was to endorse unconsciously a series of assumptions, which rested not on facts but reasons of convenience. Poincaré would later extend and complement this thesis on the hypotheses of geometry to provide a broad philosophical conception in his noteworthy *Science and Hypothesis* (1902).

Eugène Dupréel, who embraced conventionalism in the 1920s, highlights the novelty of Poincaré's proposal by pointing to the fact that no philosophical dictionary in use in the major European languages around the turn of the century even registers the term "convention"—with the sole exception of a brief reference to Hume in Baldwin's *Dictionary of Philosophy and Psychology* (Dupréel 1925: 283). Following the reception of Poincaré's ideas, lexicographers would be forced not only to take into account the recent debates on science, but also to recall what is an important philosophical concept in its own right. Hume was referring to an argument that went back to antiquity. The conventional nature of language and law had often been defended. On the basis of this thesis, the skeptic Sextus Empiricus went so far as to question dogmatic philosophy as well as science in favor of a general suspension of judgment. His views became popular again during the sixteenth century, and the founders of modern science sought to meet the challenge by conceiving various metaphysical foundations for scientific knowledge. Such foundations came once again into question at the turn of the twentieth century. One can gather from the preceding remarks that the context in which Poincaré first put forth his ideas was one of a pervasive naïve empiricism.

But Poincaré was not alone. Duhem reached independently an analogous conclusion with regard to physics in an article, "Quelques réflexions au sujet des théories physiques," published in 1892 just a few weeks after Poincaré's. He states: "Physical definitions make up a genuine vocabulary; just like a French dictionary is a set of *conventions* making a name correspond to each object, so in physical theory the definitions are a set of *conventions* making a magnitude correspond to each physical notion" (Duhem 1892: 143–4, emphasis added, trans. AB). Duhem gives the example of the notion of temperature. This notion is conceived so as to represent heat, but one should understand that between the two there is no natural relation. And Duhem proceeds to draw the implications: there are an infinite number of possible definitions. The physicist thus has freedom of choice.

But what is more significant is that two years later, in a companion article "Quelques réflexions au sujet de la physique expérimentale," Duhem went on to apply the same critical analysis to experimental physics. Here, one finds the first formulation of the claim according to which one can never condemn an isolated hypothesis but only a whole body of theory, known today as the Duhem–Quine thesis (Duhem 1894: 187). The article explores the numerous consequences of this thesis, offering a novel analysis of experiment. On this basis Duhem claims that observation implies recourse to interpretation; it is accomplished in the light of theories. Some twelve years later he would take up this analysis of experiment in conjunction with the ideas put forth in his earlier article on physical theory, in order to reach the broad conception given in *The Aim and Structure of Physical Theory* (1906).

Let us dwell on the reasons underlying Poincaré's and Duhem's foray into philosophy. Although the former made every effort to bring his study to a general readership and to introduce the necessary philosophical references, the connection with his own scientific work is conspicuous. Indeed, one of Poincaré's major scientific discoveries concerns automorphic functions and exploits the analogy with Lobatchevski's geometry. Such applications go to prove that non-Euclidean geometries are in no way a vain exercise, as some had contended.

As for Duhem, he was involved in formulating a general thermodynamics. He was thereby led to reflect on the concepts and hypotheses of this branch of physics and how they could contribute to the task of unifying physics. The reflections of Poincaré and Duhem derive from their practice as scientists. This gives substance to their views. It is perhaps the consequence of an exceptional situation: a crisis leading up to a revolution, a period of extraordinary science, to employ Thomas Kuhn's vocabulary. What had occurred was a breakdown of the rules and values, giving rise to discussions of a philosophical nature. At the same time, the link with the scientific knowledge of the time may explain some of the difficulties of their specific views. Poincaré would uphold Euclidean geometry as the language of physics; Duhem would remain satisfied with macroscopic thermodynamics.

The advent of a movement of thought

The simultaneous emphasis on conventions by an eminent mathematician and a distinguished physicist was to impress their contemporaries. Gaston Milhaud was one of the first to bring together Poincaré's and Duhem's views. In "La science rationnelle" in 1896, he argues that all mathematical systems of hypotheses involve conventions, and therefore Poincaré's thesis applies as much to geometry as to physics. Duhem's analysis of experimental testing reveals the role of hypothetical reasoning. Milhaud emphasizes the theoretical intermediaries that intervene between the fact and the law. In speaking of Kepler's third law, he writes: "It is not just the theories confessed as it were, the hypotheses clearly stated, the elaborate constructions which separate the observer from the thing observed, it is also sometimes the *conventions* or almost *unconscious definitions*, of which one would not stop to think" (Milhaud 1896: 287, emphasis added, trans. AB). Milhaud was intent on deploying fully these ideas.

What Milhaud saw in the early articles of Poincaré and Duhem was material suitable for establishing a general course of study in the philosophy of science. He went on to introduce expressions that do not occur in Poincaré or Duhem, speaking of the contingency, the creative power and the intellectual activity inherent in scientific practice. Milhaud was thereby signaling a connection with the philosophy of Émile Boutroux, who had drawn attention to such factors, in questioning various forms of determinism (Boutroux 1874). Now, this connection is all the more interesting, as Poincaré was always reserved in expressing his general philosophical views. But Boutroux was his brother-in-law, and their writings reveal a mutual influence.

Milhaud was soon joined by Édouard Le Roy. In "La science positive et les philosophies de la liberté", he likewise emphasized the relation between the assertion of the conventional nature of scientific hypotheses and the claim of the global character of testing. Le Roy did not hesitate to generalize the considerations of his predecessors. In speaking of the principle of the conservation of energy, he writes:

> Our principle presents itself . . . as the very definition of a closed system. Wherever it is found to fail, we shall say that there was an exchange between the system considered and the outside. Here again the law seems so well established only because it has become a *conventional decree* helping to fix our language.
>
> (Le Roy 1900: 318, emphasis added, trans. AB)

Le Roy endorsed a Bergsonian outlook and believed that his analysis of science favored a clear break with Kant. He perceived here the signs of a movement of thought, which he called a "new positivism" (Le Roy 1901).

In 1907 when Abel Rey published his dissertation *La Théorie de la Physique chez les physiciens contemporains*, he could attempt some sort of assessment of the discussions initiated by Poincaré, which had been pursued for well over a decade. According to Rey, these discussions renewed our understanding of scientific objectivity and could be seen as a response to the far-reaching achievements of science over the past half century, including the advent of the social sciences. But Rey adopted, more wholeheartedly than his predecessors, the atomic theory of matter as championed by Jean Perrin. He was careful to point out that this did not mean a return to classical mechanism. A considerable amount of material on the conventional aspects of science had been accumulated by then, and the term of conventionalism was soon to gain currency.

The difficulty is that one is confronted with numerous and divergent forms of this doctrine. Some commentators have gone so far as to discount Poincaré or Duhem as genuine or wholehearted conventionalists, in a conspicuous move to claim them in favor of present-day realism, while guarding against Le Roy's so-called radical conventionalism. Such a construal is deeply misleading. The situation at the turn of the twentieth century is in a sense diametrically opposed to that of today: the French thinkers under consideration were fighting the traditional view of science, imbued by naïve empiricism, mechanism, causalism, etc. We are now confronted instead with the dangers of relativism, incommensurability and social construction. Moreover, there is no evidence that Poincaré and Duhem ever abandoned the basic tenets set forth in their early writings; they continued to point out the conventions involved in science, or better, the claims implied thereby: the free choice of axioms, the multiplicity of logically possible systems of hypotheses, the decisions inherent in experimental testing, etc. Poincaré, extending the claim he had made with regard to geometry, asserted that some of the fundamental principles of physics are likewise conventions. Duhem generalized his early remarks with regard to physical magnitudes: the physicist is free to choose his definitions; several such definitions are possible, and the mathematical symbols have no natural relation to the properties signified. It appears rather that the discussions prompted the protagonists to clarify their arguments and incited them to seek viable solutions. To be sure, early conventionalism did not give rise to a homogeneous doctrine; it represented a variety of positions, which interacted closely with one another. In fact, all the thinkers mentioned—including the maverick Le Roy—defended in some way or another the value of science. Their views were more opposed to naïve empiricism than to realism and are perhaps better construed as the conception according to which our knowledge of the external world is an intricate mixture of facts and artifacts.

The reception of conventionalism

The debates we have been studying were quick to have an impact in German-speaking countries. In 1907 Philipp Frank called attention to Poincaré's basic insight in one of his first articles, "Experience and the Law of Causality," claiming that this fundamental law should be considered as a purely conventional definition. Frank's endeavor to apply Poincaré's ideas to the principle of causality elicited a response from Gerhard Hessenberg, a German mathematician based at the time in Bonn and later at the University of Tübingen. In an article under the title "Willkürliche Schöpfungen des Verstandes?" he seized the occasion of an examination of the paradoxes of set theory, including Poincaré's discussion of Richard's paradox, to take issue with Frank. Number being one of several organizing principles of reason, he proceeded to consider space, time, and causality. Hessenberg then delivered his claim: "Indeed academics of the rank of Poincaré,

Dedekind, Weber have wisely been wary of taking to extremes their 'conventionalism'" (1908: 161, trans. AB). What he objected to was the implicit criticism of Kant's conceptions that he perceived in this recent publication among others. Hessenberg's analysis is neither open-minded nor perceptive, but his article signals the appearance of the expression of conventionalism much earlier than hitherto acknowledged by historians. "Conventionalism" is then an actor's category, to use an anthropological term, and, in consequence, carries a historical significance.

Far from being shaken by such a label, Frank rose to the challenge of defending it (Frank 1908: 230). A few years later in a review of Poincaré's *Last Thoughts*, he asserts clearly his position:

> Much has been voiced against "conventionalism" in the past few years. But one needs only to call to mind Poincaré's expression, which is so clear, free from any exaggeration and ambiguity, or verbosity, to dispel any doubts, and we willingly again acknowledge "conventionalism" as an effort of clarification directed against the pseudo-problems of a merely grammatically oriented pseudo-philosophy.
>
> (Frank 1914: 54, trans. AB)

In a retrospective account, Frank would acknowledge that he was involved at the time in an informal discussion group, which included Otto Neurath and Hans Hahn (Frank 1949: 1). This discussion group has come to be called the first Vienna Circle, and historians have emphasized its importance (Haller 1991; Uebel 2000). It brings to light the slow maturation and multiple influences that contributed to the development of logical empiricism or logical positivism, as it was likewise called by its advocates. According to Frank, Poincaré and Duhem played a decisive role in initiating a new attitude; they marked a clear break with traditional theories of knowledge. But there were difficulties in Poincaré and Duhem: in particular, their ambivalence with regard to atomism, relativity, and, ultimately, quantum theory. Rey, who supported the new atomism from the outset, helped to adapt their philosophical views to the recent discoveries, and conventionalism received a new lease of life. Rey provided a synthesis of the aphoristic remarks of Poincaré and showed that Duhem's account of the structure of physical theories was a step toward an axiomatic point of view. As Frank further adds in his historical introduction: "We agreed with Rey's characterization of Poincaré's contribution as a 'new positivism'" (Frank 1949: 9). In summary, what Frank saw here was a movement of thought that was not unrelated to Mach's endeavor, and that represented a deeper understanding of scientific activity.

Neurath, while corroborating Frank's account, began to sketch his own philosophical project, in particular with regard to the holist argument: "Duhem has shown with special emphasis that every statement about any happening is saturated with hypotheses of all sorts and that these in the end are derived from our whole world-view. We are like sailors who on the open sea must reconstruct their ship but are never able to start afresh from the bottom" (Neurath 1921/1973: 199). Neurath gives to Duhem's thesis, which was based on a careful analysis of experiment in physics, a broad scope and places it within a sociological perspective. Like Frank, he rebuts those who see here merely "a new fashion" (Neurath 1916/1983: 28); a break with traditional views is necessary. And, he concludes, foreshadowing the encyclopedic project he would develop later: "The theory of systems of hypotheses has been greatly advanced by men like Mach, Duhem, Poincaré. The right moment may now have come to group the systems of hypotheses of all sciences systematically" (ibid.: 31). Here, he is bringing together several related claims made by these thinkers and putting them to work in favor of his systematization of knowledge, which involves not only physics but the whole of science.

One discovers similar interests in Frank and Neurath's German counterparts. Schlick referred significantly to Poincaré in connection with general relativity. At first blush this reference may come as a surprise, for the French mathematician had stated that if several geometries were possible, one would in the final analysis choose Euclidian geometry to describe the world. Schlick nevertheless found in Poincaré a philosophical framework conducive to the interpretation of Einstein's new theory. What he took up was the claim that Poincaré made in his later writings of the relativity of geometry.

If Schlick does not appear to have professed to be a conventionalist, he voiced his sympathy with the theory of conventions promoted by Poincaré. This would not prevent Reichenbach for reproaching him to incline too much in this direction: "[Schlick's] conception of the theory of relativity, which is related to Poincaré's conventionalism, is shared by Einstein" (Reichenbach 1922/1978: 34). The inclusion of Einstein in this category can be explained by a certain ambivalence in his early texts.

Schlick further develops the consequences of this relativity of geometry in his *General Theory of Knowledge*. But more important for him in this setting is Poincaré's analysis of definition:

> By a suitable choice it is always possible under certain circumstances to obtain an unambiguous designation of the real by means of the concept. Conceptual definitions and coordinations that come into being in this fashion we call *conventions* (using this term in the narrower sense, since in the broad sense, of course, all definitions are agreements). It was Henri Poincaré who introduced the term 'convention' in this narrower sense into natural philosophy.
>
> (Schlick 1918/1974: 71)

Likewise on the German philosophical scene, Carnap went through a conventionalist period. Thus, he writes in one of his first articles after his dissertation:

> It is the main thesis of the conventionalism expounded by Poincaré and further developed by Dingler that in the construction of physics we have to make *stipulations* [*Festsetzungen*], that are subject to our free choice. This means that the particular components of physical statements that result from these stipulations can neither be confirmed nor refuted by experience. But the choice among these stipulations ought not therefore to be made arbitrarily, rather it should follow certain methodological principles—and in the end the principle of maximal simplicity has to decide.
>
> (Carnap 1923/2019: 211, orig. emphasis)

I leave aside here the case of Dingler. Although the latter was a go-between, his own position is both confused and ambiguous, as Gereon Wolters convincingly shows (Wolters 1985). Carnap, in a manner similar to Schlick, turns around Poincaré's argument with respect to the criterion of simplicity and considers physics and geometry taken together. He is probably drawing on Duhem's holism, on which he explicitly calls with regard to the logic of measurement, that is, coordinating definitions (Carnap 1926). Conventionalism thus occupied him in several publications up to *The Logical Structure of the World* (Carnap 1928).

It is striking that several major figures of the Vienna Circle, both Austrian and German, were so attracted to conventionalism early in their careers. Conventionalism, as interpreted by the Vienna Circle, provided a stepping-stone in the elaboration of a common program, supporting anti-foundationalism, positivism, and decisionism. Thereafter, conventionalism was submerged in their new project. Carnap moved toward logical construction and logical syntax. Schlick adopted a Wittgensteinian verificationism. Neurath embarked on his *Encyclopedia of Unified Science*. These evolutions did not, however, prevent the debate from springing up again in different forms. Thus, Carnap strikingly evokes Poincaré and Duhem in connection with the "liberalization of empiricism" (Carnap 1963: 56). And Neurath brings out decisions and holism in his encyclopedic project (Neurath 1936).

Conventions and history

Poincaré, in response to Le Roy, stressed the pragmatic criteria that allow scientists to reach reliable decisions. He believed that it is possible to distinguish to some extent the artificial from the natural, the theoretical from the empirical. Poincaré went on to suggest that these criteria could be justified in a number of ways. He claimed that science aims at structures rather than things. He also referred to the intersubjective agreement of scientists. Duhem, on the other hand, claimed to find a solution to the difficulties brought up by Le Roy in his holist thesis: "Hypotheses which by themselves have no physical meaning undergo experimental testing in exactly the same manner as other hypotheses" (Duhem 1906/1991: 216).

Both Poincaré and Duhem called attention to the temporal dimension of scientific theories. In rejecting an a priori form of the intuition of space, Poincaré made legitimate the question of its origin. He pursued this issue by offering a careful analysis of the genesis of the concept of space. He resorted to mathematical methods as well as the findings of psychology and physiology, thus calling for a rigorous study of the functioning of the mind. In contrast, the founders of mathematical logic were unable to benefit from psychological research out of fear of falling into what they saw as psychologism. The later development of cognitive science and the instruments with which it has provided us has completely changed the situation. This may explain a renewed interest in Poincaré's work.

Duhem in his late *Traité d'énergétique* offers an example of the numerous conventions involved in a thorough presentation of the principle of the conservation of energy:

> To create a mathematical symbol suitable for representing the value of work [*oeuvre*], we form an expression that satisfies certain conditions that we decide to impose on it. We establish these conventions not at random; we choose them in such a way that they offer the image of the simplest and most prominent traits presented by the notion of work, or at least that they match these traits without difficulty.
>
> (Duhem 1911: I, 82, trans. AB)

Conventions, then, are those conditions that are required but not unequivocally determined. Duhem brings out no less than ten such conditions, among them the definition of kinetic energy as a magnitude that is conserved. This fact should put to rest any attempt to minimize the centrality of convention in Duhem.

In order to make himself understood, Duhem does not hesitate to mention a few historical figures: Cardanus, Descartes, and Leibniz. They represent three major stages leading to the constitution of the modern concept. Duhem is alluding here to his historical research, in particular his *Études sur Léonard de Vinci*. Cardanus was responsible for spreading Leonardo's ideas, who had taken up and refined the medieval theory of impetus (Duhem 1906–1913: I, 223–8). Benefiting from this legacy, the founders of modern science endeavored to establish an efficacious mathematical theory. What had to be sorted out were the different concepts involved (velocity, quantity of motion, and kinetic energy proper) and to settle on the most appropriate one for establishing the conservation principle. Such a task could be carried out only through a long and complex process. One may raise objections to the details of Duhem's historical interpretation, which asserts a continuity between the natural philosophy of the Middle Ages and modern physics. But there is no doubting in Duhem's case the importance of history for philosophy of science (Brenner 2014). This aspect of the work of their conventionalist predecessors found little or no echo in the logical empiricists except for Neurath and Frank.

Conclusion

The movement initiated by Poincaré and Duhem arose in a specific context: the breakdown of the Newtonian paradigm and the advent of a series of sciences devoted to the human subject and society. The development of science over the past century has given rise to new challenges; for example, those raised by cognitive science, nanotechnology, and computer science. Yet one could argue that present-day discourse on simulations continues the debates on models and idealizations of the turn of the twentieth century. The concept of virtual reality further exacerbates the issue of the natural–artificial distinction.

In endeavoring to moderate the fierce quarrel over social construction, Ian Hacking issues the following warning: "The roots of social construction are in the very logical positivism that so many present-day constructionists prefer to detest" (1999: 42–43). As we saw, Carnap among others drew on the work of the French conventionalists who had already given importance to construction. But the logical positivists were not inclined to study historical events, psychological genesis, or social processes. As we have noted, "convention" is a generic term referring to various elements of scientific activity that go beyond the axiomatic structure of theories: the heuristics, the ontologies, the rational values, etc. We may ask: Where do these other elements come from? How do they evolve over the course of history? Why do they slip into oblivion? New paths of research are needed to explore such questions. Conventionalism has not disappeared from the philosophical scene.

References

Ben-Menahem, Y. (2006) *Conventionalism*, Cambridge: Cambridge University Press.

Boutroux, E. (1874) *De la contingence des lois de la nature*, Paris: Presses universitaires de France. Repr. 1991.

Brenner, A. (2014) "Epistemology Historicized: The French Tradition," in M. C. Galavotti et al. (eds.), *New Directions in the Philosophy of Science*, Dordrecht: Springer, pp. 727–36.

——— (2015) *Les textes fondateurs de l'épistémologie française*, Paris: Hermann [This volume contains the major texts of the French authors mentioned here].

Carnap, R. (1923) "Über die Aufgabe der Physik und die Anwendung des Grundsatzes der Einfachstheit," *Kant-Studien* 28: 90–107. Trans. "On the Task of Physics and the Application of the Principle of Maximal Simplicity," in Carnap (2019), pp. 209–46.

——— (1926) *Physikalische Begrifsbildung*, Karlsruhe: Braun. Trans. "Physical Concept Formation," in Carnap (2019), pp. 340–440.

——— (1928) *Der logische Aufbau der Welt*, Berlin: Weltkreis-Verlag. Trans. *The Logical Construction of the World*, Berkeley: University of California Press, 1967.

——— (1963) "Intellectual Autobiography," in P. A. Schilpp (ed.), *The Philosophy of Rudolf Carnap*, La Salle, IL: Open Court, pp. 3–84.

——— (2019) *Collected Works, vol. 1: Early Writings* (ed. by A. W. Carus et al.), Oxford: Oxford University Press.

Duhem, P. (1892) "Quelques réflexions au sujet des theories physiques," *Revue des questions scientifiques* 31: 139–77.

——— (1894) "Quelques réflexions au sujet de la physique expérimentale," *Revue des questions scientifiques* 34: 179–229.

——— (1906) *La théorie physique: Son objet et sa structure*, Paris: Vrin. Trans. *The Aim and Structure of Physical Theory*, Princeton: Princeton University Press, 1991.
——— (1906–1913) *Études sur Léonard de Vinci*, Paris: Hermann, vol. 3.
——— (1911) *Traité d'énergétique*, Paris: Gauthier-Villars, vol. 2.
Dupréel, E. (1925) "Convention et raison," *Revue de métaphysique et de morale* 32: 283–310.
Frank, P. (1907) "Kausalgesetz und Erfahrung," *Annalen der Naturphilosophie* 6: 443–50. Trans. "Experience and the Law of Causality," in Frank (1949), pp. 53–60.
——— (1908) "Willkürliche Schöpfungen des Verstandes? Bemerkung zu dem Aufsatz von G. Hessenberg," *Jahresbericht der deutschen mathematiker-Vereinigung* 17: 227–32.
——— (1914) "[Review H. Poincarés *Letzte Gedanken*]," *Monatshefte für Mathematik und Physik* 23: Literaturberichte 54–55.
——— (1949) *Modern Science and Its Philosophy*, Cambridge, MA: Harvard University Press.
Frey, G. (1976) "Konventionalismus," in J. Ritter et al. (eds.), *Historisches Wörterbuch der Philosophie*, Basel: Schwabe, pp. 1078–9.
Hacking, I. (1999) *The Social Construction of What?* Cambridge, MA: Harvard University Press.
Haller, R. (1985) "Der erste Wiener Kreis," *Erkenntnis* 22: 341–58. Transl. "The First Vienna Circle," in T. Uebel (ed.), *Rediscovering the Forgotten Vienna Circle. Austrian Studies on Otto Neurath and the Vienna Circle*, Dordrecht: Kluwer, 1991, pp. 95–108.
Hessenberg, G. (1908) "Willkürliche Schöpfungen des Verstandes?" *Jahresbericht der deutschen mathematiker-Vereinigung* 17: 145–62.
Horwich, P. (1998) "Conventionalism," in E. Craig (ed.), *Routledge Encyclopedia of Philosophy*, London: Routledge, pp. 666–9.
Le Roy, E. (1900) "La science positive et les philosophies de la liberté," in *Premier Congrès international de philosophie*, Paris: Armand-Colin, pp. 313–41.
——— (1901) "Un positivism nouveau," *Revue de métaphysique et de morale* 9: 138–53.
Milhaud, G. (1896) "La science rationnelle," *Revue de métaphysique et de morale* 4: 280–302.
Neurath, O. (1916) "Zur Klassifikation von Hypothesensystemen," *Jahrbuch der Philosophischen Gesellschaft an der Universität Wien 1914 und 1915*, pp. 39–63. Trans. "On the Classification of Systems of Hypotheses," in Neurath (1983), pp. 13–31.
——— (1921) *Anti-Spengler*, Munich: Callway. Excerpts trans. "Anti-Spengler," in Neurath (1973), pp. 158–213.
——— (1936) "L'encyclopédie comme 'modèle'," *Revue de Synthese* 12: 187–201. Trans. "Encyclopedia as Model," in Neurath (1983), pp. 145–58.
——— (1973) *Empiricism and Sociology* (ed. by M. Neurath and R. S. Cohen), Dordrecht: Reidel.
——— (1983) *Philosophical Papers 1913–1946* (ed. by R. S. Cohen and M. Neurath), Dordrecht: Reidel.
Poincaré, H. (1891) "Les géometries non euclidiennes," *Revue générale des sciences* 2: 769–74.
——— (1902) *La science et l'hypothèse*, Paris: Flammarion, repr. 1968. Trans. "Science and Hypothesis," in Poincaré (1913), pp. 27–198.
——— (1908) *Science et méthode*, Paris: Kimé, repr. 1999. Trans. "Science and Method," in Poincaré (1913), pp. 357–546.
——— (1913) *Foundations of Science*, Lancaster, PA: The Science Press. Repr. 1946.
Reichenbach, H. (1922) "Der gegenwärtige Stand der Relativitätsdiskussion," *Logos* 10: 316–78. Trans. "The Present State of the Discussion on Relativity," in *Selected Writings* (ed. by M. Reichenbach and R. S. Cohen), Dordrecht: Reidel, 1978, vol. 2, pp. 3–47.
Rey, A. (1907) *La théorie de la Physique chez les physiciens contemporains*, Paris: Alcan.
Schlick, M. (1918) *Allgemeine Erkenntnislehre*, Berlin: Springer, 1918, 2nd rev. ed. 1925. Trans. *General Theory of Knowledge*, Lasalle: Open Court, 1974.
Uebel, T. (2000) *Vernunftkritik und Wissenschaft: Otto Neurath und der erste Wiener Kreis*, Vienna: Springer.
Verein Ernst Mach (1929) *Wissenschaftliche Weltauffassung. Der Wiener Kreis*, Vienna: Wolf. Trans. "The Scientific Conception of the World. The Vienna Circle," in O. Neurath, *Empiricism and Sociology* (ed. by R. S. Cohen and M. Neurath), Dordrecht: Reidel, 1973, pp. 299–318; rev. trans. (with orig. annotated bibliography) "The Scientific World-Conception. The Vienna Circle," in F. Stadler and T. Uebel (eds.), *Wissenschaftliche Weltauffassung. Der Wiener Kreis. Hrsg. vom Verein Ernst Mach (1929)*, Vienna: Springer, 2012, pp. 75–116.
Wolters, G. (1985) "The First Man Who Almost Wholly Understands Me: Carnap, Dingler, and Conventionalism," in N. Rescher (ed.), *The Heritage of Logical Positivism*, New York: University Press of America, pp. 93–107.

9
EINSTEIN AND LOGICAL EMPIRICISM

Fynn Ole Engler

Albert Einstein was one of the most influential and well-known physicists of the twentieth century due to his groundbreaking work in the field of relativity and quantum theory. But Einstein's philosophical considerations were also of decisive importance for the elaboration and interpretation of his physical theories as well as for the development of the modern philosophy of science (see Frank 1949; Zahar 1989; Howard 2014). Conversely, in the years following the discovery of general relativity, there were also philosophers who had their say in the field of physics, which Einstein took very seriously. These included some representatives of the new movements in scientific philosophy, above all the philosopher-physicists Moritz Schlick (1920, 1921, 1923) and Hans Reichenbach (1920, 1924, 1928).

This was no coincidence, because these philosophers also took up the enlightening and emancipatory potential of the scientific knowledge of the time and tried to advance the understanding of the new physical theories by philosophical means. Schlick and Reichenbach in particular, together with Philipp Frank (1911), Ernst Cassirer (1921), Joseph Petzoldt (1923), and Émile Meyerson (1925), played an important role on the way to a viable philosophical interpretation of Einstein's relativity theories (see Howard 1984, 1991; Hentschel 1990; Friedman 1999; Ryckman 2005). However, their work also determined Einstein's philosophical sensibility, enabling him to pursue the research program of a unified field theory based on the model of general relativity, and they also influenced his strong reluctance to a too positivistic-inspired interpretation of quantum mechanics (see Van Dongen 2010; Sauer 2014; Giovanelli 2018).

In turn, reading and discussing philosophical literature stimulated Einstein's thinking throughout his life. In the process, he developed and transformed his epistemological and methodological convictions (see, e.g., Einstein 1914a, 1916, 1918, 1921, 1933), which are difficult to assign to a certain philosophical direction, always in close connection with his physical research (Norton 2020). However, Einstein never gave up his sympathy for a kind of realism (see Fine 1986: ch. 6; Howard 1993; Zahar 2007: ch. 9; Ryckman 2017: chs. 8–9). His 1946 statement, at the age of 67, that "physics is an attempt conceptually to grasp reality as something that is considered to be independent of its being observed" (1949: 77, compare 1931: 291), is a commitment to a basic attitude that he maintained all his life, as well as reflections on methodological and heuristic issues. Such issues included the relationship between theory construction and measurement, which was at the center of the debate between conventionalism and empiricism in the early twentieth century, when Einstein sided with the founders of logical empiricism in

DOI: 10.4324/9781315650647-11

defending the empirical integrity of general relativity against the critique of neo-Kantian apriorism (see Howard 1994).

Einstein and Schlick

From the mid-1910s to the mid-1920s, Einstein saw Schlick as his favorite philosopher. The fact that Schlick in particular was able to fill this prominent role at this time had something to do with his critical attitude concerning fundamental concepts and general principles in physics (see Schlick 1915), which he shared with Einstein. In particular, both were adherents of David Hume's epistemological approach, which also obliged them to a shared opposition to any form of a priori reasoning in science and philosophy.

Just like Einstein, Schlick was an open-minded border crosser between the sciences, including empirical psychology, mathematics, and various philosophical strands (see Renn 2004, 2007b: 62–70; Engler 2009). Starting out from concrete challenges and problems, both developed their philosophy of science as a reflective enterprise, that also integrated physical theorizing, from a number of sources, in particular from a critique of neo-Kantianism, an epistemological analysis of fundamental concepts such as space, time, and causality, as carried out by empiricists (besides Hume's psychologico-critical method, Ernst Mach's historico-critical method played a central role in this by questioning the self-evidence of the foundations of classical mechanics; see Holton 1968 and Renn 2007b), but also from conventionalist aspects, as found in the works of Pierre Duhem and Henri Poincaré (see Howard 1990, 2014: sect. 2 and Ryckman 2017: ch. 7).

Einstein's preoccupation with philosophical reflections enabled him in particular to address deep-rooted conceptual tensions between the distinct domains of classical physics, mediated by an atomistic worldview (see Renn and Rynasiewicz 2014 for a discussion of his perspective on fundamental borderline problems between mechanics, thermodynamics and electrodynamics), but also to strive for a conceptual unity of physics. A first step along this path was Einstein's special theory of relativity, which he formulated in June 1905, a time in which he worked as Technical Expert, Third Class, at the patent office in Bern but was also engaged in philosophical studies, in particular of the concepts of space and time.

Special relativity

As is well known, special relativity as a "principle theory," in contrast to a "constructive theory"—the former consists of empirically obtained high-level generalizations which can be precisely formulated mathematically, while the latter provides, on the basis of some relatively simple formalisms, a model for the complex phenomena in question (see Einstein 1919). Special relativity is based on two postulates (see Einstein 1905: sect. 1). The principle of relativity states that the same physical laws apply to the behavior of bodies in reference systems moving uniformly with respect to each other. The second principle says that in all such systems, the speed of light is the same. Within the framework of classical physics, these two postulates cannot be reconciled. This is because the measured speeds also change during the transition from a stationary to a moving reference system. Thus, according to classical physics, speeds do not remain invariant when the reference systems are changed. Paradoxically, however, this is exactly what the principle of the constancy of the speed of light requires. This paradox could be clarified only by a fundamental change in the understanding of space and time, as Einstein did in his *annus mirabilis* 1905 (see Stachel 2005 and, for details, Renn and Rynasiewicz 2014). In doing so, he also had to take into account his previously gained insights into the microstructure of radiation (light quantum hypothesis) and its compatibility with the macrostructure described by the electrodynamics of James C. Maxwell and Hendrik Antoon Lorentz. Einstein assumed the validity of the relativity principle and the correctness of the Lorentz transformation equations for space and time coordinates. Moreover, he placed the auxiliary quantities for lengths and times, which Lorentz had introduced for moving bodies in the ether, at the center of special relativity; thus, he revolutionized the understanding of space and time in a first step. According to this new understanding, the measurement of space–time intervals between events depends on the relative speed of reference systems. Thus, different observers might disagree on whether two distant events really took place simultaneously.

On the way to special relativity, however, the psychological-critical analysis of the concepts of space and time, as carried out by Hume in his *A Treatise on Human Nature*, was also of decisive importance. Einstein had read Hume's work intensively together with his friends Conrad Habicht and Maurice Solovine in the so-called Olympia Academy, an informal discussion group launched around Easter 1902 (see CPAE, vol. 2: xxiv–xxv for a comprehensive list of literature that was read by Einstein and his friends). It helped him to recognize the intellectual freedom he needed to formulate new concepts of space and time that could reconcile the two aforementioned principles, which were incompatible from the point of view of classical physics (see Norton 2010: 372–81; Engler and Renn 2013: 128–37). Thus, Einstein was able to emphasize the importance of an empirically oriented philosophy in his first letter to Schlick of 14 December 1915:

> Your representations that the theory of rel[ativity] suggests itself in positivism, yet without requiring it, are also very right. In this also you saw correctly that this line of thought had a great influence on my efforts, and more specifically, E. Mach, and even more Hume, whose *Treatise of Human Nature* I had studied avidly and with admiration shortly before discovering the theory of relativity. It is very possible that without these philosophical studies I would not have arrived at the solution.
>
> (CPAE, vol. 8, Doc. 165)

Shortly before, Schlick had sent an article on the philosophical significance of the principle of relativity to Einstein (see Schlick 1915). During this time, however, he also worked on the *General Theory of Knowledge*, which he had essentially completed by the end of 1915, but whose public availability was delayed until early 1919 due to the war. To Schlick's delight, Einstein reacted with great sympathy to his article:

> Yesterday I received your paper and have studied it thoroughly already. It is among the best that has been written on relativity to date. From the philosophical perspective, nothing nearly as clear seems to have been written on the topic. At the same time, you have a complete command of the subject material. I have nothing to criticize about your representations.
>
> (Albert Einstein to Moritz Schlick, 14 December 1915 in CPAE, vol. 8, Doc. 165)

Einstein also invited Schlick to his home in Berlin, where, after posts at the University of Zurich and the Charles University of Prague, he held the directorship of the Kaiser Wilhelm Institute of Physics, a professorship at the University and a membership of the Prussian Academy of Sciences. The decisive hint that their first meeting took place only a few days later is a letter from Schlick to his father dated 14 December, in which he announced a visit to Berlin: "We will now arrive on Friday . . . at 9:54 p.m. at the Stettiner train station" (Moritz Schlick to Albert

Schlick, 14 December 1915, Noord-Hollands Archief, Schlick Estate, Inv.-No. 128). It is therefore more than likely that Schlick also visited Einstein on the weekend of 18–19 December. As it turned out, this meeting paved the way for Einstein out of a tricky situation he had gotten into with the successful formulation of the general theory of relativity in November 1915, but it also led to a brilliant confirmation of a central thesis in Schlick's *General Theory of Knowledge*.

General relativity

Since 1907, Einstein had been working on an extension of the relativity principle to nonuniform motion (see Janssen 2014: sects. 2–5; Gutfreund and Renn 2015: 7–36; Ryckman 2017: ch. 6). He first recognized that the identity of essence between inertia and gravity offered the possibility, at least locally, to mentally simulate the effects of gravity by the effects of acceleration. This not only made it possible to maintain the intimate relationship between inertial and gravitational mass of bodies, which was expressed in the fact, known since Galileo, that all bodies fall with the same acceleration towards Earth, which ultimately established the *principle of equivalence*; but also, together with the findings of special relativity, gave rise to a further revision of the concepts of space and time. These no longer served as a fixed stage for physical events but became part of these events themselves, and an expression of what was traditionally called gravity. Space and time are thus determined by the gravitational field, while the gravitational field on the basis of the equivalence principle now depends on the distribution of masses in the universe in the same way that inertial effects depend on the distribution of masses in the universe. This idea was influenced by Einstein's reading of Mach's work (see, e.g., Hoefer 1995; Renn 2007b for the role of what Einstein later called *Mach's principle* for general relativity).

Besides these two principles, Einstein's general theory of relativity is based on the *principle of general covariance*, which states that all physical laws are form-invariant, regardless of the observer's state of motion (see Stachel 1980; Norton 1984; Renn 2007a: vols. 1–2 for Einstein's struggle with general covariance). At a certain point, however, Einstein came to the conclusion that there can be no generally covariant field equations of gravitation. On this basis, in May 1913, he formulated the *Entwurf* theory with the help of his friend, the mathematician Marcel Grossmann (Einstein and Grossmann 1913). In this context, Einstein put forward two arguments that seemed to contradict a generally covariant formulation of the theory of relativity. While the first argument, which was based on the conservation laws for energy and momentum, soon turned out to be untenable (cf. Renn and Sauer 2007: Sect. 7), the second argument was based on a combination of mathematical, physical, and epistemological considerations. This was the notorious "hole argument" that Einstein (together with his close friend Michele Besso) developed in late August 1913 as a post-hoc rationality for the abandonment of general covariance (see Janssen 2007: sects. 1–2 and Stachel 2014). On the one hand, he had assumed that in a matter-free region of space–time—the "hole"—two space–times of a four-dimensional manifold can be physically distinguished from each other solely on the basis of their coordinates. On the other hand, he considered it necessary that the gravitational field equations in the matter-free region have an unambiguous solution. Assuming that space–time points can be identified independently of the metric tensor only by their coordinates, the field equations did not provide a unique solution in this matter-free region. Therefore, Einstein dropped the requirement of general covariance for his field equations in order to preserve the unambiguous determination of physical processes (see Einstein 1914b: 1066–7 for the elaborated version of the hole argument).

Schlick was quite familiar with the consequences of the hole argument. Against this background, he stated in his already-mentioned article that Mach's principle was untenable as a heuristic guideline for the relativity program (1915: 183–4). However, this probably challenged

Einstein, since he had meanwhile succeeded in finding generally covariant field equations. In addition, he did not want to shake Mach's heuristics. Nevertheless, he still lacked a convincing argument for his new understanding of space and time, which he quickly stated in his letter to Schlick, namely that "space and time lose the last vestige of physical reality" (CPAE, vol. 8, Doc. 165).

The hole argument, which initially prevented a persuasive interpretation of general relativity, was soon replaced by the "point-coincidence argument," which Einstein probably took over from Schlick at the end of 1915 (see, for details, Engler and Renn 2013: 146–52; see also Howard and Norton 1993; Howard 1999; Giovanelli 2013). In this context, Schlick had arrived at a uniform method of measurement based on spatiotemporal coincidences. In the *General Theory of Knowledge*, he stated: "Determination of space and time are always effected by means of *measurement*, and all measurements, from the most primitive to the most advanced, rest on observation of spatio-temporal coincidences" (Schlick 1918/1925: 275). Therefore, the coincidence method also made it possible to unambiguously assign a spatiotemporal schematized system of concepts, as represented by Einstein's general theory of relativity, to the physical events they designate. Furthermore, these point events proved to be invariant under arbitrary coordinate transformations and thus fulfilled the requirement for general covariance. Finally, the adoption of Schlick's method of space–time coincidences also complements the important role that Mach's principle played ontologically for the general theory of relativity, in that epistemologically material events take precedence over the hitherto-fixed structure of space–time. Thus the spatiotemporal coincidences are the only real thing in physics and can be determined objectively by measurements. Shortly after New Year's Day 1916, Einstein wrote to Besso:

> Nothing is *real* physically except for the entirety of the spatio-temporal point coincidences. If, for ex., physical events were to be constructed out of the motions of mass-points alone, then the meeting of the points, i.e., the intersection points of their world lines, would be the only real, that is, principally observable, things. These intersection points naturally remain intact in all transformations (and no new ones are added), only if certain uniqueness conditions are maintained. It is thus most natural to demand that the laws not determine more than the spatio-temporal coincidences as a whole. This is accordingly already achieved with generally covariant equations.
>
> (Albert Einstein to Michele Besso, 3 January 1916 in CPAE, vol. 8, Doc. 178, orig. emphasis)

Thus, under the condition of general covariance, the physical meaning and philosophical significance of space–time coordinates were satisfactorily clarified for the moment. Nevertheless, Einstein's further discussions with Schlick and Reichenbach about the role of conventions and experience in the 1920s, especially in the context of his strive for a unified field theory, led to differences with the founders of logical empiricism, who in turn took sides with the anti-realistic spirit of quantum mechanics (see Schlick 1936; Reichenbach 1944).

Geometry, experience and realism

It was Einstein who, in April 1920, drew Schlick's attention to Reichenbach's *The Theory of Relativity and A Priori Knowledge*. Their subsequent discussion centered on the question of whether a Kantian conception of space and time was compatible with Einstein's theory of relativity. Reichenbach attempted to reformulate Kant's doctrine of a priori principles of knowledge in a relativized version (see Friedman 2001). For him, constitutive principles continued to play a decisive role in experience in the sense of Kant, but, contrary to their originally apodictic character, they were valid only in relation to a specific theoretical framework. Thus, Reichenbach pleaded for a constitutive, contingent, and relativized a priori.

However, Schlick could not make out any objective difference between the conventions he defended and Reichenbach's constitutive principles. He wrote to Einstein: "Reichenbach does not seem to me to have done justice to Poincaré's theory of conventions; what he calls a priori principles of coordination and rightly distinguishes from empirical principles of connection seems to me to be completely identical to Poincaré's 'conventions' and not to have any meaning extending beyond that" (Moritz Schlick to Albert Einstein, 9 October 1920 in CPAE, vol. 10, Doc. 171). Assuming the inseparability between geometry and physics, in practice, which took up the abstract distinction between purely conceptual definitions and empirical knowledge in the first edition of the *General Theory of Knowledge* (1918: Sect. 8), and left no room for the category of a Kantian a priori, Schlick wanted to ensure the empirical integrity of Einstein's relativistic theory of gravitation (see Schlick [1920: ch. 5]; after an intensive exchange with Schlick in autumn 1920, Reichenbach also took a similar position).

In turn, in his widely read lecture *Geometry and Experience* delivered at a public meeting of the Prussian Academy of Sciences on 27 January 1921, Einstein defended the position (inspired by the exchange with Schlick) that what he called "practical geometry," in sharp contrast to "purely axiomatic geometry" (1921: 256), was an empirical science, and therefore, the use of non-Euclidean geometry in general relativity was not merely a matter of convention (see also Friedman 2008; Howard 2014: sect. 3; Ryckman 2017: ch. 7).

From the mid-1920s onwards, however, this tension also determined Einstein's search for a unified field theory, i.e., between the geometrization of the gravitational and electromagnetic fields, which should also take into account the fundamental properties of matter, and its compatibility with a rigid rod and clock measuring practice (see Brown 2007; Giovanelli 2014; Lehmkuhl 2014). But Einstein's goal of creating a unified foundation for all physics, as a free invention of the human mind guided by simplicity and univocalness, was also embedded in his lifelong program of conceptually grasping reality. In this context, Schlick's linguistic turn and Reichenbach's conventionalism were too positivistic (see Einstein 1949: 677–8; Reichenbach 1949; also Giovanelli 2016) as well as the quantum mechanics, which Einstein, nevertheless, was willing to integrate in his field theoretical approach in order to yield a comprehensive real, essentially "materialistic" worldview (see Gutfreund and Renn 2020: 101).

References

Brown, H. R. (2007) *Physical Relativity: Space-Time Structure from a Dynamical Perspective*, Oxford: Clarendon Press.

Cassirer, E. (1921) *Zur Einstein'schen Relativitätstheorie: Erkenntnistheoretische Betrachtungen*, Berlin: Bruno Cassirer. Trans. "Einstein's Theory of Relativity," in Cassirer, *Substance and Function and Einstein's Theory of Relativity*, Chicago: Open Court, 1923, pp. 351–456.

CPAE (1987–present) *The Collected Papers of Albert Einstein*, Princeton: Princeton University Press.

Einstein, A. (1905) "Zur Elektrodynamik bewegter Körper," *Annalen der Physik* 17: 891–921. Repr. and trans. "On the Electrodynamics of Moving Bodies," in CPAE, vol. 2, Doc. 23.

——— (1914a) "Antrittsrede," *Königlich Preussische Akademie der Wissenschaften (Berlin) Sitzungsberichte*, Halbband 2, pp. 739–42. Trans. "Principles of Theoretical Physics (Inaugural lecture)," in Einstein (1954/1994), pp. 240–4, repr. in CPAE, vol. 6, Doc. 3.

——— (1914b) "Formale Grundlage der allgemeinen Relativitätstheorie," in *Königlich Preussische Akademie der Wissenschaften (Berlin) Sitzungsberichte*, Halbband 2, pp. 1030–85. Trans. "The Formal Foundation of the General Theory of Relativity," in CPAE, vol. 6, Doc. 9.

——— (1916) "Ernst Mach," *Physikalische Zeitschrift* 17: 101–4. Repr. and trans. "Ernst Mach," in CPAE, vol. 6, Doc. 29.

——— (1918) "Motive des Forschens," in Deutschen Physikalische Gesellschaft (ed.), *Zu Max Plancks sechzigstem Geburtstag*, Karlsruhe: C. F. Müller, pp. 29–32. Trans. "Principles of Research," in *Ideas and Opinions*, New York: The Modern Library, 1994, pp. 244–8. Repr. in CPAE, vol. 7, Doc. 7.

——— (1919) "Time, Space, and Gravitation," *The Times* (London), 28 November 1919, pp. 13–4. Repr. in CPAE, vol. 7, Doc. 26.

——— (1921) *Geometrie und Erfahrung*, Berlin: Springer. Trans. "Geometry and Experience," in Einstein (1954/1994), pp. 254–68. Repr. in CPAE, vol. 7, Doc. 52.

——— (1931) "Maxwell's Influence on the Evolution of the Idea of Physical Reality," in Einstein (1954/1994), pp. 291–5.

——— (1933) *On the Method of Theoretical Physics*, The Herbert Spencer Lecture, delivered at Oxford, 10 June 1933. Oxford: Clarendon Press.

——— (1949) "Autobiographical Notes" and "Reply to Criticism: Remarks Concerning the Essays Brought Together in This Co-operative Volume," in P. A. Schilpp (ed.), *Albert Einstein: Philosopher-Scientist*, La Salle: Open Court, pp. 1–95, 665–88.

——— (1954) *Ideas and Opinions*, New York: Bonanza. Repr. New York: Modern Library, 1994.

Einstein, A. and Grossmann, M. (1913) *Outline of a Generalized Theory of Relativity and a Theory of Gravitation*, Leipzig: Teubner. Repr. in CPAE, vol. 4, Doc. 13.

Engler, F. O. (2009) "Über das erkenntnistheoretische Raumproblem bei Moritz Schlick, Wilhelm Wundt und Albert Einstein," in F. Stadler and H.-J. Wendel (eds.), *Stationen. Dem Philosophen und Physiker Moritz Schlick zum 125. Geburtstag*, Vienna: Springer, pp. 107–45.

Engler, F. O. and Renn, J. (2013) "Hume, Einstein und Schlick über die Objektivität der Wissenschaft," in F. O. Engler and M. Iven (eds.), *Moritz Schlick—Die Rostocker Jahre und ihr Einfluss auf die Wiener Zeit*, Leipzig: Leipziger Universitätsverlag, pp. 123–56.

Fine, A. (1986) *The Shaky Game. Einstein, Realism and the Quantum Theory*, Chicago: University of Chicago Press, 2nd ed., 1996.

Frank, P. (1911) "Das Relativitätsprinzips und die Darstellung der physikalischen Erscheinungen im vierdimensionalen Raum," *Annalen der Naturphilosophie* 10: 129–61.

——— (1949) "Einstein's Philosophy of Science," *Review of Modern Physics* 21: 349–55.

Friedman, M. (1999) *Reconsidering Logical Positivism*, Cambridge: Cambridge University Press.

——— (2001) *Dynamics of Reason. The 1999 Kant Lectures at Stanford University*, Stanford: CSLI Publications.

——— (2008) "Space, Time, and Geometry: Einstein and Logical Empiricism," in P. L. Galison, G. Holton and S. S. Schweber (eds.), *Einstein for the 21st Century: His Legacy in Science, Art, and Modern Culture*, Princeton and Oxford: Princeton University Press, pp. 205–16.

Giovanelli, M. (2013) "Erich Kretschmann as a Proto-Logical-Empiricist. Adventures and Misadventures of the Point-Coincidence Argument," *Studies in History and Philosophy of Modern Physics* 44: 115–34.

——— (2014) "'But One Must Not Legalize the Mentioned Sin': Phenomenological vs. Dynamical Treatments of Rods and Clocks in Einstein's Thought," *Studies in History and Philosophy of Modern Physics* 48: 20–44.

——— (2016) ". . . 'But I Still Can't Get Rid of a Sense of Artificiality': The Reichenbach-Einstein Debate on the Geometrization of the Electromagnetic Field," *Studies in History and Philosophy of Modern Physics* 54: 35–51.

——— (2018) "'Physics Is a Kind of Metaphysics': Émile Meyerson and Einstein's Late Rationalistic Realism," *European Journal for Philosophy of Science* 8: 783–829.

Gutfreund, H. and Renn, J. (2015) *The Road to Relativity. The History and Meaning of Einstein's "The Foundation of General Relativity,"* Princeton and Oxford: Princeton University Press.

——— (2020) *Einstein on Einstein. Autobiographical and Scientific Reflections*, Princeton and Oxford: Princeton University Press.

Hentschel, K. (1990) *Interpretationen und Fehlinterpretationen der speziellen und der allgemeinen Relativitätstheorie durch Zeitgenossen Albert Einsteins*, Boston: Birkhäuser.

Hoefer, C. (1995) "Einstein's Formulations of Mach's Principles," in J. B. Barbour and H. Pfister (eds.), *Mach's Principle: From Newton's Bucket to Quantum Gravity*, Boston: Birkhäuser, pp. 67–87.

Holton, G. (1968) "Mach, Einstein, and the Search for Reality," in *Thematic Origins of Scientific Thought. Kepler to Einstein*, Cambridge and London: Harvard University Press, 1988, rev. ed., pp. 237–77.

Howard, D. (1984) "Realism and Conventionalism in Einstein's Philosophy of Science: The Einstein-Schlick Correspondence," *Philosophia Naturalis* 21: 618–29.

——— (1990) "Einstein and Duhem," *Synthese* 83: 363–84.
——— (1991) "Einstein and *Eindeutigkeit*: A Neglected Theme in the Philosophical Background to General Relativity," in J. Eisenstaedt and A. J. Kox (eds.), *Historical Studies in General Relativity*, Boston: Birkhäuser, pp. 154–243.
——— (1993) "Was Einstein Really a Realist?," *Perspectives on Science* 1: 204–51.
——— (1994) "Einstein, Kant, and the Origins of Logical Empiricism," in W. Salmon and G. Wolters (eds.), *Language, Logic, and the Structure of Scientific Theories*, Pittsburgh: University of Pittsburgh Press, pp. 45–105.
——— (1999) "Point Coincidences and Pointer Coincidences: Einstein on the Invariant Content of Space-Time Theories," in H. Goenner, J. Renn, J. Ritter, and T. Sauer (eds.), *The Expanding Worlds of General Relativity*, Boston: Birkhäuser, pp. 463–500.
——— (2014) "Einstein and the Development of Twentieth-century Philosophy of Science," in M. Janssen and C. Lehner (eds.), *The Cambridge Companion to Einstein*, Cambridge: Cambridge University Press, pp. 354–76.
Howard, D. and Norton, J. D. (1993) "Out of the Labyrinth? Einstein, Hertz, and the Göttingen Answer to the Hole Argument," in J. Earman, M. Janssen and J. D. Norton (eds.), *The Attraction of Gravitation*, Boston: Birkhäuser, pp. 30–62.
Janssen, M. (2007) "What Did Einstein Know and When Did He Know It? A Besso Memo Dated August 1913," in J. Renn (ed.), *The Genesis of General Relativity. Sources and Interpretations, vol. 2: Einstein's Zurich Notebook: Commentary and Essays*, Dordrecht: Springer, 2007, pp. 785–837.
——— (2014) "'No Success Like Failure. . .' Einstein's Quest for General Relativity, 1907–1920," in M. Janssen and C. Lehner (eds.), *The Cambridge Companion to Einstein*, Cambridge: Cambridge University Press, pp. 167–227.
Lehmkuhl, D. (2014) "Why Einstein Did Not Believe That General Relativity Geometrizes Gravity," *Studies in History and Philosophy of Modern Physics* 44: 316–26.
Meyerson, E. (1925) *The Relativistic Deduction: Epistemological Implications of the Theory of Relativity*, Dordrecht: Reidel, 1985.
Norton, J. D. (1984) "How Einstein Found His Field Equations, 1912–1915," in D. Howard and J. Stachel (eds.), *Einstein and the History of General Relativity*, Boston: Birkhäuser, 1989, pp. 101–59.
——— (2010) "How Hume and Mach Helped Einstein Find Special Relativity," in M. Dickson and M. Domski (eds.), *Discourse on a New Method: Reinvigorating the Marriage of History and Philosophy of Science*, Chicago, IL: Open Court, pp. 359–86.
——— (2020) "Philosophy in Einstein's Science," in P. MacEwen (ed.), *Idealist Alternatives to Materialist Philosophies of Science*, Leiden: Brill, pp. 95–127.
Petzoldt, J. (1923) *Die Stellung der Relativitätstheorie in der geistigen Entwicklung der Menschheit*, Zweite, verbesserte und vermehrte Auflage, Leipzig: Barth.
Reichenbach, H. (1920) *Relativitätstheorie und Erknenntnis A Priori*, Berlin: Springer. Trans. *The Theory of Relativity and A Priori Knowledge*, Berkeley: University of California Press, 1965.
——— (1924) *Axiomatik der relativistischen Raum-Zeit-Lehre*, Braunschweig: Vieweg. Trans. *Axiomatization of the Theory of Relativity*, Berkeley: University of California Press, 1969.
——— (1928) *Philosophie der Raum-Zeit-Lehre*, Berlin: De Gruyter. Trans. *The Philosophy of Space and Time*, New York: Dover, 1957.
——— (1944) *Philosophic Foundations of Quantum Mechanics*, Berkeley and Los Angeles: University of California Press.
——— (1949) "The Philosophical Significance of the Theory of Relativity," in P. A. Schilpp (ed.), *Albert Einstein: Philosopher-Scientist*, Evanston, IL: The Library of Living Philosophers, pp. 289–311.
Renn, J. (2004) "The Relativity Revolution from the Perspective of Historical Epistemology," *Isis* 95: 640–8.
——— (ed.) (2007a) *The Genesis of General Relativity*, Dordrecht: Springer, vol. 4, 2007.
——— (2007b) "The Third Way to General Relativity: Einstein and Mach in Context," in J. Renn (ed.), *The Genesis of General Relativity, vol. 3: Gravitation in the Twilight of Classical Physics: Between Mechanics, Field Theory and Astronomy*, Dordrecht: Springer, 2007, pp. 21–75.
Renn, J. and Rynasiewicz, R. (2014) "Einstein's Copernican Revolution," in M. Janssen and C. Lehner (eds.), *The Cambridge Companion to Einstein*, Cambridge: Cambridge University Press, pp. 38–71.
Renn, J. and T. Sauer (2007) "Pathways out of Classical Physics. Einstein's Double Strategy in his Search for the Gravitational Field Equation," in J. Renn (ed.), *The Genesis of General Relativity, vol. 1: Einstein's Zurich Notebook: Introduction and Source*, Dordrecht: Springer, pp. 113–312.

Ryckman, T. (2005) *The Reign of Relativity: Philosophy in Physics 1915–1925*, Oxford: Oxford University Press.
——— (2017) *Einstein*, London and New York: Routledge.
Sauer, T. (2014) "Einstein's Unified Field Theory Program," in M. Janssen and C. Lehner (eds.), *The Cambridge Companion to Einstein*, Cambridge: Cambridge University Press, pp. 281–305.
Schlick, M. (1915) "Die philosophische Bedeutung des Relativitätsprinzips," *Zeitschrift für Philosophie und philosophische Kritik* 159: 129–75. Trans. "The Philosophical Significance of the Principle of Relativity," in Schlick (1979), pp. 153–89.
——— (1918) *Allgemeine Erkenntnislehre*, Berlin: Springer, 2nd rev. ed., 1925. Trans. *General Theory of Knowledge*, Vienna: Springer, 1973, repr. La Salle, IL: Open Court, 1985.
——— (1920) *Raum und Zeit in der gegenwärtigen Physik*, Berlin: Springer, 3rd ed. Trans. *Space and Time in Contemporary Physics: An Introduction to the Theory of Relativity and Gravitation*, Oxford: Oxford University Press, repr. with trans. of rev. in 4th ed., 1922, in Schlick (1979), pp. 207–69.
——— (1921) "Kritizistische oder empiristische Deutung der neuen Physik?" *Kantstudien* 26: 96–111. Trans. "Critical or Empiricist Interpretation of Modern Physics?" in Schlick (1979), pp. 322–34.
——— (1923) "Die Relativitätstheorie in der Philosophie," *Verhandlungen der Gesellschaft deutscher Naturforscher und Ärzte* 87: 58–69. Trans. "The Theory of Relativity in Philosophy," in Schlick (1979), pp. 343–53.
——— (1936) "Quantentheorie und Erkennbarkeit der Natur," in Schlick, *Gesammelte Aufsätze 1926–1936*, Vienna: Gerold, pp. 377–88. Trans. "Quantum Theory and the Knowability of Nature," in Schlick, *Philosophical Papers, Volume II (1925–1936)* (ed. by H. Mulder and B. van de Velde-Schlick), Dordrecht: Reidel, 1979, pp. 482–90.
Schlick, M. (1979) *Philosophical Papers, Volume I (1909–1922)* (ed. by H. L. Mulder and B. F. B. van de Velde-Schlick), Dordrecht, Boston and London: D. Reidel.
Stachel, J. (1980) "Einstein's Search for General Covariance, 1912–1915," in D. Howard and J. Stachel (eds.), *Einstein and the History of General Relativity*, Boston: Birkhäuser, 1989, pp. 63–100.
——— (2005) "Introduction," in J. Stachel (ed.), *Einstein's Miraculous Year: Five Papers That Changed the Face of Physics*, Princeton, NJ: Princeton University Press, pp. xv–lxxi.
——— (2014) "The Hole Argument and Some Physical and Philosophical Implications," *Living Reviews in Philosophy* 17, www.livingreviews.org.
Van Dongen, J. (2010) *Einstein's Unification*, Cambridge: Cambridge University Press.
Zahar, E. G. (1989) *Einstein's Revolution: A Study in Heuristic*, La Salle, IL: Open Court.
——— (2007) *Why Science Needs Metaphysics: A Plea for Structural Realism*, Chicago and La Salle, IL: Open Court.

10

THE FIRST VIENNA CIRCLE AND THE ERLANGEN CONFERENCE

Christoph Limbeck-Lilienau

Two groupings of philosophers can be seen as direct predecessors of the Vienna Circle (formed in 1923–24) and the Berlin group: the so-called first Vienna Circle (1907–10) and the attendees at the Erlangen Conference (1923). Both groups were initiated by later members of the Vienna Circle. The pre-World War I group in Vienna featured Philipp Frank, Hans Hahn, and Otto Neurath, while the Erlangen Conference was organized by Rudolf Carnap, together with Hans Reichenbach (the future leader of the Berlin Group). Both the first Vienna Circle and the Erlangen Conference discussed ideas which later took center stage in logical empiricism. Both groups were strongly interested in the philosophy of science and logic and tried to bring philosophy into contact with recent debates in the sciences; they felt that traditional philosophy was in a crisis and in need of reconstruction. So far, however, their reconstruction has remained somewhat vague.

The first Vienna Circle

The first Vienna Circle was absent from the standard histories of logical empiricism until the 1980s. Rudolf Haller (1982a, 1982b, 1985) and Friedrich Stadler (1982: 111–17, 1997/2001: 143–61) first pointed to a group including Frank, Hahn, and Neurath that had been mentioned by Frank (1941, 1949). Haller named it the "first Vienna Circle" and suggested that it should receive special attention, particularly if one wanted to avoid the then-dominant received view of logical empiricism. Subsequent historical investigations of the philosophy of the Vienna Circle emphasized the role of this group in pioneering ideas important for the left wing of the Vienna Circle (Uebel 2000, 2003).

The introduction to Frank's first English collection of his papers (1941) emphasizes the origin of logical empiricism in the work of Ernst Mach, despite certain shortcomings, e.g., his underestimation of the importance of logic and mathematics. In this context, Frank mentions "a group of young men," Hahn, Neurath, and himself, who tried to solve these shortcomings with, on the one hand, conventionalism (Poincaré, Duhem, and Abel Rey) and, on the other hand, new developments in logic and the philosophy of mathematics (Couturat, Schröder, and Hilbert). In a much expanded version of his historical introduction (1949), Frank is more explicit. The group, Frank argues, was actually a weekly discussion group which tried to build a "new positivism," based on the ideas of Mach and the French conventionalists.

 DOI: 10.4324/9781315650647-12

According to Frank, the group thought that the principles of science were "clear-cut mathematical relations among a small number of concepts" and not Mach's abbreviated and economical descriptions of observations. In order to allow for the relative independence of a mathematized theory from observation, one had to appeal to conventions.

> According to Mach the general principles of science are abbreviated economical descriptions of observed facts; according to Poincaré they are free creations of the human mind which do not tell anything about observed facts. The attempt to integrate the two concepts into one coherent system was the origin of what was later called logical empiricism.
>
> (1949: 11–12)

Unfortunately, Frank's report is the only source on the discussion group; yet there can be no doubt that Hahn, Frank, and Neurath knew each other at the time and did discuss, in some contexts or other, the problems mentioned by Frank. Although Neurath never speaks about a regular discussion group, he refers to his early intellectual friendship with Hahn and Frank (see Uebel 2000: 69–70; Limbeck-Lilienau 2018: 52). In a letter of 1934 to the mathematician Gerrit Mannoury, Neurath wrote: "I was acquainted with him [Hahn] for about 35 years, we discussed together Poincaré, Philipp Frank reported to us about Einstein's very first publications." Earlier that summer, after Hahn's sudden death, he wrote to Hempel: "35 years of similar endeavors in different domains. The joint youth with Poincaré, Duhem etc." And to Carnap, Neurath reported in 1945: "Hahn and I have been friends for many years—since the Gymnasium time. . . . He, the older, taught me a lot of things. We, Frank and other[s] read Spinoza in the 'Rahnhof' [a Viennese café]."

Fisette (2014) suggested that the group was just part of the discussions at or after the meetings of the "Philosophical Society" in Vienna, of which Neurath, Hahn and Frank were members at this time, and calls the first Vienna Circle a "pseudo-Circle" (see also Limbeck-Lilienau 2018; Uebel 2022). Fisette's doubts raise the question what entity the term "first Vienna Circle" is actually supposed to designate. I will focus here on the philosophical discussions in which Neurath, Hahn, and Frank were involved before Hahn's appointment at the University of Czernowitz (now in Ukraine) in 1910 and will limit my reconstruction to actually documented discussions. One thesis to be considered is whether Frank's report condenses diverse discussions into a more regular discussion group.

Neurath, Hahn, and Frank were involved in several discussions on the philosophy of science between 1907 and 1910. These discussions took place in three different settings: informal meetings of Neurath, Hahn, and Frank (perhaps with others), a reading group around Hahn and Höfler on the philosophy of mathematics, and the discussions in the "Philosophical Society."

Hahn had studied mathematics in Vienna. After postdoctoral research in Göttingen, he received his habilitation in Vienna in 1905 and started to teach there. Neurath had studied economics in Berlin since 1903 and returned to Vienna in 1906. In 1907, Neurath and Hahn met regularly in order to prepare a course on the foundations of mathematics and physics with a special focus on Poincaré, Mach, and Russell. Neurath reported in a letter to Ferdinand Tönnies in spring 1907:

> A local lecturer of mathematics plans to offer a course in the next winter term on the foundations of mathematics and mechanics (following the work of Poincaré). He asked me to teach the course together with him. . . . We meet twice a week and read

> Russell's *Principles of Mathematics*. Also Mach's *History of Mechanics* is sometimes our topic.
>
> (Neurath 1907)

Apparently that course was never given, but it may have triggered further discussions on the philosophy of science, also with Frank.

In the same year, the philosopher Alois Höfler was appointed as professor of philosophy, with a special focus on pedagogy, at the University of Vienna. Höfler had studied physics with Boltzmann before turning to philosophy under the influence of Brentano and Meinong. Hahn, Neurath, and Frank entered into close intellectual exchange with Höfler, who pursued a strong interest in the philosophy of physics and mathematics. Importantly, Höfler also was one of the organizers of the "Philosophical Society at the University of Vienna," an essential forum for the young scientists Hahn, Neurath, and Frank who gave their first philosophical lectures there (see Reininger 1938; Blackmore 1995: 289–95). A central feature of the Society were its discussion evenings, some of them led by Frank and Neurath.

In 1908–9, Höfler, Hahn and another Meinongian, Hans Pichler, met regularly in a reading group on logic and the foundations of mathematics (see Limbeck-Lilienau 2015: 45). Together, they read literature on the philosophy of mathematics, especially the recent German translation of Couturat's *Les Principes des Mathématiques*, which introduced Russell's logic and logicism to the German public. Couturat's book concluded with a long and ferocious attack on Kant's philosophy of mathematics. Neurath reviewed Couturat's book (Neurath 1909), but whether he also participated in Höfler's reading group is unclear, although some details suggest so (see Pichler 1909: 22).

Two topics were central issues in the discussions and publications of Hahn, Frank, and Neurath at that time: their views about symbolic logic and the foundation of mathematics, and their conception of scientific theories. According to Frank, the main focus of his discussions with Neurath and Hahn was an empiricist-cum-conventionalist view of scientific theories and "the primary role of mathematics and logic in the structure of science" (1949: 7). It was important to show how a complex axiomatized mathematical theory could be correlated with empirical data. Frank did not say much about logic or the foundations of mathematics, contrary to Hahn and Neurath who both showed strong interest in these topics.

The beginning of the century saw major shifts with the emergence of symbolic logic and the rejection of a psychologistic understanding of logic. At the same time, the logicist program emerged, and Neurath and Hahn were well aware of these developments, as we saw. Their sympathies for these developments can be explained by their contacts to the Viennese Meinongians, with Höfler at its center.

Neurath gained a strong interest in mathematical logic, already in Berlin through the philosopher Gregorius Itelson (Freudenthal and Karachentsev 2011). At the Second International Congress for Philosophy in Geneva in 1904, Itelson had given two talks: "The Reform of Logic" and "Logic and Mathematics." Though never published, the talks became well known through a congress report by Couturat (1904). For Itelson, "psychologistic logic" is "absolutely sterile," and logic is not about thoughts, but about the objects of thought: "Logic is the *science of objects in general*;" indeed, "Logic is the science of *all* objects, real or not, possible or impossible, in abstraction of their existence" (Couturat 1904: 1038, orig. emphasis). Itelson also emphasized a strong connection between logic and mathematics: "mathematics is a purely logical science," it is a science of ordered objects (sets, groups). This conception is close to Russell's logical realism (1904), and Meinong noticed its similarity with his own theory of objects

(1907/1973: 211–12). Back in Vienna, Neurath mentioned Itelson's conception of logic in a paper on Schröder's logic, written together with Olga Hahn (Hahn and Neurath 1909). According to notes taken by Höfler, Neurath also defended Itelson's view in a discussion on the a priori he led at the "Philosophical Society" and in discussions with Pichler (Pichler 1909: 22). Hans Hahn's sympathies for Russell inclined him to a similar view.

Later, Hahn rejected the logical realism shared by the earlier Russell, Meinong, and Itelson (see his 1929: §II) and adopted the view that logic does not say anything about the world and that the rules of logic are conventionally fixed grammatical rules of a language. This was, of course, the result of the intense discussion of the *Tractatus* in the Vienna Circle. Yet, already early on, Hahn rejected any kind of intuition (or Kantian *Anschauung*) as a justification for mathematical axioms. In geometry, this led him to accept the kind of conventionalism defended by Poincaré (Hahn 1908), but he also supported the movement for the arithmetization of analysis and the attempt to give a rigorous and axiomatic basis to arithmetic, free from any intuitions or empiricist justification (Hahn 1909). Hahn also supported Peano's rigorous axiomatic foundation of arithmetic. Later, Hahn said explicitly that a *logical* foundation of arithmetic was highly desirable (Hahn 1919). Given his interest in the logicist program of Russell (1903) and Couturat (1905), he was probably attracted to such a view much earlier.

As noted, the nature of scientific theories was a second major topic of discussion. Before 1918, Neurath and Frank published only a few papers on philosophy (and Hahn none). Their main focus was the role of theories and hypotheses in science. Recent research focused on the influence of French conventionalism on Neurath, Frank, and Hahn and their Machian background (see Brenner 2002; Neuber 2010). The philosophical papers by Frank and Neurath allow us to specify their conventionalism further. Their central problem was the status of scientific hypotheses and of theoretical terms (such as "atom," "matter," etc.). Frank (1917) shows his grounding in Mach's instrumentalism, the view that theoretical terms should not be interpreted in a realistic way and that they are only "auxiliary concepts" and instruments to order most economically our sensory impressions. But Frank wanted to give a greater autonomy to a fully mathematized theory: it was to be freely chosen and then coordinated with empirical observations.

While Hahn's acceptance of conventionalism was limited to geometry, Frank also defended the view that basic empirical principles of science are often disguised conventions, in particular the principle of causation, which he regarded as a conventional stipulation (1907). He was also sympathetic to Poincaré's structuralism, which he used in his criticism of the realistic interpretation of theoretical concepts. A scientific theory is essentially about the mathematical relations between the phenomena, and not about entities designated by theoretical terms. Like Poincaré, he held that in the history of a science the theoretical terms are replaced by others, while the mathematical structure expressed by the theory is mostly preserved (Frank 1917).

Neurath was particularly impressed by the fact, emphasized by Duhem, that hypotheses are underdetermined by observation and that alternative hypotheses are always possible. For him, the acceptance of a theoretical hypothesis requires always a decision under uncertainty. In a paper on the psychology of decisions (1913), he denied that there is a fundamental difference between practical decisions (action) and theoretical decisions (the acceptance of a hypothesis). Soon after, he developed an account of scientific theories which made it readily intelligible that there exist always alternative systems of hypotheses. The focus of his two papers on the history of optical theories (1915, 1916) lies on singling out a set of "elementary representations" or "elementary propositions" representing elementary properties of the system under analysis and showing how the possibility of alternative organization undermines crude dichotomies as between the undulatory theory and the emission theory, allowing a more fine-grained analysis and more precise and gradual distinctions. Neurath's analysis is clearly influenced on the one hand by Mach, although Neurath does not require the basic representations to be about observables, i.e., empirically given elements; on the other hand, it is influenced by the rigorous decompositional analysis defended by the Meinong school (see Meinong 1894).

Frank's claim that he, Neurath, and Hahn were strongly interested in the theories of the conventionalists and tried to integrate their ideas into the Viennese setting represented by Mach and Boltzmann thus finds clear support; yet, it may be noted that a coherent and unified philosophy cannot be constructed out of the limited set of papers they wrote at that time, or even their reviews (listed in Uebel 2003: 87–88).

The Erlangen Conference

The Erlangen Conference—which attracted less attention from philosophers (but see Thiel 1993; Carus 2007: 156–60, and various mentions in Damböck 2016)—was a meeting organized by Rudolf Carnap and Hans Reichenbach in March 1923 in Erlangen, Germany. The purpose of the conference was to assemble a group of philosophers and scientists aiming to put philosophy, especially epistemology and philosophy of science, on a more scientific basis. In a 1940 talk at Harvard, Carnap called the Erlangen Conference the "first nucleus" of the Vienna Circle (1940), and in his autobiography, he stated: "The Erlangen Conference may be regarded as the small but significant initial step in the movement of a scientific philosophy in Germany" (1963: 13). For Reichenbach and Schlick, the conference was a meeting of the "exact tendency in philosophy" (Reichenbach 1923; Schlick 1923a). Through the planning for the conference, three leading members of logical empiricism, Carnap, Reichenbach, and Schlick came into close contact for the first time and became aware of their common goal of an exact or scientific philosophy. The topic of the conference was the application of the theory of relations of *Principia Mathematica* to epistemology and philosophy of science. One topic discussed at length was the way relations and structures could be used to understand the construction of physical objects out of the data of sensory experience. The conference also afforded Carnap the opportunity to present a sketch of his *Aufbau* project and discuss the challenges of the logical construction of all scientific concepts from an empirical basis.

The idea for the conference was initially developed by Carnap and the philosopher Karl Gerhards during a meeting in October 1922. Carnap had just written a paper, "From Chaos to Reality" (1922), a brief outline of the project that later became the *Aufbau*. Several aspects of the constructive procedure in that paper are already strikingly similar to the later book. Gerhards had similar interests, having written his doctoral thesis on Mach's epistemology under the supervision of the critical realist Oswald Külpe and a habilitation on "The Sensory Basis of Physical Knowledge." Just before the meeting with Carnap, he had published a paper in which he tried to develop a mathematically precise model of the way we construct representations of three-dimensional objects from two-dimensional visual inputs (Gerhards 1922; see also Richardson 2016). Both Carnap and Gerhards wanted to answer an epistemological question in a mathematically precise way: how do we get from sensory experience to representations of physical objects and of the other entities postulated by the sciences? Both were completely indifferent to "metaphysical" questions: whether reality is fundamentally constituted by external objects or by sensory experience was irrelevant for their constructive projects. Shortly after his meeting with Gerhards, Carnap formulated the program of the conference. Its aim, Carnap said, was to use Russell's theory of relations "for a structural theory of the objects of knowledge or a theory of the orders of reality." The first part of the conference was dedicated to the formal theory of relations and structures, the second part to its application to epistemology. The themes were the "construction of reality" and a "structural theory of knowledge," more precisely the way reality is constructed out of "the given" (see Limbeck-Lilienau 2015: 84–85).

It is unclear how Reichenbach got involved into the planning of the conference, but he soon became a driving force behind it. He contacted Schlick and asked him to participate. Although Schlick strongly supported the conference, he could not leave Vienna to attend. Reichenbach also invited two psychologists, the Gestalt psychologist Kurt Lewin, and Fritz Heider. Heider was a student of Meinong and worked in Berlin in close contact with the Gestalt psychologists. The presence of the two psychologists was important for questions about the nature of sensory experience, as experience was supposed to form the basis of Carnap's constructional system. For the logical aspects of Russell's theory of relations and the theory of structures, Carnap invited the young mathematician from Göttingen, Heinrich Behmann. Behmann wrote a dissertation on *Principia Mathematica*, supervised by Hilbert, and was one of the driving forces in the introduction of Russell's ideas in Göttingen (Mancosu and Zach 2015). In 1921, Behmann wrote a habilitation on the *Entscheidungsproblem* (the decision problem) which he sent to Carnap. Another participant from Göttingen was the physicist Paul Hertz, also a close collaborator of Hilbert and co-editor with Schlick of Helmholtz's epistemological writings (1921). Besides the scientists from Göttingen and the psychologists from Berlin, there were only three philosophers at the conference, after Schlick had to cancel: Carnap, Reichenbach, and Bernhard Mertens, a friend of Carnap. Neo-Kantians, then still the dominant tendency in German philosophy, were not invited.

The conference took place from 6 to 13 March 1923 at the Philosophical Academy, a privately funded and internationally well-connected institution which had among its members philosophers such as Russell, Cassirer, Dewey, and Heidegger, though none of them were involved in the conference organized by Carnap. In the first part of the conference, Carnap and Behmann gave several talks on relations and structures, and Mertens gave one. The titles of the talks can be gleaned from Carnap's diary, here in their chronological order: Behmann on "Bernays's Axiomatics of Logic," Carnap on the "Theory of Relations," "The Importance of Structures," "The Logistic Concept of Structure," and "The General Significance of Structures," Behmann "On Symbolism" and on the "*Entscheidungsproblem*," Merten on "A Combinatorial Treatment of Structures," Carnap on "Some Remarks on Singular Characterizations" and "Characterization of Structures," Behmann on "The Ontological Proof of the Existence of God."

Three abstracts of Carnap's talks are preserved (1923). In the first, Carnap described how Russell's theory of relations can be used for a theory of structures, with the basic concepts of "constellation," "isomorphism," and "structure." A structure is defined as a class of isomorphic constellations and constellations as instances of a certain structure. Carnap also emphasized an "applied" theory, which specifies how structures can be coordinated with concrete objects, thus rendering the theory of structures useful for epistemology and philosophy of science. Carnap defended a structural view of knowledge. Only structural properties of objects can be known and communicated: "Every science is a science only insofar as it is a theory of structures." In the second talk abstract, together with Mertens, Carnap developed a theory specifying how to give a numeric characterization of structures, and the third abstract outlines the *Aufbau* project.

The second part of the conference was dedicated to talks on physics, epistemology, and psychology, with Carnap's epistemological project on center stage. Again, the titles of the talks can be gleaned from Carnap's diary: Carnap on the "Construction of Reality," Heider on "Color Perception," Lewin on "Comparative Theory of Science," Behmann on "1 + 1 = 2," Reichenbach on "Topology of Time," Hertz on "Analytic Judgments," Behmann on "Implicit Definitions." In his talk, Carnap explained his project of constructing the concepts of all objects of knowledge from a very limited number of basic concepts and relations. In the unpublished longer version of his autobiography, Carnap reported that his talk triggered an intense discussion:

> There was a heated debate on the question whether a momentary experience could contain sense-data as actual parts or not. Hertz declared actual components indispensable, while Lewin rejected them emphatically from the point of view of gestalt psychology. Reichenbach tried to reconcile the two sides by the conception that the controversy was largely a question of terminology. I tried to show that a certain method of logical analysis, which I called "quasi-analysis" did justice to the justified demands of both sides by preserving on the one hand the experiences as indivisible units and on the other hand, constructing certain complexes of experiences that correspond to the traditional components.
>
> (1956)

Through his method of "quasi-analysis" (1928: §§71–73), Carnap later sought to reconcile the opposed conceptions by constructing the "elements" out of relations between experiences which lacked such components.

The Erlangen Conference was an essential step in the genesis of the *Aufbau*, but it was also the first attempt to connect philosophers who had "a common basic attitude and the common aim of developing a sound and exact method in philosophy" (Carnap 1963: 13). Some of the talks were published later (Heider 1927; Lewin 1927), and some grew into book projects (Reichenbach 1924; Behmann 1927; Carnap 1928, 1929). Schlick himself followed the conference closely through his correspondence with Reichenbach and Carnap. After the conference and with the support of Schlick and Carnap, Reichenbach tried to found a "Journal of Exact Philosophy," a plan which eventually failed. In March 1923, Schlick wrote to Reichenbach that the closing words of Russell's *Our Knowledge of the External World* (1914) could have served as the motto of the Erlangen Conference:

> The one and only condition, I believe, which is necessary in order to secure for philosophy in the near future an achievement surpassing all that has hitherto been accomplished by philosophers, is the creation of a school of men with scientific training and philosophical interests, unhampered by the traditions of the past, and not misled by the literary methods of those who copy the ancients in all except their merits.

Schlick would become the essential link between the group of friends of pre-World War I Vienna and the group formed in Erlangen. In the summer of 1923, Schlick wrote to Russell about the Erlangen meeting, adding (1923b):

> You recommend a working union of those who have a gift for philosophical and scientific thinking as a sure means of promoting a rapid progress in philosophical research. Now there is a group of talented and enthusiastic men who have the same belief and who are willing and capable to work on a plan suggested by your words.

Shortly afterwards, Schlick formed such a "working union" of scientifically oriented philosophers in Vienna. The Erlangen Conference was the first step in the formation of such a group, perhaps more than the group known as the first Vienna Circle. Nonetheless, Neurath, Hahn, and Frank would form an essential part of the new group, while Reichenbach was to form a similar group in Berlin with Lewin as an important member.

References

ASP = Archive of Scientific Philosophy, Hillman Library, University of Pittsburgh.
VCA = Vienna Circle Archive, Rijksarchif Noord-Holland, Haarlem.
Blackmore, J. (1995) *Ludwig Boltzmann: His Later Life and Philosophy*, Dordrecht: Springer.
Behmann, H. (1927) *Mathematik und Logik*, Leipzig: Teubner.
Brenner, A. (2002) "The French Connection: Conventionalism and the Vienna Circle," in M. Heidelberger and F. Stadler (eds.), *History of Philosophy of Science. New Trends and Perspectives*, Dordrecht: Springer, pp. 277–86.
Carnap, R. (1922) "Vom Chaos zur Wirklichkeit," RC 081-05-01, Carnap Papers, ASP.
——— (1923) [Three Position Papers], RC 091-17-12, Carnap Papers, ASP.
——— (1928) *Der logische Aufbau der Welt*, Berlin: Weltkreis-Verlag. Trans. *The Logical Structure of the World*, Berkeley and Los Angeles: University of California Press, 1967.
——— (1929) *Abriß der Logistik, mit besonderer Berücksichtigung der Relationentheorie und ihrer Anwendungen*, Vienna: Springer.
——— (1940) "Über Wiener Kreis," [Talk at Harvard, Nov. 1, 1940], RC 110-08-03, Carnap Papers, ASP.
——— (1956) [Draft of Autobiography], Box 2, CM 3, Carnap Collection, Young Research Library, University of California, Los Angeles.
——— (1963) "Intellectual Autobiography," in P. A. Schilpp (ed.), *The Philosophy of Rudolf Carnap*, Library of the Living Philosophers, La Salle: Open Court, pp. 3–84.
Carus, A. (2007) *Carnap and Twentieth-Century Thought: Explication as Enlightenment*, Cambridge: Cambridge University Press.
Couturat, L. (1904) "Logique et Philosophie des Sciences," *Revue de Métaphysique et de Morale* 12: 1037–77.
——— (1905) *Les Principes des Mathématiques*, Paris: Félix Alkan. Trans. *Die philosophischen Prinzipien der Mathematik*, Leipzig: Werner Kleinkhardt, 1908.
Damböck, C. (ed.) (2016) *Influences on the Aufbau*, Dordrecht: Springer.
Fisette, D. (2014) "Austrian Philosophy and Its Institutions: Remarks on the Philosophical Society of the University of Vienna (1888–1938)," in A. Reboul (ed.), *Mind, Values and Metaphysics*, Dordrecht: Springer, vol. 1, pp. 349–74.
Frank, P. (1907) "Kausalgesetz und Erfahrung," *Annalen der Naturphilosophie* 6: 445–50. Trans. "The Law of Causality and Experience," in Frank (1949), pp. 50–60.
——— (1917) "Die Bedeutung der physikalischen Erkenntnistheorie Machs für das Geistesleben der Gegenwart," *Naturwissenschaften* 5: 65–72. Trans. "The Importance for Our Times of Ernst Mach's Philosophy of Science," in Frank (1949), pp. 61–79.
——— (1941) *Between Physics and Philosophy*, Cambridge, MA: Harvard University Press.
——— (1949) *Modern Science and Its Philosophy*, Cambridge, MA: Harvard University Press.
Freudenthal, G. and Karachentsev, T. (2011) "G. Itelson—A Socratic Philosopher," in O. Pombo, J. Symons and J. M. Torres (eds.), *The Unity of Science*, Dordrecht: Springer, pp. 109–26.
Gerhards, K. (1922) "Der mathematische Kern der Aussenwelthypothese," *Die Naturwissenschaften* 17: 423–30 and 18: 446–53.
Hahn, H. (1908) "[Review of H. Poincaré, *The Value of Science*]," *Monatshefte für Mathematik und Physik* 18: 33–34.
——— (1909) "[Review of L. Tesar *Elemente der Differential- und Integralrechnung*]," *Monatshefte für Mathematik und Physik* 20: 12–13.
——— (1919) "[Review of A. Pringsheim, *Vorlesungen über Zahlen- und Funktionslehre*]," *Göttingische gelehrte Anzeigen* 9–10: 321–38. Trans. in Hahn 1980, pp. 9–19.
——— (1929) "Empirismus, Mathematik, Logik," *Forschungen und Fortschritte* 5. Trans. "Empiricism, Mathematics, and Logic," in Hahn (1980), pp. 39–42.
——— (1980) *Empiricism, Logic, Mathematics* (ed. by B. McGuinness), Dordrecht: Reidel.
Hahn, O. and Neurath, O. (1909) "Zum Dualismus der Logik," *Archiv für Philosophie, Neue Folge* 15: 149–62. Trans. "On Duality in Logic," in J. Cat and A. Tuboly (eds.), *Neurath Reconsidered: New Sources and Perspectives*, Springer: Cham, 2021, pp. 493–505.
Haller, R. (1982a) "New Light on the Vienna Circle," *The Monist* 65: 25–37.
——— (1982b) "Das Neurath-Prinzip," in F. Stadler (ed.), *Arbeiterbildung in der Zwischenkriegszeit. Ausstellungskatalog mit Forschungsteil*, Wien: Löcker, pp. 79–87. Trans. "The Neurath Principle: Its Grounds

and Consequences," in T. Uebel (ed.), *Rediscovering the Forgotten Vienna Circle*, Dordrecht: Reidel, 1991, pp. 117–29.

——— (1985) "Der erste Wiener Kreis," *Erkenntnis* 22: 341–58. "The First Vienna Circle," Repr. in T. Uebel (ed.), *Rediscovering the Forgotten Vienna Circle*, Dordrecht: Reidel, 1991, pp. 95–107.

Heider, F. (1927) "Ding und Medium," *Symposion* 1: 109–57.

Hertz, P. and Schlick, M. (eds.) (1921) *Hermann von Helmholtz, Schriften zur Erkenntnistheorie*, Berlin: Springer.

Lewin, K. (1927) "Idee und Aufgabe der vergleichenden Wissenschaftslehre," *Symposion* 1: 61–93.

Limbeck-Lilienau, C. (2015) "Der Wiener Kreis—eine Illustrierte Geschichte des Logischen Empirismus," in C. Limbeck-Lilienau and F. Stadler (eds.), *Der Wiener Kreis. Texte und Bilder zum Logischen Empirismus*, Vienna: LIT Verlag, pp. 31–411.

——— (2018) "The First Vienna Circle: Myth or Reality?" *Hungarian Philosophical Review* 62: 50–65.

Mancosu, P. and Zach, R. (2015) "Heinrich Behmann's 1921 Lecture on the Decision Problem and the Algebra of Logic," *Bulletin of Symbolic Logic* 21 (2): 164–87.

Meinong, A. (1894) "Beiträge zur Theorie der psychischen Analyse," *Zeitschrift für Psychologie und Physiologie der Sinnesorgane* 6: 417–55. Repr. in Meinong (1968–78), vol. 1, pp. 305–95.

——— (1907) *Über die Stellung der Gegenstandstheorie im System der Wissenschaften*, Leipzig: Voigtländer. Repr. in Meinong (1968–78), vol. 5, pp. 197–365.

——— (1968–78) *Gesamtausgabe* (ed. by R. Haller et al.), Graz: Akademische Druck- u. Verlagsanstalt.

Neuber, M. (2010) "Philosophie der modernen Physik—Philipp Frank und Abel Rey," *Grazer Philosophische Studien* 80: 131–49.

Neurath, O. (1907) Letter to F. Tönnies, undated. Sig. Cb 54.56, Neurath 1907–8/Nr. 3. Tönnies Nachlass, Landesbibliothek Schleswig-Holstein, Kiel.

——— (1909) "Philosophisch-soziologische Bücherei," *Der Kunstwart* 23 (1): 138–41.

——— (1913) "Die Verirrten des Cartesius und das Auxiliarmotiv: Zur Psychologie des Entschlußes," *Jahrbuch der philosophischen Gesellschaft an der Universität Wien*. Trans. "The Lost Wanderers and the Auxiliary Motive (On the Psychology of Decision)," in Neurath 1983, pp. 1–12.

——— (1915) "Prinzipielles zur Geschichte der Optik," *Archiv für die Geschichte der Naturwissenschaft und der Technik* 5: 371–89. Trans. "On the Foundations of the History of Optics," in Neurath, *Empiricism and Sociology* (ed. by R. S. Cohen and M. Neurath), Dordrecht: Reidel, 1973, pp. 101–12.

——— (1916) "Zur Klassifikation von Hypothesensystemen," *Jahrbuch der philosophischen Gesellschaft an der Universität Wien*. Trans. "On the Classification of Systems of Hypotheses," in Neurath (1983), pp. 13–31.

——— (1983) *Philosophical Papers 1913–1946* (ed. by R. S. Cohen and M. Neurath), Dordrecht: Reidel.

Pichler, H. (1909) *Über die Erkennbarkeit der Gegenstände*, Vienna: Braumüller.

Reichenbach, H. (1923) Letter to Schlick, Feb. 3, 1923, Sig. 105/Reich-8, Schlick Nachlass, VCA.

——— (1924) *Axiomatik der relativistischen Raum-Zeit-Lehre*, Braunschweig: Vieweg. Trans. *Axiomatization of the Theory of Relativity*, Berkeley: University of California Press, 1969.

Reininger, R. (ed.) (1938) *50 Jahre Philosophische Gesellschaft an der Universität Wien 1888–1938*, Vienna: Verlag der Philosophischen Gesellschaft an der Universität Wien.

Richardson, A. (2016) "External World Problems: The Logical Construction of the World and the 'Mathematical Core of the External World Hypothesis'," in Damböck (2016), pp. 1–14.

Russell, B. (1903) *The Principles of Mathematics*, Cambridge: Cambridge University Press.

——— (1904) "Meinong's Theory of Complexes and Assumptions I," *Mind* 13: 204–19.

——— (1914) *Our Knowledge of the External World*, Chicago/London: Open Court.

Schlick, M. (1923a) Letter to Reichenbach, March 26, 1923, Sig. R 015-63, Reichenbach Nachlass, ASP.

——— (1923b) Letter to B. Russell, undated (Summer 1923), Sig. 114-Ru-6. Schlick Nachlass, VCA.

Stadler, F. (1982) *Vom Positivismus zur 'Wissenschaftlichen Weltauffassung'. Am Beispiel der Wirkungsgeschichte von Ernst Mach in Österreich von 1895 bis 1934*, Vienna: Löcker.

——— (1997) *Studien zum Wiener Kreis. Ursprung, Entwicklung und Wirkung des Logischen Empirismus im Kontext*, Frankfurt a. M.: Suhrkamp. Trans. *The Vienna Circle. Studies in the Origins, Development, and Influence of Logical Empiricism*, Vienna: Springer, 2001.

Thiel, C. (1993) "Carnap und die wissenschaftliche Philosophie auf der Erlanger Tagung 1923," in R. Haller and F. Stadler (eds.), *Wien—Berlin—Prag. Der Aufstieg der wissenschaftlichen Philosophie*, Vienna: Hölder-Pichler-Tempsky, pp. 175–88.

Uebel, T. (2000) *Vernunftkritik und Wissenschaft. Otto Neurath und der erste Wiener Kreis*, Vienna: Springer.

——— (2003) "On the Austrian Roots of Logical Empiricism: The First Vienna Circle," in P. Parrini, W. Salmon and M. Salmon (eds.), *Logical Empiricism. Historical and Contemporary Perspectives*, Pittsburgh: University of Pittsburgh Press, pp. 67–93.

——— (2022) "The First Vienna Circle: What Kind of Formation Was It—and Why Does It Matter?" in D. Romizi, M. Wulz and E. Nemeth (eds.), *Edgar Zilsel: Philosopher, Historian, Sociologist*, Cham: Springer.

11
THE VIENNA CIRCLE AND THE ERNST MACH SOCIETY

Friedrich Stadler

The history of the Vienna Circle has been researched very extensively during the last few decades. In light of the results gained, it is clear that the once-standard picture of it as a monolithic block of radical positivists aiming at making the world safe for science by either finding a foundation of knowledge claims in the indubitable experiential given or else rejecting them as meaningless pseudo-propositions can no longer be upheld. In its place we now possess a great number of studies that underline a multidimensional pluralism characterizing the membership and the work of this group throughout its relatively short existence. This chapter aims to illuminate the Circle's "public outreach" activities in connection with the Ernst Mach Society (Verein Ernst Mach), which are still not as well understood as they should be.

The Vienna Circle

In terms of both personnel and research interests, we must distinguish what we now know as the Vienna Circle as going through at least three phases. The pre-history with the so called "first Vienna Circle," with Philipp Frank, Hans Hahn, and Otto Neurath before World War I, will not be addressed explicitly here (see CH. 10). Then there is first the *pre-public phase* which extended from the formation of Moritz Schlick's by-invitation-only discussion circle in 1924 (not long after taking up his professorial appointment in Vienna) until 1929. The early core group consisted of the philosopher Schlick, the mathematician Hahn, the economist Neurath, the physicist Frank (based in Prague as successor of Einstein), and, from 1926, the philosopher Rudolf Carnap, as well as advanced students of Schlick and Hahn. During this pre-public period, Wittgenstein's *Tractatus* (1922) was extensively studied in their meetings for several years and decisively influenced their philosophy by its novel conception of logic as tautologous or analytic. They applied this conception in Russell's programmatic fashion to all of mathematics, thereby freeing empiricism from an impossible task and allowing themselves to reject Kant's synthetic a priori, recognizing only empirically testable or formally provable knowledge. This period as yet saw relatively few philosophical publications by its members, though Schlick published a revised second edition of his *General Theory of Knowledge* in 1925 (first ed. 1918) and Carnap published his first major work, *The Logical Structure of the World*, in 1928.

The start of the second, the *public phase*, in 1929 was marked by what since came to be known as the unofficial "manifesto" of the group, *Wissenschaftliche Weltauffassung. Der Wiener*

 DOI: 10.4324/9781315650647-13

Kreis ("The Scientific World-Conception. The Vienna Circle"), signed by Carnap, Neurath, and Hahn and published under the aegis of the Verein Ernst Mach (Ernst Mach Society). Celebrating Schlick's decision to remain in Vienna despite an attractive call to return to Germany, it outlined the philosophical position and aims of the group around him and also gave it the name by which it is known to posterity—in Vienna it was known as the "Schlick Circle." From the presentation of this manifesto in Prague in September 1929 at the First Conference on the Philosophy of the Exact Sciences onward, the second phase witnessed an intensive publication activity on the part of most members, aided by Carnap's co-editorship (with Hans Reichenbach) of the philosophical journal *Erkenntnis* and the editorship of two series of monographs, *Schriften zur wissenschaftlichen Weltauffassung* (*Writings on the Scientific World-Conception*) by Schlick and Frank and *Einheitswissenschaft* (*Unified Science*) by Neurath, as well as one further Conference on the Philosophy of the Exact Sciences in Königsberg in 1930 and the organization of the six International Congresses for the Unity of Science starting in Paris in 1935. During this period the Circle attracted a number of visitors from Scandinavia, Italy, Poland, Great Britain, the United States, and even China and, thanks to its radical critique of metaphysics as cognitively meaningless, also attained a level of public visibility and notoriety uncommon for academic working groups. By then the philosophical positions elaborated by its members began to differentiate themselves and two camps developed, broadly speaking, separated by their attitude towards positions they saw taken in the *Tractatus* (beyond those on the nature of logic), so much so that by the mid-1930s visitors commented on it in their reports on and/or critiques of the new philosophy.

This public phase ended abruptly in 1936 with the murder of Schlick, which robbed the group of its official head. However, by then the Viennese group had already begun to shrink due to the death of Hahn in 1934 and the emigrations of Schlick's student Herbert Feigl in 1931, of Neurath in 1934, and Carnap in 1935. Now the Circle entered its third phase, the *phase of internationalization*, with its members dispersed across the Western world where they continued their cooperation with former visitors and broadly like-minded philosophical groups like (neo-)pragmatists which further broadened the range of philosophies pursued. The International Congresses for the Unity of Science continued until 1941 and an ambitious multilayered publication project was started, the *International Encyclopedia of Unified Science*, whose run of two volumes of ten monographs each was concluded only in 1970—but not before publishing, as part of it, Thomas Kuhn's *The Structure of Scientific Revolutions*, notably to great acclaim from Carnap as one of its first readers in 1962 (for a comprehensive in-depth survey of the Circle, including discussion of some pre-history and long-term consequences, see Stadler 1997/2015).

Here, I wish to draw attention to a fact about the Vienna Circle that philosophers tend to forget perhaps a bit too easily, rightly concerned as they are to explore the subtleties and implications of the doctrines of its members that may not always have been presented in sufficient detail or, alternatively, that have become of new interest in light of much more recent philosophical developments. While they were indeed a group of mostly philosophically interested scientists or scientifically trained philosophers who sought to bring the theory of knowledge up to date—make it fit for purpose again after the conceptual upheavals since the previous turn of the century: the foundational debates in mathematics, the methodological debates in the burgeoning social sciences and the ongoing conceptual revolutions in physical science—this was not all they wished to achieve. Rather, they thought it imperative that the fundamentals of their own new conception of what knowledge is, how it is gained and how it can be extended, should be as widely disseminated as possible. It was their express intention to replace the false idols of popular imagination and authoritarian indoctrination (insight by genius or tradition upheld for its own sake), and for their replacement, the sober "scientific world-conception," to become a common good of societies deciding their own laws and institutions.

With this ambition, the members of the Vienna Circle did not stand alone. Rather they formed part of a wider trend, operative in Vienna since the final years of the nineteenth century, of a culture of independent groups and societies aiming at the promotion of particular goals to foster improvement of the physical and cultural conditions of different parts of the population or even the population as a whole. This social movement can be summarily captured by the term "late Enlightenment" (*Spätaufklärung*). It is all too easy perhaps to think of the ambition of the members of the Vienna Circle to make a difference to the world around them as, however laudable, ultimately extrinsic to their philosophy. However, the ambition to make a difference was intrinsic to the position they themselves self-consciously adopted in the formation of their own public entity, the Ernst Mach Society. This can be clearly seen at its foundation in 1928: there, the members of the Schlick's discussion circle—not yet called by its later name—responded to initiatives by existing representatives of this late Enlightenment movement and cooperated extensively (for by-now classical treatments of these Viennese movements, see Fuchs 1949; Belke 1978).

The Ernst Mach Society

Apart from a recent overview of the history and impact of the Mach Society and a comparison with the Berlin Society for Empirical (later: Scientific) Philosophy (Sandner 2019), and older work in German (Stadler 1982, 1992), the embeddedness of the Schlick Circle in the culture of Vienna and their mutual significance for each other have not received sufficient elaboration.

In his letter of protest against the government-decreed dissolution of the Mach Society in 1934 (see Stadler 1982: 196–7), Schlick, who had served as the chair of the Mach Society from beginning to end, recalled what had been the motives for its inception. Prominently among those (and typically focused upon by philosophers) was the desire that extra-mural public philosophical events—like, up to then, the open meetings of the Philosophical Society at the University of Vienna which had been founded in 1888 by students of Franz Brentano (see Reininger 1938)—should exhibit a greater focus on "scientific philosophy." According to Schlick, the plan of establishing a specific subunit for empiricist philosophy within the Philosophical Society failed. Founding the Mach Society may thus appear simply as the creation of a parallel society of their own design. But pure philosophy was not the sole mover in the developments towards the founding of the Mach Society. The contrast between the diminishing participation by Vienna Circle members in the activities of the Philosophical Society and the contemporaneous increase of their participation in various institutions of Viennese adult education and in movements for social reform outside of the university tells a different story. So does that, at the same time, certain scholars (among them Hahn, who headed the Viennese chapter of the Association of Socialist University Teachers) came under pressure due to the right-wing political developments at the university, including a marked increase of anti-Semitism, promoted by professors and students operating a successful academic policy network (Rathkolb 2013; Taschwer 2015). The growth of these proto-fascist and even national-socialist groupings since the beginning of the First Republic provides a non-negligible part of the socio-political background against which the manifesto of 1929 was written and which had prompted the foundation of the Mach Society in 1928.

On the other side, as it were, was the late Enlightenment movement. Its representatives were located politically on a spectrum ranging from liberalism to socialism. They were very loosely united after 1919 in the so-called Free Association of Cultural Societies (Freier Bund kultureller

Vereine) which included mainly the Ethical Movement (Ethische Bewegung), Monists (Österreichischer Monistenbund), Freethinkers (Freidenkervereinigung), and the association dedicated to promoting the reform ideas of Josef Popper-Lynkeus with a name taken from one of his programmatic slogans: Verein Allgemeine Nährpflicht. In addition, there was cooperation and interaction with groups which variously agitated for pacifism and different types of social reform as well as with the women's movement. Another player of an originally more academic but, later on, broader social orientation was the Vienna Sociological Society, founded in 1907 by Wilhelm Jerusalem, Rudolf Goldscheid, and others. Last but not least, there was Neurath's Social and Economic Museum in Vienna (Gesellschafts und Wirtschaftsmuseum in Wien) which he established in 1925 and led until 1934. The plethora of these groups and institutions provided the framework and occasions for the development of progressive agendas in the face of the growing threat from the political right. An integral part of this was that they also provided space for interdisciplinary discussion groups which were of interest to the members of the Schlick Circle. It is this network of late Enlightenment groups that the Circle joined through the foundation of the Mach Society.

This context was alluded to in the manifesto, but it has rarely been spelled out. To begin with, we must remember that Carnap, Hahn, and Neurath, who are commonly listed as its authors, signed the dedication to Schlick only "for the Ernst Mach Society," but that the manifesto itself (the "brochure" as they called it) was "edited by the Ernst Mach Society" (as the cover put it). Little attention has been paid to this because this Society was mainly conceived of as a mere external appendage of the academic Schlick Circle that regularly met in the Boltzmanngasse 5. Yet it is important to appreciate its separate identity. Prominent as they doubtlessly were in the life of the Mach Society, members of the Schlick Circle were but one group among others in cooperation. A handbill, titled "To All Friends of the Scientific World-Conception!" and distributed publicly around the time of its founding and reprinted at the end of the manifesto brochure, stated the aim of the cooperative venture very clearly.

> The Ernst Mach Society . . . will organize lectures and publications about the present position of the scientific worldview in order to demonstrate the significance of exact research for the social and the natural science. *In this way, the intellectual tools of modern empiricism are to be developed, tools that are needed also in shaping public and private life.* In this aim the Ernst Mach Society feels itself united with many of the lading minds of our age who live and work far away from each other, but also with broader circles who have confidence in the scientific world-conception.
>
> (Verein Ernst Mach 1929/2012: 112, orig. emphasis)

Like the manifesto, which enlarged on the themes here emphasized, the Society's handbill did not merely appeal for cooperation among the groups of Vienna's late Enlightenment, but it already represented such a cooperative venture.

Several institutional developments in the mid-1920s were strong motives for the foundation of a new joint public forum, and they brought the Vienna Circle into cooperation with groups outside academia which stood in the tradition of enlightenment broadly conceived. Recent research has emphasized the increasing nationalistic radicalization inside the University of Vienna, especially in the humanities. A substantial part of the professorship showed open sympathies for German nationalism and saw the university as an instrument in defense of "German culture," often with strong anti-Semitic connotations (Stadler 1997/2015: ch. 9). From early on, Hahn had publicly protested against these tendencies (e.g., Hahn 1924). Nevertheless, they gradually assumed dominance in two academic organizations which initially were central for Schlick and his group: the Vienna International University Lectures, and the Philosophical Society at the University of Vienna. The former was an international forum of the university with the aim of inviting foreign scholars for lectures. Initially, Schlick was strongly involved in this forum and, for example, tried several times to win Bertrand Russell for such lectures. After 1926, this forum was taken over by an increasingly dominant right-wing and nationalistic fraction of the professorship. The Philosophical Society had a similar fate (see Fisette 2014). The Society was the main public forum for philosophical lectures in Vienna since the turn of the century, and it placed a strong emphasis on the discussions that followed them. Starting in the 1910s, many of the later members of the Vienna Circle (Neurath, Hahn, Frank, Zilsel, Kraft) had given lectures there, and still in the 1920s, members of Schlick's circle regularly did so. In 1927, when the Society was taken over by the neo-Kantian Robert Reininger and turned into a local branch of the German Kant Society, discussions were restricted and the lectures increasingly gained a nationalistic touch (e.g., many lectures on German romanticism). From this point on, many members of the Vienna Circle avoided lecturing in the Society. In light of these developments, the Vienna Circle stood in need of a new public outlet. In alliance with the aforementioned social and cultural groups, the Mach Society was founded in November 1928 in a municipal building of Red Vienna, opening its initial meeting with a lecture by Neurath on "Ernst Mach and the Exact World-Conception." But the groups with whom the Circle joined forces now also stood in need of this new alliance. Let's consider some of them further.

The Schlick Circle in Vienna's late Enlightenment

The Austrian Monist Society was founded already before World War I and propagated an evolutionary and anti-metaphysical perspective on science and a non-dualist worldview in agreement with a future socialist society. It differed from the German sister society around Ernst Haeckel and Wilhelm Ostwald with its politics, namely its pacifist position and democratic attitude. In Austria-Hungary the monists found a sympathetic leader in Friedrich Jodl (1849–1914), professor of philosophy at the University of Vienna until his early death in 1914. Jodl also was an adherent of an anti-metaphysical philosophy and held leading offices in the Ethical Movement (see Belke 2021). Schlick too participated in the activities of the monists and Otto Neurath and Herbert Feigl also lectured there. Schlick had already given a lecture, "On the Meaning of Life," to the German Monist Society, where his *General Theory of Knowledge* was seen as paving the way for a monistic worldview, even if by way of a critique of crude "positivism." Schlick himself provided support for this interpretation, having expressed his preference of monism over dualism as there being "only one kind of reality" (1918/1985: 325–7). (A somewhat similar trend towards monism can be seen in Bertrand Russell, who at approximately the same time adopted a neutral monism inspired by Mach.) Schlick's favorite disciple Herbert Feigl also lectured in the Monist Society in 1930 on "Natural Law and Freedom of the Will" and later on adopted a similar position with his own monistic solution to the mind-body problem (see CH. 26). Neurath also gave a talk there on "God in History" but also showed himself to be highly skeptical of simplistic versions of monism and ideologized freethinking (see Neurath 1932). The monists' materialism seemed stuck in the nineteenth century.

One person who deserves special attention in the context is the sociologist Rudolf Goldscheid (1870–1931). As an activist of the monist and pacifist movements, he caused a heated debate, even a crisis, at the aforementioned Vienna International University Lectures in 1923 by his harsh criticism of the reactionary and anti-Semitic tendencies of the authorities. Later conflicts were foreshadowed here, for despite interventions by Schlick, the outcome was a strengthening of the German-nationalist tendency in the format of these courses. But Goldscheid was not only a fellow fighter but also a visionary in certain respects. Recent research strongly suggests that the later version of the scientific world-conception can be regarded as indebted to Goldscheid and read as a further elaboration of the anti-metaphysical, evolutionary and social reform conception of the Monist Society (Stoppelkamp forthcoming).

All these connections make clear that there existed a strong similarity and partial identity between doctrines of the monism and the aims of the Mach Society with its scientific world-conception. To be sure, the latter assumed a more scholarly format, with its commitment to modern science taking the form of a clearly formulated philosophy of science that sought to overcome traditional (historical and dialectical) materialism. The monists for their part were favorably predisposed to working with the Vienna Circle. This convergence is also suggested in Lily Herzberg's programmatic survey of the main philosophical currents in monism (1928), published in the last issue of *Annalen der Philosophie*, the predecessor of *Erkenntnis*.

Another very important ally for the emerging scientific world-conception among the late Enlightenment groups were the Freethinkers. They were explicitly anti-metaphysical, squarely basing their beliefs on natural science, and propagated schemes for social and sexual reform, including the institutionalization of the separation of church and state, school reform and a general ethical education based not on traditional religion. Importantly, it was the Freethinkers who provided the organizational framework for the founding of the Mach Society in 1928 (for instance, the Freethinkers shared their office with the newly founded Mach Society). Their journal *Der Pionier* (*The Pioneer*) publicized the activities of the Mach Society and advocated an anti-Platonist "Epicurean Marxism" not unlike that of Neurath who published in the journal of this group, e.g., on Marx and Epicurus (1928b) following his book discussing the relation of personal life to the class struggle earlier that year (1928a). It was due, then, to the significant contribution by the Freethinkers (and not the monists) that the Mach Society developed on the basis of a more professional and informed commitment to empirical and formal scientific research (even though earlier on they tended to hanker after a general dialectics of science). Already in 1926, some members of the Freethinkers sought to raise the intellectual level of their association by planning an additional forum for scientific discussions, but without success. It was only after those attempts that members of the Vienna Circle joined the cause to modify their simplistic scientific outlook in the direction of a more comprehensive and social scientific world-conception.

Importantly, value questions were not neglected in the late Enlightenment movement. Its ethical wing was institutionalized by the Ethical Community, often referred to as "Ethical Movement" (Ethische Bewegung), whose concern lay with life reform and moral education. This group propagated the naturalistic and anti-clerical ethics and moral philosophy of the already-mentioned Friedrich Jodl. The Ethical Movement had originally developed in New York City, founded by the American social reformer Felix Adler, but was strongly supported by Jodl, who co-founded its German and Austrian branches. Jodl's Ethical Movement was continued in Austria after the First World War by his follower Ludwig Börner (1882–1951) in the context of Viennese adult education, offering a combination of pacifism, monism, and free-thinking called "critical optimism." (This was another manifestation of social-liberal reformism, or "scientific humanism," to use a label employed by Carnap and Feigl in their North American emigration.) Again, it is not surprising that Moritz Schlick served as a leading member of this Ethical Movement which aimed to combine a eudemonistic ethics and worldly wisdom (*Lebensweisheit*), a combination which became manifest in the "secular Sunday celebrations" (*weltliche Sonntagsfeiern*) as an alternative to traditional church services. As a member of this group, Schlick lectured in 1928 on "Ethics of Duty and Ethics of Charity" (vs. Kant), which was elaborated later on in his book *Problems of Ethics* (1931/1939). In his obituary for Schlick,

Börner praised the assassinated philosopher as a follower of Jodl and friend of Popper-Lynkeus and expressed appreciation of his attempts to balance ethical theory and practice (1936). The Ethical Community also enjoyed the support of Rudolf Carnap and Viktor Kraft along with other later members of the Mach Society.

The Verein Allgemeine Nährpflicht also articulated a humanist-cosmopolitan program. It advocated the provision of a non-religious ethical basis for school and adult education as well as for social politics in general. The program had been developed by Josef Popper Lynkeus (1838–1921), one of the closest friends of Ernst Mach; both were revered by the young Albert Einstein as intellectual and moral role models whom he visited during his stays in Vienna in the 1920s. The Verein Allgemeine Nährpflicht professed pacifism and argued for a mixed economy which allowed for the free provision of basic food stuffs, housing, and clothing by planned production (a "mandatory nutrition service" [*allgemeine Nährpflicht*] was envisaged to replace traditional military service [*Wehrpflicht*]) while it left room for a free market in luxury goods in parallel. Neurath often referred to Popper-Lynkeus, not only during his engagement in the Bavarian revolution after World War I, and Schlick and Richard von Mises praised Popper-Lynkeus's humanism, pacifism and social reform for different reasons (see Belke 1978). Fittingly, Popper-Lynkeus was listed in the 1929 manifesto as forerunner of the scientific world-conception, together with Mach.

It is evident then that the Mach Society was deeply embedded in these, what nowadays would be called "countercultural," currents. Schlick served as chair, although he did not really like Hahn's socialism or the proletarian attitude of some other members like Neurath. The fact that he remained in this office until its forced dissolution after the Civil War in February 1934 by the Austro-Fascist regime (against which he protested strongly) is only one indicator of the importance that he too attached the common goal and that here was characterized as late Enlightenment. Naturally, the strong personal and political differences left their trace. In one of the last meetings of the Executive Committee in October 1933, Schlick proposed to change its name to "Society for Consistent Empiricism" (Gesellschaft für konsequenten Empirismus), a name which Schlick chose for his own philosophical position afterwards. But these foibles—and others—pale in comparison to the unifying purpose that brought him and his colleagues and the various late Enlightenment groups together in the Mach Society.

Conclusion

It can be taken as established, I think, that the foundation of the Mach Society was a successful attempt to complement the activities of the Schlick Circle in the academic sphere by participating in a forum offering an empirical and scientific philosophy to the wider public—far away from the Philosophical Society of the University of Vienna and, of course, in the University itself. It follows that the Philosophical Society cannot be counted as a precursor of the Schlick Circle and the Mach Society (as by Fisette 2014); rather, it provided the reason, even a cause, for the formation of the Vienna Circle and the Mach Society.

To elaborate briefly, in 1927 there existed in the Schlick Circle a shared self-awareness of being (part of) a genuine movement. The presence in the Philosophical Society of members like Hahn, Feigl, Kraft, Juhos, and Schlick was not judged to provide sufficient exposure to the public any more, given the marginalization of empiricism there and the dominance of Kant and neo-Kantianism (see Stadler 2018). This desire for a greater public exposure on the part of the Schlick Circle is also detectable in other actions they took at the time. Their steps towards the founding of a new institution were accompanied by efforts to establish a periodical of their own for their new philosophy. Both moves succeeded. The first issue of *Erkenntnis* appeared under the editorship of Rudolf Carnap and Hans Reichenbach by taking over Hans Vaihinger's *Annalen der Philosophie* in 1930. It was officially edited for the Berlin Society of Empirical (later: Scientific) Philosophy—and the Ernst Mach Society in Vienna. (The latter also ran a short-lived publication series, the manifesto being the first.)

In sum, the foundation of the Mach Society was the result of prompting from both inside and outside the Schlick Circle; it was not a spontaneous decision of its members to establish a second external discussion group for scientific philosophy. It would be naïve not to take into account the socio-cultural context of the late Enlightenment movements in Vienna after World War I, a context which contributed to and in turn was enriched—and intended to be so—by the founding and the activities of the Verein Ernst Mach during its six years of existence—with Mach as an uncontested role model for such engagement for enlightenment.

References

Belke, I. (1978) *Die sozialreformerischen Ideen von Josef Popper-Lynkeus (1838–1921) im Zusammenhang mit allgemeinen Reformbestrebungen der Wiener Bürgertums um die Jahrhundertwende*, Tübingen: J.C.B. Mohr (Paul Siebeck).

——— (2021) "Friedrich Jodl (1849)—Ein Vorgänger des Wiener Kreises," in C. Damböck, G. Sandner and M. Werner (eds.), *Empiricism, Life Reform, and the German Youth Movement*, Cham: Springer.

Börner, L. (1936) *Moritz Schlick als Ethiker*, Vienna: Nachlass Börner, Stadt- und Landesbibliothek.

Fisette, D. (2014) "Austrian Philosophy and Its Institutions: Remarks on the Philosophical Society of the University of Vienna (1888–1938)," in A. Reboul (ed.), *Mind, Values, and Metaphysics*, Dordrecht: Springer, pp. 349–74.

Fuchs, A. (1949) *Geistige Strömungen in Österreich 1867–1918*, Vienna: Globus. Repr. Vienna: Löcker, 1984.

Hahn, H. (1924) "Lehr- und Lernfreiheit an den Hochschulen," *Der Kampf* 17: 169–75.

Herzberg, L. (1928) "Die philosophischen Hauptströmungen im Monistenbund," *Annalen der Philosophie und philosophische Kritik* 7: 113–35.

Neurath, O. (1928a). *Lebensgestaltung und Klassenkampf*, Berlin: Laub. Excerpts trans. "Personal Life and Class Struggle," in Neurath (1973), pp. 249–98.

——— (1928b) "Marx und Epikur," *Der Freidenker* 32: 188–90.

——— (1932) "Die 'Philosophie' im Kampf gegen den Fortschritt der Wissenschaft," *Der Kampf* 25: 385–9. Repr. in Neurath, *Gesammelte philosophische und methodologische Schriften* (ed. by R. Haller and H. Rutte), Vienna: Hölder-Pichler-Tempsky, 1981, pp. 571–6.

——— (1973) *Empiricism and Sociology* (ed. by M. Neurath and R. S. Cohen), Dordrecht: Reidel.

Rathkolb, O. (ed.) (2013) *Der lange Schatten des Antisemitismus. Kritische Auseinandersetzung mit der Geschichte der Universität Wien im 19. und 20. Jahrhundert*, Göttingen: V&R Unipress.

Reininger, R. (ed.) (1938) *50 Jahre Philosophische Gesellschaft an der Universität Wien 1888–1938*, Wien: Verlag der Philosophischen Gesellschaft an der Universität Wien.

Sandner, G. (2019) "The Scientific World-Conception in the Making: Towards the Ideological Roots of Logical Empiricism in Berlin and Vienna," in F. Stadler (ed.), *Ernst Mach—Life, Work, Influence*, Cham: Springer, pp. 271–82.

Schlick, M. (1918) *Allgemeine Erkenntnislehre*, Berlin: Springer, 2nd rev. ed. 1925. Trans. *General Theory of Knowledge*, Vienna: Springer, 1974, repr. LaSalle: Open Court, 1985.

——— (1931). *Fragen der Ethik*, Vienna: Springer. Trans. *Problems of Ethics*, New York: Prentice-Hall, 1939.

——— (1982) *Vom Positivismus zur 'Wissenschaftlichen Weltauffassung'. Am Beispiel der Wirkungsgeschichte von Ernst Mach in Österreich von 1895 bis 1934*. Wien-München: Löcker.

——— (1992). "The Verein 'Ernst Mach'—What was It Really?" in J. Blackmore (ed.), *Ernst Mach—A Deeper Look*, Dordrecht: Kluwer, pp. 363–78.

——— (1997) *Studien zum Wiener Kreis. Ursprung, Entwicklung und Wirkung des Logischen Empirismus im Kontext*, Frankfurt a. M.: Suhrkamp. 2nd abridged ed., Cham: Springer, 2015. Trans. *The Vienna Circle. Studies in the Origins, Development, and Influence of Logical Empiricism*, Cham: Springer, 2015.

——— (2018) "Kant and Neo-Kantianism in Logical Empiricism. Elements of a Research Program," in V. Waibel, M. Ruffing and D. Wagner (eds.), *Natur und Freiheit. Akten des XII. Internationalen Kant-Kongresses*, Berlin: Walter de Gruyter, pp. 763–90.

Stoppelkamp, B. (forthcoming) *Wiener Naturalismus: Österreichische Philosophie um 1900 und die Entstehung der wissenschaftlichen Weltauffasung des Wiener Kreises*, PhD diss., University of Vienna, Vienna.

Taschwer, K. (2015) *Hochburg des Antisemitismus. Der Niedergang der Universität Wien im 20. Jahrhundert*, Vienna: Czernin.

Verein Ernst Mach (1929) *Wissenschaftliche Weltauffassung. Der Wiener Kreis*, Vienna: Wolf. Trans. "The Scientific Conception of the World. The Vienna Circle," in O. Neurath, *Empiricism and Sociology* (ed. by R. S. Cohen and M. Neurath), Dordrecht: Reidel, 1973, pp. 299–318; rev. trans. (with orig. annotated bibliography) "The Scientific World-Conception. The Vienna Circle," in F. Stadler and T. Uebel (eds.), *Wissenschaftliche Weltauffassung. Der Wiener Kreis. Hrsg. vom Verein Ernst Mach (1929)*, Vienna: Springer, 2012, pp. 75–116.

Wittgenstein, L. (1922) "Logisch-Philosophische Abhandlung," *Annalen der Naturphilosophie* 14: 185–262. Bilingual ed. trans. by F. Ramsey and C. K. Ogden *Tractatus Logico-Philosophicus*, London: Kegan Paul, Trench Trubner & Co., 1922, rev ed. 1933, repr. London: Routledge, Kegan, Paul, 1983; trans. by D. F. Pears and B. F. McGuinness, London: Routledge, Keagan Paul, 1961, repr. 1974.

12

THE BERLIN GROUP AND THE SOCIETY FOR SCIENTIFIC PHILOSOPHY

Nikolay Milkov

In the late 1920s and early 1930s, the philosopher Hans Reichenbach led a group of like-minded colleagues in Berlin that must count as an independent point of origin of the movement of logical empiricism. Like the Vienna Circle with whom they cooperated on numerous occasions, their concern was to develop a philosophy of science adequate to the latest advances in science itself. Differences of philosophical background and interests, however, resulted in a pronouncedly realist conception with distinct emphases in their account of scientific knowledge.

The Berlin Group: ideas

The Berlin Group of Logical Empiricism—to be distinguished from the Society for Empirical/Scientific Philosophy (covered later in this chapter)—originated with Hans Reichenbach's seminars, held from October 1926 at the University of Berlin. In the spring of 1928, Walter Dubislav joined the Group, and soon they met in a special colloquium, led by both of them together. After Reichenbach was forced to emigrate to Istanbul in the summer of 1933, and until the spring of 1935, Dubislav ran it alone. The Group was rather small, its core members being Reichenbach, Dubislav, Kurt Grelling, Alexander Herzberg, and, occasionally, also Kurt Lewin and Wolfgang Köhler. Carl Gustav Hempel, Olaf Helmer, Martin Strauss, and Valentin Bergmann were among its younger participants. Paul Oppenheim can be seen as its associate member.

In contrast to at least some members of the Vienna Circle, the Berlin Group considered philosophy to be a sound academic discipline. They devoted themselves to logical and epistemological criticism (analysis) of science and mathematics to determine what we really know. Their task was empirical: to isolate and identify (*herauszuschälen*) the first principles and truths of human knowledge on the basis of the latest results in mathematics and science. These principles substantiate or ground human knowledge. It is clear that they change with every new important development in science and mathematics (Dubislav 1929a). In other words, they present a kind of relative a priori. It is worth noticing that determining this relative a priori was already the main objective of Reichenbach's dissertation (1920a): the complementation of Kant's principles of science with the "principle of probability" is usefully understood in this fashion.

The next task of "natural philosophy"—a term Dubislav and Reichenbach used widely—was to organize and present the results achieved in science logically and epistemologically. First,

DOI: 10.4324/9781315650647-14

logically, in the sense that the claims of science have to be put in an impeccable logical order and that its concepts have to be correctly constructed and defined. In the 1920s, Reichenbach tried to fulfill this task by way of the axiomatization of physics, and Dubislav, with the help of formal logic, by formulating a consistent theory of definition (1928). Second, scientific results had to be organized epistemologically, in the sense that the logically well-constructed scientific theories have to be correctly connected to our perceptions and to the physical world of everyday life (Dubislav 1930).

These reorganizations were tasks for philosophers, Reichenbach argued, since "scientific research does not leave a man time enough to do the work of logical analysis" (1951: 123). Scientists are searching for ever-new discoveries and aim to explain these in new theories. The task of analyzing these theories logically and epistemologically is assigned to philosophers. Reichenbach's distinction between the context of discovery—the context in which scientists develop their theories—and the context of justification—the context in which philosophers reflect about their claims (1938: 6–7)—was introduced in this spirit.

The members of the Berlin Group further maintained that science and philosophy have the same source of knowledge: science itself. Philosophy is just general knowledge—to be more exact, knowledge of the general principles and concepts of science that are put in a logically and epistemologically correct form. Grelling presented this understanding of philosophical knowledge, regardless of whether it was obtained by a scientist or a philosopher, in its clearest form in his report on philosophy of science in Germany for *The Monist* (1928).

The precursor of this approach of the Berlin Group was the Jakob Friedrich Fries Society founded by Leonard Nelson (1882–1927). It met between 1913 and 1921 in Göttingen, though it had already existed since 1908 (Peckhaus 1994). The journal of this Society, *Abhandlungen der Fries'schen Schule, N.S.*, in which, Paul Bernays, among others, published four essays, was founded in 1906 and appeared until 1937.

The Berlin Group also followed Ernst Cassirer (as Neurath once noted [1930–1]). Some of its members developed elements of a comparative theory of science (*Wissenschaftslehre*), particularly Lewin and, partly, Reichenbach (in the 1920s, the two were in close contact). In this context, Lewin introduced new concepts into science in order to bring to light new and alternative structures. Among these, especially the concept of genidentity (*Genidentität*), i.e., the persistence of an object of both physics and biology from one point in time to another, enjoyed popularity. It was used in Reichenbach's *The Philosophy of Space and Time* (1928), in his *The Direction of Time* (1956), and in Carnap's *Aufbau* (1928) (Milkov 2021). (Later, this concept was also used by Adolf Grünbaum, Bas von Fraassen, Kevin Mulligan, and Barry Smith, among others.) Although Reichenbach's program was not identical to Lewin's (Grelling 1928: 98), he adopted some of its aspects. In particular, Reichenbach hoped that the "logical analysis" of different sciences could also bring to light connections between their ever-changing principles (Milkov 2011). Reichenbach pursued this idea especially in his manifesto *Ziele und Wege der heutigen Naturphilosophie* ("The Aims and Methods of Physical Knowledge," 1931/1978).

Another comparative theorist of science Reichenbach (and also Lewin) collaborated with in the 1920s was Paul Oppenheim. Reichenbach met him in 1921 and soon became one of the first links of a chain of Oppenheim's "scientific fellows." Later links of this chain were Hempel, Grelling, Helmer, Hilary Putnam, and Nicholas Rescher. Around 1929, when the first phase of his philosophical development came to an end (see below), Reichenbach recommended to Oppenheim the collaboration with his student Hempel.

This brief account shows that the Berlin Group did not have its roots only in Berlin and was active not only between 1926 and 1933. As noted, Reichenbach worked together with Lewin and Oppenheim between 1921 and 1926 when he was still in Stuttgart. After he left Berlin in July 1933, and until the spring of 1935, the Group continued its work under Dubislav's leadership. Then, from 1937 to 1939, Grelling led a "new Berlin Group" in which the logicians and mathematicians Franz Graf von Hoensbroech, Leopold Löwenheim, and Jürgen von Kempski were active (Peckhaus 1994). From 1938 to 1939, Oppenheim, Hempel, and Grelling formed a new chapter of the Berlin Group in Brussels. At the beginning of the 1940s, another offshoot, the so-called "H_2O Group," operated in the USA. All these circles can be seen as different formations of the Berlin Group. (Rescher [2005] even maintained that the Pittsburgh Center for Philosophy of Science had its roots in the Berlin Group.)

The profound interest in natural philosophy eventually led the Berlin Group to begin to develop a general philosophy of science of a type that was to survive the later demise of orthodox logical empiricism. Dubislav made the first step in his book *Naturphilosophie* (*Philosophy of Nature*, 1933) and Hempel, who wrote two very positive reviews of the book, was the first to understand the new elements in Dubislav's conception:

> in contrast to Schlick's philosophy of nature and also Zilsel's, problems that have the character of specific issues of natural science (albeit rather general ones), like the problem of life, are given a back seat in favor of a detailed systematic discussion of the logical and methodological problems of natural scientific knowledge.
>
> (Hempel 1934: 760, trans. NM; see also his 1933)

In short, Dubislav directed attention to the "general principles of investigation in the natural sciences" (this is the title of the central chapter 4 of Dubislav's book); he did not merely seek to demonstrate how new scientific theories were to be justified empirically, albeit with the help of logic and mathematics. Importantly enough, traces of Dubislav's influence on Hempel can be easily discovered in the latter's influential work, *Philosophy of Natural Science* (1966; see Milkov 2021).

Reichenbach was deeply impressed by the Vienna Circle's plea for the establishment of a close connection between philosophy and science, and for about ten years after the publication of its manifesto in September 1929, he was engaged in exploring its problems. It is worth remarking, however, that he was less influenced by the specific doctrines of the Vienna Circle but instead adopted its main themes: meaning, verification, truth, empiricism. Reichenbach approached these themes following his own intuitions. His explorations found expression, in particular, in the book *Experience and Prediction* (1938), in which he sharply criticized the Vienna Circle's (earlier) way of treating these themes and suggested alternative solutions. Reichenbach went back to his own pre-Vienna-Circle subject of investigation—the "logical analysis" of science—only after he relocated to the USA. He did so particularly in his work *Philosophical Foundations of Quantum Mechanics* (1944).

Working on problems set out by the Vienna Circle, Reichenbach advanced a form of "consistent empiricism" according to which human knowledge is only a product of empirical observations (1951: 259). He believed that he ultimately disproved the claims of the "rationalism" (apriorism). At the same time, encouraged by his friends Grelling and Dubislav, Reichenbach developed a strong interest in the problems of logic which found expression in his 1947 work. All these developments shaped Reichenbach's distinctive form of logical empiricism. The influence of the Vienna Circle on Dubislav was even less pronounced. It can only be discerned in Dubislav's concept of "logical behaviorism" (1933: 69), which was nothing more than his interpretation of Neurath's and Carnap's physicalism.

The Berlin Group: members

In the preceding section we attempted a comprehensive presentation of the main ideas of the Berlin Group, whose joint leaders were Reichenbach and Dubislav. Unfortunately, Reichenbach is often considered a "one-man orchestra," the only man to lead the Group. This explains why only his philosophy is well presented in the literature (e.g., Salmon 1977). Recently, however, it has been shown that as the leader of the Berlin Group, Dubislav was his equal (Milkov 2015, 2016). To bring home this point, we next offer a review of instances where Reichenbach followed Dubislav's philosophical lead.

During his short stay as a student at the University of Göttingen in the summer of 1914, the young Dubislav lived in the direct neighborhood of Leonard Nelson. Apparently, at that time, he also came into contact with Nelson's circle. This is supported by the number of references Dubislav made to Jakob Friedrich Fries and Leonard Nelson in his later writings (Dubislav 1926, 1929a). Note that this also makes it quite possible that he was associated with Grelling, who was Nelson's assistant at the time, and with Reichenbach, who had close contacts with Grelling, though as a newcomer, Dubislav was perhaps not personally introduced to them.

Be that as it may, until January 1928, Dubislav and Reichenbach still communicated by letter, even though both lived in Berlin. Their relationship changed radically after Dubislav sent Reichenbach a preliminary version of his essay "Elementarer Nachweis der Widerspruchslosigkeit des Logik-Kalküls" ("Elementary Proof of the Consistency of the Logical Calculus," 1929b). There, he developed a new kind of truth-tables in logic which deeply impressed Reichenbach, and for good reason. Dubislav's innovation helped him to advance a new probability logic, according to which the truth-values, "true" and "false," are only two points on a continuous scale of validity that measure the level of its probability. It was Dubislav who drew Reichenbach's attention to the importance of logic for his studies, in which, before 1928, he did not show a serious interest. Although Reichenbach spoke of "logical analysis" of science early on (1920b), he meant with this an axiomatization of science and its epistemological criticism—not logic proper.

Dubislav also helped Reichenbach to better understand and formulate the concept of "coordinating definition." It has to be remembered that Dubislav was a radical formalist in logic and philosophy of mathematics who closely followed David Hilbert's axiomatic method. Consequentially, he also applied this method in his philosophy of science (Milkov 2016). According to Dubislav, the task of both scientists and philosophers of science is to coordinate the "objects" (i.e., the events, facts, and phenomena) of the external world with impeccably formed calculation systems. Most importantly, they must unequivocally attach the "objects" to the calculi and also provide rules of interpretation for the resulting system.

It must also be noted that Dubislav and Reichenbach had a common program in ethics which clearly differed from that of the Vienna Circle. Both groups believed that there is no truth in ethics and that the latter does not convey knowledge. But while the Vienna Circle defended different forms of noncognitivism, according to some of which moral judgments are primarily an expression of human emotions, Dubislav and Reichenbach claimed that these are implicit commands and therefore are related by logical principles. On this score, too, Dubislav was the trendsetter: he was the first to develop this conception in print in (1937), only to be followed by Reichenbach ten years later (1947: 344, 1951: 280ff.).

Another important member of the Berlin Group was Kurt Grelling. There are good grounds to believe that Grelling inspired Reichenbach to deal with probability when both were in Göttingen in 1914. This conjecture is also supported by the fact that Grelling's early paper, "Die philosophischen Grundlagen der Wahrscheinlichkeitsrechnung" ("The Philosophical Foundations of the Probability Calculus," 1910), defended the objective interpretation of this discipline against Carl Stumpf's subjectivism. In it, Grelling also linked his thoughts on probability to the problem of induction. Importantly enough, these two conceptions were the kernel of Reichenbach's theory of probability throughout his career. Although Grelling and Reichenbach did not meet again between 1914 and 1926, they kept in touch (see ASP HR 015-54-06 and HR 044-06-21). They resumed their regular meetings when Reichenbach moved from Stuttgart to Berlin in October 1926. Grelling had already relocated from Göttingen to Berlin in 1922. In fact, these were the first steps in setting up the Berlin Group proper.

The related positions of Reichenbach and Grelling on the theory of probability were first brought to public attention at the First Conference on the Epistemology of the Exact Sciences in Prague, 1929. In the discussion (Grelling 1930), the two maintained that true science is possible only when it is based on the principle of induction: only the latter can help substantiate scientific predictions. Both resolutely rejected Carnap's and Waismann's—and, eventually, Wittgenstein's—logical conception of probability. From 1936, however, Grelling turned away from Reichenbach's interpretation of the theory of probability and came to agree more with Carnap's. At the same time, he developed original works in formal ontology (Grelling and Oppenheim 1937; Milkov 2021).

An important contribution of Grelling to the Berlin Group were his translations of four books of Bertrand Russell—*The Analysis of Mind* (1921), *The ABC of Relativity* (1925), *The Analysis of Matter* (1927), and *An Outline of Philosophy* (1929)—into German between 1927 and 1930, the period when the Berlin Group began its work. It should be noted that these books belong to a particular period of Russell's philosophical development. They are the products of the fresh start in philosophy he made in 1919 after Wittgenstein (and indirectly, also Frege) pointed out serious problems in Russell's philosophy of language (Milkov 2013a). In these four books, Russell philosophically assessed new developments in the field of psychology (notably, Watson's behaviorism) and physics (in particular, Einstein's theory of relativity). Grelling's translations were well known to the members of the Berlin Group and stimulated the discussions they led.

Another member was Alexander Herzberg (1887–1944), who received a PhD in both medicine and philosophy (Schernus 1994). For years, he was a member of the "*Deutsche Monistenbund*" (a society of freethinkers which was under the influence of Ernst Haeckel) and made numerous contributions to its journal, *Monistische Monatshefte*. As one of the co-founders of the monists' and positivists' Society for Empirical Philosophy, Herzberg was instrumental in the transition of its leadership into the hands of Reichenbach and Dubislav (see the next section). This was one reason why Reichenbach highly appreciated him. But Herzberg was also theoretically well integrated into the Berlin Group, whose philosophical program, after all, was interdisciplinary. They were interested not only in the newest developments in physics, medicine, biology, and technology but also in psychology and sociology of science.

Reichenbach, in particular, believed that we can analyze science in three different ways. First, through "logical analysis", i.e., through axiomatic, logical and epistemological analysis: this helps to "rationally reconstruct" science, and this is the path Reichenbach himself followed. Second, through psychological analysis: this shows why scientists and philosophers make theoretical mistakes. This was Herzberg's task. His book *The Psychology of Philosophers* (1926/1929) was important for the Berlin Group exactly in this context. Reichenbach knew quite well that "the psychological root of all rationalism in the wider sense is an extralogical motive . . . it is the search for certainty" (1951: 32). Purely on theoretical grounds, rationalism is not defendable, but its appeal is psychologically understandable. Finally, third, we can also make sociological analyses of science. These examine the impact of the social environment on scientific and philosophical theories.

Carl Gustav Hempel is often presented as a close associate of the Vienna Circle (Friedman 2003). This, however, is not the whole truth. Hempel spent the fall term of 1929 as a student at the University of Vienna and, thanks to a letter of recommendation from Reichenbach, attended some sessions of the Vienna Circle. However, Hempel spent much less time in Vienna than in Berlin, where he studied under Reichenbach from 1926 to 1933 and wrote a dissertation on probability, Reichenbach's specialty (Milkov 2013b). It is no coincidence, therefore, that Hempel's most influential idea, the deductive-nomological model of explanations in science, squared well with the leading endeavor of the Berlin Group—to justify (to substantiate) human knowledge. It is also no accident that he positively and extensively reviewed all of Reichenbach's books published after 1928—until 1937 in the *Jahrbuch über die Fortschritte in der Mathematik*, and then in the *Journal of Symbolic Logic*.

From 1934 through 1939, Hempel worked with Oppenheim in Brussels as his "scientific secretary." A product of this work is Hempel and Oppenheim (1936). In the fall of 1939, both emigrated to the USA. (Hempel had previously worked at the University of Chicago for nine months in 1937–1938 as Carnap's assistant.) It was in America that, between 1942 and 1944, a new cohort of the Berlin Group came to life—this time at Princeton, where Hempel joined Oppenheim and Helmer (Carnap named it the "H_2O Group"). Hempel's most influential papers, "Studies in the Logic of Conformation" (1945) and (with Oppenheim) "Studies in the Logic of Explanation" (1948), reflect this collaboration. The issues they explored were clearly close to Lewin's and Oppenheim's (and partly also to Reichenbach's) program for the comparative theory of science, ranging as they did from the logic of classification and the systematic ordering of science to taxonomy and the theory of ordering concepts, which reflects conceptual isomorphism among different sciences (Rescher 1997).

Olaf Helmer (1910–2011) was the youngest member of the Berlin Group and a close friend of Carl Hempel. He studied under Reichenbach, did his doctoral degree under Dubislav in Berlin in 1934, and then wrote a second dissertation in London in 1936 under Susan Stebbing. In 1937–8, Helmer worked, together with Hempel, under Carnap's guidance at the University of Chicago. In 1938–44, he taught mathematics, first at the University of Illinois in Urbana, and then at CUNY. In 1944–5, he collaborated with Oppenheim.

From 1946 onwards, Helmer worked at the RAND Corporation in Santa Monica (California), which he left in 1968 to co-found the Institute for the Future in Palo Alto (California). Between 1973 and 1976, he was a professor for futurology at the University of California in Los Angeles. His younger friend, Nicholas Rescher, remembered Helmer's turn to futurology in the following terms: "Once he became engrossed in matters of prediction and futurology, this replaced all other concerns. He never returned to the work on confirmation and evidentiation that characterized his early interest" (1997: 166).

It does not take much to realize that Helmer's great enthusiasm for futurology also had its roots in the discussions of the Berlin Group and in the conversations he had with his former teacher Reichenbach. Indeed, the craft of making "good predictions" was of primary importance to the latter's epistemology of science. Reichenbach maintained that "the ascertainment of a degree of probability by means of an inductive inference" (1951: 242) are free posits—and it is important that these posits are well grounded so that they can be confirmed by upcoming events.

The Society for Scientific Philosophy

Like the Vienna Circle, which had its public forum in the Ernst Mach Society, the Berlin Group led the open-to-the-public Berlin Society for Scientific Philosophy. The latter, in turn, had its roots in the Society for Empirical Philosophy but ought to be distinguished from it. The Berlin Society for Empirical Philosophy was founded by the positivist Josef Petzoldt in February 1927 as the Berlin chapter of the International Society for Empirical Philosophy. Georg von Arco, Max Deri, Alexander Herzberg, Reginald Zimmermann, Friedrich Kraus, and August von Parseval were among its founding members. The International Society for Empirical Philosophy itself was organized by the publisher Raymund Schmidt in 1925 in Leipzig, i.e., before Petzoldt started collaborating with it, in support of the journal *Annalen der Philosophie*, which already had something of a scientific orientation. Initially, that journal was launched in support of the "as if" philosophy of the neo-Kantian Hans Vaihinger.

The Berlin Society for Empirical Philosophy was formally organized, but not officially registered in court records. It defined its activity through lectures and discussions, hosting from 10 to 20 talks per year. Usually, the lectures were attended by 100 to 300 people. It met on Tuesdays at the theater of the Second Medical Clinic of the famous Charité Hospital in Berlin. (Friedrich Kraus was the director of the clinic at the time.) The very fact that its meetings took place in the medical school indicates the scientific orientation of this philosophical society. But the venue of the Society's lectures also explains why discussions of philosophical problems of medicine and psychology were prominent at its sessions (Danneberg and Schernus 1994).

In the first year and a half of Petzoldt's Berlin Society, Reichenbach remained skeptical about its viability, formally becoming its member only in October 1928. In fact, it was Dubislav who prompted Reichenbach to take this step. And it was Neurath who urged him, some months later, to take over its presidency. Neurath's idea was a reformed Berlin Society that would be a counterpart—not necessarily a satellite organization—of the Ernst Mach Society (see ASP HR 014-06-31). At just this time, in May 1929, Petzoldt fell ill and resigned as the president of the Society. Reichenbach, however, was still skeptical. Still, on 21 June, he expressed hope to be elected to the Board of the Verein Ernst Mach (ASP HR 014-06-28) which, he argued, would spread the influence of the Verein across the whole Germanophone world. Yet for some reason, he changed his mind the following week, and on 30 June 1929, he wrote to Carnap about the Berlin Society: "Recently, Dubislav and I were integrated into the Board, where we, together with Herzberg, have the real power" (ASP HR 013-39-34). To be more precise, Reichenbach was elected President (*Vorsitzender*) of the society, and Dubislav its Secretary (*Geschäftsführer*).

The next couple of years saw Reichenbach and his friends trying to transform the work of the Society. A clear indication of this is that by the end of 1931, the Society for Empirical Philosophy was renamed the Society for Scientific Philosophy. Together with the change of the Society's name came also a change in the subjects of the lectures presented at it. More specifically, after February 1932, the word "empirical," or "empiricism," didn't appear in the title of any of the 42 lectures delivered. In contrast, before that point of change, in 67 lectures the word "empirical" occurred eleven times. Apparently, the new format of the Society put greater stress on the fact that it explored philosophically problems of science than that its method was empirical. Dubislav and Reichenbach were the heart and soul of the Society for Scientific Philosophy. Altogether, Dubislav made nine presentations at it, and Reichenbach six. Herzberg, on his part, lectured three times. Grelling and Hempel, in contrast, never gave a lecture to the Society. This shows once more that the Berlin Group and the Society were two different entities.

The members of the Society largely represented the scientific elite of Berlin but also of other scientific centers in Germany. Most were seasoned researchers and respected authorities in their fields, many of them holding leading positions in prestigious academic departments and institutes. The lineup of lecturers hosted by the Society included no less than four Nobel nominees and three laureates: Friedrich Kraus, Max von Laue, Otto Meyerhof, and Wilhelm Ostwald.

The Society served as a forum for innovative scientists, like the founder of Gestalt psychology, Wolfgang Köhler, or the brain researcher, Oskar Vogt. It also attracted talented younger interdisciplinary scientists, such as the biologist and systems theorist Ludwig von Bertalanffy from Vienna. To further their own original research programs, these scientists sought out the sort of stimulus to innovative thinking that the Society's philosophically keyed interdisciplinary discussions fostered. And it is not difficult to see as well that these discussions supported the goal of the Berlin Group to determine and exhibit the relative a priori of different sciences.

References

ASP HR = Hans Reichenbach Papers, Archive of Scientific Philosophy, University of Pittsburgh.

Carnap, R. (1928) *Der logische Aufbau der Welt*, Berlin: Weltkreis-Verlag. Trans. *The Logical Structure of the World*, Berkeley: University of California Press, 1967, repr. Chicago: Open Court, 2003.

Danneberg, L., Kamlah, A., and Schäfer, L. (eds.) (1994) *Hans Reichenbach und die Berliner Gruppe*, Braunschweig: Vieweg.

Danneberg, L. and Schernus, W. (1994) "Die Gesellschaft für wissenschaftliche Philosophie," in Danneberg, Kamlah, Schäfer 1994, pp. 391–481.

Dubislav, W. (1926) *Die Fries'sche Lehre von der Begründung. Darstellung und Kritik*, Dömitz: Matting.

——— (1928) "Zur kalkülmäßigen Charakterisierung der Definition," *Annalen der Philosophie und philosophischen Kritik* 7: 136–45.

——— (1929a) *Zur Methodenlehre des Kritizismus*, Langensalza: Herman Beyer.

——— (1929b) "Elementarer Nachweis der Widerspruchslosigkeit des Logik-Kalküls," *Journal für die reine und angewandte Mathematik* 161: 107–12.

——— (1930) "Über den sogenannten Gegenstand der Mathematik," *Erkenntnis* 1: 27–48.

——— (1933) *Naturphilosophie*, Berlin: Junker und Dünnhaupt.

——— (1937) "Zur Unbegründbarkeit der Forderungssätze," *Theoria* 3: 330–42.

Friedman, M. (2003) "Hempel and the Vienna Circle," *Minnesota Studies in the Philosophy of Science* 18: 94–114.

Grelling, K. (1910) "Die philosophischen Grundlagen der Wahrscheinlichkeitsrechnung," in *Abhandlungen der Fries'schen Schule, N.S.* 3: 439–78.

——— (1928) "Philosophy of the Exact Sciences: its Present Status in Germany," *The Monist* 38: 97–119.

——— (1930) "Diskussion über Wahrscheinlichkeit," *Erkenntnis* 1: 260–85.

Grelling, K. and Oppenheim, P. (1937) "Der Gestalt-Begriff im Lichte der neuen Logik," *Erkenntnis* 7: 211–25.

Hempel, C. G. (1933) "Walter Dubislav, *Naturphilosophie*," *Jahrbuch über die Fortschritte der Mathematik* 59: 56–57.

——— (1934) "Walter Dubislav, *Naturphilosophie*," *Deutsche Literaturzeitung* 55: 759–62.

——— (1945) "Studies in the Logic of Confirmation," *Mind* 54: 1–26, 97–121. Rev. and repr. in Hempel (1965), pp. 3–46.

——— (1965) *Aspects of Scientific Explanation*, New York: Free Press.

——— (1966) *Philosophy of Natural Science*, Englewood Cliffs: Prentice-Hall.

Hempel, C. and Oppenheim, P. (1936) *Der Typusbegriff im Lichte der neuen Logik*, Leiden: Sijthoff.

——— (1948) "Studies in the Logic of Explanation," *Philosophy of Science* 15: 135–75. Repr. Hempel (1965), pp. 245–90.

Herzberg, A. (1926) *Zur Psychologie der Philosophie und der Philosophen*, Leipzig: Meiner. Trans. *The Psychology of Philosophers*, London: Kegan Paul, 1929.

Milkov, N. (2011) "Anmerkungen des Herausgebers," in H. Reichenbach (ed.), *Ziele und Wege der heutigen Naturphilosophie*, Hamburg: Felix Meiner, pp. 147–58.

——— (2013a) "The Joint Philosophical Program of Russell and Wittgenstein and Its Demise," *Nordic Wittgenstein Review* 2: 81–105.

——— (2013b) "Carl Hempel: Whose Philosopher?," in N. Milkov and V. Peckhaus (eds.), *The Berlin Group and the Philosophy of Logical Empiricism*, Berlin: Springer, pp. 293–309.

——— (2015) "On Walter Dubislav," *History and Philosophy of Logic* 36: 147–61.
——— (2016) "Walter Dubislav's Philosophy of Science and Mathematics," *HOPOS: The Journal of the International Society for the History of Philosophy of Science* 6: 96–116.
——— (2021) "Kurt Grelling and the Idiosyncrasy of the Berlin Logical Empiricism," in S. Lutz and A. Tuboly (eds.), *Logical Empiricism and the Physical Sciences*, London: Routledge, pp. 64–83.
Neurath, O. (1930–1) "Historische Anmerkungen," *Erkenntnis* 1: 311–14.
Peckhaus, V. (1994) "Von Nelson zu Reichenbach: Kurt Grelling in Göttingen und Berlin," in Danneberg, Kamlah and Schäfer 1994, pp. 53–86.
Reichenbach, H. (1920a) "A Philosophical Critique of the Probability Calculus," in Reichenbach 1978, vol. 2, pp. 312–27.
——— (1920b) *Relativitätstheorie und Erkenntnis priori*, Berlin: Springer. Trans. *The Theory of Relativity and A Priori Knowledge*, Berkeley: University of California Press, 1965.
——— (1928) *Philosophie der Raum-Zeit-Lehre*, Berlin: de Gruyter. Trans. *The Philosophy of Space and Time*, New York: Dover.
——— (1931) *Ziele und Wege der heutigen Naturphilosophie*, Leipzig: Meiner. Trans. "The Aims and Methods of Physical Knowledge," in Reichenbach 1978, vol. 2, pp. 120–225.
——— (1938) *Experience and Prediction*, Chicago: Chicago University Press.
——— (1944) *Philosophical Foundations of Quantum Mechanics*, Berkeley: University of California Press.
——— (1947) *Elements of Symbolic Logic*, New York: Macmillan.
——— (1951) *The Rise of Scientific Philosophy*, Berkeley: University of California Press.
——— (1956) *The Direction of Time*, Berkeley: University of California Press.
——— (1978) *Selected Essays: 1909–1953* (ed. by M. Reichenbach and R. S. Cohen), Dordrecht: Reidel, vol. 2.
Rescher, N. (1997) "H_2O: Hempel–Helmer–Oppenheim. An Episode in the History of Scientific Philosophy in the 20th Century," *Philosophy of Science* 64: 779–805.
——— (2005) "The Berlin School of Logical Empiricism and its Legacy," *Erkenntnis* 64: 281–304.
Russell, B. (1921) *The Analysis of Mind*, London: Allen and Unwin.
——— (1925) *The ABC of Relativity*, London: Kegan Paul.
——— (1927) *The Analysis of Matter*, London: Kegan Paul.
——— (1929) *An Outline of Philosophy*, London: Allen and Unwin.
Salmon, W. C. (1977) "The Philosophy of Hans Reichenbach," *Synthese* 34: 5–88.
Schernus, W. (1994) "Alexander Herzberg: Psychologie, Medizin und wissenschaftliche Philosophie," in Danneberg, Kamlah, and Schäfer (1994), pp. 33–51.

13
WOMEN IN LOGICAL EMPIRICISM

Frederique Janssen-Lauret

Historians of analytic philosophy often paint an all-male picture of the early stages of its development, including logical empiricism. Coffa's celebrated history of "a decade in the philosophical life of what may loosely be called Vienna" (Coffa 1991: 1), for example, mentions no female philosophers or logicians. The logical empiricist project is sometimes taken to be not just overwhelmingly male, but somehow masculine. Its aim of devising a philosophy to match the new science and mathematics—general relativity, Hilbert's program, the incompleteness of arithmetic—and the central role it gives to science as a source of knowledge seems to some feminist historians at odds with women's supposed preference for the normative: "In the twentieth century . . . women philosophers—such as Elizabeth Anscombe, Iris Murdoch, Mary Midgley, and Philippa Foot—have continued to demonstrate a strong interest in moral theory" (Broad 2006: 1069). The verification criterion of meaningfulness pronounced much of moral theory to be cognitively meaningless. Yet women were far from underrepresented in logical empiricism compared to other intellectual movements of its day. The Vienna Circle had three female members: Olga Hahn-Neurath, Rose Rand, and Olga Taussky. Relatedly, Wolenski (1989) names nine prominent female philosophers and logicians of the Lvov-Warsaw School: besides his own teacher, Izydora Dąmbska, also Eugenia Ginsberg, Janina Hosiasson, Maria Kokoszyńska, Daniela Tennerowa-Gromska, Maria Ossowska, Seweryna Łuszczewska, Halina Słoniewska, and Janina Kotarbińska. Only Ossowska worked on moral philosophy. The others published on mathematical or philosophical logic, philosophy of science, or psychology.

Logical empiricism's focus on science was not, in fact, apolitical or at odds with feminism. It was informed by philosophical reasons, but also by a drive for evidence-based policy, social progress, egalitarianism, and anti-racism (Stadler 2007). Many logical empiricists held left-wing political views. Several were Jewish, including many of the women. They knew they had reason to fear the rise of fascism and Nazism around them. Those who could, fled the continent, often to the USA or UK. Several of those who could not were cruelly murdered by the Nazis, including Hosiasson and Ginsberg. Dąmbska joined the resistance (Perzanowski 1983: 379). It has been suggested that logical empiricists' anti-metaphysical or anti-essentialist attitudes derived in part from their well-founded worries about Heideggerian, neo-Thomist, and Hegelian metaphysics informing fascist ideology (Uebel 2016: §2.3; Janssen-Lauret 2018: xviii—xix). Logical empiricists who escaped to the McCarthyite USA were under pressure to depoliticize the view or face marginalization (Reisch 2007).

 DOI: 10.4324/9781315650647-15

Among the pivotal theses of logical empiricism, we find, besides anti-metaphysics and verificationism, views of scientific theories as syntactic, axiom-based structures (or, alternatively, collections of mathematical models), confirmation theory, and the unity of science. Female logical empiricists, and female sympathetic critics of logical empiricism, made significant contributions to all of the above. Women's contributions to philosophy and logic have historically often fallen into obscurity despite their quality because of a combination of the sexism of their peers, institutional sexism affording few jobs and opportunities to women, and the sexism of historians who neglect their works compared to those of men (Janssen-Lauret forthcoming). For the women of Vienna, Berlin, Lvov, and Warsaw, these factors were present and further compounded by their being forced into exile or marginalization after fleeing fascism, and by their publications in German, French, or Polish remaining unavailable in English.

Women of the Vienna Circle

Among the members of the Vienna Circle were a female philosopher, Rose Rand, and two female mathematicians, Olga Hahn-Neurath and Olga Taussky. Olga Hahn (1882–1937) was the sister of Hans Hahn, the mathematician, and fellow member of the Vienna Circle. She studied mathematics and philosophy at the University of Vienna from 1902, and received her PhD in 1911. Hahn was triply marginalized as a woman, a Jewish scholar, and a person with a disability. She had become blind at the age of 22 as the result of an inflammation of the optic nerve. While still writing her PhD, she published her first paper on the algebra of classes in Schröder's logic, setting out to axiomatize Schröder's system (Hahn 1909). Soon afterwards, she and Neurath published a joint paper, proposing key revisions to the axioms and proof procedure of Schröder's system (Hahn and Neurath 1909). Hahn was awarded her PhD for a paper on the theory of inference (Hahn 1910). (Cat 2019: 242–3 notes that her papers received laudatory mentions in C. I. Lewis's influential bibliography on logic of 1918 and also featured in Church 1936.) Hahn married Otto Neurath in 1912 and used the name "Hahn-Neurath" from then on, later participating regularly in Schlick's seminars. Following the Austro-fascist government's crackdown on political opponents in 1934, she and Otto settled in The Hague, where she died in 1937.

Olga Taussky (1906–1995) attended Schlick's seminar from 1925 while working on her PhD in mathematics at the University of Vienna. Although Taussky was close to Kurt Gödel, whom she met there, her own academic interests were in algebraic number theory, not mathematical logic or logical empiricism. Looking back, she wrote, "I was the youngest in age in the Vienna Circle, but I was disappointed that these gatherings could not give me guidance for my work in number theory. Had I realized what Gödel would achieve later, I would not have run away" (Taussky-Todd 1988b). Upon completing her PhD, in 1931 Taussky took a job in Göttingen as co-editor of the number theory volume of Hilbert's work. At Göttingen, she met the famous algebraicist Emmy Noether, like her a rare Jewish female academic in an increasingly anti-Semitic Central Europe. Taussky fled to Noether's new department, Bryn Mawr, in 1934. She subsequently held short-term posts at Girton College Cambridge and the University of London, where she met her husband-to-be, fellow mathematician John Todd. Although she is frequently called "Olga Taussky-Todd" after her marriage, she continued to publish papers in mathematics journals as "Taussky" (e.g., Taussky 1988a). During the Second World War Taussky first taught at Queen's University, Belfast, then undertook war work at the British National Physical Laboratory. After the war ended, she and her husband moved to the California Institute of Technology. Todd took up a professorial post, but Taussky, as a woman, was only offered a research assistantship despite her strong publication record. Taussky finally became the first female professor of mathematics at her institution in 1971 (Goodstein 2020).

The work of Rose Rand (1903–1980) was more squarely philosophical. Born in Lvov (now Lviv, Ukraine) to a Jewish family, she studied philosophy in Vienna with Schlick. In 1937, Rand was awarded her PhD, on the philosophy of Tadeusz Kotarbiński, the first such study to be conducted outside Poland (Betti 2001: 102–3). She summarized her research in a German-language article (Rand 1938), which Ernest Nagel, in his review, called "a very useful summary of Kotarbiński's main work available only in Polish" (Nagel 1938: 169). Rand explained Kotarbiński's views on the three functions of language—designation, meaning, and representation—his one-category ontology, known as "reism," his resultant anti-essentialism, his views on formal logic, and their connection to Lesniewski's ontology. She also briefly sketched his views on methodology. Rand's paper on Kotarbiński remains untranslated. Soon after its appearance, Rand fled Austria for the UK but was unable to find academic employment. Over the years she taught, translated texts, and sometimes worked in a factory. Her second paper, published in 1939 and translated into English some two decades later, concerned the logic of "demand-sentences": rules, instructions, sentences containing "should" or "ought" or "must." Rand rehearsed the verificationist criterion of meaningfulness and argued that demand-sentences, while not true or false, nevertheless "make sense." We distinguish ungrammatical or contradictory demands, which cannot be complied with, from ones that can. Rand remarked that logical and syntactic rules are themselves expressible as demand-sentences, so logical empiricists would be unwise to deny that such sentences make sense. She proposed a criterion of meaning according to which declarative sentences make sense iff they are verifiable, but demand-sentences make sense iff they can be complied with. Compliance, Rand pointed out, is not itself reducible to verification: "if someone says 'Stop', what is verified by the act of stopping on the part of the person addressed, is not the sentence 'Stop', but the sentence '*x* stopped'" (Rand 1939/1962: 243). She argued that the principles of logic did not directly apply to demand-sentences but did do so by analogy, proposing several typical analogues of familiar logical forms, including *modus ponens* and *modus tollens*, and also suggesting that "No F must φ; x is an F; so x must not φ" is an additional analogue of a valid argument form for demand-sentences. The term "demand-sentences" itself Rand credits to Walter Dubislav, a member of the Berlin Circle for Scientific Philosophy (see CH. 12).

Also associated with the Berlin-based circle of logical empiricists was Wilma Papst, whose 1930 PhD dissertation offered one of the first philosophical treatments of Frege (as its title indicates). Conscious of Weyl's criticisms of Hilbert's program, Papst held that some additional philosophy was needed to avoid falling into formalism. She saw Frege's underlying philosophy as a general approach of the "Gestalt" concept (Heinemann and Reichenberger 2019). The mathematician Hilda Geiringer was also based in Berlin originally. After Hitler's rise to power, Geiringer, who was Jewish, took up a post in Istanbul alongside her mentor and later husband, Richard von Mises. Geiringer specialized in statistics and the calculus of probability, including philosophical questions about probability. Geiringer argued that we can calculate the probability of given scientific hypotheses with the help of Bayes' Theorem. But she opposed Reichenbach's view that the calculus of probability answers the philosophical problem of induction and of applying axiomatized theories to experience (Geiringer 1939).

Two further female logical empiricists from Vienna were Else Frenkel-Brunswik (1908–1958) and Käthe Steinhardt (1894–1985). A student of the first female *Dozent* (associate professor) at the University of Vienna, psychologist Charlotte Bühler, Frenkel took part in discussions of the Vienna Circle. She and her husband, fellow psychologist Egon Brunswik, fled in 1938 and found a new home in the United States, where Frenkel-Brunswik co-authored the landmark book *The Authoritarian Personality* (Adorno et al. 1950). In "Psychoanalysis and the Unity of Science," she made the case that psychoanalysis was "closer to a truly scientific theory than most observers realize" (1954: 338). She argued that psychoanalytic theoretical statements could largely be construed as Carnapian dispositional statements, just shorthand for summaries of observation statements, rather than as statements about unobserved theoretical entities. She called for "more formal attempts in the direction of an axiomatization of the psychoanalytic system" (1954: 290; see Borchers 2003). Steinhardt completed a PhD with Schlick in 1936. She worked on converting his lectures into a book—a project cut short by Schlick's death—and published a paper on the analysis of perception-statements (Steinhardt 1938).

Susan Stebbing on logical empiricism

In 1933 Susan Stebbing (1885–1943), then about to be appointed the UK's first female philosophy professor, gave the annual "Philosophical Lecture" to the British Academy. Her subject was "Logical Positivism and Analysis" (1933a, 1933b). Stebbing, though often neglected by present-day philosophers and historians, was a key figure in the development of analytic philosophy (Beaney and Chapman 2017; Janssen-Lauret 2017). While she was not herself a logical empiricist, Stebbing was a sympathetic critic who did much to promulgate logical empiricism in the UK (see also CH. 34).

Stebbing studied moral sciences at Girton College, but as a woman, she was not permitted to graduate from Cambridge. She moved to London for postgraduate study in the mid-1910s. There she met Moore, Russell, and Whitehead, shortly afterwards took up a lectureship at the London women's college, Bedford, and published papers in top journals almost every year, largely about philosophical analysis and the philosophy of science. Stebbing had met Schlick at a conference in 1930, become intrigued by logical empiricist philosophy of science, and written reviews of Jørgensen and Petzäll by the time she gave the Philosophical Lecture.

Stebbing expressed appreciation for the logical empiricists' trenchant critique of traditional metaphysics of the Hegelian and essentialist varieties and for its dedication to the analysis of language and to the "new logic" (1933b: 16). But she also pressed several objections to the logical empiricists, naming Carnap, Waismann, and Schlick, from the point of view of her own movement, the Cambridge School, of whom she mentioned Moore, Russell, and Whitehead—but not Wittgenstein, whom she tentatively grouped with the logical empiricists. In a nutshell, Stebbing held that the logical empiricist style of analysis views philosophical and empirical theories as overly syntactic, leaves too little room for semantics, and, without good reason, dismisses modest, empiricist types of metaphysics (like Stebbing's). She appealed to her own work on analysis (1932), which distinguished grammatical analysis of language from directional analysis of facts. Linguistic analysis, including definitions and the elucidation of concepts, explains a given stretch of language in terms of a further stretch of language. It may therefore be analytic or a priori. By contrast, directional analysis aims to identify what configuration of worldly objects it takes for a sentence (or proposition) to be true, and is empirical, not analytically true.

According to Stebbing, the logical empiricist's verificationist approach to meaning runs into trouble because it fails to allow for the modest metaphysics of directional analysis, according to which truth consists in agreement with reality. At the end of her lecture, she raised the question, "Can we say that there are facts which make propositions true, or can we only say that propositions are verifiable by reference to my own experience?" Her answer was, "In my opinion there are final facts, and these final facts are the facts which make propositions true (or false)" (1933b: 36). But Schlick, Waismann, Carnap, and to an extent Wittgenstein, Stebbing argued, were only able to say that propositions are verifiable by reference to each individual's experience. In her view, language, philosophy, and science cannot get off the ground without assuming that multiple minds perceive and communicate about the same worldly things (ibid.: 27). She balked at Carnap's conclusion that we cannot allow for any directional analysis of the tables in front of us other than as constructs out of our own experience (ibid.: 25).

Stebbing concluded, "Logical positivism fails, I think, in so far as it attempts to start from *a priori* assumptions with regard to the nature of language and the principles of symbolism, and, by means of these, to draw limits with regard to what we *can* think" (ibid.: 36, orig. emphasis). But her criticisms pertain only to an early stage of logical empiricism and tend towards the Wittgenstein-influenced "right wing" of the Vienna Circle. Unfortunately, Stebbing wrote no systematic treatment of later logical empiricism. By the late 1930s, increasingly horrified by the rise of fascism on the continent, she had turned her mind to writing popular books which applied her expertise on critical thinking to politics, and to helping Jewish refugees, including Neurath's Isotype Institute (Körber 2019), work which she continued until her death (Chapman 2013).

Women of the Lvov-Warsaw School

While the philosophers and logicians of Lvov and Warsaw are often collectively called "the Lvov-Warsaw School," they had no shared body of doctrine, although they all shared, and highly valued, a strong methodological focus on thorough justification and on precision and clarity. They were equally diverse with respect to gender and ethnicity. This diversity had its roots in the feminist and anti-racist views of Kazimierz Twardowski, professor of philosophy in Lvov. In 1915 a group of his former students, including Kotarbiński and Lesniewski, settled in Warsaw. Twardowski strongly encouraged and supported women as well as ethnically Jewish and Ukrainian students, all of whom faced significant discrimination in the Polish education system. Several Lvov-Warsaw women were Jewish, vulnerable qua women in the academy, but even more so, in interbellum Poland, because of their ethnicity. Two brilliant female logicians were murdered by the Nazis: Eugenia Ginsberg-Blaustein and Janina Hosiasson-Lindenbaum. Had they lived, they would probably have had long and influential careers. The great diversity of views among the women of Lvov and Warsaw means that some are closer to logical empiricism than others. It is only the former I can consider here.

Janina Kotarbińska (1901–1997) studied philosophy in Warsaw, and wrote her PhD on indeterminism in physics, biology, and the social sciences. In 1945 she became professor of epistemology, first at Poznań, then at Warsaw. Her early work remains untranslated, but it pertained to key themes of logical empiricism. It focused on the philosophy of science, including classification of physicalisms up to *Carnap's Testability and Meaning* (Hempel 1950).

Seweryna Łuszczewska-Rohmanova (1904–1978) was a mathematical logician who published primarily in Polish. She studied mathematics with Banach, and philosophy with Twardowski and Ajdukiewicz, in Lvov. She received her PhD in 1932, and soon afterwards became professor of logic in Poznań, where she spent the rest of her career (Betti 2001). Her publications largely remain untranslated, but one key paper published in English (Łuszczewska-Rohmanova 1961) sparked the small mathematical subfield of the general theory of classification, long thought relevant to the philosophy of science. Previous investigations had been restricted to finite classifications. Łuszczewska-Rohmanova's treatment set classification apart as a subject of study in its own right, capable of infinitary treatment and clearly separable from set theory. After a decades-long hiatus, the general theory of classification is now a subject of study for logicians in the twenty-first century (Parrochia and Neuville 2013). Łuszczewska-Rohmanova can therefore be seen as having achieved a key logical empiricist goal, succeeding in converting a question previously considered a philosophical one into a logico-mathematical one.

The works of Maria Kokoszyńska (1905–1981), a PhD student of Twardowski and visitor to the Vienna Circle in 1934–5, are clearly relevant to logical empiricism. Most of her papers were published either in German (Kokoszyńska 1936, 1937/38) or Polish, and remain unavailable in English in their entirety, although key passages from the Polish are translated in Brożek (2017).

While Tarski's theory of truth (1933) remains the most influential work to come out of the Lvov-Warsaw School from the point of view of analytic philosophy, it was Kokoszyńska who wrote the first paper about its philosophical implications (1936) and defended it in correspondence with Neurath (Mancosu 2008). She framed it as a direct challenge to Carnap's syntacticism, according to which truth was metaphysical and unscientific, to be replaced with syntax.

> In recent philosophical research there has been a tendency towards eliminating the absolute conception of truth, which, incidentally, conforms to the everyday conception (and according to which, roughly speaking, the truth of a sentence consists in its agreement with reality), from science or replacing it with another one. . . . [I want to] make some critical remarks about the manner in which philosophers want to liberate themselves from the absolute conception of truth, and then generally cast doubt on the idea that such a notion is unscientific.
>
> (Kokoszyńska 1936: 143, trans FJL)

Kokoszyńska then made the case that agreement between language and reality means that truth is indeed semantic, not syntactic, but that the logical method of satisfaction by sequences renders it nevertheless clear and scientific. "What is expressed in [Convention T] is that which people believed could be extracted from the classical conception of truth: to prove this, let me just point to the well-known 'definition' of truth we can find in Aristotle, especially to the words ['to say of what is that it is, or of what is not that it is not, is true']" (ibid.: 151, trans. FJL). She concluded that while "the absolute conception of truth, in relation to a particular language, can therefore in no way be reduced to a syntactic conception related to that language" (ibid.: 153, trans. FJL), this is not problematic. As absolute truth is not unscientific, the quest for a criterion of truth can be made scientific.

Kokoszyńska's argument here has some affinity with Stebbing's 1933 argument against Carnap's syntacticism, leaving room for an empiricist approach to metaphysics or semantics. Kokoszyńska appears to have been moved by similar motivations when she argued that the claim of the unity of science was not just unclear but either implausible—if it meant a strict reductive physicalism—or too weak to be of much use if it meant little more than anti-metaphysics (Kokoszyńska 1937/38). She considered the boundary between science and metaphysics more porous than the Vienna Circle, arguing that some apparently metaphysical claims are in fact highly general scientific statements which are confirmable. Her argument is very compressed, but Brożek suggests that she had in mind statements like "All events have causes," or "Mental events have physical causes" (Brożek 2017: 26).

The contributions of Hosiasson-Lindenbaum (1899–1942) to the philosophy of science deserve to be better known to those interested in logical empiricism (see Galavotti 2008, 2014). After studying mathematics and philosophy in Warsaw and receiving her PhD on inductive reasoning in 1926, Hosiasson worked as a schoolteacher for some years. She held a visiting fellowship in Cambridge in 1929–30 and contributed to *Mind* (1931). In 1935, she married fellow logician Adolf Lindenbaum and began to use a hyphenated surname. Having presented at the 1935 Paris Unity of Science Congress (1936, see Galavotti 2018), Hosiasson-Lindenbaum had planned, like Tarski, to attend the Harvard Unity of Science Congress. Tarski, who, like her, was Jewish, was saved by his presence at Harvard in early September 1939, when the Nazis invaded Poland. But Hosiasson-Lindenbaum's visa was denied, so she could not travel (Feferman and Feferman 2004: 108). She fled to Vilnius, but two years later Lithuania, too, was invaded by the Nazis. The Gestapo murdered Hosiasson-Lindenbaum in 1942.

Hosiasson-Lindenbaum sent her brilliant solution to the paradox of the ravens, the first ever published, to the *Journal of Symbolic Logic* in August 1940. Her topic is central to one of the key themes of logical empiricism: confirmation theory. Carl Hempel, who first formulated the raven paradox, held a deductive-nomological model of scientific explanation, according to which a general hypothesis is a conditional, confirmed by positive instances and disconfirmed by negative ones. Intuitively speaking, a positive instance is something satisfying the conditional's antecedent and consequent, a negative instance something satisfying the antecedent, but not the consequent. "All ravens are black," for example, is confirmed by instances which are ravens and black, and disconfirmed by ravens which aren't black. But "all ravens are black" is logically equivalent to "no non-black things are non-ravens" and to "Everything is either not a raven or black." The paradoxical consequence is that observations of non-black non-ravens—like my blue scarf—confirm the hypothesis "all ravens are black." Hempel (1945) sought to disqualify such paradoxical results, feeling that they should not count as confirming the hypothesis in any way. Hosiasson-Lindenbaum's solution, by contrast, counted "paradoxical" instances as confirming the hypothesis, but only to a negligible degree.

Hosiasson-Lindenbaum proposed a probabilistic approach to confirmation, according to which confirmation comes in degrees. She assigned numbers on a scale of 0 to 1 as values of the function $\mathfrak{c}$ (*a*,*b*), meaning the degree of confirmation of *a* with respect to *b*. Intuitively, she explained, a hypothesis is more probable the more observed instances it has. With respect to confirmation theory, this implies, for the degree of confirmation $\mathfrak{c}$, that "f4) The smaller is the a priori $\mathfrak{c}$ of a fact, the more does the $\mathfrak{c}$ of its law or hypothesis increase when this fact is observed" (Hosiasson-Lindenbaum 1940: 134). Her solution has at its heart the application of f4) to the raven paradox.

Hosiasson-Lindenbaum remarked, "we would find it rather curious if a chemist, in order to confirm ['All kitchen salt is soluble in water'], should take substances insoluble in water and then examine them to see if they are salt, instead of taking salts in order to discover whether they are soluble in water" (1940: 137). We would find this strange, she thought, because the observation of objects insoluble in water raises the degree of confirmation of "All kitchen salt is soluble in water" only by a negligible degree compared to the observation of salt. After all, the number of things insoluble in water is vast compared to the number of things which are salt, so, Hosiasson-Lindenbaum proved, "if the number of $\bar{\text{B}}$'s (non-B's) is greater than the number of A's, then the $\mathfrak{c}$ of an instance of $\bar{\text{B}}$ being $\bar{\text{A}}$ is, *ceteris paribus*, greater than the $\mathfrak{c}$ of an instance of A being B. Thus, on the strength of f4), the second instance raises the $\mathfrak{c}$ of the sentence 'Every A is B' by a greater value than the first" (Hosiasson-Lindenbaum 1940: 137, orig. emphasis). Her solution, which has not been given the attention it deserves by philosophers in part because of its thoroughgoing mathematical character, has recently been rediscovered and highly praised by authors in the field of confirmation theory (Gaifman 1979: 138; Galavotti 2007: 120).

References

Adorno, T. W., Frenkel-Brunswik, E., Levinson, D. J. and Sanford, R. N. (1950) *The Authoritarian Personality*, New York: New Harper.

Beaney, M. and Chapman, S. (2017) "Susan Stebbing," in E. N. Zalta (ed.), *The Stanford Encyclopedia of Philosophy*, https://plato.stanford.edu/archives/sum2017/entries/stebbing.

Betti, A. (2001) "Le Donne nella Filosofia Analitica Austriaca," in M. Marsonet (ed.), *Donne e Filosofia*, Florence: Erga Edizioni, pp. 98–118.
Borchers, D. (2003) "No Woman, No Try? Else Frenkel-Brunswik and the Project of Integrating Psychoanalysis into the Unity of Science," in F. Stadler (ed.), *The Vienna Circle and Logical Empiricism: Reevaluation and New Perspectives*, Dordrecht: Kluwer, pp. 323–38.
Broad, J. (2006) "Female Philosophers," in A. Grayling, A. Pyle, and N. Goulder (eds.), *The Continuum Encyclopedia of British Philosophy*, Bristol: Thoemmes Continuum, vol. 20, pp. 1066–9.
Brożek, A. (2017) "Maria Kokoszyńska: Between the Lvov-Warsaw School and the Vienna Circle," *Journal for the History of Analytical Philosophy* 5: 19–36.
Cat, J. (2019) "Neurath and the Legacy of Algebraic Logic," in Cat and Tuboly (2019), 241–338.
Cat, J. and Tuboly, A. T. (eds.) (2019) *Neurath Reconsidered*, Cham: Springer.
Chapman, S. (2013) *Susan Stebbing and the Language of Common Sense*, Basingstoke: Palgrave Macmillan.
Church, A. (1936) "A Bibliography of Symbolic Logic," *Journal of Symbolic Logic* 1: 121–216.
Coffa, J. A. (1991) *The Semantic Tradition from Kant to Carnap: To the Vienna Station*, Cambridge: Cambridge University Press.
Feferman, S. and Feferman, A. B. (2004) *Alfred Tarski: Life and Logic*, Cambridge: Cambridge University Press.
Frenkel-Brunswik, E. (1954) "Psychoanalysis and the Unity of Science," *Proceedings of the American Academy of Arts and Sciences* 80: 271–347.
Gaifman, H. (1979) "Subjective Probability, Natural Predicates and Hempel's Ravens," *Erkenntnis* 14: 105–47.
Galavotti, M. C. (2007) "Confirmation, Probability, and Logical Empiricism," in Richardson and Uebel (2007), pp. 117–35.
——— (2008) "A Tribute to Janina Hosiasson-Lindenbaum: A Philosopher Victim of the Holocaust," in R. Scazzieri and R. Simili (eds.), *Migration of Ideas*, Sagamore Beach: Watson, pp. 179–94.
——— (2014) "Probabilistic Epistemology: A European Tradition," in M. C. Galavotti et al. (eds.), *European Philosophy of Science*, Dordrecht: Springer, pp. 77–88.
——— (2018) "The Sessions on Induction and Probability at the 1935 Paris Congress: An Overview," *Philosophia Scientiae* 22 (3): 213–32.
Geiringer, H (1939) "Über die Wahrscheinlichkeit von Hypothesen," *Journal of Unified Science (Erkenntnis)* 8: 151–76.
Goodstein, J. R. (2020) "Olga Taussky-Todd," *Notices of the American Mathematical Society* 67 (3): 345–53.
Hahn, O. and Neurath, O. (1909) "Zum Dualismus in der Logik," *Archiv für systematische Philosophie* 15: 149–62. Trans. "On Duality in Logic," in Cat and Tuboly (2019), 493–505.
Hahn, O. (1909) "Zur Axiomatik des logischen Gebietkalküls," *Archiv für systematische Philosophie* 15: 345–7. Trans. "On the Axiomatics of the Logical Calculus of Domains," in Cat and Tuboly (2019), pp. 507–9.
——— (1910) "Über die Koeffizienten einer logischen Gleichung und ihre Beziehung zur Lehre von den Schlüssen," *Archiv für systematische Philosophie*, 16: 149–76.
Hempel, C. (1945) "Studies in the Logic of Confirmation," *Mind* 54: 1–26, 97–121.
——— (1950) "[Review of J. Kotarbińska 'Le Physicalisme et les Étapes de son Évolution']," *Journal of Symbolic Logic* 14: 247.
Hosiasson, J. (1931) "Why Do We Prefer Probabilities Relative to Many Data?" *Mind* 40: 23–36.
——— (1936) "La théorie des probabilités est-elle une logique généralisée? Analyse critique," in *Actes du Congrès international de philosophie scientifique, vol. IV: Induction et probabilité*, Paris: Hermann, pp. 59–64.
Hosiasson-Lindenbaum, J. (1940) "On Confirmation," *Journal of Symbolic Logic* 5: 133–48.
Heinemann, A. S. and Reichenberger, A. (2019) "Zur frühen Frege-Rezeption im deutschsprachigen Raum: Papst und Krenz," in M. Wille and V. Peckhaus (eds.), *Fregesche Variationen*, Mentis: Münster.
Janssen-Lauret, F. (2017) "Susan Stebbing, Incomplete Symbols, and Foundherentist Meta-Ontology," *Journal for the History of Analytical Philosophy* 5: 6–17.
——— (2018) "Willard Van Orman Quine's Philosophical Development in the 1930s and 1940s," in W. Carnielli, F. Janssen-Lauret and W. Pickering (eds.), *The Significance of the New Logic*, Cambridge: Cambridge University Press, 2018, pp. xiv–xlvii.
Janssen-Lauret, F. (forthcoming) "Grandmothers of Analytic Philosophy," in *Feminist Philosophy and Formal Logic* (*Minnesota Studies in Philosophy of Science*), Minneapolis: University of Minnesota Press.
Körber, S. (2019) "Thinking About the Common Reader: Neurath, Stebbing and the Modern Picture-Text Style," in Cat and Tuboly (2019), pp. 451–70.

Kokoszyńska, M. (1936) "Über den absoluten Wahrheitsbegriff und einige andere semantische Begriffe," *Erkenntnis* 6: 143–65.

——— (1937/38) "Bemerkungen über die Einheitswissenschaft," *Erkenntnis* 7: 325–35.

Lewis, C. I. (1918) *A Survey of Symbolic Logic*, Berkeley: University of California Press.

Łuszczewska-Rohmanova, S. (1961) "Classification as a Kind of Distance Function," *Studia Logica* 12: 41–66.

Mancosu, P. (2008) "Tarski, Neurath and Kokoszynska on the Semantic Conception of Truth,", in D. Patterson (ed.), *New Essays on Tarski and Philosophy*, Oxford: Oxford University Press, pp. 192–224.

Nagel, E. (1938) [Review of R. Rand, "T. Kotarbińskis Philosophie auf Grund seines Hauptwerkes: 'Elemente der Erkenntnistheorie, der Logik und der Methodologie der Wissenschaften',"] *Journal of Symbolic Logic* 3: 169.

Papst, W. (1930) *Gottlob Frege als Philosoph*, Phil. diss., Humboldt University, Berlin.

Parrochia, D. and Neuville, P. (2013) *Towards a General Theory of Classifications*, Basel: Springer.

Perzanowski, J. (1983) "Izydora Dąmbska (1904–1983)," *Studia Logica* 42: 379–81.

Rand, R. (1938) "T. Kotarbińskis Philosophie auf Grund seines Hauptwerkes: 'Elemente der Erkenntnistheorie, der Logik und der Methodologie der Wissenschaften'," *Erkenntnis* 7: 92–120.

——— (1939) "Logik der Forderungssätze," *Internationale Zeitschrift für Rechtsphilosophie* 1: 308–22. Trans. "The Logic of Demand-Sentences," *Synthese* 14 (1962): 237–54.

Reisch, G. (2007) "From 'the Life of the Present' to the 'Icy Slopes of Logic': Logical Empiricism, the Unity of Science, and the Cold War," in Richardson and Uebel (2007), pp. 58–90.

Richardson, A. and Uebel, T. (eds.) (2007) *The Cambridge Companion to Logical Empiricism*, Cambridge: Cambridge University Press.

Stadler, F. (2007) "The Vienna Circle: Context, Profile, and Development," in Richardson and Uebel (2007), pp. 13–40.

Stebbing, L. S. (1932) "The Method of Analysis in Metaphysics," *Proceedings of the Aristotelian Society* 33: 65–94.

Stebbing, L. S. (1933a) "Logical Positivism and Analysis," *Proceedings of the British Academy* 19: 53–87, published separately as (1933b) *Logical Positivism and Analysis*, London: Humphrey Milford.

Steinhardt, K (1938) "Einige Bemerkungen zum Begriffe der Wahrnehmung und zu den Wahrnehmungssätzen," *Synthese* 3: 118–27.

Tarski, A. (1933) *Pojecie prawdy w jezykac nauk dedukcyjnych*, Warsaw. German trans. "Der Wahrheitsbegriff in den formalisierten Sprachen," *Studia Philosophica* 1 (1935): 261–405; English trans. "The Concept of Truth in Formalized Languages," in Tarski, *Logic, Sematics, Metamathemathics* (ed. by J. H. Woodger), Oxford: Clarendon Press, 1956, pp. 152–278, 2nd ed. (ed. by J. Corcoran), Indianapolis: Hackett, 1983.

Taussky, O. (1988a) "How I Became a Torchbearer for Matrix Theory," *American Mathematical Monthly* 95: 801–12.

Taussky-Todd, O. (1988b) "Remembrances of Kurt Gödel," *Engineering & Science* 51: 24–28.

Uebel, T. (2016) "Vienna Circle," in E. N. Zalta (ed.), *The Stanford Encyclopedia of Philosophy*, https://plato.stanford.edu/archives/spr2016/entries/vienna-circle.

Wolenski, J. (1989) *Logic and Philosophy in the Lvov-Warsaw School*, Dordrecht: Kluwer.

PART II

Characteristic theses of and specific issues in logical empiricism

14

LOGIC AND THE FOUNDATIONS OF MATHEMATICS IN EARLY LOGICAL EMPIRICISM

Erich H. Reck

As the so-called manifesto of the Vienna Circle, *Wissenschaftliche Weltauffassung* (Verein Ernst Mach 1929), highlights, several logicians and philosophers of mathematics had a decisive influence on it, especially Bertrand Russell, David Hilbert, and Ludwig Wittgenstein. In this chapter, we consider themes associated with these three names that form a crucial part of the background, the early rise, and the legacy of logical empiricism.

Our focus is on the Vienna Circle and the period from 1905 to 1930, for reasons that will become apparent. Three of its members are central in this context: Hans Hahn, Rudolf Carnap, and Kurt Gödel. But other members play smaller roles too: Karl Menger, Otto Neurath, and Moritz Schlick; also Herbert Feigl, Philipp Frank, Olga Hahn-Neurath, Rose Rand, Kurt Reidemeister, and Friedrich Waismann. And besides the impact of Russell, Hilbert, and Wittgenstein, there are connections to other logicians and philosophers of mathematics: Bernard Bolzano, L. E. J. Brouwer, Louis Couturat, Walter Dubislav, Abraham Fraenkel, Gottlob Frege, Frank Ramsey, Ernst Schröder, and Alfred Tarski, among others.

Various members of the Vienna Circle had interests in logic from early on. Debates about the foundations of mathematics were an integral part of their discussions too, with connections to all three of the main schools at the time: logicism, formalism, and intuitionism. In addition, there was a particularly strong interest in the nature of logic, which led to novel views about both logic and mathematics. Some technical achievements resulted as well, including Gödel's famous meta-logical theorems.

Logic in the first Vienna Circle: Hahn, Frank, and Neurath

The members of what is often called the "first Vienna Circle," which met from 1907 on, were Hans Hahn, Otto Neurath, and Philipp Frank (see CH. 10). In 1902, Hahn received his PhD in mathematics in Vienna; Neurath, who had also taken classes in mathematics at the University of Vienna, went on to get a PhD in history of economics in Berlin in 1906; Frank obtained his PhD in physics in 1907, again in Vienna. Between 1902 and 1905, Hahn and Frank also spent several semesters at the University of Göttingen (Hahn as a postdoctoral researcher and Frank as a student). This was shortly after the publication of Hilbert's *Grundlagen der Geometrie*, in 1899. Hilbert and other members of his school were then busy axiomatizing other parts of

 DOI: 10.4324/9781315650647-17

mathematics and physics, including Ernst Zermelo with his trailblazing work on axiomatic set theory.

It was during their time in Göttingen that Hahn and Frank were first introduced to debates about logic and the foundations of mathematics. Hahn was collaborating with Zermelo then, by co-writing surveys on the calculus of variations (one of his areas of expertise, besides real analysis, functional analysis, and topology), and it was in this connection that he encountered Russell's logicism for the first time (Menger 1980). Frank was working with Hilbert, Felix Klein, and others on geometry and physics; but in this context too, foundational issues were not far away. When Hahn and Frank returned to Vienna, in 1904 and 1905, they brought their familiarity with these topics with them. And when Neurath joined their group, after his return from Berlin, one of the first texts they studied together was Russell's *Principles of Mathematics* from 1903 (Uebel 2000: 70).

Attention to Russell's new logic, and his associated logicism, was thus on the agenda of the Vienna Circle from the very beginning. The members' readings in this area were not restricted to Russell, however. For example, they read Louis Couturat's *Les principes des mathématiques* (1905); and for this they were joined by Alois Höfler, then the leader of the Philosophical Society at the University of Vienna and author of a logic book written with Alexius Meinong in 1890. Hahn and Höfler also collaborated on re-editing some of Bernard Bolzano's writings. One outcome was a new edition of Bolzano's *Paradoxien des Unendlichen*, published in 1920 (see CH. 6). Neurath too worked extensively on logic during these early years, publishing a series of articles on Ernst Schröder and Boolean algebra in 1909–10, several of them co-authored with Hans Hahn's sister, Olga Hahn (later Neurath's wife) (see Cat 2019).

Around 1910 the paths of Hahn, Frank, and Neurath diverged for a while. Hahn got his first professorship in mathematics at the University of Czernowitz (now Ukraine) in 1909. Frank became professor of physics at the University of Prague (as Einstein's successor) in 1912. While Neurath stayed in Vienna, he also traveled extensively in the Balkans for research purposes. In World War I, which disrupted everything, all three served in the Austrian army, some seeing active duty (like Wittgenstein, as is well known). Despite the publication of several major works in logic and related fields just before it—Whitehead and Russell's *Principia Mathematica, Vols. I—III* (1910–13), Russell's *Our Knowledge of the External World* (1914), also neo-Kantian works such as Paul Natorp's *Die logischen Grundlagen der exakten Wissenschaften* (1910) and Ernst Cassirer's *Substanzbegriff und Funktionsbegriff* (1910)—the war now overshadowed all research. This would only change again around 1921, especially in Vienna.

Schlick's and Carnap's interests in logic before Vienna

Moritz Schlick and Rudolf Carnap, who would soon move to Vienna, also developed an interest in modern logic and the philosophy of mathematics early. Focusing on physics initially, Schlick finished his dissertation under Max Planck in Berlin in 1904, but he soon pursued his interests in philosophy as well. These concerned epistemology, ethics, aesthetics, and some logic (initially traditional syllogistic logic). In his early work in epistemology, which extended into the philosophy of science, Hilbertian axiomatics was important, including the topic of "implicit definitions" (1918: Part I). His subsequent book on Einstein's theories of relativity (1920) was quickly acknowledged as authoritative (see CH. 8). Schlick received his first professorship at the University of Rostock in 1911, from where he moved on to Kiel (both in northern Germany). During the war he too had to serve in the (German) army.

In 1910–14, Rudolf Carnap studied at the University of Jena, majoring in physics, mathematics, and philosophy. He encountered modern logic right away by taking several classes with Frege (see Reck and Awodey 2004). As Frege had few students, this was unusual; he was still largely unknown at the time (although the young Wittgenstein also visited him during this period). Carnap did not immediately recognize the full potential of Frege's logic, as he noted later. And he too was drawn into World War I as a soldier. It was only in 1921 that Carnap finished his dissertation, under the neo-Kantian Bruno Bauch, on different conceptions of space (1922). While not concerned with logic directly, the use of Hilbertian axiomatics again played a crucial role in it.

Carnap started to educate himself more thoroughly in the new logic in 1920–1. He did this by studying several of Russell's books, such as *Our Knowledge of the External World* (1914) and *Introduction to Mathematical Philosophy* (1919), but also by going back to Frege's published works. Whitehead and Russell's *Principia Mathematica* was on his reading list too, but Carnap had trouble finding an affordable copy (because of severe inflation in Germany, among other reasons). He wrote to Russell directly and asked for a copy. While Russell did not have a spare one, he sent him a 30-page handwritten summary of the book's main results (Reck 2004). At the time, Carnap lived in the South-West of Germany (in Buchenbach), without an academic job, and attending classes by Edmund Husserl at the University of Freiburg (1921–2).

Carnap also corresponded with other young logicians and "mathematical philosophers" during this period, especially Hans Reichenbach in Berlin and Heinrich Behmann in Göttingen (a member of Hilbert's school). As one outcome, they organized a conference in Erlangen in 1923 (see CH. 11). In addition, Carnap and Behmann had plans to co-author an introductory book on Fregean-Russellian logic to make its techniques more widely accessible. While the two did not finish it, the project later turned into Carnap's *Abriss der Logistik* (1929), a text that had a significant influence in Vienna.

Logic and foundations of mathematics in the Vienna Circle

In 1921 Hahn returned to Vienna (from the University of Bonn), as an established professor of mathematics. He was an inspiring lecturer as well, and one topic for his classes became logic. In 1922 he offered a course on Boolean algebra; and in 1924–5 he covered *Principia Mathematica* in a series of classes co-taught with Kurt Reidemeister (another faculty member in Vienna and an expert on geometry). The detailed study of Whitehead and Russell's new logic, together with their logicism, had become a priority for him (cf. Sigmund 1995). Neurath too returned to Vienna in 1921, after having been imprisoned for his activities in the failed Bavarian revolution. And Frank began to visit frequently from Prague, so that the three friends could restart their joint activities.

Soon thereafter, in 1922, Schlick was hired as professor of philosophy in Vienna. He quickly became the figure around which the Viennese "scientific philosophers" gathered in regular evening meetings. From 1924 on, partly at the request of younger members of the group, such as Herbert Feigl and Friedrich Waismann (Schlick's assistants), a main topic in these meetings became the philosophy of logic and mathematics. In 1926, Carnap came to Vienna for his habilitation under Schlick, to be published as *Der Logische Aufbau der Welt* (1928). Schlick then offered classes on Russell's philosophy, while Carnap took over Hahn's logic course, with *Abriss der Logistik* as his textbook.

Like Hahn, Carnap was sympathetic to Russell's logicism, an earlier version of which he had encountered in Frege's works. He also remained interested in Hilbertian axiomatics, which he tried to reconcile with logicism from early on (Reck 2013). In addition, he became very interested in Fraenkel's *Einleitung in die Mengenlehre* (1923), and more specifically, in the section on "general axiomatics" in it. That section included a preliminary, informal comparison of several notions of completeness for axiom systems. Carnap tried to go further by applying formal-logical tools to the issue: he embedded Fraenkel's discussion in an early version of the simple theory of types and tried to prove corresponding theorems based on it (Reck 2007).

Carnap never fully succeeded with this project. Thus, it was largely ignored later, also in his own recollections. But what resulted at the time, around 1928, was a relatively complete book manuscript, entitled *Untersuchungen zur allgemeinen Axiomatik* (2000). This manuscript was circulated within the Vienna Circle (importantly including Kurt Gödel by then). This episode illustrates that engagement with logic and the foundations of mathematics by members of the Vienna Circle did not remain limited to a passive reception. Nor did it just concern applications of logic to philosophy and other fields, as a narrow focus on Carnap's *Aufbau* may suggest, although the emphasis on such applications was important in Vienna.

Connections to Wittgenstein and his *Tractatus*

Before expanding on the fate of Carnap's general axiomatics project, another influence on the Vienna Circle with respect to the philosophy of logic needs to be factored in: Wittgenstein's *Tractatus Logico-Philosophicus*. A partly corrupted version of this text was first published in a German journal in 1922 (see Schmidt 2016); a corrected bilingual version, with an English translation by Frank Ramsey, was published later the same year. Despite Russell's preface, the importance of the *Tractatus* was not recognized right away, also in Vienna. Two events changed that. First, in 1923 Ramsey paid an initial visit to Wittgenstein in Puchberg (where he then worked as an elementary school teacher) to talk to him about the *Tractatus*, and in 1924, he combined this with visits to Vienna where he met with Schlick, among others. Second, Kurt Reidemeister gave a presentation on the *Tractatus* to the Schlick Circle in 1924, which led to its in-depth study in the meetings of the Circle in 1925–6.

Several aspects of the *Tractatus* were received very positively in those meetings, including its focus on language to solve, or dissolve, philosophical problems. Probably the most important ingredient, however, was Wittgenstein's notion of tautology. A central goal for the Vienna Circle was always to develop an updated, fully satisfactory form of empiricism, thereby building on earlier empiricists, such as Mach, Mill, and Hume, and distancing themselves from Neo-Kantianism (Natorp, Rickert, etc.). A main reason for Hahn's and Carnap's interest in Russellian logicism was that it provided a novel perspective on mathematics directed against Kantian ideas (most explicitly in Couturat's 1905 book). Yet neither Frege nor Russell had given a fully satisfactory account of the nature of logic. Wittgenstein's notion of tautology filled that gap and did so in a deflationary way. Logical truths no longer described the "most abstract or general features" of the world, as Russell still thought only a few years earlier (1919: 169), but were empty of content. For the Viennese and other empiricists, this promised an understanding of all of mathematics along the same deflationary lines.

Wittgenstein himself came back to Vienna in 1926, involved in the design of a house for his sister Margarethe Stoneborough. Schlick got in touch with him in 1927. This led to a number of meetings with members of the Vienna Circle, including Schlick, Carnap, Waismann, and Feigl (see CH. 30). Among the topics discussed were the principle of verifiability, but also themes in logic and the foundations of mathematics. Wittgenstein was not sympathetic to the "scientific spirit" of the Vienna Circle, which contributed to a partial split within it. Members on the "right wing," including Schlick and Waismann, adopted Wittgenstein's informal approach to philosophy (with Waismann attempting to co-write a book with him). Members on the "left wing," like Carnap, Hahn, and Neurath, pushed in a more "scientific" direction and advocated the use of formal-logical tools.

Menger, Gödel, and breakthroughs in metalogic

Two of Hahn's, and partly Carnap's, students in Vienna soon played crucial roles with respect to logic and the foundations of mathematics as well. Both were mathematicians with strong philosophical interests, and neither was sympathetic to Wittgenstein's informal approach. The first was Karl Menger, who finished his dissertation in mathematics under Hahn in 1924. Menger continued his training in the Netherlands for two years, working with L. B. J. Brouwer on issues in topology and geometry (including the theory of dimensions for which Menger is well known). There he encountered Brouwerian intuitionism firsthand. He returned to Vienna in 1927, as a faculty member in mathematics. He also soon organized a regular "Mathematical Colloquium" (see Menger 2012), besides attending meetings of the Vienna Circle.

Menger was not a strict follower of Brouwer's intuitionism, but his presence in the Circle raised the profile of intuitionist or constructivist approaches to mathematics. It also made people more aware of the clashes between Brouwer and Hilbert in the 1920s. In fact, on Menger's invitation, Brouwer visited Vienna in 1928 to give lectures on mathematics, science, and language. These lectures were well attended (and proved important for Wittgenstein). Yet neither Menger, with his ties to Brouwerian intuitionism, nor Carnap and Hahn, with their sympathies for Frege-Russell logicism and their long-standing interest in Hilbertian axiomatics and formalism, were dogmatic in the ensuing debates. Carnap had tried to combine logicism and formalism for a while already; and all three now started to advocate a kind of tolerance, thus forms of pragmatic pluralism, concerning logic, the foundations of mathematics, and beyond (including Carnap's famous "principle of tolerance" from 1934; see Uebel 2005).

Hahn's second important student at the University of Vienna, several years younger than Menger, was Kurt Gödel. He had moved to Vienna in 1924 so as to study theoretical physics, then shifted towards pure mathematics, e.g., by taking classes in number theory (with Philipp Furtwänger). This led him to Menger's Mathematical Colloquium, in which he participated actively. He also took classes on logic by Carnap, on Russell by Schlick, and on philosophy more generally by Heinrich Gomperz. From 1926 on, Gödel too was invited to meetings of the Vienna Circle. Another crucial influence on him, from outside, was Hilbert. This included Hilbert's recent articles "Die logischen Grundlagen der Mathematik" (1923) and "Über das Unendliche" (1926), as well as Hilbert and Ackermann's *Grundzüge der Theoretischen Logik* (1928). Gödel also attended a lecture by Hilbert at the 1928 mathematics congress in Bologna, where the problem of the completeness of first-order logic was raised (Hilbert 1929).

Gödel's dissertation, finished under Hahn in 1929, addressed this problem head on, by providing the required proof (building on earlier proofs of completeness for sentential logic by Paul Bernays and Emil Post). This established Gödel immediately as a promising, even prominent logician. He published the core of the dissertation (1930); he also presented it at a conference in Königsberg in the fall of 1930 (more on which soon). Gödel immediately went on to other major contributions, namely his famous incompleteness theorems (1931). Hilbert's work was again an important part of the background for them; yet equally relevant was Carnap's "general axiomatics" project. Gödel knew about the latter directly by reading the manuscripts for *Untersuchungen zur allgemeinen Axiomatik* and for *Abriss der Logistik* that circulated in Vienna. In turn, Carnap was one of the first to find out about Gödel's new results.

A main target for Gödel's second incompleteness theorem (on the unprovability of consistency for arithmetic) was Hilbert's proof-theoretic program, in which proofs of consistency were a core desideratum. But both it and Gödel's first incompleteness theorem (on the incompleteness of the axioms for arithmetic and related systems) were clearly directed at Carnap's project too. They showed, among others, that Carnap's and other logicians' assumptions about the relationship between the proof-theoretic and the model-theoretic sides of logic were mistaken in crucial respects (see Reck 2007). At the time Carnap had not yet finished, much less published, his *Allgemeine Axiomatik* manuscript. After he became aware of its basic flaws, in 1930, it disappeared in a drawer. Because of that disappearance, Gödel's results came to be seen as directed mainly at Hilbert. This was reinforced by the fact that John von Neumann was present when Gödel mentioned them briefly during a discussion at the Königsberg conference in 1930; von Neumann quickly perceived their negative significance for the Hilbert program.

Important innovations, philosophical themes, and further connections

Among the contributions to logic and the foundations of mathematics in the Vienna Circle, Gödel's theorems are surely the most profound and lasting. They alone require that any comprehensive account of it, or of logical empiricism more generally, needs to cover these areas. But Gödel's theorems did not arise in a vacuum, as we saw. He was influenced by Hilbert's writings, to be sure; but his interests were shaped more directly by logic classes he took and by foundational debates he witnessed in Vienna, involving Hahn, Carnap, Menger, and others. There was the direct connection to Carnap's "general axiomatics" project; and that project was, in turn, tied to debates about "implicit definitions" in the sciences, e.g., in Schlick's works. Both Gödel's results and Carnap's work in logic also had a broader impact in philosophy.

In taking modern logic very seriously, the members of the Vienna Circle were not alone; this also happened in Hilbert's Göttingen, in Reichenbach's Berlin, and in Tarski's Warsaw (more on the latter two soon). The master text for all of them was *Principia Mathematica*. But the logic in it was overly complicated and problematic in some respects, especially as a form of logicism. This was pointed out in Frank Ramsey's article, "The Foundations of Mathematics" (1925). One of the contributions in Carnap's textbook, *Abriss der Logistik*, was to build on Ramsey by promoting a simple theory of types as the logical framework, not the ramified version of *Principia*. In fact, *Abriss* provided one of the first formal presentations of this framework (later refined by Gödel, Church, and others). Another contribution by Carnap was to suggest a variety of applications of type theory, both in *Abriss* and in *Allgemeine Axiomatik*.

Various of the logicians mentioned were then moving towards "meta-logic," i.e., the systematic study of general properties of logical systems: Carnap, with his use of type theory to advance Fraenkel's study of notions of completeness; Hilbert, with his proof-theoretic emphasis on the notion of consistency; and Gödel, with his famous theorems. The logical systems at issue included not just simple and ramified type theory, but also first-order logic and sentential logic, after being recognized as self-contained and interesting sub-systems. There was the investigation of axiomatized mathematical theories, such as arithmetic, analysis, and set theory, as part of "meta-mathematics." Besides their intrinsic significance, these projects began to be treated as models for studying other topics too, including: other forms of deductive logic (modal, many-valued, etc.); probability and inductive logic (e.g., Bayesian approaches); the content of scientific theories (their "non-observational" parts, among others); and linguistic meaning (formal syntax and semantics). From the 1930s on, Carnap in particular contributed to such topics in influential ways, thus developing the blueprint for "formal philosophy."

Two philosophical issues concerning logic and mathematics have played a prominent role in our survey so far: the clashes between logicism, formalism, and intuitionism; and the precise nature of logic. For members of the Vienna Circle intent on updating empiricism, a combination of Russell's logicism and Wittgenstein's deflation of logic was especially attractive. These thinkers acknowledged some open questions about logicism (e.g., the status of the axioms of reducibility, infinity, and multiplicity in *Principia*); they were also aware that extending the notion of tautology beyond sentential logic was not unproblematic. But they treated both as challenges rather than as looming refutations. This resulted in modified, unorthodox forms of logicism, e.g., that proposed in Carnap's presentation at the Königsberg conference in 1930. Similarly for several attempts by Carnap, from the 1930s on, to explicate the notion of "analytic truth," which were meant as a generalization of the notion of tautology. Later, these attempts were challenged by W. V. O. Quine, in debates with repercussions still felt today.

But arguably the most significant contribution to the philosophy of logic in the Vienna Circle consisted in Carnap's, Hahn's, and partly Menger's shift from a realist, absolutist view about logic to a more "tolerant" alternative. This meant giving up the assumption that there is one "correct" logic and replacing it with the study of a variety of logical systems. The main method in this context was to apply the tools of "meta-logic," as developed by Hilbert, Carnap, Gödel, etc. The resulting position was a novel kind of pragmatist pluralism, which viewed logic, mathematics, and language more generally in a conventionalist and partly formalist way. Having said that, not all members of the Vienna Circle agreed with this position. Gödel, in particular, rejected what he viewed as a problematic form of "nominalist conventionalism" in Carnap, opposing it with a new form of realism, especially concerning set theory. This too led to extended debates in philosophy, from the 1940s–1950s on, with lasting repercussions.

The Berlin Group and the Lvov-Warsaw School

Besides the Vienna Circle, there were two other influential groups of "scientific philosophers" at the time, in Berlin and Lvov-Warsaw. Both should be mentioned in our context too, at least briefly. (For more on each, see CHS. 13 and 32).

Hans Reichenbach was the central figure in the Berlin Group, formed in the 1920s. We saw that he was in contact with Carnap early, in connection with the Erlangen Conference in 1923. Reichenbach was interested in logic too, but like Schlick, he is remembered more for his work in the philosophy of science. Two other members of his group focused primarily on logic and the foundations of mathematics: Walter Dubislav and Kurt Grelling. Their works were again based on close studies of *Principia Mathematica*, on influences by the Hilbert school, and related facts. (Grelling had studied under Hilbert at the University of Göttingen. Dubislav was strongly influenced by *Principia* and by Hilbertian axiomatics as well, but his interests reached back to Frege and Bolzano.) As both died young, their impact was not felt as strongly and they tend to be neglected (but see Milkov and Peckhaus 2013). Carl Hempel should be mentioned as well. Through an extended visit to Vienna in the 1920s, he was strongly influenced by the philosophers there. He too is mostly remembered for his work in the philosophy of science (on explanation, confirmation, etc.); but like Hahn and Carnap, he developed an unorthodox form of logicism and defended it well into the 1950s (see CH. 22).

From the late 1920s on, there were contacts between the Vienna Circle, the Berlin Group, and the Lvov-Warsaw School. The latter included several productive logicians, like Alfred Tarski, Jan Łukasiewicz, and Stanisław Leśniewski. These contacts were facilitated by Rose Rand, a member of the Vienna Circle with connections to Poland. Tarski was another thinker Menger brought to Vienna, for the first time in 1930, so as to present lectures, thus serving as an emissary of "Polish logic." As Hahn, Carnap, and Gödel learned during that visit, Tarski and his colleagues had gone beyond *Principia Mathematica* to investigate "meta-logical" and "meta-mathematical" themes too, thereby achieving greater clarity, e.g., with respect to distinguishing "object-" from "meta-language." Besides Gödel's results, it was conversations with Tarski on this topic that prompted Carnap to abandon his "general axiomatics" project. Tarski and Gödel noticed several fruitful connections between their works as well; and by exploring them further, they influenced twentieth-century logic even more (Reck 2013).

Conclusion

Among the contributions by the logical empiricists surveyed in this chapter, several proved very influential philosophically: their adoption and promotion of modern logic; their combination of Russellian logicism with empiricism; and their attempts to reconcile both with Hilbertian axiomatics. Wittgenstein's notion of tautology was another main innovation, seen as leading to a deflationary conception of logic congenial to "logical" empiricism. The rise of an attitude of "tolerance" for logic and mathematic, in works by Hahn, Carnap, and Menger, was important too, with a novel form of pragmatic pluralism as the result. In terms of substantive technical contributions to logic and the foundations of mathematics, there were not only Gödel's theorems, but also the systematic adoption of simple type theory and the early exploration of metalogical issues, by Carnap, Gödel, Tarski, etc. Finally, all of these innovations had a lasting impact on philosophical methodology, by promoting novel forms of "formal philosophy."

References

Carnap, R. (1922) *Der Raum. Ein Beitrag zur Wissenschaftslehre*, Kant Studien Ergänzungshefte 56. Trans. "Space. A Contribution to the Theory of Science," in Carnap, *Collected Works* Vol. I, (ed. by A. W. Carus et al.), Oxford: Oxford University Press, 2019, pp. 22–208.

——— (1928) *Der logische Aufbau der Welt*, Berlin: Weltkreis-Verlag. Trans. *The Logical Structure of the World*, Berkeley: University of California Press, 1967, repr. Chicago: Open Court, 2003.

——— (1929) *Abriss der Logistik*, Vienna: Springer.

——— (2000) *Untersuchungen zur Allgemeinen Axiomatik* (ed. by T. Bonk and J. Mosterin), Darmstadt: Wissenschaftliche Buchgesellschaft.

Cassirer, E. (1910) *Substanzbegriff und Funktionsbegriff*, Berlin: Bruno Cassirer. Trans. "Substance and Function," in Cassirer, *Substance and Function and Einstein's Theory of Relativity*, 1923, repr. New York: Dover, 1953, pp. 3–346.

Cat, J. (2019) "Neurath and the Legacy of Algebraic Logic," in J. Cat and A. T. Tuboly (eds.), *Neurath Reconsidered: New Sources and Perspectives*, Cham: Springer, pp. 241–338.

Couturat, L. (1905) *Les principes des mathématiques, avec un appendice sur la philosophie des mathématiques de Kant*, Paris: Vrin.

Fraenkel, A. (1923) *Einleitung in die Mengenlehre*, Berlin: Springer, 3rd ed., 1928.

Gödel, K. (1930) "Die Vollständigkeit der Axiome des logischen Funktionenkalküls," *Monatshefte für Mathematik und Physik* 37: 349–60. Trans. "The Completeness of the Axioms of the Functional Calculus of Logic," in Gödel 1986, pp. 103–23.

——— (1931) "Über formal unentscheidbare Sätze der Principia Mathematica und verwandter Systeme I," *Monatshefte für Mathematik und Physik* 38: 173–98. Trans. "On Formally Undecidable Propositions of *Principia Mathematica* and Related Systems I," in Gödel 1986, 144–95.

——— (1986) *Collected Works. I: Publications 1929–1936* (ed. by S. Feferman et al.), Oxford: Oxford University Press, 1986.

Hilbert, D. (1899) *Grundlagen der Geometrie*, Leipzig: Teubner. Trans. *Foundations of Geometry*, LaSalle: Open Court, 1971.

——— (1923) "Die logischen Grundlagen der Mathematik," *Mathematische Annalen* 88: 151–65. Trans. "The Logical Foundations of Mathematics," in *From Kant to Hilbert: A Source Book in the Foundations of Mathematics* (ed. by W. B. Ewald), Oxford: Oxford University Press, 1996, vol. 2, pp. 1134–48.

——— (1926) "Über das Unendliche," *Mathematische Annalen* 95: 161–90. Trans. "On the Infinite," in *From Frege to Gödel: A Source Book in Mathematical Logic 1897–1931* (ed. by J. van Heijenoort), Cambridge, MA: Harvard University Press, 1967, pp. 367–96.

——— (1929) "Probleme der Grundlegung der Mathematik," *Mathematische Annalen* 102: 1–9. Trans. "Problems in the Foundations of Mathematics," in *From Brouwer to Hilbert: The Debate on the Foundations of Mathematics in the 1920s* (ed. by P. Mancosu), Oxford: Oxford University Press, 1998, pp. 227–33.

Hilbert, D. and Ackermann, W. (1928) *Grundzüge der Theoretischen Logik*, Berlin: Springer, 2nd ed., 1938. Trans. *Principles of Mathematical Logic*, New York: Chelsea Publishing, 1950.

Höfler, A. and Meinong, A. (1890) *Logik*, Vienna: Tempsky.

Menger, K. (1980) "Introduction," in H. Hahn, *Empiricism, Logic, and Mathematics* (ed. by Brian McGuinness), Vienna: Reidel, pp. 9–18.

——— (2012) *Ergebnisse eines Mathematischen Kolloquiums* (ed. by E. Dierker and K. Sigmund), Vienna: Springer.

Milkov, N. and Peckhaus, V. (eds.) (2013) *The Berlin Group and the Philosophy of Logical Empiricism*, Dordrecht: Springer.

Natorp, P. (1910) *Die logischen Grundlagen der exakten Wissenschaften*, Leipzig: Teubner.

Ramsey, F. P. (1925) "The Foundations of Mathematics," *Proceedings of the London Mathematical Society* 25: 338–84. Repr. in Ramsey, *Philosophical Papers* (ed. by D. H. Mellor), Cambridge: Cambridge University Press, 1990, pp. 164–224.

Reck, E. (2004) "From Frege and Russell to Carnap: Logic and Logicism in the 1920s," in S. Awodey and C. Klein (eds.), *Carnap Brought Home*, Chicago: Open Court, pp. 151–80.

——— (2007) "Carnap and Modern Logic," in M. Friedman and R. Creath (eds.), *The Cambridge Companion to Carnap*, Cambridge: Cambridge University Press, pp. 176–99.

——— (2013) "Developments in Logic: Carnap, Gödel, and Tarski," in M. Beaney (ed.), *Oxford Handbook on the History of Analytic Philosophy*, Oxford: Oxford University Press, pp. 546–71.

Reck, E. and Awodey, S. (eds.) (2004) *Frege's Lectures on Logic. Carnap's Jena Notes, 1910–1914*, Open Court: Chicago.

Russell, B. (1914) *Our Knowledge of the External World as a Field for Scientific Method in Philosophy*, LaSalle: Open Court.

——— (1919) *Introduction to Mathematical Philosophy*, London: Allen & Unwin.

Schlick, M. (1918) *Allgemeine Erkenntnislehre*, Berlin: Springer, 2nd rev. ed., 1925. Trans. *General Theory of Knowledge*, LaSalle: Open Court, 1974.

——— (1920) *Raum und Zeit in der gegenwärtigen Physik*, Berlin: Springer, 3rd ed. Trans. *Space and Time in Contemporary Physics: An Introduction to the Theory of Relativity and Gravitation*, Oxford: Oxford University Press, 1920, repr. with trans. of rev. in 4th ed. (1922) in Schlick, *Philosophical Papers, Volume I (1909–1922)* (ed. by H. L. Mulder and B. F. B. van de Velde-Schlick), Dordrecht, Boston and London: D. Reidel, 1979, pp. 207–69.

Schmidt, A. (ed.) (2016) "Metadata for LPA Ostwald Print, Wittgenstein Source Facsimile Edition of Tractatus Publication Materials," in A. Pichler (ed.), *Wittgenstein Source*, Bergen: WAB, www.wittgensteinsource.org.

Sigmund, K. (1995) "Hans Hahn and the Foundational Debate," in W. DePauli-Shimanovich et al. (eds.), *The Foundational Debate*, Dordrecht: Kluwer, pp. 235–45.

Uebel, T. (2000) *Vernunftkritik und Wissenschaft: Otto Neurath und der erste Wiener Kreis*, Vienna: Springer.

——— (2005) "Learning Logical Tolerance: Hans Hahn on the Foundations of Mathematics," *History and Philosophy of Logic* 26: 175–209.

Verein Ernst Mach (1929) *Wissenschaftliche Weltauffassung. Der Wiener Kreis*, Vienna: Wolf. Trans. "The Scientific Conception of the World. The Vienna Circle," in O. Neurath, *Empiricism and Sociology* (ed. by R. S. Cohen and M. Neurath), Dordrecht: Reidel, 1973, pp. 299–318; rev. trans. (with orig. annotated bibliography) "The Scientific World-Conception. The Vienna Circle," in F. Stadler and T. Uebel (eds.), *Wissenschaftliche Weltauffassung. Der Wiener Kreis. Hrsg. vom Verein Ernst Mach (1929)*, Vienna: Springer, 2012, pp. 75–116.

Whitehead, A. N. and Russell, B. (1910–13) *Principia Mathematica* Vols. I–III, Cambridge: Cambridge University Press.

Wittgenstein, L. (1922) "Logisch-Philosophische Abhandlung," *Annalen der Naturphilosophie* 14: 185–262. Bilingual ed. trans. by F. Ramsey and C. K. Ogden *Tractatus Logico-Philosophicus*, London: Kegan Paul, Trench Trubner & Co., 1922, rev ed. 1933, repr. London: Routledge, Kegan, Paul, 1983; trans. by D. F. Pears and B. F. McGuinness, London: Routledge, Keagan Paul, 1961, repr. 1974.

15

CONCEPTIONS OF TRUTH IN EARLY LOGICAL EMPIRICISM

Pierre Wagner

According to an old philosophical conception, truth is to be defined as an adequacy between the intellect and a thing (Thomas Aquinas) or, in a more contemporary vein, as a correspondence between language and reality. If this definition implies that truth is independent of our recognition of the correspondence, it has metaphysical implications that make it hard for an empiricist to accept. Logical empiricists have usually adhered to the view that the meaning or meaningfulness of a sentence is or is strongly dependent on the possibility of recognizing whether this sentence is true or false (see CH. 16). It is no wonder, therefore, that they rejected a metaphysically loaded interpretation of the correspondence theory of truth. Some of them have been tempted to reject the notion of truth altogether, to try to show that it is possible to do without it, or to adopt a definition which amounted to a conflation of truth and verification, obviously at odds with the traditional or the ordinary meaning of "truth."

No unified answer has been given to this predicament and there is no such thing as *the* logical empiricists' theory of truth. Although the question of the essence of truth was not a major issue in their agenda for the reason just given, truth was neither neglected nor ignored and it has even been the object of debates and disputes among members of the Vienna Circle. But they agreed that distinctions need to be made because the definition of truth depends on the kind of sentences (or propositions, or judgments) involved: logical, mathematical, empirical, universal, philosophical, and elementary sentences are not said to be true in the same sense and a separate discussion is required in each case. The result is a variegated set of issues, complicated by the key idea that a true sentence may have a conventional component in addition to an empirical one. This was especially important with respect to language: how is truth dependent on language or on grammar, and is it possible to make the linguistic or the logical form of a sentence explicit in order to clarify what it is that makes the sentence true?

Another twist in issues about truth resulted from the fact that not all grammatically correct sentences which have the form of declarative sentences were interpreted as having a truth-value. Value judgments, for example, were regarded as prescriptions or commands and philosophical statements were interpreted either as purely nonsensical, as grammatical rules or analytic sentences in disguise, as recommendations, or as mixtures of several components in need of clarification.

In their discussion of truth, logical empiricists were influenced by many authors, among whom Wittgenstein and Tarski deserve special notice. A major idea comes from Wittgenstein's

DOI: 10.4324/9781315650647-18

Tractatus (1922): the idea that any sentence is functionally composed out of basic sentences and that the truth-value of a sentence is a function of the truth-values of the basic sentences out of which it is composed. Logical empiricists did not accept this idea uncritically, but many of the questions about the truth of the so called "protocol sentences" were derived from it: are there elementary sentences, and if so, how should they be characterized, and are they incorrigible? Wittgenstein's later idea that the truth of a sentence depends on the grammar of the language to which it belongs was also influential, as well as Wittgenstein's question of whether it is possible to describe the logical form of a sentence.

Tarski's work on the definition of a truth predicate had a no-less-important impact. In his celebrated "Wahrheitsbegriff" paper (published in Polish in 1933 and translated into German in 1935), Tarski presents a method for defining a predicate "true in L" for a large class of formal languages L. Four provisos are important. First, what Tarski calls a "language" is actually equipped with axioms and rules of proof so that we would rather call it a "language-system." Second, "formal" means that the rules of formation and the rules of proof depend only on the logical form of the sentences, not on their meaning (although they do have a meaning; "formal" does not mean "non-interpreted"). Third, Tarski's technique for defining truth applies to formal constructed languages, not to natural languages. Tarski argues that no adequate definition of a truth predicate is possible for natural languages because there is no way to avoid the semantic paradoxes in them. Fourth, the definition of "true in L" is formulated in an essentially richer metalanguage so that "true in L" is not definable in L itself.

In what follows, we shall focus mainly on Schlick's, Carnap's, and Neurath's views on truth in the early period of logical empiricism and then discuss the immediate impact Tarski's work had on these. The evolution of their views was also deeply influenced by their mutual interactions, but unfortunately, this cannot be discussed here.

Schlick's conception of truth

As early as 1910, Schlick devoted a long paper (written for his habilitation) to the question of truth, in which he defends a new and original view and rejects, among others, the idea of truth as a relation of resemblance between a representation and its object; to be true is not to be a *copy* of anything. The truth-bearers are not ideas but judgments, conceived as complex signs designating existing states of affairs, and a judgment is true when the designation is *univocal*, false if it is ambiguous. If someone says "the tree is red" while the intended tree is green, it is not known whether this judgment designates a green tree or a red tree; the judgment is not univocal, so it is false. Judgments are not isolated but connected in a system: "On one side we have a system of facts, and on the other a system of judgments. Each member of the second system that is coordinated one-to-one with a member of the first is said to be *true*" (1910/1979: 96, orig. emphasis).

In his *General Theory of Knowledge* (1918), Schlick again takes up the issue of truth, but this time in the context of a theory of knowledge. Judgments, still designating the existence of a relation, are also true if the designation is univocal. But Schlick now remarks that knowledge is much more than the mere truth of a judgment; knowledge presupposes the use of signs which have already been used elsewhere in the interconnected system of judgments. Not every truth is a piece of knowledge, but only those specific truths which make new connections between "old" concepts, concepts which have already been introduced by implicit definitions, i.e., by true judgments which have the conventional character of definitions. Schlick thus makes an interesting connection between truth, knowledge, and convention, now insisting that the kind of truths which have often been mistaken for synthetic a priori judgements actually have, according to him, the epistemological character of definitions.

By 1930, Schlick had fallen under the spell of Wittgenstein's strict understanding of his motto "The sense of a proposition is the method of its verification." In consequence, Schlick realized that there are important sentences which do not have any recognizable truth-value. For example, the laws of physics cannot strictly speaking be *verified*. As Schlick put it, borrowing Wittgenstein's terms, "at bottom a law of nature does not even have the logical character of an 'assertion,' but represents, rather, a 'prescription for the making of assertions'" (1931/1979: 188). Another case of important sentences without truth-values were philosophical sentences when they essentially have an elucidatory function (as in the *Tractatus*).

In the three lectures given under the title "Form and Content" in London in 1932 and published posthumously, the paragraphs devoted to the analysis of truth show how the influence of Wittgenstein's *Tractatus* fuses with his older views. Schlick now explicitly agrees with the traditional characterization of truth as a correspondence, but the nature of this correspondence is explained as an *identity of structure* between propositions and facts: "The world consists of facts, the facts have a structure, and our propositions will picture the facts correctly, they will be *true*, if they have the same structure" (1938/1979: 348, orig. emphasis).

Schlick mentions an objection which is sometimes raised against the conception of truth as correspondence, to the effect that a comparison of a proposition with the fact it expresses is impossible, because facts are known to us by no other means than these propositions (ibid.: 349). Schlick's answer is that propositions are given to us empirically as complex signs, which are nothing but facts in the real world, and that the comparison of two facts is something we so commonly do in our daily life that nobody could reject its possibility. Striking in this reply, as in the objection, is the apparent conflation of the two questions: "What is the nature of truth?" and "How to recognize truth?" A similar conflation seems evident here: "A proposition will be verified, its truth will be established, if the structure of the sentence is the same as the structure of the fact it tries to express" (ibid.). However, once we remember that for Schlick an unverifiable sentence is (cognitively) meaningless, the close connection he establishes between the issues of truth and verification becomes understandable.

Under the influence of Wittgenstein, Schlick offered a new defense of his view of truth and verification by the middle of the decade. Defining a proposition as a sentence "*together with the logical rules belonging to them*, i.e., certain prescription as to how the sentence is to be *used*," Schlick insisted that "in order to verify the proposition I have to ascertain whether those rules have actually been obeyed—why should that be impossible?" (1935: 67, orig. emphasis) What can be seen here is that in the mid-1930s, his view of truth as correspondence was still embedded in a discussion of verification and of the disputed possibility of comparing statements with facts.

Carnap's treatment of truth up to and including *The Logical Syntax of Language*

In his 1922 dissertation on space, Carnap followed Frege in taking truth to be one of the few basic notions which are to be regarded as undefinable. First, a judgment is defined as all that is susceptible of being true or false, and the notion of judgment is then used to define propositional functions, on which logic and mathematics are built. In *The Logical Construction of the World* (1928), devoted to the project of a rational reconstruction of the empirical part of science, the reconstruction of the formal part is presupposed and no specific discussion of a conception of truth is required. It must be noted, however, that here Carnap adopted an extensionalist strategy for which the notion of a truth-value is important.

In his paper "Die Antinomien und die Unvollständigkeit der Mathematik" (1934b), the English version of which was incorporated into *The Logical Syntax of Language* in 1937, Carnap devotes a whole paragraph to an analysis of the concepts "true" and "false." His focus is now on the careful elaboration of logical frameworks in which the reconstruction of science is supposed to take place. Because the concepts *true* and *false* "are usually regarded as the principal concepts of logic" (1934a/1937: §60b), Carnap needs to explain their place in this new project. The trouble is that the customary use of these terms allows formulations such as "*A* is true" where "*A*" is a sentence, which leads to contradictions related to the well-known antinomy of the liar. His first conclusion is that the predicates "true" and "false" referring to sentences of some language *L* should be used not in *L* itself, but in a metalanguage *ML*. The *ML*-sentence "'Snow is white' is true in *L*", where "'Snow is white'" is a name in *ML* of the *L*-sentence "Snow is white," does not have the same paradoxical consequences. Carnap sketches the following strategy to be followed: the two predicates "true in L" and "false in L" are included as primitive symbols in a metalanguage *ML* and axioms such as "each sentence of *L* is either true or false" or "no sentence of *L* is both true and false," which provide rules for their use, are then formulated in *ML*. (Note that today, when axioms are given for a truth predicate, they are usually formulated in *L*, not in *ML*.)

In *Logical Syntax*, however, Carnap does not investigate such axiomatization of truth any further. The syntactical method which he favors excludes the use of semantic concepts such as *true* and *false*: "*truth and falsehood are not proper syntactical properties*" (ibid.: §60b, orig. emphasis). Whereas "consequence," "analytic," "contradictory," "provable," and other related methodological terms are definable in a syntax-language, "true" and "false" are not. Carnap's second conclusion is that in the context of the syntactic method, the adequate strategy consists in showing how to make do without these two predicates, and translations are proposed to indicate how this could be done. For example, "*A* is true" may be translated into "*A*," and "*A* is false" into "not *A*" (in these cases, *ML*-sentences are translated into *L*-ones). But this deflationist technique does not cover cases such as "all theorems are true" where the number of theorems is infinite. Distinctions need to be made between different uses of "true" and "false." In logical investigations, when the truth or falsity of *A* is determined by the axioms and the rules of the system considered, "true" may be translated as "valid," "logically valid," or "provable," and "false" as "contravalid," "contradictory," or "refutable" as the case may be, all these terms being syntactically definable (in this case, the translation takes place inside *ML*). Other cases are considered, such as "If *A* is true, then *B* is true" syntactically translatable as "B is a consequence of A."

The suggested translations, however, do not cover every case. Carnap implicitly recognizes the limits of the syntactical method on this point when he writes that "the *majority* of ordinary sentences which make use of ['true' or 'false'] can be translated either into the object-language or into the syntax-language" (ibid., emphasis added). Whereas logical truth is reduced to logical validity, factual truth is not reducible to any syntactic concept and no translation is suggested for sentences such as "Everything Anna says is true."

Carnap's deflationist eliminativist attitude with respect to truth does not imply that the concepts *true* and *false* are either condemned as unclear or rejected; neither are they reduced to processes of verification or confirmation. The indicated method of axiomatization shows that truth is recognized as a legitimate concept, although no general theory of truth is formulated or even sought for in *Logical Syntax* or in other papers in the 1930s.

Neurath's rejection of the concept of truth

Neurath had a much less conciliatory attitude with respect to "true" and "false." Sometimes he even suggested banning these terms from the language of science or, in case we would still want to use them, redefining them. In order to understand what motivates such a radical view, it is useful to go back some years and examine his comments on, and opposition to, the idea of truth as correspondence. Then he wrote, for example, that

> *statements are always compared with statements*, certainly not with some "reality," nor with "things" . . . If a statement is made, it is to be confronted with the totality of existing statements. If it agrees with them, it is joined to them; if it does not agree, it is called untrue and rejected. There can be no other concept of "truth" for science.
> (1931a/1983: 53, orig. emphasis)

What is discussed in this quotation is the process through which statements are selected and integrated into science ("the totality of existing statements"). But the idea of a comparison with reality in the first part of the quotation clearly refers to the correspondence conception of truth and the last sentence seems to suggest—misleadingly—that what has just been said is meant to characterize the concept of truth. If so, some confusion would be at work here between a *criterion* and a *definition* of truth.

The same apparent confusion recurs in other papers (1931b/1983: 66; 1934/1983: 102), in which Neurath's objection to the idea of a comparison between statements and experience, or things, or reality, is formulated in a way which easily drives the reader to the conclusion that a conception of truth is defended, different from the correspondence conception and compatible with the method of a comparison of statements with *statements*, not with *reality*. As a matter of fact, this reading was by Schlick, who interpreted Neurath as defending a coherence theory of truth. Hempel did the same when he declared that "obviously, Neurath's ideas imply a coherence theory" (1935: 51). Neurath's denial and Carnap's warning against such an interpretation were of little help, probably, at least in Schlick's case, because he was convinced that the articulation of some conception of truth was needed for the defense of empiricism, unlike Neurath. In his rejoinder to Schlick, who had criticized his radical physicalism for having "no unambiguous criterion of truth" (1934/1979: 376), Neurath explained in great detail his conception of justification, which did not require any theory of truth in any way (1934/1983: 105–7). Neurath regarded the concept of truth as a remnant of metaphysics which needed to be discarded or redefined. The same was true of other "imprecise verbal clusters" (*Ballungen*) common in the statements of everyday language which unified science was expected to transform or reinterpret. In this context, the apparent confusion between the concept of truth and the process of selecting statements for their integration into science should be understood not as the defense of a coherence theory of truth but as reflecting Neurath's willingness to replace the concept of truth and redefine "talk of truth . . . as talk of the condition of acceptability" (Uebel 2007: 233).

But what about the elementary basic statements on which other statements of science are supposed to be based? Shouldn't we regard them as indubitably true if we do not want to renounce the very idea of empiricism? Schlick believed that the assessment of observational reports depended on the possibility of comparing such basic statements with reality and, consequently, on a correspondence conception of truth. But to his notion of observational reports, the so-called "constatations" or "affirmations" (*Konstatierungen*), as indubitable, incorrigible, first-person statements which cannot even be written down, Neurath opposed his version of third-person, revisable protocol statements, which belong to the system of science and are not supposed to depend on a specific conception of truth. To Schlick's objection that empiricism demands that observational reports should be anchored into reality through a comparison of statements with facts, Neurath replied that "reality" is just another metaphysical term which should be discarded. He insisted that the possibility of discarding protocol statements be not excluded: "We shall call a statement 'false' if we cannot establish conformity between it and the whole structure of science; we can also reject a protocol sentence unless we prefer to alter the structure of science and thus make it into a 'true' statement" (1934/1983: 102). There are no unrevisable truths for Neurath, who interpreted Schlick's quest for absolute certainty (and questions such as "Why do I accept science as true?") as metaphysical and replaced it by a conception of unified science as a social practice with a multidimensional theory of the acceptability of scientific testimony at its heart (see Uebel 2009). For Neurath, we have no reason to believe that the concept of truth is indispensable for science.

Logical and mathematical truth

In *Form and Content*, Schlick declared that "one has taken *the* most important step in philosophy if one has gained a perfect understanding of the nature of logic and its relation to reality and experience" (1938/1979: 345, orig. emphasis). This step was taken with Wittgenstein's idea that logical propositions are tautologies, propositions that have no factual content and are true whatever the facts may be. To the idea that "there are *no* conditions for the validity of logic" (ibid.: 347, orig. emphasis). Schlick added that the same could be said of mathematics, something he had certainly not found in the *Tractatus*. It is not clear whether Schlick was aware of the objections which could be raised at the time against such contentions, based on Gödel's incompleteness theorem or on the plurality of existing logical systems. In 1932, Carnap was well aware of these objections, and in *Logical Syntax*, he consequently renounced not only the program of a derivation of mathematics within elementary logic, but also the very idea of correctness in logic. Yet, he was still hoping to find an explication of analyticity which would allow interpreting mathematical and logical sentences as analytic, i.e., as logically valid and thus devoid of factual content. This did not amount to see them as true *simpliciter*, because no system of objects, facts, or ideal entities was recognized, with respect to which mathematical sentences could be said "true." The project was rather to show that a rational reconstruction of science is possible, in which logic and mathematics are analytic, some acceptable explication of analyticity being given: logical and mathematical sentences were not regarded as true no matter what, but as analytically true in some proposed logical framework.

An obvious objection to this program runs as follows: is it true that there is no objective realm of logical and mathematical truths or objects? And how do you know that this is the case? To see how this objection can be met, Carnap's principle of tolerance (1934a/1937: §17)—to the effect that everybody is free to build her own system of logic (including mathematics) as she wishes—should not be read as a philosophical truth but as a decision; the decision to stop searching for the one true logic and starting investigating the "boundless ocean" of infinite possible systems (ibid.: Foreword). This first decision being taken, a second one follows: the decision to look for some specific framework in which logical and mathematical sentences can be reconstructed as analytically true sentences, for some proposed explication of analyticity such as the one Carnap exposes in *Logical Syntax*. The phrase "true by convention" has sometimes been used to characterize Carnap's conception of logical and mathematical sentences, and Carnap himself used the word "convention" (ibid.: §17). But this is misleading if it presupposes that there must be something in virtue of which these sentences should be said to be true. The radicality of Carnap's view is precisely to get rid of this very idea, logical and mathematical sentences are not *true* but *logically valid in some language L*.

The impact of Tarski's theory of truth

After Tarski explained to him his method for defining "true in *L*," Carnap quickly understood the value of his semantic approach and the use he could make of it for his own metalinguistic investigations on language-systems. For Carnap, Tarski had managed to show that "true" was a valuable scientific concept after all, and there was no reason to try to get rid of it anymore. He encouraged Tarski to present his work at the 1935 Paris Congress on Scientific Philosophy, but Tarski anticipated a negative reaction and was reluctant to do so. Carnap managed to persuade him and proposed to pave the way for Tarski's new ideas by giving a talk on truth and confirmation. In this talk, he advocated not only the distinction between true and confirmed statements but also the possibility of confronting a statement with observation (1936). A hot debate ensued among logical empiricists, between those who approved and those who rejected the use of Tarski's semantic method for defining truth. Although Neurath did not express his criticism in print, his correspondence and his interventions in private meetings, first after the 1935 Paris Congress and then again in Paris in 1937, testify of his strong disapproval (see Mormann 1999; Mancosu 2007).

Neurath had nothing to object to Tarski's logical results as such, but he anticipated that the introduction of such a metaphysically loaded term as "truth" into formal studies would have terrible effects for empiricism: philosophers would be tempted to unduly apply Tarski's method to ordinary language and be seduced into metaphysical consequences. Tarski wrote as if he regarded the classical concept of truth to mean correspondence with reality, Kokoszyńska spoke of Tarski's method as the scientific treatment of an "absolute concept of truth," and Carnap himself justified the possibility of confronting statements with facts. All this could easily look like a vindication of Schlick's views. To Neurath's objection that Tarski's definition of truth applied only to formal languages and was of no practical interest for natural languages, Tarski replied that the same remark held for the definitions of syntactical terms given in Carnap's *Logical Syntax*. But Neurath's negative reactions also had deeper reasons. For him, the need to clarify concepts did not entail that unified science had to be formulated in purely formal language-systems and the quest for a rational reconstruction of science could even be interpreted as a remnant of some metaphysical ideal. Neurath's conception of empiricism was of a different kind, which took the Encyclopedia as model and required a "Universal Jargon," in which imprecise cluster terms were inevitably mixed with scientific ones (see CH. 18). For Neurath, Tarski's definition only reflected some very specific use of the term "truth," and its use as an explication of "true" introduced more confusion than clarity.

In 1937, Neurath's suggestion was to use "accepted in the Encyclopedia" and "rejected in the Encyclopedia" instead of the mystifying "true" and "false." But he could never convince Carnap, who noted that if A is a not-yet-decided empirical sentence, "A is not accepted in the Encyclopedia" is certainly not equivalent to "A is false" (Mancosu 2007). "True" and "false" as used in ordinary language were not eliminable after all and were in need of an explication, for which Tarski's definition was a useful candidate.

However, Carnap's enthusiasm for Tarski's semantics does not mean that he adopted it as it stood and simply followed his lead. Their outlooks and goals were actually very different. For one thing, whereas Tarski was interested in the methodology of the deductive sciences, Carnap's language-systems were aimed at the reconstruction of science, including empirical science. For another, whereas Tarski very cautiously provided an explicit definition of a satisfaction relation and a truth predicate for *L* which presupposed no other previously accepted semantic concepts, what interested Carnap was the possibility of liberalizing metalanguages so as not to be restricted to the syntactic method anymore. For him, "true in *L*" was hardly more than a new metalinguistic tool for the explication of methodological concepts that were discussed in the Vienna Circle, especially the distinction between formal truth and empirical truth and the concept of analyticity; by contrast, Tarski was very skeptical about the possibility of finding a definite and precise explication for these concepts. Later in the 1940s, Carnap would actually realize that the framework of Tarskian semantics was not adequate for his own projects and could not help integrating modal, inductive, and intensional logic into one single framework. He consequently turned to other semantic methods based on ideas found in Wittgenstein, Waismann, Church, and others, in which the specifically *Tarskian* concept of truth did not play any central role anymore. What would soon be central in this new setting (and in most of Carnap's later publications) is the notion of a *state description* and the idea of a sentence "holding" in a state description, not to be confused with the now-common model-theoretic (and Tarskian) conception of a sentence being "true in a structure." In this non-Tarskian framework, a true state description is just one which correctly describes the real world and the focus is on L-truth, not on truth itself, a sentence being L-true if it "holds" in all state descriptions (see Wagner 2017).

References

Carnap, R. (1922) *Der Raum. Ein Beitrag zur Wissenschaftslehre*, Kant Studien Ergänzungshefte 56. Trans. "Space. A Contribution to the Theory of Science," in Carnap, *Collected Works* (ed. by A. W. Carus et al.), Oxford: Oxford University Press, 2019, vol. 1, pp. 22–208.

——— (1928) *Der logische Aufbau der Welt*, Berlin: Weltkreis-Verlag. Trans. *The Logical Structure of the World*, Berkeley: University of California Press, 1967, repr. Chicago: Open Court.

——— (1934a) *Logische Syntax der Sprache*, Vienna: Springer. Rev. ed. trans. *The Logical Syntax of Language*, London: Kegan, Paul, Trench Trubner & Cie, 1937, repr. Chicago: Open Court, 2002.

——— (1934b) "Die Antinomien und die Unvollständigkeit der Mathematik," *Monatshefte für Mathematik und Physik* 41: 263–84.

——— (1936) "Wahrheit und Bewährung," *Actes du Congrès International de Philosophie Scientifique, Sorbonne, Paris, 1935*, fasc. 4, pp. 18–23.

Hempel, C. G. (1935) "On the Logical Positivists' Theory of Truth," *Analysis* 2: 49–59.

Mancosu, P. (2007) "Tarski, Neurath and Kokoszynska on the Semantic Conception of Truth," in D. Patterson (ed.), *New Essays on Tarski*, Oxford: Oxford University Press, pp. 192–224.

Mormann, T. (1999) "Neurath's Opposition to Tarskian Semantics," in J. Wolenski and E. Köhler (eds.), *Alfred Tarski and the Vienna Circle*, Dordrecht: Kluwer, pp. 17–26.

Neurath, O. (1931a) "Physikalismus," *Scientia* 50: 297–303. Trans. "Physicalism," in Neurath (1983), pp. 52–57.

——— (1931b) "Soziologie im Physikalismus," *Erkenntnis* 2: 393–431. Trans. "Sociology in the Framework of Physicalism," in Neurath (1983), pp. 58–90.

——— (1934) "Radikaler Physikalismus und 'wirkliche Welt'," *Erkenntnis* 4: 346–62. Trans. "Radical Physicalism and the 'Real World'," in Neurath (1983), pp. 100–14.

——— (1983) *Philosophical Papers 1913–1946* (ed. by R. S. Cohen and M. Neurath), Dordrecht: Reidel.

Schlick, M. (1910) "Das Wesen der Wahrheit nach der modernen Logik," *Vierteljahresschrift für wissenschaftliche Philosophie und Soziologie* 34: 386–477. Trans. "The Nature of Truth According to Modern Logic," in Schlick (1979), vol. 1, pp. 41–103.

——— (1918) *Allgemeine Erkenntnislehre*, Berlin: Springer, 1918, 2nd rev. ed. 1925. Trans. *General Theory of Knowledge*, Vienna: Springer, 1974, repr. LaSalle, IL: Open Court, 1985.

——— (1931) "Die Kausalität in der gegenwärtigen Physik," *Die Naturwissenschaften* 19: 145–62. Trans. "Causality in Contemporary Physics," in Schlick (1979), vol. 2, pp. 176–209.

——— (1934) "Über das Fundament der Erkenntnis,", *Erkenntnis* 4: 79–99. Trans. "The Foundation of Knowledge," in Schlick (1979), vol. 2, pp. 370–87.

——— (1935) "Facts and Propositions," *Analysis* 2: 65–70. Repr. in Schlick (1979), vol. 2, pp. 400–4.

——— (1938) "Form and Content," in Schlick, *Gesammelte Aufsätze 1926–1936*, Vienna: Gerold, pp. 151–250. Repr. in Schlick (1979), vol. 2, pp. 285–369.

——— (1979) *Philosophical Papers* (ed. by H. L. Mulder and B. van de Velde-Schlick), Dordrecht: Reidel, vol. 2.

Tarski, A. (1935) "Der Wahrheitsbegriff in den formalisierten Sprachen," *Studia Philosophica* 1: 261–405 (trans. of Polish original in *Trav. Varsovie* 34 (1933)). Engl. trans. "The Concept of Truth in Formalized Languages," in Tarski, *Logic, Semantics, Metamathematics*, Oxford: Clarendon Press, 1956, pp. 152–278.

Uebel, T. (2007) *Empiricism at the Crossroads: The Vienna Circle's Protocol Sentence Debate*, Chicago: Open Court.

——— (2009) "Neurath's Protocol Statements Revisited: Sketch of a Theory of Scientific Testimony," *Studies in History and Philosophy of Science* 40: 4–13.

Wagner, P. (2017) "Carnapian and Tarskian Semantics," *Synthese* 194: 97–119.

Wittgenstein, L. (1922) *Tractatus Logico-Philosophicus* (bilingual ed. trans. by F. Ramsey and C. K. Ogden), London: Kegan Paul, Trench Trubner & Co., 1922, rev. ed. 1933, repr. London: Routledge, Kegan, Paul, 1983; trans. by D. F. Pears and B. F. McGuinness, London: Routledge, Keagan Paul, 1961, repr. 1974.

16
VERIFICATIONISM

James Justus

Verificationism was the linchpin of logical empiricism. It sustained the highly contentious judgments that traditional metaphysics was a fraud, and that philosophical methodology should much more closely emulate scientific (and mathematical) practice. It also formed the common thread uniting the remarkably diverse epistemological views of the logical empiricists. For its critics, in sharp contrast, verificationism was an irredeemably flawed doctrine, a blinkered account of what makes statements meaningful, and ultimately responsible for logical empiricism's demise.

Concerns about verification did not emerge *sui generis* among logical empiricists. The idea that sensory experience is the ultimate arbiter of knowledge, and perhaps distinguishes sense from nonsense, has shaped much of the modern history of philosophy. Its significance is especially conspicuous in the empiricist tradition: Berkeley's attack on Locke's notion of material substance, Hume's deflation of causation and admonition to commit sophistry and illusions to the flames, John Stuart Mill's unwavering naturalism, Russell's principle of acquaintance, and, on the scientific side, Mach's scientific phenomenalism and the conceptual and operational acuity that facilitated Einstein's monumental discoveries and so impressed the logical empiricists. But more than any other group of thinkers, insights and concerns about verification compelled and confirmed the systematic, scientific worldview they built. That philosophical edifice largely stands or falls with verificationism.

As a benchmark of cognitive evaluation, verificationism is seemingly straightforward to state: non-analytic statements with no connection to experience are spurious. The apparent simplicity is, however, illusory. Verificationism—what counts as verifi*able* in particular—is embedded in a web of confusions and controversies about how "experience" should be understood, what exact connection is required, and the nature of the resulting illegitimacy when the connection is absent, among others. Perhaps most damaging to the historical reputation of logical empiricism is the charge that the verifiability requirement is self-refuting. It is neither analytic nor an empirical proposition upon which experience bears; it therefore glaringly fails its own cognitive standard. Like many of the supposedly decisive shortcomings of logical empiricism, this criticism falls short. It does considerable historical injustice to what logical empiricists actually said about how they understood the verifiability requirement, and the cogency of that understanding. That history, like many contemporary critical appraisals of logical empiricism, merits a second look.

 DOI: 10.4324/9781315650647-19

Verificationism as theory of meaning

"Verificationism" is associated with two distinct but related tenets held by logical empiricists: a verificationist theory of meaning, and a verifiability criterion of meaningfulness. The former, logically stronger view entails the latter, but only the latter is rightly considered non-negotiable for card-carrying logical empiricists. And only the latter was universally adopted among them. As a full theory of meaning, verificationism was endorsed by a proper subset, and a set whose cardinality diminished as the influence of Wittgenstein waned.

Schlick (1936: 341) gave the view its most succinct expression: "The meaning of a proposition is the method of its verification." With the only exception being Waismann, no logical empiricist had more interactions with Wittgenstein than Schlick, and Schlick largely credited their discussions for this account of meaning. The sentence immediately preceding the canonical slogan, for example, states what Schlick considered an equivalent formulation: "Stating the meaning of a sentence amounts to stating the rules according to which the sentence is to be used." (It's unclear the formulations are equivalent. In natural languages, usage and the verification conditions that determine truth-values can diverge.) The rules of "logical grammar" in question include explicit and ostensive definitions of terms. By the latter, Schlick included not only simple acts of verbally labeling objects, but more complicated cases in which "the meaning of the words is defined by the way we use them" in certain "complex situations" (ibid.: 342). How 'chance' or 'because' would be used in such situations were cited as examples. With words defined, rules of logical grammar then specify the verification conditions under which sentences are true or false, which determines their meaning.

This account of meaning seemed grounded in scientific practice, especially the epoch-making developments in physics of the time. As an exemplar, Schlick mentioned Einstein's characterization of simultaneity. He also approvingly cited Bridgman's *The Logic of Modern Physics*. Schlick's recognition of affinities between operationalism and the verificationist theory of meaning was not idiosyncratic. Both G. Bergmann and Hempel thought the approaches were "closely akin." Nor is the similarity superficial. Within the domain of physics, Bridgman's almost exclusive focus, what else could verification methods amount to than precisely the operational measurement procedures (and standardizations) that proved so scientifically fruitful for so many concepts: length, mass, simultaneity, time, etc.?

What became clear was that both views deliver an indefensibly impoverished account of meaning. Meaning as expressed and communicated in natural languages is simply too rich, multifarious, and amorphous to be reliably rendered with the early-Wittgensteinian resources that inspired the "method of verification" view. Even within physics, operationalism seemed myopic. Operational procedures seemed incapable of comprehensively rendering the full meaning of many (if not all) important concepts of physics, and in fact, many useful concepts seemed to resist any (useful) operational characterization (Chang 2009). And besides Wittgenstein's evolution in outlook, many logical empiricists increasingly disenchanted with his views began exploring very different accounts of meaning in natural language (e.g., Carnap 1955). What was never abandoned, however, was the commitment to verifiability as a criterion of meaningfulness.

Verifiability as a criterion of meaningfulness

Some early attempts to develop such a criterion were steeped in verificationist approaches to meaning. In "The Elimination of Metaphysics by Logical Analysis of Language," Carnap developed a verifiability criterion that focused foremost on characterizing the meaningfulness of words, rather than statements. Language syntax provided the demarcation (1932/1959:

§§2–3). Specifically, meaningful words must satisfy two conditions: (1) conform to the syntax of the language containing them; and (2) be deducible (via statements exemplifying them) from "protocol" statements, the evidence statements of science (the form, content, and status of which remained under discussion). Explaining these conditions, Carnap defined the "elementary sentence form" of a word as the simplest sentence form in which it can occur. A meaningful word must have a syntactically fixed mode of occurrence in its elementary sentence form, and instances of this form must be deducible from protocol statements. By late 1932, Carnap had realized that protocol statements were corrigible, but the deducibility focus of (2) seems to deny that inductive arguments can show words are meaningful. Carnap also did not specify the form of the required deduction, but this was almost certainly intentional. He likely realized the challenge this posed, as criticisms of Ayer's criteria later revealed.

Sequences of meaningful words can nonetheless be meaningless if they violate the language's grammatical syntax or "logical syntax." "Caesar is and" is not significant because it violates (English) grammatical syntax. According to Carnap, "Caesar is a prime number," though grammatically sound, is not meaningful because it violates logical syntax; "Caesar" and "prime number" belong to different logical categories. (Four years later, Carnap's evaluation [1936–7: 169] of these kinds of statements had changed: he then held they were meaningful but false.)

Grammatical syntax varies across languages. Its role in Carnap's 1932 criterion linguistically relativizes meaningfulness. This does not compromise the criterion's critical force against metaphysics since, Carnap emphasized, metaphysical statements are meaningless because they violate logical, not grammatical syntax. The concept of logical syntax, therefore, constitutes the normative basis of his criterion. Carnap (1932/1959: §4) thought, for instance, that meaningless statements could not be formed in languages where grammatical and logical syntax coincide. Carnap later abandoned the idea of a uniquely correct, universal "logical syntax," most clearly in the *Logical Syntax*, and this catalyzed a radically different approach to the search for a verifiability criterion in 1936.

Not all early verifiability criteria were wedded to Wittgenstein-inspired verificationism about meaning. In *Der Logische Aufbau der Welt*, Carnap claimed statements were verifiable, and thereby meaningful, if they could be translated—via basic statements about physical objects or "elementary experiences"—into a constructional system. Questions without verifiable answers were meaningless pseudo-questions (1928a/1967: §§179–80). However, despite countervailing claims in Carnap's autobiography and some suggestive passages in the *Aufbau* (esp. ibid.: 291–2), strong verificationism was not central to the *Aufbau* (Creath 1982). Its primary concern was the notion of a constructional system, which may explain why a precise verifiability criterion was not presented.

Carnap proposed the first explicit, semi-formal criterion in *Scheinprobleme in der Philosophie* (*Pseudoproblems in Philosophy*). Intending to demonstrate the meaninglessness of the realism-idealism debate and other philosophical controversies, Carnap presented a criterion of "factual content":

> If a statement *p* expresses the content of an experience E, and if the statement *q* is either the same as *p* or can be derived from *p* and prior experiences, either through deductive or inductive arguments, then we say that *q* is 'supported by' the experience E . . . A statement *p* is said to have "factual content," if experiences which would support *p* or the contradictory of *p* are at least conceivable, and if their characteristics can be indicated.
>
> (1928b/1967: §7)

Only statements with factual content, Carnap emphasized, are meaningful. This entails, strictly speaking, that factual mathematical and logical statements are meaningless, but Carnap's choice of examples strongly suggests its intended application was in synthetic, not analytic statements.

For Carnap, one strength of the 1928 criterion was its weakness. Conceiv*able*, not necessarily actual, connections to "experiences" make statements meaningful. With the term "supported" and reference to inductive arguments, moreover, Carnap only required experiences confirm or disconfirm statements, rather than conclusively establish their truth or falsity; seemingly a weaker requirement than the deducibility condition four years later. Carnap proposed this criterion shortly after writing the *Aufbau*, and its weak requirement accords well with the weak form of verificationism permeating that work.

The reflexivity problem

Carnap and logical empiricists generally believed verifiability criteria, unlike anti-metaphysical critiques that judged metaphysics false, justified eliminating metaphysics as a meaningless philosophical hoax. Two serious challenges confronted such a strong claim:

The reflexivity problem—a verifiability criterion (or statement thereof) is not analytic, does not satisfy itself, and is therefore meaningless.

The adequacy problem—any criterion will inevitably evaluate obviously meaningful statements as meaningless and vice versa.

It is not widely appreciated that Ayer was the first to address the reflexivity problem. There seemed to be two thoroughly unpalatable options. One, misconstrue verifiability criteria as (highly dubious) empirical claims about how meaning or meaningfulness is actually understood. Or, two, misconstrue them as dogmatic stipulations about how meaning or meaningfulness should be characterized. Although Ayer eventually recognized they were neither, his initial position was confused. A criterion was a "significant empirical proposition" (1934: 344), but also "something that is presupposed in any enquiry into the meaning of meaning, or any other philosophical enquiry, and therefore cannot appear as the conclusion of such an enquiry" (ibid.: 343). But presuppositions necessarily presupposed in enquiry seem a priori if not analytic, and certainly not empirical.

Two years later Ayer appreciated the difficulty and suggested such a criterion is a "definition" that must satisfy an adequacy condition: it must classify statements unequivocally considered meaningful as meaningful and statements unequivocally considered meaningless as meaningless (1936b). The intended sense of "definition" here is anything but lexical and, as Hempel (1951) later clarified, is better captured by Carnap's notion of explication. A verifiability criterion explicates the vague idea of meaningfulness by pinpointing and precisifying what it is that makes statements meaningful. It is therefore neither analytic nor synthetic but a bit of proposed conceptual engineering. The precisification must accord with basic commitments about the meaningfulness of common statements in natural language and show, exactly, how theoretical claims in science, which mathematical abstraction often takes far from the observable domain, nevertheless derive their meaningfulness from that domain. With these desiderata met, a criterion as explication would precisely characterize what (nonanalytic) meaningfulness requires.

Another supposed pillar principle of logical empiricism might seem to preclude this response to the reflexivity problem: commitment to an exhaustive analytic-synthetic distinction. It should be more widely recognized that quite early on, several logical empiricists (e.g., Carnap, Hempel, Neurath, Reichenbach) abandoned such an absolutist distinction. Carnap, of course, never wavered from the conviction that the distinction can be drawn within a particular language if its semantics are systematic and clear enough to determine meanings (see Carnap 1955). But explications concern "external" choices between languages: the original language containing the vague notion and the regimented language with the more precise explication. Unlike questions "internal" to a language that admit of definite cognitive answers (analytic or synthetic, true or false, five-sigma supported or not), at the heart of Carnap's internal–external methodology described in "Empiricism, Semantics, and Ontology" is the belief that choices between languages are "noncognitive," only evaluable on pragmatic grounds. As an explication of meaningfulness, the primary pragmatic concern for a criterion is the kind of adequacy condition Ayer identified. With that condition satisfied, metaphysicians could still reject a proposed explication that deemed their subject empirically inert, but only at a steep price. To vindicate their discipline a more liberal criterion must be formulated that judges commonplace, scientific, *and* metaphysical statements meaningful in a non-*ad hoc* manner while judging clearly nonsensical or vacuous assertions meaningless. In this way the reflexivity problem is ancillary to the adequacy problem.

The adequacy problem

The adequacy problem, however, was much more challenging. Identifying what type of modality meaningfulness requires was the problem (recall "conceiv*ability*" in Carnap's first criterion). Unsurprisingly, criticisms often uncharitably exploit the modal ambiguity of proposed criteria. Stace (1935), for instance, argued that early verifi*ability* criteria make non-analytic statements about the past meaningless because the past cannot "in principle" be accessed to verify them. This involved two misconceptions. First, Stace thought verifiability required the possibility of *conclusively* establishing the truth-values of statements. By 1932, however, Schlick had rejected this strong requirement, and Carnap (1928b) had also rejected it: meaningful statements are "supported" by conceivable experiences, not necessarily conclusively demonstrated.

The second misconception traded upon a more intractable modal ambiguity of "possibility of verification." For logical empiricists, "possibility" had three viable senses: practical, "physical" (Carnap 1936–7: 423), or logical. Practical possibility was obviously not the intended sense. Schlick's (1932) discussion of measuring mountains on the far side of the moon made that clear. But Schlick (1932), Carnap (1928b), and Ayer (1934, 1936b) were silent about whether logical or "physical" possibility was needed, the latter being Carnap's later label for nomological possibility. Stace's criticism broached an important general question. What is the status of statements about physically inaccessible events—in the past, in remote regions of the universe, in causally hostile environments in which no detection equipment (or humans) could exist, etc.—for which no present evidence is available and no future evidence ever would be? Without pinpointing exactly what kind of possibility is necessary, facile appeals to verifiability or confirmability failed to render satisfactory guidance about meaningfulness, and in general seemed doomed to indefensible vagueness.

Schlick (1936: 349) was one of the first to address this problem by arguing for logical, not physical possibility of verification: "when we speak of verifiability we mean logical possibility of verification, and nothing but this." For Schlick (and Carnap), physical possibility meant nomological possibility—consistency with the "laws of nature"—not the stronger notion of consistency with all facts about the past (temporal possibility). If nomological possibility is required, Schlick argued, which statements are meaningful is only ascertainable with empirical inquiry, about what the laws of nature are in particular. Schlick believed this conflates meaning with truth-value. Influenced by Wittgenstein, Schlick (1936: 349) thought statement meaning is determined a priori by "rules of logical grammar of language," and only with meaning fixed a priori is truth-value assessable. Since many nomological generalizations remain undiscovered and none have been established with absolute certainty, sharply delineating the physically possible from the impossible, and thus the meaningful from the meaningless, cannot be done.

But if logical possibility marks the bounds of empirical intelligibility, meaningful questions can concern states of affairs inconsistent with extremely well-confirmed nomological generalizations. Questions about velocities greater than 3×10^8 m/s, for example, may be significant even though they exceed the constraint of consistency with relativity theory. Thus, only statements whose confirmation conditions are logically impossible fail Schlick's criterion, which seemed (and seems) unduly permissive. More wedded to scientific practice and less aligned with Wittgenstein, Carnap (1936–7: 133) favored nomological possibility: "the physical possibility of the process of confirmation," determines empirical significance. But beyond the alignment with scientific practice, Carnap offered no explicit rebuttal to Schlick's argument.

In fact, without the clarity formal precision provides, simply (informally) specifying that the relevant modality must be nomological or logical adequately grounds neither Carnap nor Schlick's agenda. With logical possibility, logical empiricists could not reveal metaphysical statements as meaningless by demanding specification of their confirmation conditions. Metaphysicians could elide the request by responding that conceiving of these conditions may surpass human cognition. After all, they involve empirically impossible logical possibilities that might be conceptually inaccessible. Furthermore, the most sophisticated semantics for modal statements available to logical empiricists, Carnap's *Meaning and Necessity*, sanctioned this response. If the set of logical possibilities were recursive, a decision procedure would determine whether any putative conceptually inaccessible possibility alluded to by metaphysicians was genuinely logically possible. For Carnap (1947), however, logical possibilities are state-descriptions, and in both his "object" language S_1 and modal language S_2, the set of state-descriptions is non-denumerable and hence neither recursive nor recursively enumerable.

A similar stratagem exists for nomological possibility. Confirmation conditions for metaphysical claims are empirically possible, the line would go, but limited cognitive capabilities might prevent even the currently metaphysically savvy from clearly ascertaining what they are. As Berlin (1939: 293) keenly warned about the vague insistence that meaningfulness requires observation be "relevant" to a statement's truth-value: "Relevance is not a precise logical category, and fantastic metaphysical systems may choose to claim that [physically possible] observation data are 'relevant' to their truth." Avoiding these evasive responses and cogently addressing the adequacy problem required a more rigorous approach to formulating a verifiability criterion. The precision mathematical logic afforded and remarkable achievements it had recently facilitated made formalization the obvious strategy.

Ayer's criteria

A formal characterization of what it is for a statement to "say something" about, or "bear upon" a set of statements was needed. This characterization would then constitute the basis of a verifiability criterion, or criterion of empirical significance as it came to be called: of non-analytic statements, only those saying something about observation statements are meaningful. (A full criterion of cognitive significance, which would delimit meaningful from meaningless statements, would include an empirical significance criterion and an analyticity criterion delimiting analytic from synthetic statements.) In the first edition of *Language, Truth, and Logic*, Ayer proposed the first formal criterion: "the mark of a genuine factual proposition. . . [is that] some experiential propositions can be deduced from it in conjunction with certain other premises without being deducible from those other premises alone" (1936a/1946: 38–39). "Experiential propositions" designated observation statements. With this criterion, seemingly straightforward but problematically vague notions of "relevance" were explicated, with explicit deducibility relations specifying the required connection with observation exactly.

But the clarificatory power of formal precision can expose deficiencies that vague representations conceal. Such a deficiency was almost immediately apparent: it seemed to judge almost any statement meaningful. Consider any statement S and observation statement O. S and $S \rightarrow O$ entail O and $S \rightarrow O$ alone almost never does (Berlin 1939). According to the criterion, S and $S \rightarrow O$ are therefore significant unless $S \rightarrow O$ entails O, which holds just when $S \vee O$ is logically true (Lewis 1988). Ayer missed this point in the second edition and misinterpreted Berlin's criticism as showing that any statement is significant (1936a/1946: 11). This nuance notwithstanding, there was nothing to salvage. Lewis (1988) proved Ayer's criterion *does* judge any statement significant if there are at least two contradictory observation statements, which there surely are.

To rectify this shortcoming, in the second edition Ayer (1936a/1946: 13) developed a more complex, recursive criterion that distinguishes *directly* from *indirectly* verifiable statements:

> (DV) A statement S is *directly verifiable* iff: with a set of observation statements Ω, S entails at least one observation statement not entailed by Ω, alone.
>
> (IV) S is *indirectly verifiable* iff:
>
> > (i) with a set of statements Π, S entails at least one directly significant statement not entailed by Π alone; and,
> > (ii) Π contains only analytic, directly significant, or statements that can be shown indirectly significant independently.

Although not verifiable, analytic statements were considered meaningful. (Ayer [1936a/1946: 78] declared a statement analytic if "its validity depends solely on the definitions of the symbols it contains.")

The counterexample was more sophisticated, but Alonzo Church (1949) showed Ayer's second criterion fell as definitively as the first. Assume three logically independent observation statements O_1, O_2, O_3 exist and S is any statement. $(\neg O_1 \wedge O_2) \vee (\neg S \wedge O_3)$ is directly verifiable by (DV), since with O_1 it entails O_3. Also, $(\neg O_1 \wedge O_2) \vee (\neg S \wedge O_3)$ and S together entail O_2. Hence, by (IV) S is indirectly verifiable, unless $(\neg O_1 \wedge O_2) \vee (\neg S \wedge O_3)$ alone entails O_2. But if so, $\neg S \wedge O_3$ entails O_2 so that $\neg S$ is directly verifiable. Thus, any statement is indirectly verifiable or its negation is directly verifiable. On the reasonable requirement that $\neg S$ is verifiable iff S is, Ayer's criterion judges any statement verifiable.

The received view is that Church's criticism was the death knell of an already moribund project (e.g., Soames 2003: 391). The lively and technically sophisticated interplay between well-motivated amendments and intricate counterexamples that continued for several decades, however, belies this negative assessment (see Justus 2006). Church's criticism, for instance, exploits the internal structure of molecular statements. But that vulnerability, and many others, are easily and intuitively avoided by additional recursive requirements, just as in the transition from Ayer's first to second verifiability criterion (see Makinson 1965). To illustrate, consider the first iteration of the "patch and puncture industry." Since the lax attitude of Ayer's criterion towards the internal structure of $(\neg O_1 \wedge O_2) \vee (\neg S \wedge O_3)$ precipitates the problem, the straightforward fix that immediately suggests itself is restricting which molecular statements can be used to establish indirect significance via condition IV (Nidditch 1961). Requiring that all statements in Π of IV be analytic, directly significant, or independently indirectly significant, *and composed only of such statements* blocks the second inference in Church's analysis. Ayer's original criteria were much too slack; specifying verifiability with full recursive precision renders the criterion impervious to criticisms like Church's (see Wright 1986).

In one respect Ayer's criteria were atypical for logical empiricists, and clearly deficient for at least one. Unlike most logical empiricists, especially Carnap, Ayer had an affinity for an "ordinary-language" philosophical methodology, and his criteria were intended for natural languages. Carnap, in sharp contrast, thought the poor structure and insufficient systematicity of natural languages was primarily responsible for the existence and persistence of metaphysics. For Carnap, verifiability criteria—or "empirical significance" criteria in his preferred terminology—that *bear on* statements of natural languages need not be formulated *within* them. By developing criteria for suitably constructed artificial languages, Carnap thought portions of natural languages could be revealed to be significant (or not).

Empirical significance in a "language of science"

In an influential review, Hempel (1951) expressed pessimism that an adequate criterion would ever be found. Early attempts, such as Ayer's, were deemed deficient, and he correctly argued Carnap's (1936–7) construction of an artificial language for science via reduction postulates could not in fact accommodate most scientific theorizing. Instead, Hempel thought cognitive significance was a matter of degree and, crucially, only holistically assessable at the system level: "Significant systems range from those whose entire extralogical vocabulary consists of observation terms, through theories whose formulation relies heavily on theoretical constructs, on to systems with hardly any bearing on potential empirical findings" (1951: 74). Comparing theories by clarity, predictive and explanatory power, or simplicity, Hempel suggested, may be the only way to manifest the failings of metaphysics.

Carnap's last criterion, the most sophisticated of all the logical empiricists' attempts, specifically rejects the suggestion that empirical significance is a matter of degree, but it incorporates Hempel's insight that it should be relativized to theoretical context. Carnap (1956) formulated the criterion in a constructed language which divides into a theoretical language L_T and observation language L_O. V_O is the class of non-logical constants of L_O, and V_T is the class of *primitive* non-logical constants of L_T. Members of V_O designate observable properties and relations such as "hard" and "in physical contact with." Variables of L_O range over "concrete, observable entities (e.g., observable events, things, or thing-moments)" (1956: 41). L_O has at least one finite model and (for reasons of expositional ease only) is strictly extensional and constructive; that is, it contains no modalities and every variable value is designated by a term of L_O.

Members of V_T, called theoretical terms, designate unobservable properties and relations such as "electron" or "magnetic field." L_T, according to Carnap (1956: 43), "includes all of mathematics that is needed in science and also all kinds of entities that customarily occur in any branch of empirical science." Carnap defined a theory as a finite set of statements within L_T represented by their conjunction T. T was intended to represent the "fundamental laws of science, and not other scientifically asserted sentences, e.g., those describing single facts" (1956: 51). A set of correspondence rules, represented by their conjunction C, connects terms of V_T and V_O.

Besides requiring correspondence rules to contain V_T and V_O terms, Carnap rightly did not restrict their form. Hempel's (1951) criticism that so-called reduction sentences could not accommodate most scientific theorizing convinced him to adopt a more liberal view about the relationship between theoretical and observable terms: reduction postulates were only one of many different ways in which theory and observation could be connected.

With this framework, Carnap (1956) presented the criterion in three "definitions":

(D1) A theoretical term M is *empirically significant relative* to a class K with respect to $L_T, L_O, T,$ and $C =_{df}$

(i) $K \subset V_T$;

(ii) $M \notin K$; and,

(iii) there are statements $S_M, S_K \in L_T$, and $S_O \in L_O$ such that:

(a) S_M contains M as the only non-logical term;

(b) the non-logical terms in S_K belong to K;

(c) $(S_M \wedge S_K \wedge T \wedge C)$ is consistent;

(d) $(S_M \wedge S_K \wedge T \wedge C) \models S_O$; and

(e) $\neg[(S_K \wedge T \wedge C) \models S_O]$.

(D2) M_n is empirically significant with respect to $L_T, L_O, T,$ and $C =_{df}$ there is a sequence of theoretical terms $\langle M_1, \ldots, M_n \rangle (M_i \in V_T)$ such that every M_i is *significant relative* to $\{M_1, \ldots, M_{i-1}\}$ with respect to $L_T, L_O, T,$ and C.

(D3) An expression A of L_T is an empirically significant statement of $L_T =_{df}$

(i) A satisfies the rules of formation of L_T; and,

(ii) every non-logical term in A is significant as in (D2).

Similar to Ayer's criteria, these definitions, especially (D1[d],[e]), explicate the idea that empirical significance requires making a predictive difference. For a term M to be empirically significant, there must be a statement S_M containing M as its only theoretical term (D1[a]) in which S_M is indispensable in the nontrivial derivation of at least one observation statement (D1[c–e]). The indispensability may be mediated: the derivation can depend on other theoretical terms K (D1[ii, iii]), those occurring in S_K (D1[b]). Note, similar to the priority in his 1932 criterion, that the primary focus is on *term* significance; statement significance depends upon the significance of the terms they contain. With such a focus, Carnap was able to avoid the criticisms made of Ayer's criterion. By making statement significance parasitic in this way, Carnap precludes criticisms exploiting the internal structure of non-atomic statements, such as Church's.

Following Hempel's (1951) suggestion, Carnap also relativized empirical significance to particular languages *and* theoretical contexts (the theory and correspondence rules T and C). This was intended to reflect scientific methodology: theoretical terms and statements often bear on observations only in interdependent groups. Testing hypotheses in most special sciences requires methods of experimental design, data acquisition, and statistical analysis that are in turn informed by other theoretical concepts and results. And since theoretical terms often only bear on observation in conjunction with other theoretical terms, (D2) implicitly distinguishes theoretical terms whose significance depends on other theoretical terms from those that acquire their significance directly when $<M_1, \ldots, M_{n-1}>$ is empty. For any theoretical terms to be significant, there must be at least one of the latter (directly gaining significance via reduction sentences, for instance).

Despite its philosophical significance—for logical empiricists, of course, but frankly for almost all of contemporary analytic philosophy—Carnap's (1956) criterion garnered very little attention. Only a few critical responses were published. Kaplan's criticisms (published in [1975]) involving "deoccamization" and definitional extensions of theories, along with similar criticisms in Rozeboom (1960), are the most discussed. Deoccamization targets the requirement for at least one directly significant *M*, but the criterion is easily generalized to avoid the objection (Creath 1976) and its critical force is dubious regardless (see Justus 2014). The other criticisms are similarly ineffective.

Take theory extension. The supposed problem is that adding statements (even definitions) to *T* may change the significance of theoretical terms. Kaplan thought this was counterintuitive because definitions are "ordinarily thought of" as not adding empirical content to theories. But what Kaplan missed is that *modal* claims are Carnap's target of assessment: *could* theoretical terms make predictive differences in the observable realm? That assessment ranges over *all* circumstances, including those far from the ordinary, and far from what's probable or plausible. Kaplan's intuition may be worth respecting when S_K and S_M represent true, probable, or otherwise epistemically defensible conditions, but Carnap was concerned with whether implausible, even absurd counterfactual S_K and S_M far outside "ordinary" contexts could help establish connections between theoretical terms and observation. Kaplan's examples are of this sort and so reveal precisely how definitional extensions can change whether theoretical terms *potentially* bear on observation statements. In addition, Kaplan's and Rozeboom's objections in effect trade on an alteration of the logico-linguistic framework under analysis and so would not render unequivocal judgments even if the previous problem were somehow remedied (see Justus 2014).

Until recently, the deficiencies of these criticisms were lost in the high tide of repudiation of logical empiricism. Searching for such a criterion was just another philosophical dead end of a misguided movement. But as the work of the logical empiricists is increasingly appreciated rather than caricatured, the tide appears to be turning.

References

Ayer, A. J. (1934) "Demonstration of the Impossibility of Metaphysics," *Mind* 43: 335–45.

——— (1936a) *Language, Truth, and Logic*, London: Gollancz, 2nd ed., with an added introduction, New York: Dover, 1946.

——— (1936b) "The Principle of Verifiability," *Mind* 45: 199–203.

Berlin, I. (1939) "Verification," *Proceedings of the Aristotelian Society* 39: 225–48.

Carnap, R. (1928a) *Der logische Aufbau der Welt*, Berlin: Weltkreis Verlag. Trans. *The Logical Structure of the World*, Berkeley: University of California Press, 1967, repr. Chicago: Open Court, 2003.

——— (1928b) *Scheinprobleme in der Philosophie*, Berlin: Weltkreis Verlag. Trans. "Pseudoproblems in Philosophy," in Carnap (1928a/1967), pp. 301–43.

——— (1932) "Überwindung der Metaphysik durch logische Analyse der Sprache," *Erkenntnis* 2: 219–41. Trans. "The Elimination of Metaphysics Through Logical Analysis of Language," in A. Ayer (ed.), *Logical Positivism*, Glencoe: Free Press, 1959, pp. 60–81.

——— (1936–7) "Testability and Meaning," *Philosophy of Science* 3: 419–71 and 4: 1–40.

——— (1947) *Meaning and Necessity*, Chicago: University of Chicago Press, 2nd ed. 1956.

——— (1955) "Meaning and Synonymy in Natural Languages," *Philosophical Studies* 6: 33–47. Repr. in Carnap, *Meaning and Necessity*, Chicago: University of Chicago Press, 1956, 2nd ed., pp. 233–47.

——— (1956) "The Methodological Character of Theoretical Concepts," in H. Feigl and M. Scriven (eds.), *The Foundations of Science and the Concepts of Psychology and Psychoanalysis*, Minneapolis: University of Minnesota Press, pp. 38–76.

Chang, H. (2009) "Operationalism," in E. N. Zalta (ed.), *The Stanford Encyclopedia of Philosophy*, https://plato.stanford.edu/archives/fall2009/entries/operationalism/.

Church, A. (1949) "[Review of Ayer, *Language, Truth and Logic*, 2nd ed.]," *Journal of Symbolic Logic* 14: 52–53.

Creath, R. (1976) "On Kaplan on Carnap on Significance," *Philosophical Studies* 30: 393–400.

——— (1982) "Was Carnap a Complete Verificationist in the *Aufbau*?" *Philosophy of Science Association PSA* 1: 384–93.

Hempel, C. (1951) "The Concept of Cognitive Significance: A Reconsideration," *Proceedings of the American Academy of Arts and Sciences* 80: 61–77.

Justus, J. (2006) "Cognitive Significance," in S. Sarkar and J. Pfeifer (eds.), *Philosophy of Science: An Encyclopedia*, New York: Routledge, pp. 131–40.
——— (2014) "Carnap's Forgotten Criterion of Empirical Significance," *Mind* 123: 415–36.
Kaplan, D. (1975) "Significance and Analyticity," in J. Hintikka (ed.), *Rudolf Carnap, Logical Empiricist*, Dordrecht: D. Reidel.
Lewis, D. (1988) "Ayer's First Empiricist Criterion of Meaning: Why Does It Fail?" *Analysis* 48: 1–3.
Makinson, D. (1965) "Nidditch's Definition of Verifiability," *Mind* 74: 240–7.
Nidditch, P. (1961) "A Defense of Ayer's Verifiability Principle Against Church's Criticism," *Mind* 70: 88–89.
Rozeboom, W. (1960) "A Note on Carnap's Meaning Criterion," *Philosophical Studies* 11: 33–38.
Schlick, M. (1932) "Positivismus und Realismus," *Erkenntnis* 3: 1–31. Trans. "Positivism and Realism" in Schlick 1979, pp. 259–84.
——— (1936) "Meaning and Verification," *Philosophical Review* 45: 339–69. Repr. in Schlick 1979, pp. 456–81.
——— (1979) *Philosophical Papers, vol. 2 (1925–1936)* (ed. by H. L. Mulder and B. van de Velde-Schlick), Dordrecht: Reidel.
Soames, S. (2003) *Philosophical Analysis in the Twentieth Century*, Princeton: Princeton University Press.
Stace, W. (1935) "Metaphysics and Meaning," *Mind* 44: 417–38.
Wright, C. (1986) "Scientific Realism, Observation and the Verification Principle," in G. Macdonald and C. Wright (eds.), *Fact, Fiction and Morality*, Oxford: Blackwell, pp. 247–74.

17
NONCOGNITIVISM

Anne Siegetsleitner

In recent years, it has become accepted that there is more to logical empiricism than philosophy of science and mathematical logic. Some logical empiricists were even interested in morality as a common practice of daily life and in ethics as the philosophical inquiry of morality, but all can be said to share a belief in what Rudolf Carnap called "scientific humanism": the belief that it is the task of humanity itself to improve human life conditions and that science is one of the most valuable means to this improvement (1963a: 83). Some engaged personally with a liberal or socialist agenda. Nevertheless, most were reluctant to investigate these topics or other areas of value-oriented or normative practices in their own philosophical work. The reason is often held to be the movement's purported commitment to noncognitivism regarding value statements and norms. Some critics even allege that the logical empiricists were caught in a paradox. How could they be noncognitivists and at the same time engage in moral, social, and political activities?

In order to clarify these issues, the first part of this chapter will deal with the meaning of noncognitivism and how this relates to the logical empiricist verifiability principle of meaningfulness. The second part lays out how some important logical empiricists positioned themselves in the matter. Rudolf Carnap, Alfred J. Ayer, Hans Reichenbach, Friedrich Waismann, and Otto Neurath are well known for their noncognitivist stance, but the contributions of Moritz Schlick and Victor Kraft seem to challenge this consensus.

Noncognitivism and the verifiability principle of meaningfulness

"Noncognitivism" is a highly contested term. To stay close to contemporary philosophy of values and norms, I take the following theses as the core ideas of noncognitivism: (1) Value statements and/or norms have no truth conditions, i.e., they are not apt for truth or falsity in any substantial sense and cannot be confirmed or disconfirmed, and (2) value statements and/or norms express attitudes. These claims come in many variants but, in essence, for a statement to be cognitively meaningful, it has to be truth-valuable.

The commitment by logical empiricists to noncognitivist accounts of value statements and norms is generally considered to follow from their commitment to the verifiability principle of meaningfulness. This principle considers a statement as cognitively meaningful if and only if either: (1) one can state exactly under which empirical conditions the statement

DOI: 10.4324/9781315650647-20

can be confirmed or disconfirmed (i.e., the statement is in principle empirically verifiable); or (2) it is analytically true. If one holds that value statements are neither empirically verifiable nor analytically true and therefore not cognitively meaningful, one is committed to noncognitivism.

The verification principle never received a definite and uncontested formulation, and different versions come with varying sorts of radicality and range of application (see CH. 16). In his "The Elimination of Metaphysics through Logical Analysis of Language," Carnap seems to restrict meaningfulness in general to empirical or tautological statements. Philosophy of value and normative theory are explicitly mentioned: "In the domain of *metaphysics*, including all philosophy of value and normative theory, logical analysis yields the negative result *that the alleged statements in this domain are entirely meaningless*" (1932/1959: 60–61, orig. emphasis). There is no reason to assume that the same does not hold of value statements and norms outside philosophy, especially morality. It is clear from the context, however, that only cognitive meaninglessness was intended to be predicated of them and that only unconditional value statements and norms are targeted (see Carnap 1934). Ayer, in his influential *Language, Truth and Logic*, also separated factual from non-factual statements and allowed for emotional significance (1936/1971: 16). Logical empiricists always allowed for other kinds of meaningfulness like "emotional," "emotive," "instrumental," or "imperative" meaning. Some, like Karl Menger, rejected the talk of meaninglessness altogether (1934/1974: 185). Some formulations of the verifiability principle were extremely weak, especially Schlick's. Nonetheless, in most versions, unconditional value statements and norms turn out to be cognitively meaningless because they allow no interpretation as empirically testable or as tautological.

Noncognitivism in Carnap, Ayer, Reichenbach, Waismann, and Neurath

Logical empiricists are commonly regarded as proponents of a strong kind of emotivism (e.g., Urmson 1968: 15). Emotivism claims that value statements are (primarily) expressions of a pro or con attitude of the speaker towards various objects (and aim at evoking similar attitudes in the addressee). Strong emotivism claims that value statements or norms are nothing but expressions of such attitudes. A different variant of noncognitivism is prescriptivism which regards value statements and norms as a species of prescriptions or commands (see Hare 1952). Ethical emotivism and prescriptivism exclude metaethical positions like naturalism and intuitionism. Both of these hold that moral statements can be true or false, and that their truth or falsity can be known. Whereas naturalists believe that goodness or rightness can be identified or reduced to "natural" properties like happiness or the will of God, intuitionists think that there are special moral facts or properties which are to be known by intuition.

One of the main points of reference for the close link between logical empiricism and emotivism is Carnap. From his Viennese time onwards he viewed the content of morality as determined by individual decision (in his earlier work, traces of neo-Kantian and phenomenologist value theories were detected by Mormann 2006). For Carnap, metaphysical statements and affirmations of unconditional values "serve for the *expression of the general attitude of a person towards life*" (1932/1959: 78, orig. emphasis). In his London lectures *Philosophy and Logical Syntax*, Carnap offered a variant of this and stated: "But actually a value statement is nothing else than a command in a misleading grammatical form. It may have effects upon the actions of men, and these effects may either be in accordance with our wishes or not; but it is neither true nor false. It does not assert anything and can neither be proved nor disproved" (1935: 24). What value statements express for Carnap are permanent emotional or volitional dispositions (ibid.: 29). In these publications, Carnap presented himself as a noncognitivist of an emotivist kind with certain prescriptivist leanings.

In later years, Carnap allowed for the logical treatment of value statements and norms. In his reply to Kaplan on value statements published in the Schilpp volume, he developed a formal language for pure optatives and introduced the optative-operator "utinam" (meaning "wish that"). Conditional value judgments are expressively interpreted as factually meaningful as they express means-ends relations and are open to empirical scrutiny and confirmation (1963b: 999). While working on this reply in 1958, Carnap made additional notes on value concepts, which were published in 2017. There he further considered the embedding of value statements and norms in wider contexts and discussed standards of rational acceptance not only for factual statements connected with valuations but also for pure optatives (1958; for discussion, see Carus 2017). Even so, Carnap continued to hold to an individualistic version of moral decisionism or voluntarism according to which individual decisions determine the content of a person's morality and ethics. As Carnap put it in an interview in 1964, fundamental values are up to "the individual decision of any human according to his conscience or value feeling or however one may call it" (1993: 146, trans. AS).

In the English-speaking world, a similar perspective was promulgated by Ayer's *Language, Truth, and Logic*:

> The presence of an ethical symbol in a proposition adds nothing to its factual content. Thus if I say to someone, "You acted wrongly in stealing that money," I am not stating anything more than if I had simply said, "You stole that money." . . . I am simply evincing my moral disapproval of it. It is as if I said, "You stole that money," in a peculiar tone of horror, or written it with the addition of some special exclamation mark. The tone . . . adds nothing to the literal meaning of the sentence. It merely serves to show that the expression of it is attended by certain feelings in the speaker.
>
> (1936/1971: 110)

Ayer's formulations have been widely taken as paradigmatic for the metaethical position of logical empiricism in general. He interpreted ethical and moral terms as "expressions and excitants of feeling" (ibid.: 113). Where Carnap spoke of permanent emotional or volitional dispositions, Ayer used "disapproval," "sentiments," or "feeling." Like Carnap, Ayer also mentioned further addressee-oriented purposes moral and ethical terms serve: "They are calculated also to arouse feeling, and so to stimulate action. Indeed some of them are used in such a way as to give the sentences in which they occur the effect of commands" (ibid.: 111). Moral and ethical statements of this kind are not truth-apt because they do not express genuine propositions (ibid.: 112). In Ayer's approach, as in Carnap's, no genuine moral disagreements are possible, only clashes of emotions or at best disagreements about questions of fact relevant to moral claims. Nevertheless, arguments about the assumptions made by moral framework are possible even within Ayer's approach (ibid.: 115–16). This is a point we will need to consider further.

Reichenbach presented his noncognitivist account forcefully in *The Rise of Scientific Philosophy*. Interpreting central moral and ethical statements (utterances) as directives which express volitional decisions of the speaker (1951: 291), Reichenbach shifted the focus from the speaker's emotion to her volitional decisions. Imperatives are a special kind of directives, which are used for the direction of persons other than the speaker (ibid.: 280). Combined with his version of the verifiability principle, this committed Reichenbach to noncognitivism with regard to fundamental decisions or goals. However, he stated explicitly that empirical knowledge and logic may inform whether a goal adopted by a fundamental decision is attainable and compatible with other goals. Moral or political considerations concerned with such means-ends relations are of the cognitive type. Therefore, knowledge can resolve disputes over means, if the disputants agree on ends. "If some fundamental aims are the same, quite a few moral issues are transformed into logical issues" (ibid.: 297). According to Reichenbach, there may be some truth or falsity in moral and political reasoning, but not with regard to the fundamental values or goals expressed.

Reichenbach's noncognitivist account of morality stresses that moral agents have to decide which imperatives they take to be moral ones, but notes that at least some of these concern what is necessary for social life, i.e., rules the compliance with which is mandatory. In this context, he proposed the democratic principle "*Everybody is entitled to set up his own moral imperatives and to demand that everyone follows these imperatives*," but also stressed that this did not give carte blanche to individuals but required them to secure consensus with others (ibid.: 295, orig. emphasis). Reichenbach's account makes clear that noncognitivism in theoretical ethics does not commit one to value neutrality in the practical field of morality. On the contrary, it is the task of a moral agent (although not of the scientific philosopher) to set up moral imperatives. Reichenbach drew the important conclusion: "Whenever there comes a philosopher who tells you that he has found the ultimate truth, do not trust him" (ibid. 302). For him, the essence of noncognitivism lies precisely in the "recognition of the non-logical relation between knowledge and action" (1979: 478).

Waismann held a similar position in his short but clearly written lecture dating from 1938 or 1939, "Ethics and Science." Having rejected naturalism and intuitionism, he declared that ethics and morality concern the will and that precepts or prescriptions are not true or false: approving something means affirming something, taking a position, and making a commitment. "In the final analysis it always comes down to the question: Do you will the world to be like this? Or do you will it be like this?" (1989/1994: 49). Awareness of the need to choose provides morality with a new meaning and seriousness. "The tissue of half-truths woven by human imagination around moral laws vanishes before his eyes; he stops asking for the truth of morality and starts choosing and deciding. And does this not give ethics a new dignity alien to the old conception?" (ibid.: 51).

Neurath only ever issued passing comments on traditional ethics (e.g., 1932/1959: 305–8), but it is clear that in order to constitute a moral system, for him this choice has to be made by a group and not a single individual. For this reason, as soon as a moral system is constituted, it provides a shared set of standards for evaluation or action guidance (see Siegetsleitner 2014: 219). Neurath opted for a system which aims at improving the living conditions of men and women, and devoted his many-faceted academic work to providing tools for the social sciences in order to describe and finally improve these conditions. Importantly, his sociological and economic modelbuilding eschewed the many of the limiting conditions associated with "positivist" neo-classical welfare economics, his observance of the fact-value distinction notwithstanding (see O'Neill and Uebel 2008).

Note that none of the hitherto-presented logical empiricists denies the possibility of reasoned argument relevant to value statements or norms. How far this is possible depends on the precise character of the expressed *Lebensgefühl*, sentiment, approval, or decision. Unfortunately, none of the discussed philosophers was very clear about this. If it is about an evaluative or normative framework, which in itself allows for empirical criteria and rationality, there is much room for rational deliberation and empirical scrutiny. Take utilitarianism as an example for a normative theory which can be rendered perfectly compatible with such a noncognitivist account (we treat its basic premises as adopted conventionally). After people have chosen or accepted this moral framework, they share a standard for evaluation and action guidance which provides them with an empirical criterion in the determination of what is right and what is wrong, namely, "maximizes general utility." At least in some versions of utilitarianism, the decision becomes an empirical question. Evaluation and action guidance present themselves as questions of empirical knowledge, *nota bene*, within the chosen framework. The important role science may play in finding out what will be the right thing to do is obvious too.

How reasoned ethical argumentation is possible within a noncognitive metaethical theory has become a core issue in recent theories of expressivism, especially hybrid expressivism or neo-expressivism. According to ecumenical expressivism developed by Ridge (2006), moral evaluations and/or norms express two closely related mental states. One state is a belief that certain things have a certain property F, and the other a noncognitive state like approval or disapproval of things that have F. The descriptive belief may explain the inconsistency and disagreement relations between moral evaluations and/or norms involving them. These hybrid theories share important features with pure expressivist as well as with cognitivist theories (see Schroeder 2010: 191 ff.). Similar features can be found explicitly in the noncognitivist accounts of Schlick and Kraft.

Noncognitivism in Schlick and Kraft

Some contributions to ethics by logical empiricists raise the question to what extent if any they can be called noncognitivist. This holds for the metaethical approaches by Schlick and Kraft. For both, ethics played an important part in their philosophy.

Schlick's *Problems of Ethics* mostly deals with issues in moral psychology like psychological hedonism, but metaethical questions play an important role as well. Schlick maintained that the general value predicates "valuable" and "good" have the same extension as "bringing pleasure" (1930/1939: 102). Like "x is bringing pleasure," the statement "x is valuable" describes a state of affairs, an experience of pleasure, and claims that it is either true or false. Thus, value statements have a cognitive dimension sufficient for truth-aptness. Whether further linguistic functions may be fulfilled by such statements is in his view irrelevant as regards their verifiability. According to Schlick's account, value statements are verifiable, namely by the occurrence of a certain experience, a feeling of pleasure (ibid.: 105). So if the denial of truth-aptness is a necessary component of noncognitivism, Schlick was no noncognitivist.

But what does Schlick's general view about valuability mean for the special case of "morally good"? Following Schlick's definition of social morality, the statement "x is morally good (from a social point of view)" translates into "x is approved by society, because society believes that x is most advantageous to its general welfare" (ibid.: 195). (The social inclinations of altruism and love serve as prominent examples.) Although Schlick did not explain how this valuation relates to experiences of pleasure, which only can be experiences of individuals, he did ascribe a cognitive function and truth-aptness to socially interpreted value statements as far as these are verifiable by the occurrence of such an approval. The same holds for Schlick's account of individualistic morality. As far as individualistic morality is concerned, "x is morally good" is equivalent to "x is conducive to the capacity for happiness of the individual" (ibid.: 182–3). In Schlick's optimistic outlook, following the demands of social morality is also morally good in the individualistic sense (see, e.g., the case of love; ibid.: 190). For him, what used to be an ethics of duty becomes an ethics of kindness. In this respect, he simply ignores the tension between social and individualistic morality which is not solved by wishful thinking. To be sure, for Schlick, the predicate "morally good" also has a formal component: the morally good is what is demanded (ibid.: 11). Yet this deontological residue is fairly inconsequential in the special case of individual morality because it only asks to do what an individual desires to do anyway.

Schlick was an empiricist of the hedonist school but not a naturalist. On the one hand, there are moral facts, and these are facts concerning pleasurable experiences and standards of social approval: no intuitions are needed for knowing them. On the other hand, there is also the formal component of value predicates. Like contemporary hybrid theories, Schlick allowed value predicates to refer to natural properties and at the same time to fulfill a further function due to an additional component of their meaning. So Schlick might even be called a cognitivist because he holds that value statements are truth-apt and that value knowledge is possible. However, this pertains only to statements within the framework of a hedonistic account of morality, which has to be chosen in the first place. Therefore, on closer inspection, these value statements turn out to be framework-relative and conditional.

This issue becomes more explicit in the work of Victor Kraft. In 1937, he published the first edition of his *Foundations for a Scientific Value Theory*, with a second, much-revised edition in 1951. Already in the first edition, Kraft distinguishes between the factual content (*Sachgehalt*) and the value character of value concepts used in value judgments. The value character is part of every value concept and does not express an individual preference but an impersonal and general command: "*A value judgment means . . .: the command to adopt an attitude towards an object, [and to do so] generally and anonymously*, not as a particular person with regard to particular persons" (1937: 164, orig. emphasis, transl. AS). Value judgments are not individual statements but impersonal commands to statements.

According to Kraft, there are pure value concepts like "good" or "bad." However, in most value concepts an additional component is present: a factual content. This component provides the criteria for the impersonal judgment; of these, Kraft speaks as specialized value concepts (aesthetic concepts also belong to this category). In these cases a competent addressee of a value judgment is (implicitly) informed about the criteria and told that the speaker thinks that these criteria are fulfilled by the object judged. If the criteria for these impersonal judgments are empirical ones, depending on the value-oriented practice, there is, again, plenty of room for rational argumentation and knowledge. This means that for Kraft, different kinds of value statements and judgments must be distinguished. If factual content is present as well, then there may be room for knowledge and logic. (A similar position is found in Joergensen [1937/38] and Rand [1939].) To be sure, objective criticism of value judgments is possible only given a shared value standard: any objectivity of value judgments is always derivative and conditional, as Kraft makes plain (1937: 171).

In Kraft's account, value judgments are meaningful, but due to the value character of the value concepts used in them, he recoils from ascribing truth or falsehood to them and prefers to speak of their "validity" (ibid. 165). (On this point there is agreement with Feigl [1952].) Following the definition of noncognitivism used here, Kraft must be characterized as a noncognitivist. However, if one wants to stress the role of knowledge due to the factual content of specialized value concepts like in moral ones, Kraft could be called a cognitivist.

Concluding remarks

Noncognitivism holds that value statements and/or norms are not apt for truth or falsity in any substantial sense and that they express noncognitive attitudes. Regarding value statements and/or norms which are not framework-relative, conditional, or instrumental, logical empiricists are noncognitivists of various kinds.

All things considered, noncognitivism leaves no room for dogmatism or privileged access to ultimate moral truth, as none is recognized. In the end, values and norms are up to choice and decision. However, this does not logically commit theorists to the claim that all value statements or norms are beyond the scope of rational discussion. This is most certainly not true for conditional and instrumental statements. Nor does noncognitivism logically require a retreat from substantive concerns. A noncognitivist position in metaethics is unproblematically compatible with a utilitarian outlook or a socialist political or scientific agenda, as can be found in Neurath's writings. Thus, the alleged logical empiricists' paradox of being noncognitivists and at the same time engaging in moral, social, and political activities simply is a misconception of their accounts. However, noncognitivism by itself is conceptually neutral with regard to a theorist's socio-political orientation.

Noncognitivist theories have become more subtle and complicated over the years (see the survey in Schroeder 2010), but core ideas of contemporary noncognitivism can be found already in versions put forth by logical empiricists. If their writings on values and norms had been read with less negative preconceptions, perhaps some of the new directions of investigation and developments, which have been pursued in value theory, especially ethics, in recent decades, could have been pursued earlier.

References

Ayer, A. J. (1936) *Language, Truth, and Logic*, London: Gollancz, 2nd ed., 1946. Repr. London: Penguin Books, 1971.

Ayer, A. J. (ed.) (1959) *Logical Positivism*, New York: Free Press.

Carnap, R. (1932) "Überwindung der Metaphysik durch logische Analyse der Sprache," *Erkenntnis* 2: 219–41. Trans. "The Elimination of Metaphysics Through Logical Analysis of Language," in Ayer (1959), pp. 60–81.

——— (1934) "Theoretische Fragen und praktische Entscheidungen," *Natur und Geist* 2: 257–60. Repr. in C. Damböck (ed.), *Der Wiener Kreis. Ausgewählte Texte*, Stuttgart: Reclam, pp. 175–9.

——— (1935) *Philosophy and Logical Syntax*, London: Kegan Paul, Trench, Trubner & Co.

——— (1963a) "Intellectual Autobiography," in P. A. Schilpp (ed.), *The Philosophy of Rudolf Carnap*, Chicago and La Salle, IL: Open Court, pp. 1–84.

——— (1963b) "Reply to Abraham Kaplan on Value Judgment," in P. A. Schilpp (ed.), *The Philosophy of Rudolf Carnap*, Chicago and La Salle, IL: Open Court, pp. 999–1013.

——— (1993) "Interview mit Rudolf Carnap," in Carnap, *Mein Weg in die Philosophie*, Stuttgart: Reclam, pp. 134–48 (orig. 1964).

——— (2017) "Value Concepts," *Synthese* 194: 185–94 (orig. 1958).

Carus, A. (2017) "Carnapian Rationality," *Synthese* 194: 163–84.

Feigl, H. (1952) "Validation and Vindication: An Analysis of the Nature and the Limits of Ethical Arguments," in W. Sellars and J. Hospers (eds.), *Readings in Ethical Theory*, New York: Appleton-Century-Crofts, pp. 667–80.

Hare, R. M. (1952) *The Language of Morals*, Oxford: Clarendon Press.

Joergensen, J. (1937/38) "Imperatives and Logic," *Erkenntnis* 7: 288–96.

Kraft, V. (1937) *Die Grundlagen einer wissenschaftlichen Wertlehre*, Vienna: Springer, 2nd ed., 1951. Trans. *Foundations for a Scientific Analysis of Value*, Dordrecht: Reidel, 1981.

Menger, K. (1934) *Moral, Wille und Weltgestaltung: Grundlegung zur Metaphysik der Sitten*, Vienna: Springer. Trans. *Morality, Decision and Social Organization: Toward a Logic of Ethics*, Dordrecht: Reidel, 1974.

Mormann, T. (2006) "Werte bei Carnap," *Zeitschrift für philosophische Forschung* 60: 169–89.

Neurath, O. (1932) "Soziologie im Physikalismus," *Erkenntnis* 2: 393–431. Trans. "Sociology and Physicalism," in Ayer 1959, 282–320, and "Sociology in the Framework of Physicalism," in Neurath, *Philosophical Papers 1913–1946* (ed. by R. S. Cohen and M. Neurath, Dordrecht: Reidel, 1983, pp. 58–90.

O'Neill, J. and Uebel, T. (2008) "Logical Empiricism as Critical Theory: The Debate Continues," *Analyse & Kritik* 30: 379–98.

Rand, R. (1939) "Logik der Forderungssätze," *Internationale Zeitschrift für Theorie des Rechts/ Revue internationale de la théorie du droit* 1: 308–22.

Reichenbach, H. (1951) *The Rise of Scientific Philosophy*, Berkeley: University of California Press.

——— (1979) "On the Explication of Ethical Utterance," in Reichenbach, *Selected Writings: 1909–1953* (ed. by M. Reichenbach and R. S. Cohen), Dordrecht: Reidel, vol. 1, pp. 474–79 (orig. 1952).

Ridge, M. (2006) "Ecumenical Expressivism: The Best of Both Worlds?" in R. Shafer-Landau (ed.), *Oxford Studies in Metaethics 2*, Oxford: Oxford University Press, pp. 302–36.

Schlick, M. (1930) *Fragen der Ethik*, Vienna: Springer. Trans. *Problems of Ethics*, New York: Prentice-Hall, 1939.

Schroeder, M. (2010) *Noncognitivism in Ethics*, New York: Prentice-Hall.

Siegetsleitner, A. (2014) *Ethik und Moral im Wiener Kreis. Zur Geschichte eines engagierten Humanismus*, Vienna: Böhlau.

Urmson, J. O. (1968) *The Emotive Theory of Ethics*, London: Hutchinson University Library.

Waismann, F. (1989) "Ethik und Wissenschaft," in Waismann, *Wille und Motiv* (ed. by J. Schulte), Stuttgart: Reclam, pp. 158–83 (orig. 1938–9). Trans. "Ethics and Science," in F. Waismann, J. Schächter and M. Schlick, *Ethics and the Will. Essays* (ed. by B. McGuinness and J. Schulte), Dordrecht: Kluwer, 1994, pp. 35–52.

18
THE UNITY OF SCIENCE

Jordi Cat

Reductionism and its epistemological expression, foundationalism, are familiar tropes in narratives about logical empiricism. A minimal standard of reductionism requires relating all relevant concepts and generalizations to a privileged fundamental level of concepts and statements. The relation of reduction, understood as either conceptual or ontological dependence, typically rests on salient features of the fundamental level, such as generality, simplicity, and elementarity. Foundationalism, moreover, privileges certain claims from the perspective of their epistemological function, in its most radical formulations as absolute or unrevisable, and providing the grounds of the epistemic reliability and justification of all others. Of such reductionist foundationalism, logical empiricism has widely been supposed to be possessed. Contrary to this, however, alternatives to reductionism and foundationalism originated at the center of the movement with leading figures such as Neurath, Carnap, and Reichenbach.

With the turn of the twentieth century, both mathematics and physics underwent radical transformations. Logical empiricists pursued the implications that followed from mathematics turning out not to be synthetic and physics not a priori. Adopting a scientific world-conception meant investigating the possibility of objective empirical knowledge with a role for the constructive element of cognition, yet without universality or necessity, avoiding Kantian apriorism and radical positivism. The formulation of the new unity of objective, empirical knowledge was to express the new epistemological assumptions.

Unity and epistemology up to the Vienna Circle's manifesto

By the end of the nineteenth century, different neo-Kantian schools had introduced a distinction between the natural sciences (*Naturwissenschaften*) and the cultural, or human, sciences (*Geisteswissenschaften*) and sought to articulate the kind of concept formation that was distinctive of each. According to the Marburg School, led by Hermann Cohen and Paul Natorp, the mode of conceptualization of the natural sciences required general laws, and its certainty and objectivity relied on its precise mathematical form. Meanwhile the Southwestern School, led by Wilhelm Windelband and Heinrich Rickert, sought to identify the distinctive kind of theory and concept formation in the historical or cultural sciences as concerned with individuality and the pursuit of values. The opposing program of monism was led by scientists like Ernst Haeckel

DOI: 10.4324/9781315650647-21

and Wilhelm Ostwald and popularized through literature and associations, apart from strictly academic venues.

Neurath entered the discussion of the unity of science as a political economist with a historical, empiricist perspective and early on rejected the distinction between the natural and the human sciences and insisted that any classification was historically contingent (see his 1910). He urged attention to the languages employed in the formulation of scientific theories, explored new symbolisms for economics with comparative but non-measurable quantities, and stressed the need for the systematic examination of all possible cases. Neurath also insisted on attention to the relation which individual inquiries bore to the whole of science and demanded transparency of the resulting order. At this point Neurath's conception of unity required simplicity, not reductionism. His ideal required scientists to cooperate as a whole rather than act as a mere collective. He distinguished between mere separation of labor and proper division of labor, in which specialized work took place in conscious relation to a surveyed whole, connecting two levels of scientific representation (ibid.).

Moritz Schlick's *Allgemeine Erkenntnislehre (General Theory of Knowledge*, 1918/1974) presented a generalization of Planck's and Hilbert's commitment to a holistic structural unity in mathematics and physics. Empirical knowledge resulted in bringing together concepts and intuitions. Conceptual knowledge is organized systematically, for instance, in axioms expressible symbolically and through implicit definitions. These are structural descriptions, internal to the formal system and representing the meaning of terms, such as point and line in Hilbert's system of geometry, as determined by their occurrence in different axioms. By contrast, intuitions are private and imprecise ideas sometimes expressed in explicit definitions that characterize ordinary experience with practical value. The coordination between the two modes of cognition is effected through representations of coincidences of sensations from different sense modalities common to different individuals, e.g., touching and seeing a sphere. Schlick declared that the modes of cognition are ideally rendered comparable through the quantitative means of measurement, that is, exact, objective determinations of coincidences. This was what gives us scientific knowledge of a transcendent reality. Metaphysics was, by contrast, the pursuit of intuitive knowledge of the transcendent, a nonsensical project given that intuition is only acquaintance with the immanent content of experience that cannot be shared (Schlick 1926/1979: 109).

Schlick's Einsteinian turn was followed by a Tractarian turn which focused his interest on the limits and nature of philosophy, the role of language and logic in determining the scope of science, and the source of meaning in the verification of elementary statements. All meaningful questions can be answered in principle by science; there are no philosophical questions as such, only the philosophical activity of clarification of scientific propositions (Schlick 1930/1979). Unity, in the form of reduction, remained part of the image of science. It was a mark of its progress towards the complete world-picture, knowledge of one world: "Science, therefore, is a unity. It is not a mosaic, or a grove in which various kinds of tress stand side by side, but *one* tree with many branches and leaves" (Schlick 1934/1979: 141).

As it had been for Schlick and Neurath earlier, philosophy's task for Carnap was the comprehensive construction of a map in the form of a "system of knowledge." For this task, philosophy had recourse to the new resources of formal logic. Carnap's project extended beyond the general concepts of the natural sciences to the cultural objects/concepts of the human sciences, but he renounced the neo-Kantian doctrine of conceptual discontinuity between them. Carnap's *Der Logische Aufbau der Welt* (*The Logical Structure of the World*) developed a comprehensive system of scientific concepts on the assumption that "there is only one domain of objects and therefore only one science," thereby providing a model of the unity of science (1928/1967: 9). The system was a model of vertical integration. Scientific objects/concepts could be given a structural definite description in terms of formal relations to others. Key among their mutual relations was their hierarchical order so that the constitution relation between the types of objects involved the step-by-step derivation or construction from a fundamental level. Upwards, the derivation was a constitution; downwards, a reduction. From a linguistic, formal standpoint, Carnap required that all statements about a reduced or constructed object would be translated into—represented coextensively by—the ones about the more fundamental reducing or constructing objects. The structural character of the unified system would guarantee its objectivity, or at least its intersubjectivity. The system's interconnected and comprehensive character would also do the exclusionary boundary work: only statements translatable into statements within the system would locate a concept within the structure and identify its acceptable meaning. Now, whether the conditions of a phenomenalistic reduction were met or not, as Quine later objected (1951), the nature of Carnap's reductionism remains under discussion. Altogether more problematic, as recent scholarship has shown (Friedman 1987; Richardson 1998), is the foundationalism that also the *Aufbau* was accused of, most prominently again by Quine (1951).

What is often called the Vienna Circle's "manifesto" was a pamphlet aimed at persuading Schlick not to take a university position elsewhere and institutionalizing his informal circle by attributing a shared program. Unity was officially adopted as a goal, with multiple levels of representation and purposes: "the goal ahead is unified science" (Verein Ernst Mach 1929/1973: 306). At one level, the manifesto endorsed what it called the Enlightenment spirit. Its renewed relevance resided in its intellectual and social values and the new intellectual resources for their implementation. The more specific emphasis was on the value of the so-called scientific attitude, which involved pursuing the constructive and critical tasks called for by the so-called scientific world-conception. At the service of unification, the manifesto endorsed related standards: collective efforts at clarity and precision, a comprehensive conceptual system of concepts, an integrated structure for the individual disciplines within unified science with individual theories axiomatized based on conventional definitions and coordinating definitions. The manifesto stressed the social significance of the project consistent with the Enlightenment spirit as endorsed by Neurath and expressed also in the *Aufbau* (ibid.: 305, 317–18).

After the manifesto: between program and debate

The manifesto had established a consensus around the linguistic turn and the linguistic dimension of the epistemic unity of the sciences. As a result, the discussion turned to epistemic norms and took the form of the question of the language of science and its fundamental sentences: the protocol sentence debate and the linguistic demarcation between science and metaphysics. The epistemic burden lay on the basic evidence statements, or "protocol statements." The protocol-sentence debate was prompted by Neurath's criticism of Carnap's choice of the private phenomenalist basis of the given in the *Aufbau* and raised the question of the appropriate form and epistemic status of the basic evidence statements of science (see Uebel 2007a). The search for an alternative led to different conceptions of "physicalism," a weakening of the reductionism and strong challenges to foundationalist assumptions. The different positions adopted under the former term (which Schlick disliked) illustrated different metatheoretical perspectives. Schlick emphasized the significance of personal activities of meaning determination (elucidation) and justification; Carnap's logic of science aimed at a structural and formal logical rational reconstruction of scientific discourse; and Neurath's empirical behavioristics of science pursued a materialistic, constructive, contextual, and pragmatic type of naturalism, neither strictly descriptive nor reducible to the natural sciences. Notably, the attention to unified science was distinctive of the Viennese group among Central European empiricists. By contrast, in Berlin, Hans

Reichenbach's Berlin Group and the Society for Empirical Philosophy paid more attention to specific issues in the foundations of special sciences such as mathematics, physics, and psychology, while in Poland, the focus was on the foundations of logic and mathematics.

For Neurath the task of unification became the predominant focus, and he placed a distinctive emphasis on the collective scientific character of the process, its instrumental value for social life and, along the way, on the implications for the social sciences. Social unity or coordination required intellectual forms of unification acting as tools not only in planning schemes or international organization but also within science itself. In particular, Neurath stressed that coordination was required between scientific and ordinary languages with different standards of perspicuity: formally precise technical or constructed languages (as investigated by Carnap) must be shown to link up somehow with imprecise natural languages—according to the motto "metaphysical terms divide, scientific terms unite" (Neurath 1933/1987: 23). In addition, Neurath developed a visual language for social educational purposes; here, the motto was "words divide, pictures unite" (Neurath 1931d/1973: 217).

The term "scientific world-conception" similarly designated a collective historical task of scientific unification in contrast to traditional philosophical world-pictures or worldviews: it was built bottom-up, from partial and contingent empirical inquiry, rather than top-down from first principles. The purpose of the physicalist basis he argued for was to provide the inter-subjective, public basis for empirical investigations. Another objective, prominently promoted by Carnap, was the adoption of a comprehensive system of precise conventional symbolism, constructed with the aid of logic and mathematics, to enable the purification of language from metaphysics and the connection between the symbolic expressions of results. Importantly, Neurath's unity (and, increasingly, Carnap's) was not the reductive unity of the *Aufbau*'s hierarchical system or even a materialist variant thereof. As he was careful to note, "unified science is physics in *its largest aspect*, a tissue of laws expressing space-time linkages" (1931a/1983: 49, emphasis added). It privileged statements about material objects or events in space and time with broad references to "spatio-temporal order" or with "spatio-temporal data" (1931b/1983: 54), but suggested no hierarchy. Neurath often spoke of those material events and processes that constituted the empirical objects of interest quite indiscriminately as "behaviors." Different sciences concerned themselves with the behaviors of different kinds of systems (or the same), whether it was human behavior or the behavior or stars and stones. Likewise, the science of science that took over from philosophy in the discussion of conceptual foundations of the sciences he called the "behavioristics of scholars" (1936a).

For Neurath, the universal physicalist language was not a constructed, ideal language. To the extent that it was derivative of ordinary language and in the human sciences concerned actual human behavior, it was not neatly precise with terms laid out atomistically and defined exactly in formulas. It also contained imprecise terms (*Ballungen*) for complex "cluster concepts" whose metaphysical meanings can be eliminated, at least in principle. When interpreted physicalistically, they often, as a matter of actual linguistic behavior, remained in use. The universal language of unified science was at best a "universal slang" or "jargon" (1932a/1983: 64; 1932b/1983: 91–92). Its significance lies in its holistic character, informing all the key concerns of the scientific world-conception: meaning, demarcation, prediction, testing, growing interconnectedness, and application in interventions, e.g., in engineering or social planning. The value of the task of unification was far more than systematically or historically descriptive; it was also methodologically and pragmatically normative.

The negative, demarcating function of unity is discharged by the universality of physicalism. As one of its positive functions, holism allowed the determination of the correctness of a statement due to its internal place in the totality of statements (1932a/1983: 66). Coherence as a criterion of acceptance of individual statements and groups thereof (theories) is a weaker criterion than the deductive integration in Carnap's *Aufbau* or any correspondence conception of truth. Whether the latter in particular offered epistemic advantages seemed doubtful to him. Typically, Neurath also appealed to and extended Duhem's underdetermination thesis to reject the idea—Quine-style (1951)—that individual statements could be tested in isolation (1921/1973; 199; 1936a/1983; 161). Predicting, testing, and planning require more than appealing to facts or experiences (that is, data or observation or factual reports). It required the auxiliary hypotheses that themselves may need to be revised sooner or later (1921/1973: 203). Eventually, for Neurath, the acceptance (not truth) of a hypothesis—whether a statement of singular fact or an inductive generalization—became a pragmatic matter of decision, without certainty, informed by what he called extralogical factors or auxiliary motives such as accepted purposes and conventions (1934/1983: 104 and 106).

For the project of unification, this methodological holism had two consequences: antifoundationalism and antireductionism. It eliminated a strong asymmetry within the system of scientific statements that privileged protocol statements absolutely, on account of their certainty and unrevisable status. As a consequence, in the process of establishing coherence within the system, this allowed, in a situation of conflict, for the possibility of rejecting the protocol statements considered rather than introducing changes in the system in order to accommodate them. Neurath also extended Duhem's holistic diagnoses from physics to all the sciences. His example, showing the pressing practical need for unification, was a forest fire as an object for meteorology, biology, chemistry, and psychology, all of them combining to investigate the possibility of engaging in some kind of intervention (1932a/1983: 59). Neurath expressed the cognitive syntheses engendered by the holistic antifoundationalism and antireductionism he promoted with his famous simile of the ship: "We are like sailors who have to rebuild their ship on the open sea, without ever being able to dismantle it in dry-dock and reconstruct it from the best components" (1932b/1983: 92).

In the picture of unity Neurath defended, there was little room for actual reductions or the imperative of reductionism. Typically, for the study of the behavior of machines, animals, and humans, "a reduction to a law concerning atoms or other elements [was] unnecessary" since laws of macrostructures were sufficient, though he admitted that sometimes the actual "macro-laws" could not account for irregularities (1932a/1983: 68 and 75). That said, Neurath pointed out intrinsic limitations in the predictive capacity of the social sciences. Limits in scope may stem from historical contextuality or immanence of scientific practice itself, its dependence on current social conditions (1936c/2004: 506). In addition, some events cannot be predicted at all. Neurath was responsible for drawing early attention to self-reflexivity, either self-fulfilling or self-defeating, recognizing that predictions may act as co-determinants of or obstacles for what they describe (e.g., 1931c/1973: 404–5; see Uebel 2007b). Finally, he pointed to unpredictability in principle (both in natural and human sciences) associated with instability or dependence on initial conditions—so-called chaotic behavior (1944: 28).

Carnap's overarching interests concerned formal logic (first only deductive, later also inductive logic), the foundations of mathematics—especially the conflict between logicism, formalism, and intuitionism—and the reform of philosophy itself. It culminated in *Logical Syntax of Language* with the formulation of the program of philosophical reform in logical-linguistic terms announced in the manifesto: "*Philosophy is to be replaced by the logic of science*—that is to say, by the logical analysis of the concepts and sentences of the sciences, for *the logic of science is nothing other than the logical syntax of the language of science*" (1934/1937: xiii, orig. emphasis). After 1935, following his acceptance of Tarski's semantic theory of truth, Carnap loosened the strictures on this logic of science, first admitting semantics and later also pragmatics in order to provide a variety of rational reconstructions, from individual contested concepts to entire theories. These were to be developed according to pragmatic criteria in line with whatever were the purposes for which the reconstruction was undertaken. Carnap abjured absolutism with his embrace of logical tolerance: "*In logic, there are no morals*. Everyone is at liberty to build up his own logic, i.e. his form of language as he wishes. All that is required is that, if he wishes to discuss it, he must state his methods clearly" (ibid.: 52, orig. emphasis).

All along, Carnap continued to pursue the formal aspect of unification. Early on, he declared that the new logic served the unity of science by specifying logical relations between concepts and statements of different sciences and the concepts of the given (1930: §8). Neurath's critique prompted Carnap to revision the unity on a different basis. His final answer in light of the principle of tolerance was that there was no uniquely correct answer as to the logical form the rational reconstruction of scientific discourse had to adopt. Yet, for the purpose of reconstructing our language of science, Carnap was adamant that a physicalist language with terms for intersubjectively observable objects and properties was basic (Carnap 1936–7). After 1932 Carnap was no longer concerned to defend the individualistic standpoint of traditional empiricism, thereby agreeing with Neurath. But Carnap's physicalist unity wasn't Neurath's either; for him, physicalism meant first the translatability of all languages into the language of physics, later the equipollence—sameness of nonanalytic consequences—of their statements. He did not ground the unity of science on the putative fact that the test of all scientific statements had to invoke conditions specifiable in the ordinary language speaking of material objects and properties (Schlick's coincidences intersubjectivized).

The unity of science movement and the encyclopedia model

The project of the *International Encyclopedia of Unified Science* was emblematic of the circumstances, ambitions, and challenges of the unity of science project. Neurath had discussed the encyclopedia project in the 1920s with Frank and Einstein, and an official proposal was finally launched at the 1935 Paris Congress. An encyclopedia committee was established (Carnap, Frank, Jørgen Jørgensen, Charles Morris, Neurath, and Louis Rougier), and three editors were named (Neurath, Carnap, and Morris). The initial publication began in 1938. Only ten monographs were published by 1945, by the end, in 1970, they numbered twenty and included Thomas Kuhn's *Structure* (1962) to replace the history of science volume originally envisioned to be written by Federigo Enriques. Neurath had initially envisioned 26 volumes with 260 monographs and an accompanying 10-volume Visual Thesaurus (Reisch 1995). To point to the larger, collective project growing from particular areas of actual individual research, the contributions had the form of monographs focused on single scientific disciplines or particular kinds. The titles of contributions by leading members of the encyclopedia project (Morris, Carnap, Frank, and Neurath himself) featured the term "foundations," but the overall aim was far from epistemologically foundationalist.

From the mid-1930s Neurath grew increasingly restive when confronted with instances of what he considered undue rigidity or over-systematization in the treatment of scientific reasoning by colleagues. To the ideal of "THE system" he opposed what he called "the encyclopedia as 'model'" (1936b/1983: 145). In more concrete terms, to the image of the pyramid he opposed the images of the mosaic (1938: 3) and the "orchestration of the sciences" (1945–46: title). The encyclopedia stood for a collective scientific project of democratic, pluralistic, comprehensive, horizontal integration.

From a linguistic standpoint, he admitted to the ideal but limited role for precise ideal language and stressed the role of a universal jargon, acknowledging the irreducible presence of vague ordinary terms. From a logical standpoint, he acknowledged the unifying role for symbolic terminology and calculi but noted the local scope of both isolated contradictions and axiomatizations. From a methodological standpoint, he stressed the role for empirical control of hypotheses while noting the indispensability of decisions to be made by investigators. From social and disciplinary standpoints, he spoke of "systematization from below" as a form of collective cooperation among individuals working within particular, more specialized research, so that division of labor did not degenerate into isolating separation. In this sense, the encyclopedia was a model of society and a collection of tools at its disposal. Neurath defended the democratic and pluralistic character of the project as a process of collective unity of action without authoritarian integration.

Carnap added his own, more formally oriented take in his contribution to the first installment of the collective *Encyclopedia of Unified Science*. For Carnap, terms of the social sciences were reducible to terms of psychology, biology, physics, and the thing-language, provided one adopted some form of methodological individualism. The linguistic unity, the reduction of terms of one science (language) to those of another and ultimately to the thing-language, however, didn't guarantee the reduction of their associated laws, e.g., the derivation of laws of biology from laws of physics (1938: 60). In fact, no such reduction had been achieved. The situation was worse in the case of psychology and the social sciences, where the reduction to biological or physical laws seems remote. But this lack of actual unity in reductive terms was not an impossibility in principle, Carnap noted. In practice, however, it represented a practical obstacle to successful prediction, since (now echoing Neurath) in very many decisions in individual and social situations, "we need a prediction based upon a combined knowledge or concrete facts and general laws belonging to different branches of science" (ibid.: 62). For this the unity of language was necessary, but it was not sufficient.

In postwar philosophy of science, the ideal of unity of science survived, on one hand, in the discussions of scientific explanation, notably by Carl Gustav Hempel, Paul Oppenheim, and Ernst Nagel. They adopted Carnap's criterion of explanation by the subsumption of descriptions of the phenomena to be explained under empirical generalizations and required it in the form of a deductive argument: the deductive-nomological or covering-law conception of explanation. The ordering of varying scopes of generality over a class of phenomena determined the relations of reduction, down to the most general theory (Hempel and Oppenheim 1948). For Hempel, the natural sciences exhibited the application of this criterion of explanation, which played both unifying and demarcating roles; by the same criterion, the human sciences barely passed muster, typically contributing only explanation sketches that stood in need of development (1942). Nagel, also adopting a logical and linguistic standpoint, represented relations of reduction between scientific theories in terms of deductive relations between laws or hypotheses, and distinguished between different situations depending on the differences in the respective vocabularies (Nagel 1961). Notably, Oppenheim also collaborated with Hilary Putnam (who had studied under Reichenbach) on a formulation of the unity of science thesis that integrated Carnap's strict early standard of reductionism with the empiricist revisability of a working hypothesis (Oppenheim and Putnam 1958). In their paper, they argued that the sciences have been consistently aiming at an ideal of unity and have been succeeding. In their formulation of this ideal, the sciences are divided up into levels on a hierarchy, with particle physics at the bottom, and took the view that terms and laws are connected across levels and, more ontologically, that entities described at higher levels are composed and causally explained by entities at lower levels, all the way to elementary particles. Needless to say, this view conflicts with Putnam's later anti-positivist views and other more recent views on the disunity of science. Ironically, significant elements of the latter (see Galison and Stump 1996) can be recognized in Neurath's conception of unity.

References

Ayer, A. J. (ed.) (1959) *Logical Positivism*, New York: Free Press.

Carnap, R. (1928) *Der logische Aufbau der Welt*, Berlin: Weltkreis-Verlag. Trans. *The Logical Structure of the World*, Berkeley: University of California Press, 1967, repr. Chicago: Open Court, 2003.

——— (1930) "Die alte und die neue Logik," *Erkenntnis* 1: 12–26. Trans. "The Old and the New Logic," in Ayer 1959, pp. 60–81.

——— (1934) *Logische Syntax der Sprache*, Vienna: Springer. Rev. ed. trans. *The Logical Syntax of Language*, London: Kegan Paul, Trench, Trubner & Cie, 1937, repr. Chicago: Open Court, 2002.

——— (1936–7) "Testability and Meaning," *Philosophy of Science* 3: 419–71 and 4: 1–40.

——— (1938) "Logical Foundations of the Unity of Science," in Neurath et al. (1938), pp. 42–62.

Friedman, M. (1987) "Carnap's *Aufbau* Reconsidered," *Nous* 21: 521–45. Repr. in Friedman, *Reconsidering Logical Positivism*, Cambridge: Cambridge University Press, 1999, pp. 89–113.

Galison, P. and Stump, D. (eds.) (1996) *Disunity and Contextualism: New Directions in the Philosophy of Science*, Stanford: Stanford University Press.

Hempel, C. G. (1942) "The Function of General Laws in History," *Journal of Philosophy* 39: 35–48. Repr. in Hempel (1965), pp. 231–43.

——— (1965) *Aspects of Scientific Explanation*, New York: Free Press.

Hempel, C. G. and Oppenheim, P. (1948) "Studies in the Logic of Explanation," *Philosophy of Science* 15: 135–75. Repr. in Hempel (1965), pp. 245–90.

Kuhn, T. S. (1962) *The Structure of Scientific Revolutions*, Chicago: University of Chicago Press, 2nd ed., 1970.

Nagel, E. (1961) *The Structure of Science*, New York: Harcourt, Brace & World.

Neurath, O. (1910) "Zur Theorie der Sozialwissenschaften," *Jahrbuch für Gesetzgebung, Verwaltung und Volkswirtschaft im Deutschen Reich* 34: 37–67. Trans. "On the Theory of the Social Sciences," in Neurath (2004), pp. 265–91.

——— (1921) *Anti-Spengler*, Munich: Callway. Excerpts trans. "Anti-Spengler," in Neurath (1973), pp. 158–213.

——— (1931a) "Physicalism: The Philosophy of the Vienna Circle," *The Monist* 41: 618–23, Repr. in Neurath (1983), pp. 48–51.

——— (1931b) "Physikalismus," *Scientia* 50: 297–303. Trans. "Physicalism," in Neurath (1983), pp. 52–57.

——— (1931c) *Empirische Soziologie. Der wissenschaftliche Gehat der Geschichte undf Nationalökonomie*, Vienna: Springer. Partly trans. "Empirical Sociology," in Neurath (1973), pp. 391–421.

——— (1931d) "Bildhafte Pädagogik im Gesellschafts- und Wirtschaftsmuseum in Wien," *Museumskunde* 3: 125–9. Trans. "Visual Education and the Social and Economic Museum in Vienna" in Neurath (1973), pp. 215–8.

——— (1932a) "Soziologie im Physikalismus," *Erkenntnis* 2: 393–431. Trans. "Sociology and Physicalism," in Ayer (1959a), pp. 282–320, and "Sociology in the Framework of Physicalism," in Neurath (1983), pp. 58–90.

——— (1932b) "Protokollsätze," *Erkenntnis* 3: 204–14. Trans. "Protocol Sentences," in Ayer (1959a), pp. 199–208, and "Protocol Statements," in Neurath (1983), pp. 91–99.

——— (1933) *Einheitswissenschaft und Psychologie*. Vienna: Gerold. Trans. "Unified Science and Psychology," in B. McGuinness (ed.), *Unified Science*, Dordrecht: Reidel, 1987, pp. 1–23.

——— (1934) "Radikaler Physikalismus und 'wirkliche Welt'," *Erkenntnis* 4: 346–62. Trans. "Radical Physicalism and 'the Real World,'" in Neurath (1983), pp. 100–14.

——— (1936a) "Physikalismus und Erkenntnisforschung," *Theoria* 2: 97–105, 234–7. Trans. "Physicalism and the Investigation of Knowledge," in Neurath (1983), pp. 159–67.

——— (1936b) "L'encyclopédie comme 'modèle'," *Revue de Synthèse* 12: 187–201. Trans. "Encyclopedia as Model," in Neurath (1983), pp. 145–58.

——— (1936c) "Soziologische Prognosen," *Erkenntnis* 3: 398–405. Trans. "Sociological Predictions," in Neurath (2004), pp. 506–12.

——— (1938) "Unified Science as Encyclopedic Integration," In Neurath et al. (1938), pp. 1–27.

——— (1944) *Foundations of the Social Sciences*, Chicago: University of Chicago Press.

——— (1945–46) "The Orchestration of the Sciences by the Encyclopedism of Logical Empiricism," *Philosophy and Phenomenological Research* 6: 496–508. Repr. in Neurath (1983), pp. 230–42.

——— (1973) *Empiricism and Sociology* (ed. by M. Neurath and R. S. Cohen), Dordrecht: Reidel.

——— (1983) *Philosophical Papers 1913–1946* (ed. by R. S. Cohen and M. Neurath), Dordrecht: Reidel.

——— (2004) *Economic Writings: Selections 1904–1945* (ed. by T. Uebel and R. S. Cohen), Dordrecht: Kluwer.

Neurath, O., Bohr, N., Dewey, J., Russell, B., Carnap, R. and Morris, C. (1938) *Encyclopedia and Unified Science*, Chicago: University of Chicago Press.

Oppenheim, P. and Putnam, H. (1958) "The Unity of Science as a Working Hypothesis," in H. Feigl, G. Maxwell and M. Scriven (eds.), *Concepts, Theories and the Mind-Body Problem*, Minneapolis: University of Minnesota Press, pp. 3–36.

Quine, W. V. O. (1951) "Two Dogmas of Empiricism," *Philosophical Review* 60: 20–43. Rev. and repr. in Quine, *From a Logical Point of View*, Cambridge, MA: Harvard University Press, 1953, rev. ed., 1980, pp. 20–46.

Reisch, G. (1995) *A History of the International Encyclopedia of Unified Science*, PhD diss., University of Chicago, Chicago.

Richardson, A. (1998) *Carnap's Construction of the World*, Cambridge: Cambridge University Press.

Schlick, M. (1918) *Allgemeine Erkenntnislehre*, Berlin: Springer, 2nd rev. ed., 1925. Trans. *General Theory of Knowledge*, Vienna: Springer, 1974, repr., Lasalle, IL: Open Court, 1985.

——— (1926) "Erleben, Erkennen, Metaphysik," *Kantstudien* 31: 146–58. Trans. "Experience Cognition, Metaphysics," in Schlick (1979), pp. 99–111.

——— (1930) "Die Wende der Philosophie," *Erkenntnis* 1: 4–11. Trans. "The Turning Point in Philosophy," in Ayer (1959a), pp. 53–59, and in Schlick 1979, pp. 154–60.

——— (1934) "Philosophie und Naturwissenschaft," *Erkenntnis* 4: 370–96. Trans. "Philosophy and Natural Science," in Schlick (1979), pp. 139–53.

——— (1979) *Philosophical Papers vol. 2 (1925–1936)* (ed. by H. L. Mulder and B. van de Velde-Schlick), Dordrecht: Reidel.

Uebel, T. (2007a) *Empiricism at the Crossroads: The Vienna Circle's Protocol-Sentence Debate*, Chicago: Open Court.

——— (2007b) "Philosophy of Social Science in Early Logical Empiricism," in A. Richardson and T. Uebel (eds.), *Cambridge Companion to Logical Empiricism*, Cambridge: Cambridge University Press, pp. 250–77.

Verein Ernst Mach (1929) *Wissenschaftliche Weltauffassung. Der Wiener Kreis*, Vienna: Wolf. Trans. "The Scientific Conception of the World. The Vienna Circle," in Neurath 1973, pp. 299–318; rev. trans. (with orig. annotated bibliography) "The Scientific World-Conception. The Vienna Circle," in F. Stadler and T. Uebel (eds.), *Wissenschaftliche Weltauffassung. Der Wiener Kreis. Hrsg. vom Verein Ernst Mach (1929)*, Vienna: Springer, 2012, pp. 75–116.

19
THE DEDUCTIVE-NOMOLOGICAL MODEL OF EXPLANATION

Stathis Psillos

The key to understanding the logical positivist view of scientific explanation is in the thought that explanations are arguments. In particular, an explanation is a deductive argument whose premises state particular matters of fact (such as initial and boundary conditions) *and* general laws and whose conclusion is a statement of the event to be explained. Hence, an explanation is a deductive argument, one premise of which is nomological, i.e., it states a law of nature. As Carl Hempel put it (1965a: 299), explanation amounts to "deductive subsumption under general laws."

The DN Schema

Schematically put, to offer an explanation of an event *e* is to construct a valid deductive argument of the following form:

(DN) Antecedent/Initial Conditions $C_1, \ldots, C_i$
Lawlike Statements $L_1, \ldots, L_j$
Therefore, e (event/fact to be explained)

The premises are called (collectively) "explanans," and the conclusion "explanandum." But there can be no explanation without one or more general laws; hence, the argument should be such that the explanandum does not follow if the law-statements were to be excluded from the explanans.

Though Hempel should be credited with the name "deductive-nomological," the idea behind the name first appeared in Rudolf Carnap's work. Carnap suggested that an application of the logical calculus to factual statements is the derivation of a singular statement from a general one (which, given its descriptive terms, expresses a law of nature) and another singular one. His example was:

Premises: 1. c is an iron rod
2. c is now heated
3. For every x, if x is an iron rod and x is heated, x expands
Conclusion: 4. c now expands.

 DOI: 10.4324/9781315650647-22

This kind of argument, Carnap suggested, can be seen either as an *explanation* of (4) (by reference to other facts (1 and 2) and a law (3); or as a *prediction* of (4) (1939: 36). An advantage of this account was conceptual clarity. The issue of explanation was split up into two parts: one logical (is the argument deductively valid?); and another factual (is the argument sound?).

Hempel systematized the deductive-nomological model by offering four conditions of adequacy for an explanation.

1 The argument must be deductively valid.
2 The explanans must contain essentially a lawlike statement.
3 The explanans must have empirical content, i.e., they must be confirmable.
4 The explanans must be true.

The first three conditions are called "logical" (Hempel and Oppenheim 1948/1965b: 247), because they pertain to the *form* of the explanation. The fourth condition is "empirical." Hempel noted that it was an empirical matter whether the premises of an explanation were true or false. He called a DN argument that satisfies the first three conditions a "potential explanation," viz., a valid argument such that, if it were also sound, it would explain the explanandum. He contrasted it with an "actual explanation," which is a sound DN argument. The fourth condition separates a potential from an actual explanation, the correct explanation of an event. With the fourth condition, the issue of "the logical structure of explanatory arguments" (ibid.: 249, note 3) was separated from the empirical issue of what is the correct explanation of an event.

Explanation and causation

Hempel first presented the essential elements of this model in "The Function of General Laws in History" of 1942. There, he took it that laws function in the natural sciences as the links of events "in patterns which are usually referred to as *explanation* and *prediction*" (1942/1965b: 232, orig. emphasis) and aimed to show that historical explanation, qua scientific explanation, should be no less nomological: "scientific explanation can be achieved only by means of suitable general hypotheses, or theories, which are bodies of systematically related hypotheses" (ibid.: 239). The target of the proposed account of scientific explanation was *causal explanation*. Insofar as there is causal explanation in science, it should be an "explanation by scientific laws" (ibid.: 233, note 1). What does the explaining are not *sui generis* relations of cause and effect. For Hempel, as for Hume, it was untenable to think that "a careful examination of two specific events alone, without any reference to similar cases and to general regularities, can reveal that one of the events produces or determines the other" (ibid.: 241, note 7).

The DN model was offered as an account of causal explanation in that all causal explanation is DN explanation: "causal explanation is a special type of deductive nomological explanation" (1965a: 300). In his early writings, Hempel contrasted causal explanation, which requires universal laws, to probabilistic explanation, which relies on "certain probability hypotheses" (1942/1965b: 237). In later writings, he formulated the inductive-statistical model of explanation, the key idea of which is that the explanatory argument is inductive and that the law referred to in it is probabilistic. If all causal explanation is DN, the converse is not the case. Here, Hempel has in mind the explanation of less fundamental laws by being deduced from more fundamental ones. The point is taken to be that though Newton's law of gravity (together with the three laws of motion) entail (and hence explain) Kepler's laws of planetary motion, they do no cause them (1965a: 301).

The fact that explanations are arguments is meant to capture the view that causation is nomological. Causal relations among singular events are mirrored in the deductive relations between the premises and the conclusion of a suitable DN argument: "a certain event or set of events can be said to have caused a specified 'effect' only if there are general laws connecting the former with the latter in such a way that, given a description of the antecedent events, the occurrence of the effect can be deduced with the help of the laws" (ibid.: 300–1). Given the concomitant thesis that there is symmetry between explanation and prediction, it follows that when there is causation and knowledge of laws, the effect can be predicted. If explanation amounts to "nomic expectability," so does causation. Moritz Schlick was forthright about all this: "the criterion of causality is successful prediction" (1932/1979: 254; see also 1931/1970: 184). Carnap concurred: "causal relation means predictability" (1966: 192).

Part of the rationale for developing a deductive-nomological account of explanation was precisely an attempt to demystify causation. Already in the *Aufbau*, Carnap stressed that "causality means nothing but a functional dependency of a certain sort" (1928: §165). The functional dependency is between two states of a system, and it can be called a "causal law" if the two states are in temporal proximity, and one precedes the other in time. Schlick expressed this idea succinctly: "the difference between a mere temporal sequence and a causal sequence is the regularity, the uniformity of the latter. If *C* is regularly followed by *E*, then *C* is the cause of *E*; if *E* only 'happens' to follow *C* now and then, the sequence is called mere chance" (1932/1979: 239). The problem with causation was that it was too metaphysical. Carnap (1928: §20) distinguished between the "correlation problem" and the "essence problem." The essence problem is metaphysical because it purports to investigate what the alleged essence of causation is, beyond regularity. As such, it relies on the "erroneous assumption" that there is something in causation beyond correlation ("i.e. beyond mathematical function"). The correlation problem, on the other hand, is empirical. It investigates what events are correlated, where correlation is understood as subsumption under a "general functional law." Carnap immediately added that the problem of correlation is none other than finding "the laws of nature."

The success or failure of the DN model is tied to two issues. The first: what are the laws of nature? The second: are all facts about causation captured by the DN model, and are there bona fide causal explanations which are not DN? Let's start with the second issue.

Explanatory asymmetries

It has been a standard criticism of the DN model that it fails as a model of scientific explanation precisely because it leaves out of the explication of the concept of explanation important considerations about the role of causation in explanation. There is more to the concept of causation than what can be captured by DN explanations.

One type of counterexample has aimed to show that the DN model is insufficient as an account of scientific explanation. Its legendary form is the flagpole-and-shadow case. Suppose that we construct a DN explanation of why the shadow of a flagpole at noon has a certain length. Using the height of the pole as the initial condition, and employing the relevant nomological statements of geometrical optics (together with elementary trigonometry), we can construct a deductively valid argument with a statement of the length of the shadow as its conclusion. So we can DN-explain why the length of the shadow at noon is what it is. But as Sylvain Bromberger observed (1966/1992: 83; see Hempel 1962: 109–10), we can reverse the order of explanation: we can explain the height of the flagpole, using the very same nomological statements, but (this time) with the length of the shadow as the initial condition. Surely, it has been argued, this is not a bona fide explanation of the height of the pole: although it satisfies the DN model, it is not a *causal* explanation of the height of the pole. The height of the pole is the cause of its shadow at noon, but the shadow does not cause the flagpole to have the height it does.

This counterexample can be generalized by exploiting the functional character of some lawlike statements in science: in a functional law, we can calculate the values of each of the magnitudes involved in the equation that expresses the law by means of the others. Hence, given some initial values for the known magnitudes, we can calculate, and hence DN-explain, the value of the unknown magnitude. Suppose, for instance, that we want to explain the period T of a pendulum. This relates to its length l by the functional law $T = 2\pi\sqrt{l/g}$. So, we can construct a DN argument whose conclusion is some value of the period T and whose premises are the aforementioned law statement together with some value l of the length as our antecedent condition. Suppose, instead, that we wanted to explain the length of the pendulum. We could construct a DN argument similar to the previous one, with the length l as its conclusion, using the very same law statement but this time conjoined with a value of the period T as antecedent condition. If, in the former case, it is straightforward to say that the length of the pendulum causes it to have a period of a certain value, in the latter case, it seems problematic to say that the period causes the pendulum to have the length it does.

Note that this kind of counterexample does not dispute the main thesis that causal explanation is a subset of DN explanation. What they aim to show is that the DN model licenses seemingly inappropriate explanations, their inappropriateness being that they fail to capture the right causal order. The key point here is that unless causal considerations are imported into DN explanatory arguments, they fail to distinguish between legitimate (because causal) and illegitimate (because non-causal) explanations.

How telling is this point? Hempel, who noted the example of the pendulum, pointed out that functional laws are *not* causal laws but laws of coexistence, since they do not have built into them the temporal asymmetry between cause and effect. Still, it might be argued, functional laws can have their symmetry broken when a variable is manipulated: the length of the pendulum causes the period because the length can be manipulated independently of the period. Since manipulation is a causal concept, it seems that this kind of reaction to the symmetry problem is not open to Hempel. But Hempel perceptively noted that the so-called independent variable (the length) cannot be manipulated independently of the period, since if we want to change the length of a given fixed pendulum, we have first to stop it, and hence change its period (1965a: 353).

Can the DN model escape the criticism that there are adequate causal explanations which are not DN? This addresses the heart of the problem: are laws required for explanation? And, consequently, should explanations be arguments? The opposite view is that explanations are statements, and in particular causal statements of the form: *Y* because *X*. On this view, advocated by Michael Striven, law-statements play a subsidiary role in explanation by offering grounds for the connection between *X* and *Y*. Hence, on Scriven's view, "*Y* because *X*," if true, is a complete answer to the question "Why *Y*?" thereby being a complete explanation of *Y*. But there may be a second question to ask, viz., what is the ground of the claim that *X* explains *Y*? (In virtue of what facts does *X* explain *Y*?) This second question might invite stating relevant laws of nature. But, on Scriven's view, the second is not a why question: "Producing a law is one way, not necessarily more conclusive, and usually less easy than other ways of supporting the causal statement" (1958: 194). He supported this point by the famous example of the explanation of the ink stain on the carpet. Citing the fact that the stain on the carpet was caused by inadvertently knocking over an ink bottle from the table, Scriven (1962: 90) argues, "is the explanation of the state of affairs in question, and there is no nonsense about it being in doubt because you cannot quote the laws that are involved, Newton's and all the others."

If explanations are arguments, they are neither true nor false—they are valid and, if the premises are true, sound. If explanations are statements, they are true or false. The extension of this thought is that explanations are (true) causal stories, that is, stories that give causally relevant information about how an effect was brought about, without having the form of a deductive argument. Alternatively, an explanation is nothing linguistic but the things in the world that cause something to happen. The so-called "ontic conception" of explanation, advocated by Wesley Salmon, is based on a denial of the claim that explanations are linguistic devices. For the ontic conception, citing a causal mechanism can be a legitimate explanation of an event without having the form of a Hempelian DN argument (Salmon 1989: 24).

The key issue at stake has to do with how causation is understood. As noted, for the logical empiricists (at least for Schlick, Carnap, and Hempel) causation is general and hence nomological: there is no singular causation. Hence, where there is causal explanation, there is nomological explanation. Carnap insisted: "When someone asserts that A caused B, he is really saying that this is a particular instance of a general law that is universal with respect to space and time" (1966: 204).

Take the statement "*X* causes *Y*," where *X* and *Y* are individual events. This has the appearance of "individual causation" (aka singular causation). Concerning this case, Hempel noted: "the given causal statement must be taken to claim by implication, that an appropriate law or set of laws holds by virtue of which *X* causes *Y*" (1965a: 350). Since the individual events are unrepeatable, the law connects event-types. Hence, the singular statement relies on a general statement concerning kinds of events, *X* and *Y* being tokens of them. Given this, the only real option for causal explanation is to search for laws. And this is a broadly empirical issue. Carnap concurred: "When a causal relation is mentioned, there is always an implicit reference to unspecified laws of nature" (1966: 194).

Explanation, for Hempel, offers understanding. An adequate explanation of event *Y* (that is, of why *Y* happened) should offer an adequate understanding of this happening. Just citing a cause would not offer an adequate understanding, unless it was accompanied by the citation of a law that connects the two events. For just citing that *X* is the cause of *Y* does not remove the surprise of *Y's* happening, since it does not make *Y* expected given the law. Though Scriven agreed that explanation is related to understanding, he took it that understanding does not require knowing laws. So he proposed (1962: 95): "a causal explanation of an event of type *Y*, in circumstances *R* is exemplified by claims of the following type: there is a comprehensible cause *X* of *Y* and it is understood that *Xs* can cause *Ys*." Hempel's point was that for it to be understood that *Xs* can cause *Ys*, the law that connects them should be assumed, sought after, and ideally become known.

Explanation and laws

But what are the laws of nature? Schlick adopted the view that law statements are inference-tickets. The thrust of this view is that law statements should not be seen as expressing propositions, and hence as being amenable to claims of truth and falsity. Rather, they must be seen as disguised *rules of inference*. We cannot validly move from the singular claim that "*a* is *F*" to the singular claim (perhaps, prediction) that "*a* is *G*," unless we use the sentence "All *Fs* are *Gs*." On the inference-ticket view, the function of law statements is exactly this: they entitle us to make inferences such as the one just presented. Schlick's endorsement of this view was motivated by the thought that nomological statements are, strictly speaking, meaningless, because they are unverifiable. He thought, however, that though meaningless, a nomological statement represents "a direction for the formulation of propositions" (1931/1970: 98), meaning that it could license the transition from a (singular) verifiable proposition to another. Carnap's reaction to this view was based on the principle of tolerance. As he noted, treating law-like statements "not as proper sentences of the object-language, but merely as directions for the construction of sentences" is not "inadmissible." Yet, he took it that a form of language in which "the laws are treated as equally privileged proper sentences of the object-language, is, as it appears, much simpler and better adapted to the ordinary use of language in the actual sciences than the first form" (1934/1937: 321). Note that on Schlick's account, explanations are not arguments, since the seeming law statements are really rules of inference. For the same reason, explanations are not statements either. Hempel too rejected Schlick's view of laws. His chief objection was that in most typical cases, a cluster of laws (as opposed to a single law) will be necessary for making an inference from a singular statement to another one, and hence the whole cluster should be taken as an extralogical inference rule. In this case, as he noted, "the simplest way" is to treat laws as genuine statements functioning as premises in arguments and to "use only purely logical rules of inference" (1965a: 358).

From early on the logical empiricist tradition took it that the laws of nature are those regularities that are expressed by *true* universally qualified statements. In fact, the characterization of a statement as lawlike was taken to be, to a first approximation at least, purely *syntactic*: a lawlike statement is a universally quantified statement of the form *All Fs are Gs*. For a lawlike statement, then, to express a law of nature, it had to be true. But purely formal considerations cannot distinguish between genuine laws and accidentally true generalizations. Drawing this distinction was imperative for the success of the DN model, since only genuine laws explain. Accidents (though true) cannot explain. With Hempel, suppose we want to explain why John Jones is bald. To this purpose, we can construct a DN argument whose explanans are as follows: "John Jones is a member of the Greenbury School Board for 1964" and "All members of the Greenbury School Board for 1964 are bald" (ibid.: 339). Precisely because the major premise of the argument is true but not lawlike (i.e., precisely because it is an accident) this argument lacks explanatory force. In contradistinction to this, the lawlike statement "All gases expand when heated under constant pressure" can adequately explain why a certain quantity of gas in a container expanded when it was heated under constant pressure. The relevant DN argument has explanatory force because the lawlike statement expresses a genuine law. Without a robust distinction between laws and accidents, the DN model loses most of its putative force as a correct account of explanation.

Hempel tried to draw this distinction in terms of the scope of the generalization: genuine lawlike statements have "nonlimited scope" (ibid.: 267); or, as he put it elsewhere, they must be of "essentially generalized form" (ibid.: 340); that is, they should not be paraphrasable into conjunctions of their instances. But Kepler's laws are limited in scope (since they refer to the planets of the solar system) *and* genuinely explanatory. How about then taking genuine lawlike statements not to make any reference to particular objects? That is, to have "purely qualitative predicates"? Galileo's law of free fall would be a counterexample to this move. Hempel took to heart Nelson Goodman's point (1954) that whether or not a generalization will count as lawlike will depend on what kinds of predicates are involved in its expression. He even played with the idea that the suitable predicates should be "projectable" to hitherto unexamined instances, where this idea of *projectability* was couched in terms of the relative "entrenchment" of the predicates that feature in a generalization, the degree of entrenchment of a predicate being a function of its past uses in projected generalizations. But Hempel was overall skeptical of defining "entrenchment" and preferred to stick to the criterion of essential generality of lawlike statements (1965a: 342–3). He took it that Kepler's laws are explanatory, though of limited scope, since they are derivable from more fundamental laws, such as Newton's.

Scientific explanation is centrally concerned with explaining regularities, not only particular facts. Attempting to extend his DN model to the explanation of laws, Hempel encountered the following difficulty (ibid.: 273). Suppose one wants to explain a low-level law L_1 by a DN argument. One might want to achieve this by subsuming L_1 under the "more comprehensive regularity" $L_1\&L_2$, where L_2 may be any other law one likes. Although such a construction would meet all the requirements of the DN model, it wouldn't count as an explanation of L_1. Saying that the conjunction $L_1\&L_2$ is not more fundamental than L_1 would not help. The issue at stake is precisely what makes a law more fundamental than another one. Intuitively, it is clear that the laws of the kinetic theory of gases are more fundamental than the laws of ideal gases. But if what makes them more fundamental just is that the latter are derived from the former, the conjunction $L_1\&L_2$ would also count as more fundamental than its components. Hempel admitted that he did not know how to deal with this problem, which came to be known as "the problem of conjunction." In fact, the problem was not adequately addressed before Michael Friedman's (1974) account of explanation as unification.

A thought that became available regarding the distinction between laws and accidents, motivated once more by Nelson Goodman, was that laws support counterfactual conditionals, while accidents do not. Reichenbach's (1947: 368) example is very instructive. Compare the following two general statements: "All gold cubes are smaller than one cubic mile" and "All plutonium cubes are smaller than one cubic mile." Both are similarly unrestricted, and both involve projectable and purely qualitative predicates. Yet, we rightly feel that only the latter generalization is a genuine law of nature, because although it is nothing but contingent lack of resources and technical means that do not allow us to build a gold cube larger than one cubic mile, we could not *possibly* put together a plutonium cube of this (and of much less) size, even if we had the necessary quantity and means. The construction of this plutonium cube is (physically) impossible because any amount of plutonium over the critical mass would lead to an explosion detrimental to humankind. The plutonium-cube generalization has (while the gold-cube one lacks) modal force. This difference can easily be shown if we take counterfactuals into account. "If this had been a plutonium cube, then it would not have been larger than one cubic mile" is a true counterfactual, while "If this had been a gold cube, then it would not have been larger than one cubic mile" is false. The suggestion, then, is laws are the true lawlike statements that support counterfactuals.

But both Hempel and Carnap thought that counterfactuals are particularly elusive; their explication faces "notorious philosophical difficulties" (Hempel 1965a: 339). The chief one, it should be noted, is that on Goodman's account of when a counterfactual is true, laws are presupposed. Hence, counterfactuals cannot be used, non-circularly, to distinguish between laws and accidents. How about the distinguishing between lawlike statements and accidents by reference to explanation? An account put forward by A. J. Ayer and Richard Braithwaite, among others, was that the difference between laws and accidents is a difference in the epistemic attitude of scientists towards them. Laws are those generalizations which are used in explanation and prediction and are confirmed by their instances. As Ayer (1956/1963: 230) summed it up: "the difference between [laws and accidents (what he called "generalisations of fact")] lies not so much on the side of the facts that make them true or false, as in the attitude of those who put them forward." Hempel was naturally reluctant to accept this move, since if the concept of explanation is used in the clarification of the concept of law, then the DN model would be hopelessly circular (1965a: 339–40).

Hempel reluctantly admitted that he has not ended up "with a fully satisfactory general characterisation of lawlike sentences and thus of laws" (ibid.: 343). Carnap himself insisted on characterizing laws as statements which have "nomic form" and "are true" (1966: 213) and took it that the basic logical empiricist commitment is that laws do not hold with any kind of necessity. Hence, when one adds to a true lawlike statement of the form "All *F*s are *G*s," the qualifier "and this holds with necessity," one adds nothing of cognitive value since the alleged intrinsic necessity in any observed causal sequence cannot be observed, nor can it be verified in any other way (ibid.: 201). All is not lost, however, since the so-called Mill-Ramsey-Lewis account of laws is suitable for an empiricist account of lawhood (see Psillos 2002: ch. 5).

It is important to stress that Hempel offered the DN model as an *explication* of the concept of explanation, as this is used in science. As he put it, "explicating the concept of scientific explanation is not the same thing as writing an entry on the word 'explain' for the Oxford English Dictionary" (1965a: 413). Following Carnap's (1950) account of explication of the concept of probability, he noted that he aimed to make "reasonably precise" the logical structure of explanation in the sciences. As such, the project of explication requires some rational reconstruction of actual uses. Hence, if the fact that there are proof-sketches as opposed to fully developed proofs does not discredit the concept of proof, the fact that that there are elliptical explanations (or explanation-sketches) does not discredit the DN model. The analogy with the formal explication of the concept of proof in mathematics was meant to show that the DN explication of explanation can provide standards of critical appraisal and be fruitful. Indeed, the extension of this model to the explanation by means of statistical laws (see Hempel 1962) is testimony to its fruitfulness.

References

Ayer, A. J. (1956) "What Is a Law of Nature?" *Revue Internationale de Philosophie* 10: 144–65. Repr. in *The Concept of a Person and Other Essays*, London: Palgrave Macmillan, pp. 209–34.

Bromberger, S. (1966) "Why-Questions," in R. G. Colodny (ed.), *Mind and Cosmos: Essays in Contemporary Philosophy of Science*, Pittsburgh: Pittsburgh University Press, pp. 86–111. Repr. in Bromberger, *On What We Know We Don't Know*, Chicago: University of Chicago Press, 1992, pp. 75–100.

Carnap, R. (1928) *Der logische Aufbau der Welt*, Berlin: Welykreisverlag. Trans. *The Logical Structure of the World*, Berkeley: University of California Press, 1963, repr. Chicago: Open Court, 2003.

——— (1934) *Logische Syntax der Sprache*, Vienna: Springer. Rev. ed. trans. *The Logical Syntax of Language*, London: Kegan Paul, Trench, Trubner & Cie, 1937, repr. Chicago: Open Court, 2002.

——— (1939) *Foundations of Logic and Mathematics*, Chicago: University of Chicago Press.

——— (1950) *Logical Foundations of Probability*, Chicago: University of Chicago Press.

——— (1966) *Philosophical Foundations of Physics*, New York: Basic Books. Repr. as *An Introduction to the Philosophy of Science*, New York: Dover, 1974.

Feigl, H. and Maxwell, G. (eds.) (1962) *Scientific Explanation, Space, and Time*, Minneapolis: University of Minnesota Press.

Friedman, M. (1974) "Explanation and Scientific Understanding," *Journal of Philosophy* 71: 5–19.

Goodman, N. (1954) *Fact, Fiction and Forecast*, Cambridge, MA: Harvard University Press, 4th ed., 1983.

Hempel, C. G. (1942) "The Function of General Laws in History," *Journal of Philosophy* 39: 35–48. Repr. in Hempel (1965b), pp. 231–43.

——— (1962) "Deductive-Nomological vs. Statistical Explanation," in Feigl and Maxwell (1962), pp. 98–169.

——— (1965a) "Aspects of Scientific Explanation," in Hempel (1965b), pp. 331–496.

——— (1965b) *Aspects of Scientific Explanation and Other Essays*, New York: The Free Press.

Hempel, C. G. and Oppenheim, P. (1948) "Studies in the Logic of Explanation," *Philosophy of Science* 15: 135–76. Repr. in Hempel (1965b), pp. 245–90.

Psillos, S. (2002) *Causation and Explanation*, London: Acumen.

Reichenbach, H. (1947) *Elements of Symbolic Logic*, New York: Macmillan.

Salmon, W. (1989) *Four Decades of Scientific Explanation*, Minneapolis: University of Minnesota Press.
Schlick, M. (1931) "Die Kausalität in der gegenwärtigen Physik," *Die Naturwissenschaften* 19: 145–62. Trans. "Causality in Contemporary Physics," in S. Toulmin (ed.), *Physical Reality*, New York: Harper Torchbooks, 1970, pp. 83–121 and in Schlick (1979), pp. 176–209.
——— (1932) "Causality in Everyday Life and Recent Science," *University of California Publications in Philosophy* XV: 99–125. Repr. in Schlick (1979), pp. 238–58.
——— (1979) *Philosophical Papers* (ed. by B. van de Velde-Schlick and H. Mulder), Dordrecht: Reidel, vol. 2.
Scriven, M. (1958) "Definitions, Explanations and Theories," in H. Feigl and M. Scriven (eds.), *Concepts, Theories, and the Mind-Body Problem*, Minneapolis: University of Minnesota Press, pp. 99–195.
——— (1962) "Explanations, Predictions and Laws," in Feigl and Maxwell (1962), pp. 170–230.

20
THE PARTIAL INTERPRETATION OF SCIENTIFIC THEORIES

William Demopoulos

The logical empiricists articulated what is arguably the first systematic development of a theory of theories. This approach derived from the perspective afforded by modern logic and Hilbert's work on the axiomatic method. It was designed to address two key developments, the first concerned with abstract principles, the second with existence claims. In the case of abstract principles, Einstein's theories of relativity demonstrated the need to revise our conception of the epistemological status of the geometry of space and the chronometry of time; to understand the nature and epistemological significance of such a revision was pressing. As for existence claims, the early twentieth-century experimental success of atomism established the appeal to unobservable entities as a permanent feature of physical theory and raised the question of the nature of the support that attaches to claims which purport to be about entities that transcend observation.

The logical empiricist reconstruction of relativity was a corrective to the pretensions of the rationalist adherence to the synthetic a priori and the naïveté of an earlier empiricism; it was distinguished by its conventionalist interpretation of geometry and chronometry. The search for an account of theoretical claims which appeal to unobservable entities culminated in the *partial interpretation* view of theories, one of whose principal goals was to address the prima facie challenge to empiricism that such claims represent by clarifying their empirical status.

My focus in this chapter is the partial interpretation account of theories, and its elaboration by Carnap in his various Ramsey-sentence and related reconstructions of theoretical knowledge. I will articulate a central difficulty for the partial interpretation account that consists in the failure to provide an adequate account of our epistemic access to theoretical domains. To fully address this issue, it is necessary to delve into the methodology by which such access has in fact been secured. My modest goal in this chapter is to motivate the importance of the problem of epistemic access.

The partial interpretation account of theories holds that the vocabulary of a theory consists of an observational and a theoretical component, which are distinguished whether a vocabulary item applies to entities in the intended domain of the theory which are observable and so belongs to the *observation vocabulary*, or unobservable and so belongs to the *theoretical vocabulary*. In Carnap's classical formulation (1956: 41–42, 46–48), this distinction partitions vocabulary items into just these two classes. (The possibility of a third class of terms—*mixed vocabulary* items that apply to both observable and unobservable entities—will be addressed later.)

DOI: 10.4324/9781315650647-23

The sentences of the language of a theory are then divided into three classes: one generated from just the observation vocabulary, another generated from just the theoretical vocabulary, and a third generated from the combined observation and theoretical vocabularies. These are, respectively, the *observation sentences*, *theoretical sentences*, and *correspondence rules* of the language. A *theory* is the conjunction of a selection of theoretical sentences and correspondence rules. There are no special restrictions on the logic of a theory, and it may be either first-order or higher-order; the notion of a theory may also be generalized in various ways that are irrelevant to the conceptual issues on which I intend to focus.

Historically, the partial interpretation account of theoretical knowledge derives from the idea that theoretical terms are introduced by sentences which, taken by themselves, are indistinguishable in their epistemic status from the statements of a pure mathematical theory. Especially influential was Hilbert's contention that the primitives of a mathematical theory are whatever satisfies its axioms. This contention—that the postulates of a theory "implicitly define" its primitive notions—swept away the subjective associations that characterized an older tradition's understanding of a mathematical theory's primitives, even in the case of geometry, where they were thought to have a familiar "intuitive" content. The partial interpretation account sought to extend Hilbert's analysis to physics by providing an account of the empirical content of its theoretical statements that is based on the connections between theoretical terms and observation terms that are expressed by correspondence rules.

The two characteristic claims of the partial interpretation view are, first, that only the observation vocabulary is completely understood, and second, that the interpretation of the theoretical vocabulary is limited by constraints which depend only on the logical category of the theoretical terms and whatever restrictions the true observation sentences impose on the domain of unobservable entities over which the theoretical sentences and correspondence rules are evaluated. I will refer to this second claim as the *structuralist thesis*. We have yet to explain how the partial interpretation view conceives the relation between interpretations, true interpretations, and truth.

Carnap's mature reconstruction of the language of science (see his 1963, extended in 1961) builds on and extends the partial interpretation view of theories. The central notion of this account is the *Ramsey sentence* of a theory: the sentence formed by replacing theoretical terms by (new) variables of the appropriate logical category, then closing the resulting formula by adding an existential quantifier for each of the new variables. It is a very short step from the two characteristic claims of the partial interpretation account to the notion that a partially interpreted theory's Ramsey sentence captures its "factual content": the Ramsey sentence is observationally equivalent to the theory in the sense that any argument from the partially interpreted theory to a sentence of the observation language can be recovered using the Ramsey sentence instead. (The Ramsey sentence of a theory, like one of its Craig transcriptions, eliminates theoretical vocabulary, but unlike a Craig transcription, it retains the connections between observable properties and relations which are mediated by their association with theoretical properties and relations, thereby avoiding the difficulties adduced by Hempel [1958/1965: 214–16].). The Ramsey sentence's use of variables in place of uninterpreted theoretical terms simply makes explicit the commitment of the partial interpretation account to the structuralist thesis (compare Ramsey 1931: 232).

Carnap's mature reconstruction refines the doctrine of partial interpretation in two principal respects. As noted, Carnap explicates the factual content of a partially interpreted theory in terms of its Ramsey sentence. But he took things further by combining his account of the factual content of a theory with an explication of theoretical analyticity—analyticity relative to a theory—in terms of what has come to be known as the *Carnap sentence* of a theory: the conditional whose antecedent is the theory's Ramsey sentence and whose consequent is the partially interpreted theory. Before Carnap, the distinction between the factual and analytic (and hence, non-factual) components of a theory followed the distinction between postulates and definitions. But since this distinction is inherently arbitrary, its utility for a dichotomy that is supposed to reveal our factual commitments may be doubted. The Carnap sentence is justifiably regarded as analytic because it is a kind of "implicit definition" of the theoretical vocabulary, one that is provably non-factual in the sense that the only observation sentences it logically implies are logical truths. And as Winnie (1970) later showed, the Carnap sentence, like a proper definition, satisfies a special noncreativity condition that is similar to the noncreativity condition that is customary for proper *explicit* definitions (see Demopoulos 2007; Gupta 2009).

Carnap advanced the Ramsey sentence not just as a clarification of the partial interpretation view of theories, but as a correct representation of how scientists understand their theoretical claims. They intend, Carnap held, an "indeterminate" claim, one that may have many interpretations under which it comes out true. As scientists understand them, theoretical claims are indeterminate as to the interpretation of their theoretical vocabulary and *any* representative class or relation which makes true the Ramsey sentence of the theory to which the claim belongs is as acceptable as any other.

In one of his last papers on theoretical terms (1961), Carnap converts the *implicit* definition of theoretical terms by the Carnap sentence into a sequence of *explicit* definitions of them. But these explicit definitions do not eliminate—and were not intended to do so—the indeterminateness of his earlier account. Indeed, Carnap formulates his explicit definitions in what he calls a "logically indeterminate" language. The language L_{ε} which he employs is a standard first- or higher-order language enriched with Hilbert's ε-operator and the extensional axioms which govern its use. There are two such axioms. Given a formula *Fx* with one free variable, the first axiom tells us that if there is something satisfying *Fx*, then there is an "ε-representative" of *F*, denoted "$\varepsilon_x(Fx)$," that is selected by the choice function which interprets the ε-operator. The second axiom tells us that if the formulas *Fx* and *Gx* are extensionally equivalent, their ε-representatives are the same. That it should be possible to apply Hilbert's ε-operator to the Ramsey-sentence reconstruction of theories is a consequence of Carnap's observation that the Carnap sentence of a theory can be derived from a sentence that is in the same form as the first of the axioms for the ε-operator.

For Carnap, the principal virtue of this proposal is that it incorporates the convenience of having the use of a theoretical vocabulary while retaining all the characteristic indeterminateness of that vocabulary that is the hallmark of the partial interpretation view and of his mature reconstruction in terms of Ramsey and Carnap sentences. Thus, he writes that the theoretical postulates and correspondence rules

> are intended *by the scientist who constructs the system* to specify the meaning of [a theoretical term] to just this extent: if there is an entity satisfying the postulates, then [the term] is to be understood as denoting one such entity. Therefore the definition [of a theoretical term by means of Hilbert's ε-operator] gives to the indeterminate [theoretical term] just the intended meaning with just the intended degree of indeterminacy.
> (Carnap 1961: 163, emphasis added)

Not long after Carnap wrote this, John Winnie argued against the partial interpretation account—and by implication, against its elaborations by Carnap—on the basis of two theorems proved in his (1967): modulo the conceptually unimportant technical restriction that not all theoretical properties and not all theoretical relations are universal, Winnie showed that on the partial interpretation view of theories, if a theory has one realization, there is always another; and if a theory is realizable at all, it is arithmetically realizable.

In his exposition of the partial interpretation reconstruction, Winnie included in addition to the observation and theoretical predicates of a theory a separate category of *mixed predicates*, predicates that apply to both observable and unobservable entities. But on Winnie's account of the view, only the interpretation of the observation predicates must be the same in any realization of a partially interpreted theory. No such requirement applies to the interpretation of theoretical predicates; nor does it apply to mixed predicates. (For a discussion of Winnie's treatment of mixed predicates and the account proposed by Lewis [1970], see Demopoulos [2017 and 2022: ch.1], which expand on the present paper.)

To establish that a partially interpreted theory has many models if it has one model, Winnie introduced a construction based on a mapping from and onto the domain of a model of the theory that permutes at least one pair of unobservable entities so that it changes the image, under the mapping, of the interpretation of at least one theoretical predicate. By the structuralist thesis—and this is the observation Winnie's proof rests on—the theoretical predicates and relations can always be understood so that the relations which interpret them are *defined* by their images under an arbitrary one-one mapping from and onto the unobservable part of the domain. As for the properties and relations which interpret mixed predicates, their images are unchanged only by the action of the mapping from and onto the observable part of the domain. Such a construction defines the theoretical and mixed properties and relations of a new structure as the images of the properties and relations of a structure we know to be a model of the theory. Since the new properties and relations are *by construction* isomorphic to the properties and relations of a model of the theory, the structure which they define must also be a model of the theory.

Carnap and the advocates of partial interpretation take it as a desideratum of an adequate reconstruction of theoretical knowledge that it should address the empirical basis of theoretical claims. On the partial interpretation reconstruction, this problem is addressed by the provision of an explanation of our understanding of theoretical claims in terms of the connection correspondence rules establish between theoretical and observation vocabulary. Carnap's Ramsey-sentence reconstruction dissolves the problem of how we come to understand the meanings of terms which apply to unobservable entities by *eliminating* theoretical terms in favor of variables. But this dissolution of the problem is merely an emendation—not a rejection—of the partial interpretation view, an emendation that preserves the structuralist thesis. Indeed, Carnap's transition to the Ramsey sentence rests on the recognition that the partial interpretation view subscribes to this thesis. For if, in addition to whatever restrictions the true observation sentences impose on the domain of a model of a partially interpreted theory, the constraints on the interpretation of theoretical vocabulary appeal only to the logical category of the theoretical terms, then there can be no objection to their replacement by variables, and the problem of accounting for how theoretical terms are understood then simply disappears. As for the empirical basis for theoretical *claims*—as opposed to our understanding of the theoretical *vocabulary*—and the explanation of their difference from the claims of pure mathematics, modulo the elimination of theoretical vocabulary, these issues are addressed by Carnap's Ramsey-sentence reconstruction much as they are by the partial interpretation reconstruction. Instead of appealing to the association of theoretical claims with observation sentences by the mediation of correspondence rules, the empirical basis of such claims is accounted for by the association of the Ramsey-sentence transforms of theoretical claims with observation sentences that is effected by the Ramsey-sentence transforms of correspondence rules.

The centrality of the problem of the empirical basis of theoretical claims to the logical empiricists' reconstruction of theories shows why it is so misleading to characterize their account to have incorporated a "syntactic view" of theories. For the logical empiricists, theories are indeed linguistic objects. But to call the logical empiricist account "syntactic" misses the fact that it is first and foremost a reconstruction of theoretical knowledge that purports to show how observation bears on the empirical character and evidential support of theoretical claims. Evidently, neither goal can be successfully addressed without going beyond syntax—as indeed the principal logical empiricist reconstructions do when they assume that the observation language is interpreted.

This contrasts with the situation in mathematical logic where a theory is defined as a set of sentences in a language about which we assume *only* that its syntax and underlying logic are completely explicit. The logical tradition is also motivated by an epistemological problem which is no less fundamental to its notion of a theory than the epistemological motivation underlying logical empiricism is to *its* conception of a theory. The logical tradition sought to show that a *finitary* notion of proof suffices for the reconstruction of all mathematical reasoning—even reasoning within theories whose intended interpretation is over an infinite domain of arbitrarily large cardinality. But here the restriction to syntax in the characterization of a theory is entirely natural, since it is essential to the successful positive solution of the problem which motivates the logical tradition that a theory should be represented as a purely syntactic object. Although the logical empiricist approach to theories was profoundly influenced by the logical tradition, its goals were different: it sought to build a platform for the representation of the theoretical claims of physics that would be capable of illuminating their content and the basis on which they are understood and evaluated. In particular, it sought to show how observation must be a central component of an adequate empiricist solution to this problem. The nature of the questions the logical empiricist approach sought to address demanded—and were recognized as demanding—a notion of theory that includes more than the purely syntactic conception of the logical tradition.

Now, as we have seen, Carnap was prepared to *accept* multiple realizability as a point *in favor* of partial interpretation and Ramsey-sentence reconstructions. He even went so far as to endorse an arithmetical interpretation of the Ramsey sentence of a theory as the *correct* understanding of what the sentence asserts:

> I agree with Hempel that the Ramsey sentence does indeed refer to theoretical entities. . . . However, it should be noted that these entities are not unobservable physical objects like atoms, electrons, etc., but rather (at least in the form of the theoretical language which I have chosen in [1956: sect. VII] purely logico-mathematical entities, e.g. natural numbers, classes of such, classes of classes, etc.
>
> (1963: 963)

Thus, Carnap appears to have *also* anticipated and embraced the content of the second of Winnie's two theorems as an acceptable consequence of his view of the factual content of theories. In light of these facts, we would do well to set to one side the issues connected with multiple realizability and arithmetical interpretations and turn our attention to examining a different question. How do theories, whose characteristic feature is that their theoretical claims transcend observation, acquire their empirical status?

There is an argument which bears on the answer to this question that was favored by partial interpretation and related views, namely Hilary Putnam's model-theoretic argument (1977). I should emphasize that my interest is the actual argument, rather than Putnam's uses of it; these are all more various than the application I will isolate. As I understand its significance, Putnam's argument shows that the answer to our question given by the doctrine of partial interpretation and its close descendants is incompatible with the thesis that when theories which transcend observation are true, they express salient truths about unobservable entities. The fault with all these views stems from their failure to satisfactorily address the basis for our epistemic access to theoretical domains.

The model-theoretic argument consists of a simple technical argument and an observation. The technical argument establishes that any model of a theory's observational consequences can be extended to a model of the theory's theoretical sentences and correspondence rules, where the domain of this extension is the standard domain of observable and unobservable entities. The argument which establishes this conclusion also supports an observation, namely, that on the partial interpretation reconstruction of theories, as well as Carnap's various refinements of it, the conditions under which a partially interpreted theory can be shown to be *satisfiable* suffice to show that the theory is *true*.

The simple technical argument exploits a construction like Winnie's. A well-known folklore result reported in van Benthem (1978: Lemma 3.2) assures us that any model of a theory's observational consequences can be extended to a model of the theory. This is true, in particular, when the model of the theory's observational consequences is defined over the standard domain of observable entities. But so far as the folklore result is concerned, the "theoretical" or "unobservable" part of the extension might well be "abstract"—even number-theoretic. Nevertheless, if we are given such a "partially abstract" model, we can use a construction which, like Winnie's, is based on a one-one onto mapping which is the identity on the observable part of the domain of this model and is arbitrary from its theoretical part onto the theoretical domain of non-abstract unobservable entities. To do so we require the additional and contingent assumption that the cardinalities of the theoretical domains of the two models are the same. Relative to this assumption, Putnam's argument proceeds by defining the theoretical properties and relations over the standard domain of observable and unobservable entities as the images of the properties and relations of the abstract model under a mapping that is the identity map on the observable part of the domain of the partially abstract model, and is arbitrary from its theoretical part onto the theoretical part of the domain of non-abstract theoretical entities. But since the defined properties and relations meet all the conditions that the partial interpretation account is capable of imposing on theoretical properties and relations, the argument shows not merely that the theory is satisfiable, but that, on the partial interpretation account, it is *true*.

Winnie and Putnam exploit the same technical idea in their respective definitions of the theoretical relations which interpret the theoretical predicates of a partially interpreted theory. But their arguments support what are conceptually very different difficulties for the view. Putnam's argument is not directed at the existence of *multiple* realizations, nor does it concern the existence of an *arithmetical* model of a partially interpreted theory. Rather, Putnam's argument takes us from a cardinality assumption, and the existence of what might well be an arithmetical model of the kind explored by Winnie, to the conclusion that, on the partial interpretation view, the fact that a theory is *satisfiable* over the intended domain of observable and theoretical entities suffices to show that it is *true*. But it is clear that the notion that the truth of what is asserted about unobservable entities might depend only on their *number* runs counter to one of the simplest and least contentious convictions of "realism" and, indeed, of common sense. This is the conviction that if a theory is true, this is because its theoretical claims have captured a salient aspect of the reality they seek to describe, an aspect that goes beyond any mere question of cardinality.

The partial interpretation account of theories claims to reconstruct the empirical status of a theory's theoretical statements using only the theory's logico-mathematical framework and the apparatus of correspondence rules. But when we are restricted to just these resources, the truth of theoretical claims reduces to their satisfiability in any sufficiently large model of the true observation sentences. This shows that the reconstruction has failed to correctly represent the nature of the epistemological status of a theory's theoretical claims. It has failed because the epistemic basis for such an assertion of *satisfiability* is entirely different from what is required by an assertion of *truth*. The idea that the claim that a theory is true should depend only on a cardinality constraint and a logical argument fails to adequately separate the epistemic basis for the truth of the theoretical assertions of an empirical theory from the epistemic basis for the mere satisfiability of the "abstract" assertions of a purely mathematical theory over a given domain.

This conclusion should perhaps have been anticipated, given the origin of the partial interpretation view in Hilbert's conception of the foundations of geometry. In his correspondence with Frege, Hilbert defended the idea that satisfiability in a sufficiently large domain is a suitable surrogate for the "truth" of a mathematical theory. But whatever its plausibility for theories of pure mathematics, the methodological demands we impose on the theoretical claims of physics cannot be captured by so weak a requirement, at least not if we wish to preserve the methodological difference between physics and pure mathematics. An advocate of partial interpretation might respond to this objection by recalling that a physical theory will qualify as true not if it is merely satisfiable, but only if it is satisfiable in a model which is an extension of a domain that forms the basis for a model of the true observation statements. By contrast, the domains which bear witness to the "truth" of a mathematical theory need not have any connection with such a model. For an advocate of partial interpretation, the theories of physics are true because they are *empirically adequate* in the sense that they have observational consequences, all of which are true; but a theory of pure mathematics is not necessarily associated with *any* observation language and is not required to be empirically adequate.

However, this response misses the point of the model-theoretic argument as we have presented it: provided the domain over which a partially interpreted theory involving unobservable entities is interpreted includes the domain of the model of the true observation sentences, it is a consequence of the partial interpretation view that the method of argument by which we are able to establish the "truth" of a purely mathematical claim over a given domain also suffices to establish the truth of a theoretical claim.

The model-theoretic argument puts us in a position to see why the multiple realizability which afflicts partial interpretation and Ramsey-sentence reconstructions is largely tangential to the acceptability of these reconstructions. For suppose we are given a realization of the sort that Putnam's argument shows is possible. Even if we are able to rule out alternative realizations, we must still confront the fact that these accounts allow that a theory can be true only because it has a realization that models its observational consequences and is the right size. This is contrary to our conviction that if our physical theories are true, this is because they succeed in isolating salient truths about the entities with which they deal.

David Lewis is most closely identified with the suggestion that to address Putnam's argument, we should distinguish among possible realizations of a theory (see his 1983, 1984). The core assumption of Lewis's response is that a theory is *true* only if it is true relative to a realization whose properties and relations are *natural*. Since there is nothing in Putnam's construction of his interpretation of theoretical predicates which requires that they should be natural properties and relations, Lewis argued that the construction fails to show that the theory's theoretical claims are, in the relevant sense, true. We might adapt Lewis's reply to the model-theoretic argument and supplement the partial interpretation account by restricting the class of admissible realizations to those that involve natural relations, thereby distinguishing true theories from theories that are merely satisfiable in a domain that extends a model of the true observation statements. Lewis's distinction might be further exploited to characterize true *empirical* theories as those that are not merely satisfiable in some such realization or other, but are true because they are true in a realization whose relations are natural. But we should be cautious about accepting Lewis's idea as an adequate response to the model-theoretic argument or as a guide for emending the partial interpretation view.

To begin with, Lewis's reply to Putnam leaves unresolved the problem of how we are able to make significant claims about relations that are *not* "natural." Even if we have no interest in theorizing about such relations, an adequate response to the model-theoretic argument should nevertheless explain how it is *possible* to do so without the assertion of the truth of such a theory collapsing into an assertion of its satisfiability over a domain—even a domain that extends the model of the observational consequences of the "theory" of such a natural relation. Indeed, as Fraser MacBride has remarked, on the assumption that we achieve knowledge of natural relations only with the progress of science—and perhaps only after many distractions involving nonnatural relations—anyone following Lewis's suggestion *must* have an interest in how we manage to make significant, but as it happens, misguided claims about nonnatural relations.

But secondly, and more importantly, addressing Putnam's argument by appealing to Lewis's proposal obscures the difficulty the model-theoretic argument raises for partial interpretation and Ramsey-sentence reconstructions. The problem with these approaches is not their failure to designate certain properties and relations as natural, but the fact that they are too weak to explain the difference between the epistemological status of theoretical and purely mathematical claims. But then the partial interpretation framework for addressing how theories are warranted must *also* fail to capture the methodology by which claims about unobservables are established, and this shows that a different approach to these two issues is required. An emendation of the view based on Lewis's reply to Putnam succeeds only in recording the fact that we *do* distinguish mathematical claims from the theoretical claims of physics; it has nothing to contribute to our understanding of the methodology by which we make this distinction. Nor does it contribute to our understanding of how we successfully gain epistemic access to theoretical domains in order to warrant our claims about them.

In conclusion, I have argued that there are two questions that an adequate account of theoretical knowledge must address: (1) "What is the methodology by which we gain epistemic access to theoretical domains?" and (2) "How are the physical claims about such domains distinguished from the claims of pure mathematics?" Insofar as partial interpretation and Ramsey-sentence reconstructions are unable to adequately address the second of these two questions, it is evident that they have failed to marshal the resources necessary for addressing the first.

References

Carnap, R. (1956) "The Methodological Character of Theoretical Concepts," in H. Feigl and M. Scriven (eds.), *The Foundations of Science and the Concepts of Psychology and Psychoanalysis*, Minneapolis: University of Minnesota Press, pp. 38–76.

——— (1961) "On the Use of Hilbert's ε-Operator in Scientific Theories," in Y. Bar-Hillel et al. (eds.), *Essays in the Foundations of Mathematics Dedicated to A. A. Fraenkel*, Jerusalem: The Magnus Press of Hebrew University, pp. 156–64.

——— (1963) "Replies and Systematic Expositions," in P. A. Schilpp (ed.), *The Philosophy of Rudolf Carnap*, La Salle, IL: Open Court, pp. 859–1013.

Demopoulos, W. (2007) "Carnap on the Rational Reconstruction of Scientific Theories," in M. Friedman and R. Creath (eds.), *The Cambridge Companion to Carnap*, Cambridge: Cambridge University Press, pp. 248–72.

——— (2017) "Logical Empiricist Reconstructions of Theoretical Knowledge," in H. Leitgeb et al (eds.), *Logic, Methodology and Philosophy of Science. Proceedings of the 15th International Congress*, London: College Publications, pp. 189-206.

——— (2022) *On Theories. Logical Empiricism and the Methodology of Modern Physics* (ed. by M. Friedman), Cambridge, MA: Harvard University Press.

Gupta, A. (2009) "Definitions," in E. N. Zalta (ed.), *Stanford Encyclopedia of Philosophy*, Spring 2009 ed., https://plato.stanford.edu/entries/definitions/.

Hempel, C. G. (1958) "The Theoretician's Dilemma: A Study in the Logic of Theory Construction," in H. Feigl and M. Scriven (eds.), *Concepts, Theories, and the Mind-Body Problem*, Minneapolis: University of Minnesota Press, pp. 37–98. Repr. in Hempel, *Aspects of Scientific Explanation*, New York: The Free Press, 1965, pp. 173–226.

Lewis, D. (1970) "How to Define Theoretical Terms," *Journal of Philosophy* 67: 427–45.

——— (1983) "New Work for a Theory of Universals," *Australasian Journal of Philosophy* 61: 343–77.

——— (1984) "Putnam's Paradox," *Australasian Journal of Philosophy* 62: 221–36.

Putnam, H. (1977) "Realism and Reason," *Proceedings and Addresses of the American Philosophical Association* 50: 483–98.

Ramsey, F. P. (1931) "Theories," in Ramsey, *The Foundations of Mathematics and other Logical Essays* (ed. by R. Braithwaite), Paterson, NJ: Littlefield and Adams, pp. 212–36.

van Benthem, J. (1978) "Ramsey Eliminability," *Studia Logica* 37: 321–36.

Winnie, J. (1967) "The Implicit Definition of Theoretical Terms," *British Journal for the Philosophy of Science* 18: 223–9.

——— (1970) "Theoretical Analyticity," in R. Cohen and M. Wartofsky (eds.), *Boston Studies in the Philosophy of Science*, Dordrecht: Reidel, vol. VIII, pp. 289–305.

21
THE RELATIVE A PRIORI

David J. Stump

The idea of a relative a priori goes back to Reichenbach's briefly held view in *The Theory of Relativity and A Priori Knowledge* (1920) and was developed by Carnap as a theory of linguistic frameworks in *Logical Syntax of Language* (1934). Paolo Parrini (1976; 1998: 59–60) and Michael Friedman (1991, 1999) each revived Reichenbach's idea of a relative a priori and made a good case for its importance. Right away we must note an underlying tension here. On the one hand, the official position of the Vienna Circle and later logical positivists and logical empiricists is that there is no synthetic a priori at all, because all knowledge can be shown to be either empirical or simply a matter of definition. On the other hand, while synthetic a priori knowledge was officially rejected by the Vienna Circle, several topics considered to be a priori by Kant—for example, space and time and causality—remained central topics in twentieth-century philosophy of science. Indeed, the focus of Alberto Coffa's groundbreaking book is a priori knowledge and the search for an alternative to Kant: "It would be hard to find a more crucial epistemological problem than that of the character of a priori knowledge" (1991: 1). One may think that Quine's critique of the analytic/synthetic distinction would have put the final nail in the coffin of a priori knowledge and that a holism in which all scientific statements are justified empirically has replaced the notion of any special status for what was formerly considered to be a priori. But in fact, replacement theories of the a priori were developed throughout the twentieth century. While Kant thought of a priori knowledge as fixed and absolutely certain, these replacements see the (former) a priori as changing during scientific revolutions.

Reichenbach

A look at the positions of individuals involved in the rejection of the Kantian a priori shows a rather nuanced and complicated story. Coffa, Friedman, and others have argued that Schlick went through a neo-Kantian phase before becoming a strict empiricist, although this reading of Schlick has been challenged. Of course, in his later writings, Schlick totally rejects the Kantian idea of the synthetic a priori, arguing that all synthetic statements are empirical and all a priori statements are tautologies. As Coffa tells the story, it is Einstein's theories of relativity that convinced Schlick to change his mind (Coffa 1991: 189 ff.). Reichenbach's briefly held theory of the relative a priori is seen as a last ditch attempt to make the theories of relativity compatible with Kant.

 DOI: 10.4324/9781315650647-24

Reichenbach begins with a famous distinction between two meanings of "a priori" in Kant, in order to introduce the idea of the relative a priori: "First, it means 'necessarily true' or 'true for all times' and secondly, 'constituting the concept of object'" (1920/1965: 48). Reichenbach suggests giving up on the idea of necessity, because Euclidean geometry has been shown not to be necessary (ibid.: 3), but we should hold on to the second meaning of the a priori, that of constitution. To explain this idea, Reichenbach focuses on coordinating principles that connect scientific theories expressed in mathematical terms to physical reality (ibid.: 37). Their role "is precisely to relate abstract mathematical concepts (such as the Newtonian concepts of absolute space, time, and motion) to concrete empirical phenomena (such as the observed motions of the heavenly bodies in the solar system" (Friedman 2012: 48).

Principles of coordination have a special status in scientific theories, and this is what Reichenbach means by constitution. Coordinating principles do not make factual claims but, rather, connect empirical phenomena to scientific concepts.

> [T]his view is distinct from an empiricist philosophy that believes it can characterize all scientific statements indifferently by the notion "derived from experience." Such an empiricist philosophy has not noticed the great difference existing between specific physical laws and the principles of coordination and is not aware of the fact that the latter have a completely different status from the former for the logical construction of knowledge. The doctrine of the a priori has been transformed into the theory that the logical construction of knowledge is determined by a special class of principles, and that this logical function singles out this class, the significance of which has nothing to do with the manner of its discovery and the duration of its validity.
>
> (Reichenbach 1920/1965: 93–94)

While still advocating for the relative a priori, Friedman now rejects calling the constitutive elements "coordinating principles," because this ties them too closely to an outdated theory (2010: 697–8, 777 n. 253, 781 n. 268): "The solution I am now exploring involves replacing the Kantian faculty of sensibility with what we now call physical frames of references—ostensively introduced and empirically given systems of coordinates (spatial and temporal) within which empirical phenomena are to be observed, described, and measured" (2012: 48). Related issues have also been discussed by Thomas Oberdan (2009: 197–8) and Flavia Padovani (2011, 2015). So questions remain regarding the exact nature of constitution.

In fact, there is some ambiguity in the discussion of what gets constituted by the relative a priori. On the one hand, we have the object (of knowledge) being constituted a priori by the mind, which is a very Kantian conception that may skid into idealism. This is a concern in Schlick's discussion of Reichenbach's early view (Oberdan 2009: 189), and it faces us again in Kuhn's idea of world change and "constituting nature" (1962: 110). If, on the other hand, it is scientific concepts that are constituted, we do not need to worry about idealism. I suggest that what makes something constitutive is the role that the a priori plays in scientific theory. There is no special content that makes something a priori, as analytic epistemologists are likely to argue. Rather, it is the role that something plays as a necessary precondition for further inquiry that is central. This is a general and pragmatic criterion not wedded to any particular view of theories or of semantics. Indeed, it is a criterion of constitution that focuses on scientific practice, rather than on theories.

All of the theories of the relative a priori distinguish between empirical laws and constitutive (a priori) principles, as in the quote from Reichenbach just presented. Parrini generalizes the point:

> A closer reading of the texts, now abundantly confirmed by archival research, has shown the neopositivists' full awareness of the role played in scientific knowledge not only by logic and mathematics, but also by other "forms of thinking" (as Frank called them) that rely on the exercise of a spontaneous inventive capacity, largely autonomous of experience.
>
> (1998: 25)

In some sense, all of the logical empiricists describe a form of constitutive theories, and in that sense a form of the a priori.

The example that Friedman has worked through in great detail is the history of mechanics and gravitational theory from Newton to Einstein. Newton's laws (or their relativistic counterparts) and the mathematics used in physics are the constitutive elements, while Newton's law of gravity and Einstein's field equations are empirical. The constitutive elements, of course, changed from Newtonian theory to Einstein's. Space is no longer Euclidean but, rather, has variable curvature, and Newton's second law receives a relativistic "correction" so that it is consistent with the principle that nothing can travel faster than the speed of light.

> Thus, for example, whereas Euclidean geometry and the Newtonian laws of motion were indeed necessary presuppositions for the empirical meaning and application of the Newtonian theory of universal gravitation (and they were therefore constitutively a priori in this context), the radically new mathematical and physical framework consisting of the Riemannian theory of manifolds and the principle of equivalence defines an analogous system of necessary presuppositions in general relativity.
>
> (Friedman 2008: 251)

Some constitutive principles must necessarily be in place, but they can change rather dramatically from one scientific theory to another, hence the *relative* a priori. Friedman defends the idea that conceptual revolutions occur in science when there is a change in what had been taken to be a priori knowledge: "the relativized and dynamical conception of the a priori developed by the logical empiricists appears to describe these conceptual revolutions far better than does Quinean holism" (2002: 181).

Reichenbach was right to highlight the constitutive role of certain principles and presuppositions in science. One could say that the term "a priori" is misleading in this context given that these constitutive elements are not a priori in the traditional sense. Instead, we have various theories of the constitutive elements in science: Kant's, in which the constitutive elements really are a priori in the traditional sense, and those endorsed by others such as C. I. Lewis (1923, 1929), Ernst Cassirer (1910), Arthur Pap (1946), Thomas Kuhn (1962, 2000), Michael Friedman (2001a), Ian Hacking (1982, 2012), and Hilary Putnam (1962a, 1962b, 1979). Note that putting the word "relative" or "dynamic" or "pragmatic" in front of the word "a priori" changes the meaning of the latter completely. These kinds of a priori are not fixed, are not universal, are not necessary, and are not known by any sort of intuition. Nevertheless, for the purposes of discussing Reichenbach, Carnap, and others, I generally maintain the term "a priori," even though I would prefer to give it up for the sake of clarity and replace it with "constitutive elements." That is, I would prefer to think of these philosophers as having theories of the role played by presuppositions underlying empirical inquiry in the physical sciences. Following the Kantian tradition, we can call these presuppositions constitutive of particular inquiries, but only with the qualification indicated.

Discussions with Schlick

Reichenbach's views on the relative a priori changed after an exchange of letters with Schlick, whose views of conventions changed as well (Friedman 2004: 104–5). Reichenbach discussed his changed view in print, noting that he had taken Schlick to be ignoring the constitutive significance of some concepts used in science, but now he saw their dispute as terminological (1922/1978: 37–39). Friedman rejects the idea that the dispute between Reichenbach and Schlick is merely terminological, and he describes it as follows: "Schlick argues that we should no longer characterize constitutive principles—for example, and especially, the principles of geometry—as a priori at all: we should rather characterize them as conventions in the sense of Henri Poincaré" (1994/1999: 63). Of course, Schlick's interpretation of Poincaré's conventions is quite a bit more general than Poincaré's own. The bottom line is that Schlick rejects the synthetic a priori, instead characterizing all synthetic statements as empirical. The constitutive principles are taken to be analytic and conventional.

> From Hilbert Schlick adopts the idea that the axioms of geometry "implicitly define" the primitive terms of that science. This explains why the axioms of geometry are both nonempirical and conventional: Alternative systems of geometry—Euclidean or non-Euclidean simply count as different definitions of "point," "line," "between," and so on. Hilbert's view thus accounts for the nonempirical status of pure geometry. But we need to add Poincaré's ideas to account for applied geometry. Here Schlick reasons as follows: In applying such a purely formal system of implicit definitions to our actual experience of nature, no merely empirical considerations can force us to adopt one system rather than another; rather, only experience plus the requirement of overall *simplicity* of the laws of nature yield a determinate such system.
>
> (Friedman 1994/1999: 64)

As Friedman notes, on Schlick's interpretation, conventionalism is equivalent to the Duhem–Quine thesis and is more general than Poincaré's conventionalist thesis. It is hard to explain how Poincaré can limit his conventionalism, as he clearly does, if he is interpreted as basing it on underdetermination (Stump 1989: 348; Friedman 1995, 1999: 73). Indeed, Schlick's interpretation of Poincaré leads directly to an empiricist refutation of conventionalism. Interpreted as so-called "implicit definitions" conventions are not genuinely constitutive. Either they are mere definitions and hence analytic—this was dubbed "trivial semantic conventionalism" in the literature—or they are empirical. As Friedman points out, Reichenbach's early critique of conventionalism is important, and lost in his later work where he adopts Schlick's point of view. Reichenbach's early point is that the general theory of relativity (GTR) showed that metric conventionalism is false (see Ryckman 1992).

> Reichenbach cannot accept Poincaré's (and Schlick's) conventionalism as a general philosophical doctrine about geometry as such, independent of any specific theoretical context, and for precisely this reason Reichenbach explicitly rejects conventionalism in the "Introduction" to his book [1920] . . . For Schlick, by contrast,

> geometry remains conventional or nonempirical in the context of the general theory of relativity. Duhemian holistic considerations still apply, and all that general relativity actually shows is that the *simplest* total system of natural laws employs non-Euclidean geometry: Euclidean geometry thus remains an equally "correct" option. Moreover, Reichenbach (1922) comes to agree with Schlick on this crucial point.
>
> (Friedman 1994/1999: 66–67)

Although the underlying mathematical structure of GTR does not have a determinate metric, metric is determined empirically in GTR by the distribution of mass. This was not widely understood until the discussion of spacetime theories in the 1970s and 80s.

Later in the decade, Reichenbach voices the standard logical empiricist view that there is no a priori knowledge. He refers to his early work but is now fully critical of Kant, saying that logic and definitions are presupposed but "empty," while "physics contains no principles that are a priori in the sense of 'independent of experience.' We hold to even the most general principles of knowledge solely because they prove themselves in experience" (1928/1958: 172).

Carnap

Midway in his career Carnap adopts a view that is different from the strict empiricist rejection of the Kantian a priori. According to Carnap, natural science is to be represented in a formal language. In *The Logical Syntax of Language* (1934), he divides a scientific language into a logical part and a physical part; that is, he distinguishes between what he calls L-rules and P-rules. The logical part is analytic and conventional, while the physical part is synthetic and empirical. Many different languages are possible, and the choice between them is practical, rather than cognitive. Carnap calls the choice of language an external question, and it is guided by the suitability of a language for a given purpose. There is no right or wrong language choice in an objective sense but only one relative to our purposes. Carnap's principle of tolerance famously held: "*In logic, there are no morals*. Everyone is at liberty to build up his own logic, i.e. his own form of language, as he wishes. All that is required of him is that, if he wishes to discuss it, he must state his methods clearly, and give syntactical rules instead of philosophical arguments" (1934/1937: 52, orig. emphasis).

It is worth noting that physical theory itself can determine the distinction between logical and empirical content. Section 50 of the *Logical Syntax* gives the example of metric being determined conventionally in one theory but empirically in another, implicitly agreeing with Reichenbach's assessment that Einstein's general theory of relativity disproves metric conventionalism. Importantly, for Carnap the relative a priori was analytic. Comparison of Friedman's view with Carnap's raises the question of whether the relative a priori can remain so (see Uebel 2012).

The specific context for which Carnap developed his theory was the philosophy of mathematics (Goldfarb and Ricketts 1992; Friedman 2001b; Ricketts 2007). Carnap was well aware of the competing schools of formalism, intuitionism, and logicism and saw advantages and limitation in each. His conception of logical syntax as providing alternative languages provided a way to diffuse the philosophical conflict between the schools by transforming the issues into one of language choice. As Friedman puts it, Carnap showed that the competing schools

> are not really debating about the "true nature" of mathematical objects but merely proposing different language forms, each having various advantages and disadvantages, for the total language of science. It cannot be stressed too much, I think, that this

> diagnosis and transformation of philosophical problems constitutes the main point of both the principle of tolerance and the method of logical syntax more generally.
> (2001b: 233)

Although he did not discuss historical changes in science, there is clearly a place for revolutions in his outlook. A change of language for Carnap amounts to a conceptual scientific revolution, with the change in language being a formal equivalent of a change in conceptual scheme or paradigm. Even though Carnap does not link his analysis of science to the relativized a priori, several authors have noted the similarity between Carnap's views and those of the early Reichenbach (Creath 2004: 292, Friedman 2004: 112). As Parrini puts it, "Carnap can be seen as the philosopher who made the most significant contribution to the development of the Neoempiricistic theory of the *relativized a priori*, by systematizing the standard *linguistic* version of this conception" (2009: 127, orig. emphasis).

Carnap seems to have been endlessly optimistic about the power of formal methods, but his approach leaves itself open to the critiques developed by Quine (1951) and Gödel (1995). (Hudson [2010] argues that Carnap cannot escape the critique of his principle of tolerance brought forth by Gödel, despite defenses of Carnap by Ricketts, Friedman, and Awodey and Carus [2004].) Briefly summarizing Quine's arguments here, in "Two Dogmas of Empiricism" (1951), he criticizes two ideas central to the program of the logical empiricists. First, he claims that the analytic/synthetic distinction cannot be drawn. Since this distinction is needed to even state what "modern empiricism" is (i.e., to reject the existence of synthetic a priori statements), Quine's critique is fundamental. Second, he rejects reductionism, the idea that every meaningful statement can be seen as a logical construction of observation predicates. It has been argued, however, that Quine's resulting naturalism overlooks both the historical development of science and the special role that constitutive or a priori principles play (Coffa 1986: 55–59).

Conclusion

As with the Kantian a priori, the relative a priori can be said to be constitutive in function; unlike Kant, relative a priori knowledge is contingent and has changed historically. Reichenbach distinguishes the constitutive and necessary or universal elements of the a priori, thus dividing the term "a priori" into two elements. It is certainly true that the constitutive elements in science frequently turn out to be things that were once considered to be a priori knowledge in a full-bodied sense. The dynamic, pragmatic, functional, and relative a priori are not theories of a priori knowledge but theories of the constitutive elements in science (see Stump 2015). They give a proper epistemological status to things that had formerly been called a priori.

Nevertheless, there are three ways that the theories of the constitutive elements in science are connected to the traditional idea of a priori knowledge. First, many of the examples come from things that were considered a priori, especially by Kant. Our conception of space and time, our knowledge of mathematics, and a few fundamental principles of physical theory fall into this category. Second, those advocating theories of the constitutive elements in science have emphasized that some knowledge must be in place before we can conduct scientific research in a particular way. These are necessary preconditions for the possibility of conducting a scientific inquiry, to put it in Kantian language. For example, the mathematics has to be known before the empirical inquiry can begin, given that theories are stated in mathematical language and that problems are solved using the tools of mathematics. Third, many aspects of scientific knowledge are taken for granted as established, and furthermore, some of these aspects can be taken as a criterion for further inquiry. These principles are not only "hardened" so that no one would seriously doubt them, but they also play a special role in categorizing phenomena. The constitutive elements in science play a role that is somewhat different from the usage in Kant, therefore. The focus lies on what makes a particular scientific practice possible, on necessary preconditions that are constitutive.

References

Awodey, S. and Carus, A. W. (2004) "How Carnap Could Have Replied to Gödel," in Awodey and Klein 2004, pp. 203–23.

Awodey, S. and Klein, C. (eds.) (2004) *Carnap Brought Home: The View from Jena*, Chicago: Open Court.

Carnap, R. (1934) *Logische Syntax der Sprache*, Vienna: Springer. Rev. ed. trans. *The Logical Syntax of Language*, London: Kegan Paul, Trench, Trubner, & Co., 1937, repr. Chicago: Open Court, 2002.

Cassirer, E. (1910) *Substanzbegriff und Funktionsbegriff: Untersuchungen über die Grundfragen der Erkenntniskritik*, Berlin: Bruno Cassirer. Trans. "Substance and Function," in Cassirer, *Substance and Function and Einstein's Theory of Relativity*, Chicago: Open Court, 1923, pp. 1–350.

Coffa, J. A. (1986) "From Geometry to Tolerance: Sources of Conventionalism in 19th Century Geometry," in R. G. Colodny (ed.), *From Quarks to Quasars: Philosophical Problems of Modern Physics*, Pittsburgh: University of Pittsburgh Press, pp. 3–70.

——— (1991) *The Semantic Tradition from Kant to Carnap: To the Vienna Station*, Cambridge: Cambridge University Press.

Creath, R. (2004) "Carnap's Program and Quine's Question," in Awodey and Klein (2004), pp. 279–93.

Friedman, M. (1991) "The Re-evaluation of Logical Positivism," *Journal of Philosophy* 88: 505–19. Repr. in Friedman 1999, pp. 1–14.

——— (1994) "Geometry, Convention, and the Relativized A Priori," in W. Salmon, G. Wolters (eds.), *Logic, Language, and the Structure of Scientific Theories*, Pittsburgh: University of Pittsburgh Press, pp. 21–34. Repr. in Friedman 1999, pp. 59–70.

——— (1999) *Reconsidering Logical Positivism*, Cambridge: Cambridge University Press.

——— (2001a) *Dynamics of Reason: The 1999 Kant Lectures at Stanford University*, Stanford: CSLI Publications.

——— (2001b) "Tolerance and Analyticity in Carnap's Philosophy of Mathematics," in J. Floyd and S. Shieh (eds.), *Future Pasts: The Analytic Tradition in Twentieth-Century Philosophy*, Oxford: Oxford University Press, pp. 223–55.

——— (2002) "Kant, Kuhn, and the Rationality of Science," *Philosophy of Science* 69: 171–90.

——— (2004) "Carnap and the Evolution of the A Priori," in Awodey and Klein (2004), pp. 101–16.

——— (2008) "Ernst Cassirer and Thomas Kuhn: The Neo-Kantian Tradition in History and Philosophy of Science," *Philosophical Forum* 39: 239–52.

——— (2010) "Synthetic History Reconsidered," in M. Domski and M. Dickson (eds.), *Discourse on a New Method: Reinvigorating the Marriage of History and Philosophy of Science*, Chicago and La Salle, IL: Open Court: 571 813.

——— (2012) "Reconsidering the Dynamics of Reason: Response to Ferrari, Mormann, Nordmann, and Uebel." *Studies in History and Philosophy of Science* 43: 47–53.

Gödel, K. (1995) "Is Mathematics Syntax of Language?" in S. Feferman et al. (eds.), *Collected Work, vol. 3: Unpublished Essays and Lectures*, Oxford: Oxford University Press, pp. 334–62.

Goldfarb, W. and Ricketts, T. (1992) "Carnap and the Philosophy of Mathematics," in D. Bell and W. Vossenkuhl (eds.), *Science and Subjectivity: The Vienna Circle and Twentieth Century Philosophy*, Berlin: Akademie Verlag, pp. 61–78.

Hacking, I. (1982) "Language, Truth, and Reason," in M. Hollis and S. Lukes (eds.), *Rationality and Relativism*, Oxford: Basil Blackwell, pp. 48–66.

——— (2012) "'Language, Truth and Reason' 30 Years Later," *Studies in History and Philosophy of Science* 43: 599–609.

Hudson, R. (2010) "Carnap, the Principle of Tolerance, and Empiricism," *Philosophy of Science* 77: 341–58.

Kuhn, T. S. (1962) *The Structure of Scientific Revolutions*, Chicago: University of Chicago Press.

——— (2000) *The Road Since Structure*, Chicago: University of Chicago Press.

Lewis, C. I. (1923) "Pragmatic Conception of the A Priori," in C. I. Lewis, *Collected Papers* (ed. by J. D. Goheen and J. L. Mothershead), Stanford: Stanford University Press, pp. 231–9.

——— (1929) *Mind and the World Order: Outline of a Theory of Knowledge*, New York: Charles Scribner's Sons. Repr. New York: Dover, 1956.

Oberdan, T. (2009) "Geometry, Convention, and the Relativized A Priori: The Schlick—Reichenbach Correspondence," in F. Stadler, H. J. Wendel and E. Glassner (eds.), *Stationen. Dem Philosophen und Physiker Moritz Schlick zum 125. Geburtstag*, Vienna: Springer, pp. 186–211.

Padovani, F. (2011) "Relativizing the Relativized A Priori: Reichenbach's Axioms of Coordination Divided," *Synthese* 181: 41–62.

——— (2015) "Measurement, Coordination, and the Relativized A Priori," *Studies in History and Philosophy of Science Part B: Studies in History and Philosophy of Modern Physics* 52: 123–8.

Pap, A. (1946) *The A Priori in Physical Theory*, New York: King's Crown Press.

Parrini, P. (1976) *Linguaggio e teoria*, Firenze: La Nuova Italia Editrice.

——— (1998) *Knowledge and Eeality: An Essay in Positive Philosophy*, Dordrecht: Kluwer.

——— (2009) "Carnap's Relativised A Priori and Ontology," in M. Bitbol, P. Kerszberg and J. Petitot (eds.), *Constituting Objectivity: Transcendental Perspectives on Modern Physics*, Berlin: Springer, pp. 127–43.

Putnam, H. (1962a) "It Ain't Necessarily So," *Journal of Philosophy* 59: 658–71. Repr. in Putnam, *Mathematics, Matter and Method*, Cambridge: Cambridge University Press, 1975, pp. 237–49.

——— (1962b) "The Analytic and the Synthetic," in H. Feigl and G. Maxwell (eds.), *Scientific Explanation, Space, and Time: Minnesota Studies in the Philosophy of Science,* Minneapolis, MN: University of Minnesota Press, vol. 3, pp. 358–97. Repr. in Putnam, *Mind, Language and Reality*, Cambridge: Cambridge University Press, 1975, pp. 33–69.

——— (1979). "Analyticity and Apriority: Beyond Wittgenstein and Quine," in P. A. French, T. E. Uehling and H. K. Wettstein (eds.), *Studies in Metaphysics: Midwest Studies in Philosophy*, Minneapolis: University of Minnesota Press, vol. 4, pp. 423–45. Repr. in Putnam *Realism and Reason*, Cambridge: Cambridge University Press, 1983, pp. 115–38.

Quine, W. V. O. (1951) "Two Dogmas of Empiricism," *Philosophical Review* 60: 20–42. Repr. in *From a Logical Point of View*, Cambridge, MA: Harvard University Press, 1953, pp. 20–46.

Reichenbach, H. (1920) *Relativitätstheorie und Erkenntnis Apriori*, Berlin: Springer. Trans. *The Theory of Relativity and A Priori Knowledge*, Berkeley: University of California Press, 1965.

——— (1922) "Der gegenwärtige Stand der Relativitätsdiskussion," *Logos*: 316–78. Trans. "The Present State of the Discussion on Relativity," in Reichenbach, *Selected Writings 1909–1953* (ed. by M. Reichenbach and R. S. Cohen), Dordrecht: D. Reidel, 1978, vol. 2, pp. 3–47.

——— (1928) *Philosophie der Raum-Zeit-Lehre*, Berlin: de Gruyter. Trans. *The Philosophy of Space and Time*, New York: Dover, 1958.

Ricketts, T. (2007) "Tolerance and Logicism: Logical Syntax and the Philosophy of Mathematics," in M. Friedman and R. Creath (eds.), *The Cambridge Companion to Carnap*, Cambridge: Cambridge University Press, pp. 200–25.

Ryckman, T. A. (1992) "'P(oint)-C(oincidence) Thinking': The Ironical Attachment of Logical Empiricism to General Relativity (and Some Lingering Consequences)," *Studies in History and Philosophy of Science* 23: 471–97.

Stump, D. J. (1989) "Henri Poincaré's Philosophy of Science," *Studies in History and Philosophy of Science* 20: 335–63.

——— (2015) *Conceptual Change and the Philosophy of Science: Alternative Interpretations of the A Priori*, London: Routledge.

Uebel, T. (2012) "De-Synthesizing the Relative a Priori," *Studies in History and Philosophy of Science* 43: 7–17.

22
NONSTANDARD LOGICISM

Georg Schiemer

Logicism was a dominant position in the foundations of mathematics of the late nineteenth and early twentieth centuries. Roughly put, it is the view that pure mathematics is reducible to higher-order logic. More specifically, the logicist thesis is usually taken to consist of two claims. First, all primitive terms of an axiomatized mathematical theory can be explicitly defined by using only logical vocabulary. Second, all axioms of the theory can be deduced from purely logical principles. It follows from these two claims that all theorems of a mathematical theory are also derivable from purely logical principles. Let us call this the classical or standard logicist thesis.

It is well known that the pioneering logicists Frege, Russell, and Whitehead as well as subsequent philosophers such as Ramsey and Carnap defended variants of this view. However, the contributions of second-generation logicists often differed from classical logicism in important respects, in particular concerning: (1) the mathematical theories considered; (2) the logical principles adopted; and (3) the very concept of a logicist reduction. Thus, based on different accounts of what is meant by "logic," "mathematics," and "reducible," one can identify a number of nonstandard theories of logicism developed in the 1920s and later on.

Logicism should thus not be viewed as a monolithic research program, but rather as a family of different approaches on how the general project of reducing mathematics to logic can be made precise. The focus of this chapter will be on different theories of logicism developed in the heyday of logical empiricism, that is, roughly between 1920 and 1940. The central aim here is to analyze how Frege's and Russell's logicist programs were modified in the period in question. The changes concern not only formal details of the underlying logic such as the adoption of a simple theory of logical types, but also the kind of mathematical theories considered for the reduction to logic. Whereas classical logicism focused mainly on arithmetic, logical empiricists such as Carnap and Hahn were interested in a generalized logicist thesis which is applicable to any axiomatic theory of pure mathematics. (See also Schiemer 2022 for a survey of different versions of logicism in logical empiricism.)

Logicism and type theory

The origins of classical logicism can be traced back to foundational work in nineteenth-century mathematics, in particular on the arithmetization and rigorization of analysis. Frege's own work on the logicist reduction of arithmetic is usually considered as a natural consequence of this line

 DOI: 10.4324/9781315650647-25

of foundational research. As is well known, Frege's logicism is developed most systematically in *Grundgesetze der Arithmetik* (1893). Part 1 of the book—titled "Exposition of the Begriffsschrift"—contains a description of the logical system used for the reduction of arithmetic. This is, roughly put, a higher-order logic complemented by a naïve theory of sets or, in Frege's terminology, a theory of concept extensions. Frege's central objective in the book was to show that a version of the Dedekind–Peano axioms of arithmetic can be derived from his logical principles and that explicit logical definitions can be given for the primitive arithmetical terms. Unfortunately, this project was doomed to failure given the fact that his naïve theory of classes turned out to be inconsistent. In his famous 1902 letter, Russell informed Frege that a class theoretic paradox can be derived from his logical system containing the infamous basic law V.

Subsequent research on logicism in the twentieth century was driven by the attempt to block Russell's paradox as well as related paradoxes based on a theory of logical types. Roughly put, this is a system of higher-order logic that describes the stratification of the logical universe into a hierarchy of typed objects. The inventor and main proponent of such a logic was, of course, Russell. A systematic development of his new system was given in the first edition of *Principia Mathematica* (Russell and Whitehead 1910–13). The ramified type theory presented there was modified substantially in work by a second generation of logicians, including Hilbert and Ackermann, Carnap, Gödel, Tarski, Ramsey, Chwistek, and Quine. Their work led to the *simplification* of type theory as well as to a purely *extensional* treatment of types. Another important modification of Russell and Whitehead's original framework concerns the distinction between the syntax and the semantics. The picture emerging in work by Carnap, Tarski, and others is that of type theory as a formal set theory, i.e., a theory describing a rich and stratified universe of objects.

Simple type theory came to serve as the standard logic in the 1920s and 1930s (see Ferreirós 1999 for a detailed overview). Type theoretic systems usually discussed at the time usually contained two important higher-order axioms, namely an axiom scheme for comprehension and an axiom scheme of extensionality which states that properties are identical if they are co-extensional. In Russell's original presentation, three other axioms were taken to belong to the logical principles. The first one is the multiplicative axiom which is roughly equivalent to the set-theoretic axiom of choice. The second one is an axiom of infinity, which states that the there is a countable infinite number of objects in the individual domain of the language. A third axiom, relevant only for the original ramified treatment of types, is the axiom of reducibility.

With the adoption of type theory as a way to block the set-theoretic paradoxes deducible from Frege's logical system, it is easy to see how classical logicism can be reformulated in this new framework. A natural way to specify the logicist thesis can be given in terms of the notion of an *interpretation* of a theory in another one. Roughly put, given two axiomatic theories S and T (expressed in languages L_S and L_T, respectively), an interpretation of T in S is given by a translation of the formulas in L_T into formulas of L_S that preserves: (1) the logical structure of L_T sentences; and (2) the theorems of T. The second condition states that the translation of every theorem of the interpreted theory should also be *provable* from the axioms of the interpreting theory. Applied to the reduction of arithmetic to logic, one can show that the theory of Dedekind–Peano arithmetic (expressed in the second-order language) can be interpreted in type theory (expressed in the purely logical language). More specifically, there exists a translation of all arithmetical statements into purely logical statements based on Frege's logical definitions of the primitive arithmetical terms. This translation is theorem-preserving in the sense that for any arithmetical statement that is provable from the Dedekind-Peano axioms, its purely logical translation is derivable from the principles of type theory (see Schiemer 2022).

Logical truth in logical empiricism

Logical empiricism took shape as a philosophical movement in direct reaction to the foundational debates in mathematics at the turn of the last century. In particular, Frege's and Russell's thesis that mathematics is a branch of logic played a central role in the revival of empiricist philosophies in Vienna and elsewhere. While thinkers such as Hahn and Carnap took simple type theory to be the standard logical system for the logicist reduction of mathematics, their accounts differed from classical logicism in central respects. This is mainly due to the fact that, at their time, the conception of logic was subject to significant revision, largely in response to Wittgenstein's *Tractatus Logico-Philosophicus* (1922). Wittgenstein proposed a new analysis of logical truth in terms of the notion of a tautology. Such truths do not assert facts about the world but concern only the logical form of statements. This new conception of logic as a system of tautologies marks a sharp break with previous accounts, in particular the logical universalism shared by Frege and Russell.

As is well known, the members of the Vienna Circle wholeheartedly adopted the *Tractatus* conception of logic. How did the new understanding of logic as a set of tautologies transform the way in which logicism was understood by them? On first glance, both Carnap and Hahn seem to have embraced classical logicism in the sense outlined earlier. This is evident in a number of publications from the time, for instance, in several articles written by Carnap around 1930 that explicitly discuss Frege's and Russell's logicism (e.g., 1931). However, an important difference to their program becomes visible if one considers how the "fundamental logical sentences" are understood in these writings. Instead of characterizing logical laws as universal truths, they are conceived by Carnap as tautologies in Wittgenstein's sense. Consequently, assuming that mathematics is reducible to logic, it follows that all mathematical theorems are also purely tautological in character. This account of the nature of higher mathematics was widely shared among members of the Vienna Circle, as the following passage in the circle's official manifesto of 1929 indicates: "The conception of mathematics as tautological in character, which is based on the investigations of Russell and Wittgenstein, is also held by the Vienna Circle" (Verein Ernst Mach 1929/2012: 85).

Now, both Hahn and Carnap were aware that this revised logicist thesis built on the notion of logical tautology is problematic. It is difficult to see how logical type theory (let alone theories of higher mathematics) can be tautological in character, that is, without any ontological commitments. How did the logical empiricists aim to vindicate the logicist thesis in light of this fact? Two lines of reasoning can be mentioned here. The first concerns different attempts to *generalize* the notion of a tautology in order to make it applicable to type-theoretic logic and a fortiori also to mathematics. Hahn fully embraced the *Tractatus*-style conception of logic in his philosophical writings from the 1920s and early 1930s (e.g., 1929). In particular, he defended the view that logical laws concern only the logical form of statements and have no representational function. However, in contrast to Wittgenstein's "thin" account of tautologies, Hahn was interested in formulating a "wider" conception of tautologies (1933). This is based on the fact that he, in contrast to Wittgenstein, adopted an early conventionalism about the choice of logical principles (see Uebel 2005). Thus, Hahn argued that one can freely adopt our logical system for the study of inferences in our language. This includes the possibility to adopt set-theoretic principles such as the axioms of infinity or choice. Consequently, what counts as a valid tautological transformation is specified relative to the particular choice of a logical framework.

A second strategy to vindicate logicism is also based on the reassessment of the logical status of certain axioms of type theory. As is well known, Russell's and Whitehead's axioms of choice, infinity, and reducibility were viewed critically by many proponents of logicism, including also by logical empiricists. Consider, for instance, Carnap's *Abriss der Logistik*, where the status of these axioms is discussed in detail. As Carnap points out, the axioms of choice and of infinity "should not be included among the basic principles of logic, since its admissibility has been problematic" (1929, §24b). The decidedly existential and thus non-logical status of these axioms thus presented a central problem to type-theoretic logicism. It was clear to Carnap and others that these axioms are independent of the other logical principles of type theory but at the same time indispensable for the logicist reduction of mathematics. A possible solution to this problem was to view these axioms not as proper logical principles but rather as hypothetical assumptions in logical reasoning. More specifically, the central idea here was to reformulate those mathematical statements whose proof depends on these axioms in terms of conditional statements where the axioms in question occur in the antecedent.

This "conditional" logicism presents a weakened form of classical logicism that can be found in several works of the time (see Musgrave 1977; Coffa 1981). The first systematic formulation of this approach was given in Russell's *Introduction to Mathematical Philosophy* (1919). This approach was also adopted by several members of logical empiricism. Carnap, in particular, formulated variants of the strategy to conditionalize mathematical statements in his writings from the period in question. Thus he wrote: "[Russell] . . . transformed a mathematical sentence, say S, the proof of which required the axiom of infinity, I, or the axiom of choice, C, into a conditional sentence; hence S is taken to assert not S, but $I \supset S$ or $C \supset S$, respectively. This conditional sentence is then derivable from the axioms of logic" (1931/1983: 45).

The central motivation for this logical reconstruction was to reduce mathematics to logic without having to assert the logical truth of existential axioms such as choice and infinity. Applied to the program of reducing arithmetic to logic, this method yields a nonstandard form of logicism: arithmetical statements are also translated into purely logical statements here, but not based on explicit "logicist" definitions of the primitive terms. Instead, they are translated into conditional statements in the language of type theory.

If-thenism and general axiomatics

Logical empiricists including Hahn and Carnap formulated variants of a "conditional logicism" based on the critique of the non-tautological nature of axioms such as infinity and choice. Interestingly, the if-thenist reconstruction of mathematical statements is also adopted to generalize the logicist thesis in a different way, namely, to make it applicable to non-arithmetical theories in mathematics. This approach is again rooted in Russell's foundational work.

In his *Principles of Mathematics* (1903), the method of conditionalization was originally introduced in the discussion of non-Euclidian geometries (see Gandon 2009). Russell argued that the axioms of mutually inconsistent geometrical theories should not be viewed as assertive statements but rather as hypothetical claims about possible structures of space. Geometrical theorems, in turn, are to be expressed as quantified conditional statements that contain the *ramsified* axioms as the antecedent. This logical reconstruction has an important consequence: a geometry, conceived now as class of conditional statements, is expressible in a pure logical language and thus does not express any factual content about the world. Russell was well aware of this fact and suggested if-thenism as a natural approach to describe theories of pure mathematics: "Pure mathematics is the class of all propositions of the form 'p implies q,' where p and q are propositions containing one or more variables, the same in two propositions, and neither p nor q contains any constants except logical constants" (1903: 3). This logical reconstruction exercised an important influence on the logical empiricists. In particular, independent of the classical logicist project to reduce arithmetic to a firm logical basis, Russell's account was viewed by Carnap and Hahn, among others, as a way to capture modern axiomatic reasoning in non-arithmetical branches of mathematics.

Carnap, in particular, did not view logicism and formal axiomatics as opposing programs in the foundations of mathematics. One can view his early work on the philosophy of mathematics as an attempt to reconcile modern axiomatic mathematics with a generalized version of the Fregean or Russellian logicist thesis (compare Awodey and Carus 2001; Reck 2004). This is most explicit in his work on "general axiomatics" from the late 1920s. In particular, in Part II of his *Abriss*—titled "Applied Logistic"—Carnap suggests the following type-theoretic formalization of axiomatic theories: the primitive vocabulary of a theory is expressed as free variables (each of a specified arity and type). Axioms and theorems are expressed as sentential functions, that is, as open formulas in a modern sense. Carnap argues that an axiomatic theory gives not only an *implicit definition* of the primitive terms occurring in the axioms in the sense specified by Hilbert, but also an explicit definition of a higher-order concept, the so-called *Explizitbegriff* of an axiom system. More specifically, he holds that:

> For instance, if $x, y, \ldots \alpha, \beta, \ldots P, Q, \ldots$ are the primitive variables of the AS and if we name the conjunction of axioms (that is a propositional function) AS($x, y, \ldots \alpha, \beta, \ldots P, Q, \ldots$), then the definition of the explicit concept of this AS is:
>
> $$\hat{x}, \hat{y}, \ldots \alpha, \beta, \ldots \hat{P}, \hat{Q}, \ldots \{AS(x, y, \ldots \alpha, \beta, \ldots P, Q, \ldots)\}$$
>
> (1929: 72)

How is this approach of formalizing axiomatic theories related to the if-thenism described in Russell's *Principles of Mathematics*? Interestingly, Carnap's understanding of mathematical statements is highly similar to Russell's in this respect. While the if-thenist reconstruction is not mentioned in *Abriss*, Carnap explicitly discusses it in a related paper titled "Proper and Improper Concepts" (1927). He argues there that the mathematical content of a theorem is best expressed by a closed formula, namely a quantified conditional statement that contains the "logical product" of the axioms of a given theory in the antecedent. The theorems of a given theory are thus to be translated into purely logical statements of the form:

$$\forall x,y,\ldots\alpha,\beta,\ldots P,Q,\ldots\left[AS\left(x,y,\ldots\alpha,\beta,\ldots P,Q,\ldots\right)\rightarrow\varphi\left(x,y,\ldots\alpha,\beta,\ldots P,Q,\ldots\right)\right]$$

where variables $x, y, \ldots \alpha, \beta, \ldots P, Q, \ldots$ present the primitive vocabulary of the theory, AS presents the axioms of a theory, and φ the ramsified theorem in question. (This if-thenist construction can also be found in the writings of other logical empiricists; e.g., Hempel [1945].)

A central philosophical motivation underlying Carnap's adoption of this Russellian if- thenism was to defend some form of non-classical logicism. This generalized version of a logicist reduction is usually characterized in the modern literature in terms of two conditions (Musgrave 1977: 117–18):

1. All mathematical statements have the logical form of conditional statements with the logical product of the axioms in the antecedent and a ramsified theorem in the consequent.
2. All true mathematical statements are derivable from logical axioms.

The first condition states that all mathematical statements can be reformulated in purely logical terms. This language logicism corresponds to the weak logicist thesis discussed in Carnap's

Abriss. Given that the explicit concept of an axiom system can be expressed in purely logical terms, it follows that any mathematical theory (including non-arithmetical ones such as geometry or topology) "can be represented as a branch of logistic itself" (1929: §30b). Moreover, any theorem of a given theory can be translated into a purely logical sentence based on the if-thenist reconstruction.

As pointed out earlier, this condition is usually accompanied by a second thesis, namely that all true mathematical statements so construed become derivable from the logical (i.e., type theoretic) axioms in question. Expressed in modern logical terminology, (2) states that the if-thenist translation is also theorem-preserving, that is, it induces an *interpretation* of a mathematical theory in type theory in the sense specified earlier. Although this second thesis of conditional logicism is usually not described explicitly in published work from the 1920s and early 1930s, it is likely that this view was shared by Carnap and his fellow logical empiricists. (A version of this thesis can be found in his 2000. See Schiemer 2022.) Obviously, this if-thenism presents a weaker form of logicism than Frege's and Russell's original programs. In particular, what is missing here are explicit logicist definitions of the primitive terms of a mathematical theory. Moreover, mathematical axioms are not supposed to be derived from purely logical principles in the present account. What is derived from the principles of type theory are the ramsified conditional statements described earlier. Thus, the conditional logicism shows that all proofs of theorems can be formalized within a general type-theoretic system. This is given by the fact that for any axiomatic mathematical theory A and every statement φ in the language L_A, the following equivalence holds:

$$TT \cup \{\mathrm{A}\} \vdash \varphi \leftrightarrow TT \vdash \forall \vec{X}\left(\mathrm{A}\left(\vec{X}\right) \rightarrow \varphi\left(\vec{X}\right)\right).$$

Thus, whenever a statement is derivable from theory A (plus the logical axioms of TT, then the universal ramsification of ($A \rightarrow \varphi$) is derivable from the logical axioms alone (compare again Musgrave 1977; Coffa 1981).

Logical pluralism

Logicism lost much of its philosophical significance in the course of the 1930s, mainly as a result of Gödel's incompleteness results. Roughly put, Gödel's results show that arithmetical truth cannot be identified with logical provability. This was a serious blow for the traditional logicist thesis that arithmetic is reducible to higher-order logic. A second reason for the gradual demise of Frege's and Russell's program was that the scope of logic changed significantly in the period in question. First-order logic was eventually established as the standard logical system and replaced the logical theory of types. Moreover, logic also underwent a metatheoretic turn in work by Tarski, Carnap, and Gödel (among others). The new metalogical approach and the clear syntax-semantics distinction implied by it was clearly incompatible with the logical universalism present in the work of Frege and Russell, but also in Wittgenstein's *Tractatus*.

While logicism was challenged by these developments, it would be wrong to conclude that it was given up at the time. In fact, it remained a central position in work by philosophers affiliated with logical empiricism well after the 1930s. This is true, in particular, of Carnap's work. His project on general axiomatics—originally devised as two volumes of *Untersuchungen zur allgemeinen Axiomatik*—was eventually abandoned in 1930, mainly in response to Tarski's metatheoretic definition of notions such as categoricity, truth, and logical consequence (see Awodey and Carus 2001). As is well known, Tarski emphasized the distinction between the formulation of axiomatic (or "deductive") theories in an object language and the specification of their metatheoretic properties in a separate and richer metalanguage. Given this background, Carnap eventually developed a similar approach that presented a sharp break with Wittgenstein's position. Moreover, influenced by correspondence with Gödel in 1932, he also adopted a purely syntactic conception of logic and, more importantly, the idea of logical syntax as the study of metalinguistic properties of "logico-mathematical" languages. This new approach culminated in his *Logical Syntax of Language*, first published in 1934.

Carnap's work on logical syntax is marked by a number of important innovations. First, his account of logico-mathematical systems is decidedly metatheoretic: mathematical theories such as arithmetic are presented axiomatically in a fully specified object language. In the case of Peano arithmetic, this is the higher-order language of simple types L_{II}. As in Tarski's work, the formation and transformation rules of this language are expressed in a separate syntax language. Given this setup, a central syntactic concept introduced in *Logical Syntax* is the notion of analyticity which is explicitly introduced by Carnap as an explication of logical truth. Thus, Wittgenstein's notion of tautological truth is replaced here by a decidedly metatheoretical concept. Roughly put, analyticity for sentences in L_{II} is defined analogously to Tarski's treatment of formal truth, namely in terms of several recursive clauses for the valuation (or satisfaction) of open formulas.

The most significant innovation in Carnap's book is the fact that logical universalism of the traditional logicists is replaced by a form of logical pluralism. According to this view, there exists no unique or correct logic. Rather, according to Carnap, one can freely choose between different logical systems for the task of formalizing mathematics or the sciences. The different frameworks are equally valid or acceptable. Moreover, the choice between them should be based purely on pragmatic or instrumental considerations. Carnap's adoption of this logical pluralism is best expressed in his famous remark on the principle of tolerance: "*In logic, there are no morals*. Everyone is at liberty to build up his own logic, i.e. his own form of language, as he wishes. All that is required of him is that, if he wishes to discuss it, he must state his methods clearly, and give syntactical rules instead of philosophical arguments" (1934/1937: 52, orig. emphasis). The principle expressed here also presents a fundamental break with the *Tractatus*-style conception of logic. It is no longer the case that the nature and role of logic is ultimately grounded in metaphysical considerations concerning the relation between our language and the world. Instead, different logical systems can be adopted for different theoretical purposes and studied by metatheoretical means.

Given Carnap's new framework for the study of logic and mathematics, the question arises as to what residual role is assigned to logicism in *Logical Syntax* (see Friedman 1999 and the papers in Wagner 2009). Frege's and Russell's original program of reducing mathematics to higher-order logic is discussed in §84 of the book. However, Carnap clearly does not ascribe to it the importance he did in his pre-*Syntax* work. In fact, the study of the logical syntax of formal languages is viewed here as a way to reconcile logicism with other foundational views, in particular Hilbert's program. As a consequence, the particular understanding of the classical logicist project is significantly changed.

Notice first that Carnap is still working with a simplified version of Russell's logical theory of types in 1934. Type theory is expressed here in language L_{II} which, in contrast to his previous work on general axiomatics, is explicitly treated as an object language now. Surprisingly, L_{II} is no longer considered for the logical reduction of arithmetic. This is due to the fact that all arithmetical terms are already contained as primitive signs in the language. Moreover, the axioms of Peano arithmetic are not supposed to be deducible from the logical principles of type theory here, but they already belong to the axiom base of the logical system (1934/1937: §30). Thus, Peano arithmetic is no longer interpreted in type theory; it is now taken to be a part of it.

Given Carnap's account of "logico-mathematical systems," classical logicism obviously becomes irrelevant in this context. The logicist project of reducing mathematics to logic is replaced here by the more general project of showing that both logic and mathematics can be characterized as analytic in Carnap's new sense of the term.

A second issue that distinguishes Carnap's approach in *Logical Syntax* from classical logicism relates to his logical pluralism. We have seen that the definition of the notion of analyticity given in 1934 is a relative one: analytic truth, understood as an explication of logical or tautological truth, is specified relative to a particular language or logical system. Which mathematical statements count as analytic therefore depends on the prior choice of a logical system with its formation and transformation rules. For instance, statements of classical analysis will turn out as analytic relative to the type theoretic system L_{II}, but not analytic relative to the weaker language L_{I} of primitive recursive arithmetic also discussed in the book. Since one is free to choose between such logical frameworks, it follows that whether certain branches of mathematics count as logical also becomes a question of pragmatic choice. Thus, given Carnap's new principle of logical tolerance, the logicist reduction of mathematics to logic is no longer "a question of philosophical significance, but only one of technical expedience" (ibid.: §84). To show that the logicist thesis holds, it suffices to adopt a sufficiently strong background system that either (1) contains the mathematical axioms and primitive terms in question; or (2) that allows one to deduce these axioms in terms of sufficiently strong transformation rules.

Acknowledgement

Sections 1-3 of the present entry are based on Schiemer 2022. Research on this project received funding from the European Research Council (ERC) under the European Union's Horizon 2020 research and innovation program (grant agreement No. 715222).

References

Awodey, S. and Carus, A. (2001) "Carnap, Completeness, and Categoricity: The *Gabelbarkeitssatz* of 1928," *Erkenntnis* 54: 145–72.

Carnap, R. (1927) "Eigentliche und uneigentliche Begriffe," *Symposion* 1: 355–74.

——— (1929) *Abriss der Logistik*, Vienna: Springer.

——— (1931) "Die logizistische Grundlegung der Mathematik," *Erkenntnis*, 2: 91–105. Trans. "The Logicist Foundations of Mathematics," in P. Benacerraf and H. Putnam (eds.), *The Philosophy of Mathematics*, Englewood Cliffs: Prentice-Hall, 1964, 2nd ed., 1983, pp. 41–52.

——— (1934) *Logische Syntax der Sprache*, Vienna: Springer. Rev. ed. trans. *The Logical Syntax of Language*, London: Kegan Paul, Trench, Trubner & Cie, 1937, repr. Chicago: Open Court, 2002.

——— (2000) *Untersuchungen zur allgemeinen Axiomatik*, Darmstadt: Wissenschaftliche Buchgesellschaft.

Coffa, A. (1981) "Kant and Russell," *Synthese* 46: 247–63.

Ferreirós, J. (1999) *Labyrinth of Thought—A History of Set Theory and its Role in Modern Mathematics*, Basel: Birkhäuser.

Frege, G. (1893) *Grundgesetze der Arithmetik. I. Band*, Hildesheim: Georg Olms Verlagsbuchhandlung. Trans. *The Basic Laws of Arithmetic*, Oxford: Oxford University Press, 2013.

Friedman, M. (1999) "Tolerance and Analyticity in Carnap's Philosophy of Mathematics," in M. Friedman (ed.), *Reconsidering Logical Positivism*, Cambridge: Cambridge University Press, pp. 198–233.

Gandon, S. (2009) "Toward a Topic-Specific Logicism? Russell's Theory of Geometry in the *Principles of Mathematics*," *Philosophia Mathematica* 17: 35–72.

Hahn, H. (1929) "Empirismus, Mathematik, Logik," *Forschungen und Fortschritte* 5. Trans. "Empiricism, Mathematics, Logic," in Hahn, *Empiricism, Logic and Mathematics* (ed. by B. McGuinness), Dordrecht: Reidel, 1980, pp. 20–30.

——— (1933) *Logik, Mathematik, Naturerkennen*, Vienna: Gerold. Trans. "Logic, Mathematics and Knowledge of Nature," in B. McGuinness (ed.), *Unified Science*, Dordrecht: Reidel, 1987, pp. 24–45.

Hempel, G. (1945) "On the Nature of Mathematical Truth," *The American Mathematical Monthly* 52: 543–56.

Musgrave, A. (1977) "Logicism Revisited," *British Journal of Philosophy of Science* 28: 99–127.

Reck, E. (2004) "From Frege and Russell to Carnap: Logic and Logicism in the 1920s," in S. Awodey and C. Klein (eds.), *Carnap Brought Home: The View from Jena*, Chicago: Open Court, pp. 151–80.

Russell, B. (1903) *The Principles of Mathematics*, London: Routledge.

——— (1919) *Introduction to Mathematical Philosophy*, London: George Allen & Unwin.

Russell, B. and Whitehead, N. A. (1910–13) *Principia Mathematica*, Cambridge: Cambridge University Press, vol. 3, 2nd ed., 1927.

Schiemer, G. (2022) "Logicism in Logical Empiricism" in F. Boccuni and A. Sereni (eds.), *Origins and Varieties of Logicism*, Routledge, pp. 243–266.

Uebel, T. (2005) "Learning Logical Tolerance: Hans Hahn on the Foundations of Mathematics," *History and Philosophy of Logic* 26: 175–209.

Verein Ernst Mach (1929) *Wissenschaftliche Weltauffassung. Der Wiener Kreis*, Vienna: Wolf. Trans. "The Scientific Conception of the World. The Vienna Circle," in O. Neurath, *Empiricism and Sociology* (ed. by R. S. Cohen and M. Neurath), Dordrecht: Reidel, 1973, pp. 299–318; rev. trans. (with orig. annotated bibliography) "The Scientific World-Conception. The Vienna Circle," in F. Stadler and T. Uebel (eds.), *Wissenschaftliche Weltauffassung. Der Wiener Kreis. Hrsg. vom Verein Ernst Mach (1929)*, Vienna: Springer, 2012, pp. 75–116.

Wagner, P. (ed.) (2009) *Carnap's Logical Syntax of Language*, Basingstoke: Palgrave Macmillan.

Wittgenstein, L. (1922) *Tractatus Logico-Philosophicus*, bilingual ed. trans. by F. Ramsey and C. K. Ogden, London: Kegan Paul, Trench Trubner & Co., 1922, rev. ed. 1933, repr. London: Routledge, Kegan, Paul, 1983; trans. by D. F. Pears and B. F. McGuinness, London: Routledge, Keagan Paul, 1961, repr. 1974.

23
PROBABILITY IN LOGICAL EMPIRICISM

Marta Sznajder

While logical empiricism was not a single doctrine, what brought its representatives together was a commitment to empiricism, even if understood differently. Around 1929–30, their empiricism focused on a criterion that required that meaningful statements were only the ones that could in principle be verified by observation or deduction. A sentence is verifiable when there is, in principle, a procedure that would determine whether the sentence is true or false.

Sooner or later, anyone who submitted to this idea had to concede that strict verificationism is impossible to uphold: actual science uses terms that mix empirical with theoretical content, or terms that refer to dispositions, as well as universally quantified statements. Hence, this strict notion of verification had to be replaced with something that would reflect this more complicated reality: confirmation.

Confirmation and probability

Confirmation is a relation between evidence and a hypothesis. Through this concept, we recognize the evidential support that the evidence provides to the hypothesis without straightforwardly verifying or falsifying it. Observing more and more black ravens gives more weight to the hypothesis that all ravens are black—confirms it to higher and higher degree—but no finite number of observations can fully verify this sentence. The shift of focus towards confirmation emphasizes the essentially inductive character of scientific reasoning.

Just as in this raven example, confirmation is easily seen as a gradable notion. Degrees of confirmation can be thought of as probabilities: more or better evidence confirms a given hypothesis to a higher degree, which means making it more probable that the hypothesis is true. However, not everybody within the logical empiricism movement considered confirmation and probability to be so inextricably linked. For instance, Hempel explicitly detached confirmation from probability and focused on a qualitative concept of confirmation (see CH. 25). Carnap and Reichenbach, the main figures of the movement who engaged with the topic and whose proposals are the focus of this chapter, did explicate confirmation in terms of probability. Others, like Richard von Mises or Friedrich Waismann, while working with similar interpretations of probability as, respectively, Reichenbach and Carnap, did not model confirmation in terms of probability.

Connecting confirmation to probability does not automatically lead to a clear account of confirmation: one still needs a clear conception of probability. There were two main interpretations

DOI: 10.4324/9781315650647-26

of probability within the logical empiricism movement: the frequentist one, developed most intensively by Reichenbach, and the logical one, which Carnap focused on. The frequentist one derives probabilities from sequences of observations, and the logical one sees them as meaning relations between sentences.

Reichenbach's approach to the impossibility of full verification was a radical reconsidering of all semantics, disposing with the notion of truth altogether and replacing it with a probabilistic continuum of degrees of truth. He chose the underlying notion of probability to be the limiting relative frequency in infinite sequences of events. This choice was largely motivated by his work on causality and his rejection of the epistemic views on probability early on in his career. Probability was taken to be a property of sequences of events; it applied to the observed world and not to the language that the world was described in.

Carnap's approach was in line with the rest of his work at the time, with his focus on syntax and semantics of the scientific language. Carnap took confirmation to be a purely semantic notion, with sentences that state the degree to which one statement confirms another coming out as analytic. In this view, sentences about degrees of confirmation do not describe facts about the relations between observed events, but rather objective meaning relations between propositions. The notion of probability that Carnap chose with which to explicate this notion of confirmation was the logical probability.

Probability in the Vienna Circle

Those two main views came about in a specific historical context. Probability was an important topic in the Vienna Circle from its beginning. The main focus of their probability discussions at the time was the logical conception of probability, explored by Wittgenstein and Waismann. Wittgenstein's brief remarks on probability are located in the proposition 5.15 of the *Tractatus*. There, he sketched a simple picture of conditional probability as a relation between ranges of propositions: the proportion of the number of cases (worlds, states) that make the propositions true. In 1929, Waismann gave a few talks about probability to Schlick's Circle. His work on the topic was an elaboration and elucidation of Wittgenstein's basic idea. At the same time, Waismann did not completely shun the frequentist conception, but rather called for a future account of logical probability that will also clarify its relationship with the frequentist one (1930).

In the spring of 1929, Eino Kaila visited Vienna (see CH. 33). Kaila and Carnap met a number of times during that visit to discuss the *Aufbau*. An important part of Kaila's critique of Carnap's book was the lack of any treatment of probability. The criticism did push Carnap to consider the issue more seriously, which eventually led to "Testability and Meaning." While Carnap introduced the concept of confirmation there, he did not yet draw the full connection between confirmation and probability. Instead, he wrote that he considered it as rather impossible to explicate degree of confirmation as "the degree of probability in the strict sense which this concept has in the calculus of probability, i.e. as the limit of relative frequency" (1936–7: 427). It was only at the beginning of the 1940s that he started to work on explicating degrees of confirmation using the concept of probability, but this time choosing the logical rather than the frequentist conception.

While Carnap was almost coerced to working on confirmation and probability—by Kaila's criticisms, as well as the very nature of his interest in the language of science—Reichenbach's way to it was more straightforward. He was interested in the topic from the beginning of his career, with his doctoral dissertation being on the use of the concept of probability in scientific descriptions of the world (1915). When it came to his own explication of probability, his main influence was his Berlin colleague Richard von Mises, who developed a strictly frequentist conception of probability based on random sequences.

Besides Waismann, Carnap, and Reichenbach, other members of the movement spent parts or the whole of their careers working on confirmation and probability. Herbert Feigl's 1927 doctoral dissertation investigated the role of induction and probability in science. He argued that inductive reasoning cannot be shown to be valid, but should rather be sought to be pragmatically vindicated. Janina Hosiasson made a number of contributions to confirmation theory and the logic of inductive reasoning (1931, 1940), as well as a criticism of Reichenbach's probability logic (1936). Ernest Nagel also criticized Reichenbach's conception and argued for a truth-frequency theory of probability, focused on sequences of sentences rather than events (1939).

Two concepts of probability

While there were some efforts to offer alternatives to the two dominant positions, as well as reconciliations, in practice most of the work on probability and confirmation within the logical empiricism movement revolved around the frequentist and the logical conceptions of probability. Extreme subjectivism in Ramsey's and de Finetti's style, while developed roughly at the same time, did not properly enter those discussions until the 1950s, when it was popularized by Leonard Savage.

This dualism of conceptions was summarized by Carnap in his distinction between the two concepts of probability: probability$_1$ and probability$_2$. Probability$_2$ is the physical probability: probability manifested in sequences of observations and explicated using the frequency conception. Probability$_1$ is the logical, semantic one, explicated as the degree of confirmation. Sentences about probability$_1$ are analytic and express a logical relation between evidence and hypothesis. (However, one needs to tread with caution here when it comes to this objective, analytic character of the concept: this is post-*Syntax* Carnap, who assumes implicit relativization to a conceptual framework.) Later on, Carnap elaborated on this idea, writing that statements about probability$_2$ values occur within science, and statements about probability$_1$, belong to inductive logic, which provides rules for operations on the statements within science (1953: 192).

According to Carnap, the two concepts are not incompatible. It is a mistake to insist that only one of them is the correct probability concept, trying to dismiss or reduce the other interpretation. Both concepts—both explicanda—are legitimate objects of formal explication efforts, simply used in different contexts and for different purposes. Reichenbach's stance on that distinction was very different. He maintained that in every context where there is talk of probability, the frequency interpretation can be used. It was part of his program to show how this can be done in the epistemic context.

In spite of this "monist" attitude of Reichenbach, Carnap was positive about his work, although they did not interact on the topic a lot. By the time Carnap was working on probability full time, Reichenbach's theory was fully developed. Moreover, Reichenbach's premature death in the early 1950s prevented him from seeing Carnap's later work.

Carnap's probability: the logical interpretation

In "Testability and Meaning," Carnap was still skeptical about the possibility of having a quantitative explication of the concept of confirmation. In the five years following that paper, he seems to have changed his mind. In his diaries from the time, there are multiple reports of working on "weight" and "confirmation." Inspired by the Vienna Circle discussions, in early 1941 he was led to re-examine the work of other logical probability authors like J. M. Keynes and H. Jeffreys. Soon, an idea for a large monograph was formed, which was published in 1950 as *Logical Foundations of Probability*. From then on, Carnap spent almost all of his time on developing his inductive logic; that is, the theory of logical probability.

The probability that Carnap aims to explicate in *Foundations* is probability$_1$: the objective, semantic concept of quantitative confirmation. He proceeds according to his own explication procedure (1950: 3–8). First, one specifies the explicandum, choosing those meanings or uses of the original concept that will be the target of the explication. Only when that is done is the formal explicatum constructed. The descriptions of the explicandum that Carnap initially focused on were probability as: measure of evidential support, fair betting quotient, and estimate of relative frequency (i.e., estimate of probability$_2$). Carnapian inductive logic was a normative project, aimed to provide standards of rationality for beliefs which are influenced by evidence.

Carnap's formal explicata are the confirmation functions, or c-functions. They are functions of pairs of sentences, one representing the available evidence, and the other the hypothesis. The value of the c-function for a given pair of evidence and hypothesis statements represents the degree to which the evidence confirms the hypothesis. The formal languages that the c-functions are defined on are first-order languages with unary predicates, which restricted the applicability of inductive logic.

The c-functions are conditional probability functions, defined in the standard way on the basis of unconditional probability. The latter is in turn defined as a measure on the space of models for the language. However, the axioms of probability calculus alone do not provide specific numerical values for contingent propositions: any such sentence can in principle have any probability assigned to it by a particular function. Carnap considered the resulting theory too weak to be useful as a theory of confirmation: for successful applications, it had to offer the scientist specific confirmation, workable values for the kind of statements that she would be interested in, i.e., empirical rather than purely logical ones.

Hence, Carnap proposed further axioms, regularity and symmetry. Regularity ensures, roughly, that logically possible sentences are assigned positive probability. Symmetry requires a priori probabilities for every atomic sentence to be equal; it is a version of the infamous principle of indifference, which says that in absence of reasons to do otherwise, one should assume equal probabilities for all possible outcomes. It was Carnap's continuous reliance on versions of this principle—and his insistence that they were rationally required—that earned him the bulk of the criticism that his inductive logic received over the years. He continued to defend his use of the principle as applicable in the kind of situation he was modeling, i.e., under the assumption of the lack of any knowledge beyond the knowledge of the structure of the object language. Any further updates from this initial position of complete ignorance were to be made by conditionalizing on all of the information received. The latter condition is known as the requirement of total evidence, which was often criticized as unrealistic.

After *Foundations*, Carnap focused on the search for axiomatic representations of further classes of rationally admissible c-functions, with the additional axioms justified by considerations concerning the rationality of inductive reasoning. At first, the class of admissible confirmation functions was parametrized using a single real-valued parameter λ (1952). The λ expresses the rate at which a c-function is influenced by the observations, as opposed to being tied to the a priori assumptions about the possible observations. In the limit, with more and more observations, the confirmation values for observed properties converge to the empirical frequencies.

The last stage of the development of Carnap's inductive logic was the *Basic System of Inductive Logic* (1971a, 1980). Instead of proposing a single parametrized family, Carnap considered a range of possibilities in terms of new axioms and parameters. He relaxed the previous symmetry requirement, allowing for the prior probabilities to not be distributed evenly across all predicates of the language. He also explored ways to formalize possible statistical dependencies between different predicates.

Another development within the inductive logic program which sparked a lot of debate, was the bringing in of a new description of probability$_1$ in terms of rational decision making. The new description was of probability$_1$ as the rational probability value to be used when calculating rational expected utility of an action; Carnap eventually dropped the confirmation theory angle entirely. This change was likely brought in under the influence of John Kemeny, who was Carnap's close collaborator in the early 1950s and introduced him to the developments in the subjective epistemic probability interpretation.

Carnapian inductive logic received a lot of criticism over the years. At first it was because Carnap was mistakenly thought to have endorsed a single rational confirmation function, which was perceived as far too strict and aprioristic. Additionally, his continuing reliance on different versions of the principle of indifference was never accepted. Finally, as Carnap investigated wider and wider classes of admissible confirmation functions, his project was interpreted as having moved from explicating an objective to a subjective concept of probability—with Carnap himself consistently denying this interpretation, insisting that nothing that he ever said made his concept of probability "subjectivistic" (see, e.g., 1963: 972).

Reichenbach's probability: the frequency interpretation

After having rejected both the subjective and objective epistemic interpretations of probability in his doctoral dissertation, Reichenbach turned to an account informed primarily by scientific practice. His first book-length treatment of probability was *Wahrscheinlichkeitslehre* (1935; revised for the English translation of 1949). It was not received uncritically, with prominent figures like C. I. Lewis and Russell expressing serious disagreement. However, the basic conceptual features of his interpretation of probability, as well as his pragmatic justification of induction, have secured a lasting place in the history of philosophy and shaped many of the subsequent discussions.

Reichenbach's approach to probability is expressly empirical. Instead of coming up with rationally justified or self-evident first principles of confirmation, we are to look at our most successful inductive practices and make their methods explicit, by formalizing their assumptions into an axiomatic system. Such an approach cohered with his account of deductive logic, which was a formalist one: he saw logic as an axiomatic calculus, with the degree to which the axioms correspond to the features of the real world as a matter of "coordination," which was not an a priori issue. Reichenbach insisted that in every context where the concept of probability is used, it can be modeled using his frequency conception; this was to be true also of epistemic contexts, or contexts in which we talk about probabilities of general sentences, like the scientific laws.

As an objective feature of the world, probability is considered by Reichenbach to be a property of sequences of events: the probability of an event is the limiting relative frequency of events of its type in an infinite sequence of events. While this formulation is simple, it is a challenge to clearly spell out the details. First of all, what kind of infinite sequences of events are ones that determine probabilities? They cannot be just any sequences, but they have to be random in some way. Compare the following sequence of coin tosses: H(eads)T(ails)HTHTHT. . ., where the relative frequency of each outcome approaches 0.5, but intuitively, the outcome of the next toss becomes almost certain as the sequence grows. Reichenbach, however, was skeptical about the possibility of a successful definition of truly random sequences, something which von Mises aspired to. Hence, Reichenbach restricted himself to so-called normal sequences of events, which is a weaker concept than random sequences.

Regardless of their exact specification, in real life we do not have access to infinite sequences of observed events. This means that the probability values used in practice must be approximations, or estimations, of the ones provided by idealized, infinite sequences of events. The method of approximation that Reichenbach proposes is the Straight Rule, which says that as the probability value, one should take the empirical frequency in the actually observed sequence of events. The Straight Rule provides us with workable basic probability values for contingent statements, in the same way the symmetry axiom gave Carnap some basic probability values to start with, so to speak.

The Straight Rule needed justification. Reichenbach's argument for it was that it converges to the limiting relative frequency in the long run, i.e., that it gives values that can be arbitrarily close to the actual probability, as the number of observations increases (provided that there is a limit). However, not only does this hold for many methods other than the Straight Rule, but it does not guarantee uniform convergence, which in turn means that in any particular case we cannot know how many observations (at most) it will take to get close to the limiting frequency.

Finally, there is the issue of what events count as events of the same type: this is the reference class problem. This problem comes up especially strongly when it comes to calculating probabilities of one-off, or singular, events, for which there is no natural sequence of previous observations, like the death of a particular person or the occurrence of a previously unknown disease. According to Reichenbach, probabilities are assigned to such events in an extended sense, as posits. We are to choose the narrowest reference class of events similar to the one in question, for which we have stable statistics, and posit the frequency in that class as the one-off probability. For instance, we can assign the person in question the class of people of the same age who suffer from the same diseases, and check the death rates in that class. This simple idea, however, is problematically circular: the availability of "stable statistics" presupposes an inductive procedure to have been there already in the first place.

Posits are divided into blind and appraised ones, depending on whether there was any data to go by—in the form of statistics concerning an adequate reference class—when deciding on the value for a posit. When more information becomes available, blind posits become appraised through integrating this information into the new probability estimation. The same kind of dynamics occurs when we move from primitive to appraised knowledge, as more and longer sequences of observations become available as bases for probability estimates. Hence, the knowledge of probabilities for Reichenbach is deeply rooted in the available experience, and grows together with it.

As Reichenbach took the idea that no empirical statements can be fully verified or falsified to its logical end, he disposed of the classical notion of truth. As a result, the binary truth-values were replaced by a range of values, and the new probability logic was continuum-valued. Under the frequency interpretation, probabilities are assigned to sequences of events. However, the probability logic is defined on a formal language, which requires a propositional representation of the sequences of events. This is achieved by creating sequences of propositions describing the events in the sequence, creating an isomorphic sequence within the language.

Reichenbach's probability logic is the logic of those sequences of propositions. It is based on classical deductive logic, with additional axioms that capture the idea that probabilities are properties of infinite sequences. These axioms were: univocality, normalization, addition, and multiplication. They form, in Reichenbach's own language, an axiomatization of finitely additive probability. The final element is the Straight Rule, or rule of induction, which allows for the derivation of specific numerical values for probabilities of contingent propositions. The exact details of the syntax and semantics of the actual probability logic were left unclear or underdeveloped in many places, which makes it hard to evaluate the extent to which Reichenbach achieved formally what he was aiming at conceptually.

Apriorism, theories, induction

Reichenbach distanced himself explicitly from any kind of apriorism, both in logic and in epistemology. Hence, in his conception, specific degrees of probability can be known only a posteriori, based on enumerative induction on observations, formulated as the Straight Rule. Carnap's logical probability, on the other hand, was explicitly aprioristic, as he sought to explicate degree of confirmation as a purely semantic relation between propositions. The justification that Carnap offered for those of his axioms that extended the basic probability calculus was based on normative considerations of rationality. These two conceptions of probability lead to different stances on important topics within epistemology and philosophy of science, just as they emerge from different conceptions of semantics or logic.

The problem of single case probabilities arises differently for the two interpretations. For frequentism, this issue is immediate, since probability itself is defined as a property of sequences of events. I explained earlier how Reichenbach's solution to assign probability to singular events in a special, extended way led to the reference class problem. Carnap did not address this problem explicitly, because it did not arise so starkly for his approach. His inductive logic was syntax-based in the sense that (with some exceptions in the Basic System where he considers partially interpreted languages) he did not focus at all on what the basic predicates of the object language were supposed to mean, and whether the individual constants referred to any specific individuals that did not belong to any more general types. His inductive logic provided a priori probability values for any kind of event, as long as it could be described using one of the predicates of the object language, regardless of how singular the events could have been.

Assigning confirmation values to scientific theories, as opposed to simple observation statements, poses a significant challenge for both conceptions, albeit for different reasons. In Carnap's inductive logic, universally quantified sentences always have zero probability, which is a straightforward consequence of his measure-theoretic approach to unconditional probabilities. This means that, under his interpretation, no scientific theory could be confirmed to a positive degree. He chose to bite the bullet on this issue and to focus on one-step predictive probability: confirmation functions tell us the probability of the next observed object being of a certain kind, rather than specifying the probability for any object being of that kind. He argued that in ordinary discourse, when confirmation of theories is discussed, it is meant that they hold only of a finite, rather than infinite, number of instances (1950, §110G).

When it comes to the impact that the observations have on scientific theories, or hypotheses, Reichenbach's view was Bayesian: the a posteriori probabilities of theories were to be calculated according to the Bayes formula. The prior probabilities were calculated objectively from the empirical frequencies using the Straight Rule. However, under the frequency interpretation, it only makes sense to ascribe probability values when there is an appropriate sequence of events available. This means that to assign probabilities, or degrees of confirmation, to scientific theories, one needs some sort of sequence of theories "of the same kind," and an interpretation of how the truth frequency in such a sequence would be determined. Reichenbach suggested a solution along those lines, and the exact details were worked out by Wesley Salmon (1967). The kind of procedure envisioned by Reichenbach turned out to be essentially the same as the modern hierarchical Bayesian picture.

A full theory of confirmation and the effect that evidence has on theories requires a general justification of inductive reasoning, explaining why observations can be rationally expected to have a predictive value over the future ones. As the answer to that problem, Reichenbach offered his pragmatist vindication of induction in terms of success: if the world is such that any prediction method can succeed in it, then his inductive method will be successful (see CH. 24).

Carnap's relationship with this problem was more complicated. In *Foundations* (1950: §41f), he offered an unsatisfactory treatment of the problem of induction, claiming that in his inductive logic one can analytically prove the sentence that states that conditional on our experience, the degree of uniformity of the world is high (which is a presupposition of induction). After being repeatedly criticized on this, he continued working on a better answer, which was never published (for background details, see Carus 2017).

Later, he argued that justifying the inductive method reduces to the question of justification of the axioms for confirmation functions, which is a higher-level issue than choosing a particular function from the set of admissible confirmation functions. In the 1960s Carnap offered two kind of answers to that problem. On the one hand, he attempted to justify his axioms in terms of rationality considerations about decisions that the beliefs following those axioms would lead to (1971b). On the other hand, he made what detractors considered a controversial claim about inductive intuition: that it allows us to directly "see" the correctness of the basic axioms of inductive logic in the same way that we "see" the correctness of the basic axioms of deductive logic (1968). Defenders argue that no special faculty was meant to be invoked (Wagner 2011).

After the big two

The philosophical study of probability is currently dominated by subjective Bayesianism, which is somewhat distant from Carnap's more aprioristic approach focused on finding more and more formal constraints on rational credences. Carnap's program was continued after his death, among others by his long-time collaborator Richard Jeffrey (1973) (see also Zabell 2011). Carnapian confirmation functions were also shown by Brian Skyrms to be natural counterparts of certain classes of Bayesian priors (1996). Currently, the inductive logic program is also continued under the label of objective Bayesianism (Williamson 2016); there is also work done in the pure inductive logic tradition, which focuses on extending Carnap's functions to richer, but still uninterpreted, languages (Paris and Vencovská 2015).

Reichenbach should be given credit for integrating the frequency conception of probability into a full philosophical system and constructing a complete epistemology that took seriously the idea that all knowledge is eventually probabilistic (Eberhardt and Glymour 2011). Several of Reichenbach's own ideas were elaborated on by his students like Salmon (1967), but his probability logic did not enjoy the kind of following that Carnap's did, in terms of an organized research program continuing for decades afterwards.

References

Carnap, R. (1936–7) "Testability and Meaning," *Philosophy of Science* 3: 419–71 and 4: 1–40.

——— (1950) *Logical Foundations of Probability*, Chicago: University of Chicago Press, 2nd ed., 1962.

——— (1952) *The Continuum of Inductive Methods*, Chicago: University of Chicago Press.

——— (1953) "Inductive Logic and Science," *Proceedings of the American Academy of Arts and Sciences* 80: 189–97.

——— (1963) "Replies and Systematic Expositions," in P. A. Schilpp (ed.), *The Philosophy of Rudolf Carnap*, La Salle, IL: Open Court, pp. 859–1012.

——— (1968) "Inductive Logic and Inductive Intuition," in I. Lakatos (ed.), *The Problem of Inductive Logic*, Amsterdam: North-Holland, pp. 258–67.

——— (1971a) "A Basic System of Inductive Logic, Part I," in Carnap and Jeffrey (1971), pp. 33–165.

——— (1971b) "Inductive Logic and Rational Decisions," in Carnap and Jeffrey (1971), pp. 5–31.

——— (1980) "A Basic System of Inductive Logic, Part II," in R. Jeffrey (ed.), *Studies in Inductive Logic and Probability*, Berkeley: University of California Press, vol. 2, pp. 7–155.

Carnap, R. and Jeffrey, R. C. (eds.) (1971) *Studies in Inductive Logic and Probability*, Berkeley: University of California Press, vol. 1.

Carus, A. W. (2017) "Carnapian Rationality," *Synthese* 194: 163–84.

Eberhardt, F. and Glymour, C. (2011) "Hans Reichenbach's Probability Logic," in Gabbay, Woods and Hartmann (2011), pp. 356–89.

Feigl, H. (1927) "Zufall und Gesetz," in *Wissenschaftlicher Jahresbericht der Philosophischen Gesellschaft zu Wien*, Wien: Verlag der Philosophischen Gesellschaft an der Universität zu Wien. Repr. in R. Haller and T. Binder (eds.), *Zufall und Gesetz. Drei Dissertationen unter Moritz Schlick: F. Feigl—M. Natkin—Tscha Hung*, Amsterdam: Rodopi, 1999, pp. 2–192.

Gabbay, D. M., Woods, J. and Hartmann, S. (eds.) (2011) *Handbook of the History of Logic,* Amsterdam: Elsevier, vol. 10.

Hosiasson, J. (1931) "Why Do We Prefer Probabilities Relative to Many Data?" *Mind* 40: 23–36.

——— (1936) "La théorie des probabilités est-elle une logique généralisée? Analyse critique," in *Actes du Congrès International de Philosophie Scientifique, Facs. IV Induction et probabilité*, Paris: Hermann & Cie, pp. 58–64.

Hosiasson-Lindenbaum, J. (1940) "On Confirmation," *The Journal of Symbolic Logic* 5: 133–48.

Jeffrey, R. C. (1973) "Carnap's Inductive Logic," *Synthese* 25: 299–306.

Nagel, E. (1939) *Principles of the Theory of Probability*, Chicago: The University of Chicago Press.

Paris, J. and Vencovská, A. (2015) *Pure Inductive Logic*, Cambridge: Cambridge University Press.

Reichenbach, H. (1915) *Der Begriff der Wahrscheinlichkeit für die mathematische Darstellung der Wirklichkeit*, Leipzig: Barth. Trans. *The Concept of Probability in the Mathematical Representation of Reality*, Chicago: Open Court, 2008.

——— (1935) *Wahrscheinlichkeitslehre: Eine Untersuchung über die Logischen und Mathematischen Grundlagen der Wahrscheinlichkeitsrechnung*, Leiden: Sijthoff. Trans. and rev. *The Theory of Probability: An Inquiry into the Logical and Mathematical Foundations of the Calculus of Probability*, Berkeley: University of California Press, 1949.

Salmon, W. (1967) *The Foundations of Scientific Inference*, Pittsburgh: University of Pittsburgh Press.

Skyrms, B. (1996) "Carnapian Inductive Logic and Bayesian Statistics," *Statistics, Probability and Game Theory*, IMS Lecture Notes—Monograph Series, 30: 321–36.

Wagner, P. (2011) "Carnap's Theories of Confirmation," in D. Dieks et al. (eds.), *Explanation, Prediction, and Confirmation*, Dordrecht: Springer, pp. 477–86.

Waismann, F. (1930) "Logische Analyse des Wahrscheinlichkeitsbegriffs," *Erkenntnis* 1: 228–48. Trans. "A Logical Analysis of the Concept of Probability," in Waismann, *Philosophical Papers* (ed. by B. McGuinness), Dordrecht: Reidel, 1977, pp. 4–21.

Williamson, J. (2016) *Lectures on Inductive Logic*, Oxford: Oxford University Press.

Zabell, S. L. (2011) "Carnap and the Logic of Inductive Inference," in Gabbay, Woods and Hartmann (2011), pp. 265–309.

24

REICHENBACH AND THE PROBLEM OF INDUCTION

Flavia Padovani

Few philosophical problems occupied Reichenbach throughout his entire career as much as the problem of the justification of induction. How can we avoid the potential circularity of attempts that end up involving inductive reasoning—the difficulty that lay at the core of Hume's skeptical reasoning? We commonly use induction when we state the relationship between observed and unobserved regularities to obtain general laws, or when we predict future events on the basis of observed ones. In both cases, the inferences use probability, and this makes the justification dependent upon probability and its interpretation. As is well known, Reichenbach conceives of the probability of an event as the limit of the relative frequency of events of the same type. The underlying idea is that if we want to grant our knowledge of the physical world an *objective* value, we have to provide an objective interpretation of probability. This is the reason why Reichenbach repeatedly opposed any logical or subjective probability interpretation.

The interpretation of probability loomed large in Reichenbach's philosophy from his dissertation onwards (1915). He also always emphasized the tight connection of epistemology with the practice of science and its methodology. In his early writings, he expressed this by means of arguments about principles regarded as necessary preconditions of any empirical knowledge, in a Kantian framework. He considered the principle of induction to be on a par with those principles, thus having an a priori nature. In his later writings, the principle of induction lost its aprioricity and was transformed into a pragmatic one. This marked Reichenbach's shift to what he called "a consequent empiricism" (1936: 33), i.e., the empiricism resulting once a solution to the problem of induction was found.

The solution to this problem envisaged by Reichenbach centers around the following idea. Clearly, one cannot justify the inductive inference by relying on a factual proposition appealing to the principle of uniformity of nature (as Hume showed). Neither can one justify induction based on its reliability a priori, as Reichenbach himself tried to do in his doctoral thesis. Yet, even if the reliability of induction cannot be proved, one cannot operate without using induction. So, Reichenbach's reasoning goes, if nature is uniform, induction will be certainly successful, while if nature is not uniform, no other method would be successful in any case. Therefore, the most favorable method we have is induction, which justifies its use. This is generally known as Reichenbach's "vindication of induction" (Feigl 1950: 212).

 DOI: 10.4324/9781315650647-27

Even granting the overall wisdom of this pragmatic defense, Reichenbach's optimality argument has largely been considered insufficient (Salmon 1991: 100ff.; Niiniluoto 1998: 78ff.; Eberhardt and Glymour 2011). However, it has been recently revisited and its spirit revived (see Schurz [2019] and literature therein, as well as Beisbart [forthcoming] for its critical discussion). Here I will reconstruct the salient moments of Reichenbach's attempts to provide a solution to this problem, from his early Kantian interpretation to his later pragmatic justification.

Reichenbach's early conception

Already Reichenbach's doctoral dissertation shows how his philosophical approach arose from concern with scientific methodology, although it contains various (pseudo-)Kantian elements (Eberhardt 2011; Padovani 2011) in support of its core argument for the necessity of introducing a "principle of probability" to complement the principle of causality. In its traditional sense, the principle of causality appears to Reichenbach insufficient to provide a satisfactory (mathematical) representation of reality. In scientific practice, we use two distinct types of judgment: mathematical and physical. These two types differ essentially because in the case of the latter, we need to employ a notion of approximation that enables us to apply (that is, to "coordinate") mathematical judgements to reality, turning them into physical judgments. The possibility to use approximation in this framework relies on what Reichenbach calls the "principle of lawful distribution of values," which is supposed to guarantee, with certainty, that the frequency distribution converges to a limit—a condition without which we would not be able to conceive of any physical law. In Reichenbach's dissertation, this principle is identified with the "principle of probability" and justified via a so-called "transcendental deduction" (1915/2008: 105ff.), with all the problematic consequences that derive from it. Incidentally, in his dissertation the frequency interpretation of probability is merely suggested: the explicit equation of probability statements with limiting frequencies begins only with his "The Physical Presuppositions of the Calculus of Probability" (1920a/1978).

In autobiographical notes from 1927 (Eberhardt 2011: 135), Reichenbach sketched what he considered to be the four main results of his dissertation. Two of them were that the probability claim could be reduced to a claim of certainty, and that the principle of lawful distribution was to be considered synthetic a priori. This meant that to be justified, any probability claim should lead to conclusions considered to be reliable *with certainty*, and that the firm foundation of this "certainty" was rooted in synthetic a priori reasoning. By the end of the 1920s, Reichenbach rejected those claims. With the "comfortable resource" of synthetic a priori reasoning no longer available (1934/35: 369), he was forced to ground the reliability of probability claims differently. Admittedly, convergence could be guaranteed only with probability. However, he still regarded the other two dissertation results as successful: that for a certain class of problems, "the assumption of equiprobability can be reduced to a continuity assumption" (crucial in avoiding any appeal to subjective considerations such as the principle of indifference), and that the continuity assumption is presupposed by any physical (and thus also causal) claim. Reichenbach considered the second result to be his own discovery—and the most significant contribution to the problem of probability since Hume. Even in his later writings, he emphasized that this result could be detached from the 1915 Kantian backdrop against which it was conceived, and incorporated it into his mature view, stating that "the application of the calculus of probability contains no assumptions different from those presupposed in the application of the principle of causality" (1949: 355).

Normal induction, coordination, and simplicity: Reichenbach's approach in the 1920s

In the early 1920s, Reichenbach reflected on the status of cognitive principles in scientific theory change, principles whose traditional (Kantian) interpretation had been challenged by the theory of relativity. Reichenbach addressed this issue in his *The Theory of Relativity and A Priori Knowledge* of 1920 with an account of cognition as coordination between a set of formal statements (the mathematical machinery of a theory) and what this set is supposed to represent in reality. In Reichenbach's view, there are specific theory-relative principles that prescribe the conditions of applicability of a theory to reality, but these are not fixed once and for all. Rather, they are subject to revision if contradictions among different sets of principles emerge in the development of science (1920b/1965: sects. II and III). The distinctive function of these principles is to ground objective knowledge by providing the conditions for univocal coordination (i.e., true, objective knowledge). To be accessible in a law-like structure, the manifold of imperfect sense perception must be translated through measurement and cast into formal language. For this, we need to be able to make extrapolations of observational data. The selection of the most probable extrapolations can be carried out using the "principle of normal induction," which is independent of experience and in this framework still has an a priori (albeit contingent) character (ibid.: 14–15 and 44ff.). Ultimately, every empirical statement, every empirical law will imply the principle of induction, which ensures that it would be the last one to be discarded in any conflict with experience (ibid.: 67).

In *Axiomatization of the Theory of Relativity* (1924), Reichenbach provides another glimpse into his new epistemological approach, progressively detached from his initial Kantian leanings. Some of the principles that he originally considered as synthetic a priori now appear under the heading of "epistemological principles." In discussing the necessary presuppositions of experimental work, he emphasizes, for instance, the "assumption that the experiment, if subsequently repeated, will always yield the same result," and that an empirical law is the result of connecting "certain values of measurement . . . by a simple curve," which implies the use of "the principle of induction" (1924/1969: 5).

Reichenbach does not discuss the justification of this fundamental principle until his clear rejection of its synthetic a priori status in 1925. The principle of induction implies an assertion about perceptions, but the content of perceptions is independent of the subject, so probability can by no means be derived from a subjective source. The principle of probability does state that the content of perceptions can be assigned a specific statistical regularity. In this sense, probability appears to be an intrinsic feature of the world, hence a "metaphysical axiom," as he puts it, capturing the belief in the uniformity of nature as well as the belief in the existence of the external world. Although the validity of both beliefs can be questioned, they are equally essential as we cannot do without them (1925a/1978: 292ff.).

The same question reverberates in "The Causal Structure of the World" of 1925, in which Reichenbach develops a "topology of probability implication" grounded in an analysis of interacting causal chains (Padovani 2013). In Reichenbach's view, we use probability inferences, both when we establish with probability general statements based on a finite number of observations, and when we ascribe a probability to an individual event and from this infer the probability of its respective occurrence in a finite number of cases. Anticipating potential objections to the lack of justification for this inference, Reichenbach reiterates that it is meaningful because "it is also invariably made in science and in daily life" but does not defend it further (1925b/1978: 90). In 1929, he will eventually refer to it as an axiomatic assumption, acknowledging that the concept of probability is a "basic logical concept, the meaning of which we accept as given axiomatically" (1929/1978: 151).

In a paper on the principle of causality and its various meanings (1932, written in 1923) and in other writings of the period, including (1924) and (1929), Reichenbach introduces a crucial distinction between descriptive and inductive simplicity. While the first is a principle of economy, a property of the description alone and not of the objects described by a law, the second involves the objects under investigation. In the case of inductive simplicity, the hypothesis is that if "there exists a random selection of measured values, it is improbable that it represents an exceptional selection" and that "for a given finite number of measured points, it is the simplest curve upon which future measured values will lie" (1932: Sect. 5). So, for Reichenbach, the simplest description is the most probable (1929: Sect. 11). As we shall see, this distinction will play an important role when Reichenbach motivates the selection of his so-called straight rule based on descriptive simplicity (1949: 447).

At the end of this decade, Reichenbach still struggles with the justification of induction and recognizes that even though relying on probability and induction is a condition of knowledge of nature, "we have no means of proving that natural scientific knowledge, even if possible heretofore, will always be possible" (1929/1978: 193).

Reichenbach's attempt to justify induction: from metaphysics to pragmatism

In the 1930s, Reichenbach devotes several parts of his writings to the nature and the centrality of induction, culminating in two major publications on the topic, *The Theory of Probability* (1935a, translated and revised in 1949) and *Experience and Prediction* (1938a). In "Causality and Probability" (1930), he considers the possibility that induction is either a convention, or some sort of practical, extra-scientific rule, only to dismiss both options as inadmissible. If the principle were a convention, and thus some sort of "ordering principle," we would not be able to explain why, in order to enable predictions, science reasonably and consistently chooses the simplest curve connecting measured points, assuming that this will approximate the outcomes of future measurements. But this principle also cannot be extra-scientific, for it represents the only tool we have to differentiate between empirically well-grounded theories and arbitrary interpretations that do not properly apply to the physical world. In contrast to his previous conception of truth entailed by univocal coordination, relative to a given stage of knowledge, Reichenbach now abandons the bivalence of classical logic in favor of a continuous scale of probability, adding that, since any prediction presupposes the principle of induction, the probability inference would have no meaning without it. For him, probability statements simply cannot be interpreted within a two-valued logic; hence, "it is not possible to justify the system of scientific statements simply on the basis of deductive logic together with observational reports" (1930/1978: 342–4). The same applies to logical statements, also not justifiable in terms of deductive logic.

The construction of a probability logic, construed as a conceptual framework for any empirical knowledge, is the next step Reichenbach undertakes. The outline of the axiomatic construction of the theory of probability he presents now is meant to capture the attitudes of both the layman and the scientist towards probability, and, most importantly, to provide a solution to Hume's problem of induction. Some of the intricacies of Reichenbach's account are related to the fact that in nature, no probability sequence is derived by a rule, i.e., can be defined intensionally; such sequences can be defined only by the enumeration of their elements, i.e., extensionally. We can base our understanding of a probability sequence only on the portion that is observed

(1933a/1949: 309), that is, by counting the relative frequency in that finite section, and then suppose that this relative frequency will apply, with sufficient approximation, to the rest of the sequence. The assumption here is "that the observed value represents, within certain limits of exactness, the value of the limit for the whole sequence. This inference is called the *inductive inference*" (1949: 351, orig. emphasis). A more precise formulation of this inference is Reichenbach's famous "rule of induction" or "straight rule" (ibid.: 445), which states that the frequency observed in a finite sequence will approximately correspond to the long-run frequency, i.e., will successfully enable us to ascertain its limit *if* a limit exists and *if* the rule is used persistently. As is clear from this approach, Reichenbach defends an empirical view according to which probability values can only be established a posteriori, through the method of enumerative induction.

The degree of probability obtained through this method ultimately relies on the truth-values of individual statements. Truth is therefore a property of statements, while probability is conceived as a property of sequences of individual statements, that is, of propositional sequences (ibid.: sect. 77). In this sense, the frequency interpretation "reduces the concept of probability to the concept of truth. This is the reason"—Reichenbach argues—"why the degree of probability cannot be conceived as a predicate of individual statements but has to be referred to more general logical constructions, which are built up from individual statements in a similar way as a sequence is constructed from individual elements" (1933a/1949: 311–12). It remains unclear whether probability consists in a syntactic and semantic relationship between those individual statements, and not "between what is signified by statements," as Nagel pointed out (1936: 503). Reichenbach's reply to this criticism fell short of clarity (1938b).

The question arises as to how we can determine the degree of probability for a single case in a frequentist world. This is where Reichenbach's famous recourse to the notion of "posit" comes in. A "posit" is a statement that we consider—basically pretend—to be true, although the truth-value is not known. As he explains, the method of positing is vital both in our lives and in science, as it "serves to utilize probability statements for decisions in regard to single cases. It plays an important role in all practical applications. The merchant who stores a great amount of merchandise for the season, the farmer who wants to get in his crop, the physician who prescribes a cure—all must make decisions though they know only probability statements about the factors determining success: the merchant about the prospective demand, the farmer about the prospective weather, the physician about the illness that presumably confronts him." Like gamblers, they make predictions by betting on the most probable case (1949: 373).

Whenever we have only "primitive knowledge" of an event—i.e., "a state of knowledge that does not include a knowledge of probabilities" (ibid.: 364)—we can ascertain probabilities only by making a "blind posit," a simple anticipative hypothesis without a weight, and applying the rule (ibid.: Sect. 81, 1938a: Sect. 34). Then, when we have obtained new observational data, we can correct our first hypothesis and see whether this revised posit can lead to successful results. This secondary posit will supply a weight to the primary posit while becoming an anticipative posit in turn (1949: 461). By applying the rule again, if new evidence leads to different outcomes, we will also be able to revise the secondary posit and replace it with a new one, thus improving the degree of convergence. At each step, the observed frequency represents the best posit to be possibly updated in a subsequent step. It is in this sense that Reichenbach envisages the inductive procedure to be a "*self-corrective method*, or an *asymptotic method*" that will "automatically lead to success in a finite number of steps" (ibid.: 446, orig. emphasis).

All forms of inference made in a primitive state of knowledge are reducible to induction by enumeration. These inferences, applied to one sequence horizontally, can be then assessed through the same rule by inferences that take sequences as elements, along a vertical direction. This practice results in a system of higher-level inductions that Reichenbach calls "cross induction," which eventually leads to an advanced state of knowledge where inferences are not merely formed on the basis of observational data, but also by combining several other inferences in a network. This "method of *concatenation*" among inferences is what grants the success of induction (ibid.: 430–1, orig. emphasis). However, by what techniques this integrated assessment is supposed to take place is not fully elucidated in Reichenbach's framework (Eberhardt and Glymour 2011: 374).

For Reichenbach, the rule of induction is the "condition of possibility" of the scientific method. Scientific theories are established by what he calls "explanatory induction," which "consists in the inference from certain observational data to a hypothesis, or theory, from which the data are derivable and which, conversely, is regarded as being made probable by the data" (1949: 431). This procedure recalls the idea of the mutuality of coordination he introduced in the 1920s (see Padovani 2015, 2017, 2021), but now Reichenbach deems his justification of induction "pragmatic" because "it demonstrates the usefulness of the inductive procedure for the purpose of acting" (1949: 481). The centrality of the notion of weight in his pragmatic approach is stressed in *Experience and Prediction*, which aims at developing a theory of weight "identical with the theory of probability" (1938a: 24, and esp. Sect. 34; 1949: Sect. 81). The practical efficacy of induction in guiding our actions not only makes it indispensable, but also explains why this is the best method we have (1940: 103).

Reichenbach's justification of induction is rooted in the assumption that if there exists a limit of the observed frequency, this limit will be found by applying the straight rule. The simple idea is that this inductive method is convergent, that is to say, that the repeated use of induction will eventually detect convergence, to a sufficient degree of approximation, if a limit exists (1949: 474). One of the most crucial issues in Reichenbach's proposed solution to Hume's problem arises here. Reichenbach ineffectively appeals to descriptive simplicity to ground the selection of the straight rule (1949: 447; 1938a: 373). For him, even if all asymptotic rules are empirically equivalent in converging to the same limit in the long run, the rule of induction should be preferred because, from a descriptive point of view, it is the simplest curve in the set of asymptotic rules. However, whether these asymptotic rules are actually equivalent is not at all granted, so Reichenbach's attempt to vindicate his straight rule has not been considered successful (Salmon 1991; Galavotti 2011).

The discussion with Popper

Reichenbach's approach to probability and induction has raised several other objections, but the ones expressed by Popper are perhaps the most famous. Their debate started in the early 1930s, initially with a note by Popper to the editors of *Erkenntnis* (Popper 1932–3) in which he argued that Hume's problem of induction appeared to contradict what he deemed the fundamental tenet of logical empiricism, namely that experience alone would tell us whether statements and laws of nature are true or false. For Popper, since laws of nature cannot be reduced to observational statements, they would have to be dismissed as meaningless, if one accepts the verifiability criterion of empirical significance. Therefore, falsifiability is the only effective measure to state the empirical character of theoretical systems. Reichenbach replied (1932/33b: 428) that the idea that natural laws should be understood as representing sentences whose truth or falsehood is "unilaterally decidable" is misleading, and that a certain amount of uncertainty must be presupposed. More precisely, "any law of nature, formulated as a statement about real things," ultimately consists "in a probability statement," and as such can be properly interpreted only within a probability logic (as he outlined in his 1932/33a).

Popper wholly discarded this type of logic in his *The Logic of Scientific Discovery* (1934) along with any form of induction, whose "scientific logical function" he deemed "superfluous" (1935: 170). For him, any attempt to justify induction leads to "insurmountable" difficulties, that is, either to a form of apriorism, or to infinite regress. Additionally, there is also nothing to gain in considering the principle of induction as being merely probable and not true, as Reichenbach suggests. Like Reichenbach, Popper believes that we could never establish whether scientific theories are actually true, only whether they are falsified and thus discarded. Differently from Reichenbach's approach to all-statements (1949: 436), Popper's argument relies on the idea that even a single counterexample could falsify a theory, since we postulate the "invariance of natural laws" and that they admit of no exception (1934/2002: sect. 79). Yet, not only does he not believe anything would actually confirm a theory (which, for Reichenbach, is a logical asymmetry), but he also believes we cannot assign to theories a positive probability (ibid.: sect. 80). Contrary to Popper, Reichenbach argues that both the construction of a new theory and the process of falsification do imply the use of probability. Furthermore, he claims that we can rely on past observation to determine which hypotheses are more probable, and eventually come up with a ranking (1935b: 375).

To perform this ranking, one could construe a hypothesis as a sequence of statements, each of which consists of singular statements individually testable. The probability of this hypothesis would be decided on the basis of the relative frequency of the statements confirming the hypothesis in that sequence. For Popper, this approach "leads nowhere." Even if every second individual statement in that sequence was refuted, the hypotheses would still have a probability of 1/2 (cf. also Nagel 1936: 507). One could also try an estimate of the ratio of all confirmed statements to all those who have not been tested yet, but the result of the probability here would always be zero. According to Popper, any other attempt to deal with this construal of hypotheses would eventually lead to subjective considerations involving a measure of the confidence of the experimenter, which is obviously unacceptable in a frequency interpretation (1934/2002: 255).

For Reichenbach, another way to rank hypotheses would be to consider a hypothesis not as a sequence of statements, but as being itself an element of a sequence of hypotheses. The problem in this case is how to determine the unique reference class against which the probability of a hypothesis should be assessed. Reichenbach does not provide a clear and non-question-begging account of how this class is to be selected (see his discussion on Newton's law of gravitation in 1949: 438ff.). Even if this were possible, Popper would disagree, as to ascertain a hypothesis' probability, we must consider the relative frequency of true hypotheses within that class. Yet, if we were able to establish which hypotheses are true or false in that sequence, we would not need to use any probabilistic reasoning to start with (1934/2002: 257). (For discussion of criticisms of Reichenbach's denial that probabilities require to be grounded in certainties after all, see Atkinson and Peijnenburg 2006: 446 ff.; Eberhardt and Glymour 2011: 378).

Both ways to determine the probability of hypotheses are thus rejected by Popper. Reichenbach's reply emphasizes that looking at the probability of theories "fits naturally" into his probability logic, and that limitations due to insufficient availability of historical material to develop a way to confirm theories is not a serious objection to his theory (1935b/1978: 381). To his mind, favoring instead the notion that in pursuing knowledge we only "guess," guided by the "unscientific, metaphysical . . . faith in laws," as Popper claims (1934/2002: 278), misrepresents scientific practice. Reichenbach's view that his account remained faithful to it, and thereby was superior to Popper's, was, in the end, not widely shared.

Conclusion

In sum, from his doctoral dissertation to its culmination twenty years later in the publication of *The Theory of Probability*, Reichenbach aimed to provide mathematical and methodological foundations for his interpretation of probability and thereby furnish a solution to Hume's problem of induction. The benefits of a pragmatic approach notwithstanding, that his proposal is unavailing is out of question, but so it is that the discussions it triggered have inspired a rich amount of literature until the present day.

References

Atkinson, D. and Peijnenburg, J. (2006) "Probability Without Certainty: Foundationalism and the Lewis-Reichenbach Debate," *Studies in History and Philosophy of Science* 37: 442–53.

Beisbart, C. (forthcoming) "Lässt sich die Induktion doch rechtfertigen? Eine kritische Diskussion von neuen Ansätzen zum Induktionsproblem." *Zeitschrift für philosophische Forschung*.

Eberhardt, F. (2011) "Reliability via Synthetic A Priori: Reichenbach's Doctoral Thesis on Probability," *Synthese* 181: 125–36.

Eberhardt, F. and Glymour, C. (2011) "Hans Reichenbach's Probability Logic," in D. Gabbay, S. Hartmann and J. Woods (eds.), *Handbook of the History of Logic X: Inductive Logic*, Amsterdam: Elsevier, pp. 357–89.

Feigl, H. (1950) "De Principiis non disputandum," in M. Black (ed.), *Philosophical Analysis*, Ithaca, NY: Cornell University Press, pp. 119–56. Repr. in Feigl, *Inquiries and Provocations, Selected Writings 1929–1974* (ed. by R. S. Cohen), Dordrecht: Reidel, pp. 237–68.

Galavotti, M. C. (2011) "On Hans Reichenbach's Inductivism," *Synthese* 181: 95–111.

Nagel, E. (1936). "[Review of *Wahrscheinlichkeitslehre* by Hans Reichenbach]," *Mind* 45 (180): 501–14.

Niiniluoto, I. (1998) "Truth, Probability, and Simplicity—Comments on Hans Reichenbach's Probabilistic Empiricism," in H. Poser and U. Dirks (eds.), *Hans Reichenbach. Philosophie im Umkreis der Physik*, Berlin: Akademie Verlag, pp. 69–87.

Padovani, F. (2011) "Relativizing the Relativized A Priori: Reichenbach's Axioms of Coordination Divided," *Synthese* 181: 41–62.

——— (2013) "Genidentity and Topology of Time: Kurt Lewin and Hans Reichenbach," in N. Milkov and V. Peckhaus (eds.), *The Berlin Group of the Philosophy of Logical Empiricism*, Dordrecht: Springer, pp. 97–122.

——— (2015) "Reichenbach on Causality in 1923: Scientific Inference, Coordination, and Confirmation," *Studies in the History and Philosophy of Science* 53: 3–11.

——— (2017) "Coordination and Measurement: What We Get Wrong About What Reichenbach Got Right," *European Studies in Philosophy of Science* 5: 49–60.

——— (2021) "From Physical Possibility to Probability and Back. Reichenbach's Account of Coordination," in S. Lutz and A. T. Tuboly (eds.), *Logical Empiricism and the Physical Sciences*, London: Routledge, pp. 336–53.

Popper, K. (1932/33) "Ein Kriterium des empirischen Charakters theoretischer Systeme," *Erkenntnis* 3: 426–7.

——— (1934) *Logik der Forschung*, Vienna: Springer. Trans. *The Logic of Scientific Discovery*, London: Hutchinson, 1958, repr. London: Routledge, 2002.

——— (1935). "'Induktionslogik' und 'Hypothesenwahrscheinlichkeit'," *Erkenntnis* 5: 170–2.

Reichenbach, H. (1915) *Der Begriff der Wahrscheinlichkeit für die mathematische Darstellung der Wirklichkeit*, Leipzig: J.A. Barth. Trans. *The Concept of Probability in the Mathematical Representation of Reality*, Chicago: Open Court, 2008.

——— (1920a) "Die physikalischen Voraussetzungen der Wahrscheinlichkeitsrechnung," *Die Naturwissenschaften* 8: 46–55. Trans. "The Physical Presuppositions of the Calculus of Probability," in Reichenbach 1978, vol. 2, pp. 293–309.

——— (1920b). *Relativitätstheorie und Erkenntnis apriori*, Berlin: Springer. Trans. *The Theory of Relativity and A Priori Knowledge*, Berkeley: University of California Press, 1965.

——— (1924) *Axiomatik der relativistischen Raum-Zeit-Lehre*, Braunschweig: Vieweg & Sohn. Trans. *Axiomatization of the Theory of Relativity*, Berkeley: University of California Press, 1969.

——— (1925a) "Metaphysik und Naturwissenschaft," *Symposion* 1: 158–76. Trans. "Metaphysics and Natural Science," in Reichenbach 1978, vol. 1, pp. 283–97.

——— (1925b) "Die Kausalstruktur der Welt und der Unterschied von Vergangenheit und Zukunft," *Sitzungsberichte der Bayerische Akademie der Wissenschaften*, November: 133–75. Trans. "The Causal Structure of the World and the Difference between Past and Future," in Reichenbach 1978, vol. 2, pp. 81–119.

——— (1929) "Ziele und Wege der physikalischen Erkenntnis," in Geiger-Scheel (ed.), *Handbuch der Physik*, Berlin: Springer, vol. 4. Trans. "The Aims and Methods of Physical Knowledge," in Reichenbach 1978, vol. 2, pp. 120–225.

——— (1930) "Kausalität und Wahrscheinlichkeit," *Erkenntnis* 1: 158–88. Partly trans. "Causality and Probability," in Reichenbach 1978, vol. 2, pp. 333–44.

——— (1932) "Die Kausalbehauptung und die Möglichkeit ihrer empirischen Nachprüfung," *Erkenntnis*, 3: 32–64. Trans. "The Principle of Causality and the Possibility of Its Empirical Confirmation," in Reichenbach 1978, vol. 2, pp. 345–71.

——— (1932/33a) "Die logischen Grundlagen des Wahrscheinlichkeitsbegriffs," *Erkenntnis* 3: 401–25. Trans. (with revisions) "The Logical Foundations of the Concept of Probability," in H. Feigl and W. Sellars (eds.), *Readings in Philosophical Analysis*, New York: Appleton-Century-Crofts, 1949, pp. 305–23.

——— (1932/33b) "Bemerkung," *Erkenntnis* 3: 427–8.

——— (1934/35) "Sur les fondements logiques de la probabilité," *Recherches Philosophiques* 4: 361–70.

——— (1935a). *Wahrscheinlichkeitslehre*, Leiden: Sijthoff.

——— (1935b) "Über Induktion und Wahrscheinlichkeit. Bemerkungen zu Karl Poppers 'Logik der Forschung'," *Erkenntnis* 5: 267–84. Trans. "Induction and Probability," in Reichenbach (1978), vol. 2, pp. 372–87.

——— (1936) "L'empirisme logistique et la désagrégation de l'a priori," in *Actes du Congrès international de philosophie scientifique, Sorbonne, Paris, 1935*, Facs. I "Philosophie Scientifique et Empirisme Logique," Paris: Herman & Cie, vol. 1, pp. 28–35.

——— (1938a) *Experience and Prediction*, Chicago: University of Chicago Press.

——— (1938b) "On Probability and Induction," *Philosophy of Science* 5: 21–45.

——— (1940) "On the Justification of Induction," *Journal of Philosophy* 37: 97–103.

——— (1949) *The Theory of Probability*, Berkeley: University of California Press [Enlarged translation of (1935a)].

——— (1978) *Selected Writings: 1909–1953* (ed. by R. S. Cohen and M. Reichenbach), Dordrecht: Reidel, vol. 2.

Salmon, W. C. (1991) "Hans Reichenbach's Vindication of Induction," *Erkenntnis* 35: 99–122.

Schurz, G. (2019) *Hume's Problem Solved: The Optimality of Meta-Induction*, Cambridge, MA: MIT Press.

25

SCHLICK, CARNAP, AND FEIGL ON THE MIND-BODY PROBLEM

Sean Crawford

Moritz Schlick, Rudolf Carnap, and Herbert Feig are the most prominent of the positivists to formulate views on the mind-body problem. (C. G. Hempel's one-off treatment [1935] is set aside here; see Kim [2003] and Crawford [2013].) While their views differed from each other and changed over time, they were all committed to some form of scientific physicalism, though a linguistic or conceptual rather than ontological form of it. In focus here are their views during the heyday of logical positivism and its immediate aftermath, though some initial scene-setting of Schlick's and Carnap's pre-positivist views will help to understand the final opposing positions of Carnap and Feigl. All three philosophers are largely—entirely, in the case of Schlick and Feigl—concerned with sensations and sensory consciousness, or what Feigl came to call (after the psychologist E. C. Tolman) "raw feels" and what are now usually called "qualia" or "phenomenally conscious" mental states.

The pre-positivism of Schlick and Carnap

Schlick's and Carnap's pre-positivist views on the mind-body problem reflect a view dominant at the time in both psychology and philosophy: psychophysical parallelism—or more accurately, psycho*physiological* parallelism. In his *General Theory of Knowledge* Schlick explicitly endorses it, though his aim is to interpret it properly. As already mentioned, the mentality in question, for Schlick, is qualitative or sensory consciousness—"mental qualities"—which for him are "immediately experienced reality," the "directly given," the "content of consciousness" (1918/1985: 289). In order to avoid a variety of intractable complications involved in a metaphysically dualistic parallelism (ibid.: 299), especially ones involving causal dependency, Schlick proposed what is often called an "identity theory" but is better described as a kind of "double language" or "double access" monism (Feigl [1975: 14] suggested the term "twofold access"). This theory is part of Schlick's overall "critical realist" philosophy and especially his structuralist view of science, both of which rely on two crucial Kantian distinctions (though Feigl spoke of it as "Schlick-Russell-Eddington structuralism"): first, that between form and content, and second, that between cognition or conceptual knowledge (*Erkennen*) and intuitive experience or acquaintance (*Kennen, Erleben*). The idea is that science provides knowledge only of the structure of reality, which for Schlick takes the form of abstract uninterpreted axiomatic systems, which include scientific quantitative terms, laws, and equations, and in which

DOI: 10.4324/9781315650647-28

the concepts required for genuine knowledge are defined by Hilbertian implicit definitions that display their formal logico-mathematical relations to each other. The content of this relational structure consists of the intrinsically qualitative natures of the things related by the structure. These qualities are the "things-in-themselves," the "transcendent" objects of reality, which, contrary to Kant, are knowable indirectly through designation by the (implicitly defined) quantitative scientific concepts. But we can also directly experience a tiny subset of this intrinsically qualitative transcendent reality, namely, the subset containing the subjective mental qualities, which are mental by dint of composing parts of our individual brains, which makes them accessible to us via introspection.

Metaphysically speaking, then, the intrinsic nature of the universe is ultimately purely qualitative, a web of mostly unexperienceable intrinsic qualities (1918/1985: 283–92). We acquire knowledge of the "essence" of any one of these intrinsic qualities by "incorporating [it] into the quantitative conceptual system and thus reducing it to the fundamental intensities selected as a basis" (ibid.: 285). This is done through the "method of coincidences" (ibid.: §31), part of which involves the coordination of the intrinsic qualities with quantitative physical concepts. A small minority of these intrinsic qualities—the subjective or mental ones—are thus doubly designated: directly by intuitive, often imagistic, psychological concepts (or representations), on the one hand, which are qualitative (e.g., *yellow*); and indirectly by unintuitive, physical concepts, on the other hand, which are quantitative (e.g., *wavelength of 590–560 angstroms*). Only the latter provide knowledge of reality; the former give us mere acquaintance with some of it. "The same reality," he says—namely, that which is immediately experienced (i.e., some of the all-pervading qualities that constitute reality)—"can be designated both by psychological concepts and by physical ones" (ibid.: 310).

Even though for him mental/subjective and extra-mental/objective qualities "do not differ fundamentally" (ibid.: 328), Schlick is adamant that his monism is not a form of panpsychism, as the qualities that constitute the world are not all mental. Only the ones that constitute (some of) our brain, and with which we are directly acquainted by immediate experience, are mental, so that mind is "but a sector of the totality of the natural" (ibid.: 296). Moreover, since "reality is called 'physical' in so far as it is designated by means of the spatio-temporal quantitative conceptual system of natural science" (ibid.: 294), and all of reality can be so designated, Schlick's structuralist qualitative monism is a form of physicalism (somewhat akin to what Davidson [1970] called "nomological monism" but also with similarities to Jackson's [1998] "Kantian physicalism"). With the formation of the Vienna Circle and especially under the influence of Wittgenstein and Carnap, Schlick's double-access quality monism is left behind as his thinking takes an anti-metaphysical turn.

Turning to Carnap, we find that his *Aufbau*, written largely before he joined the Vienna Circle, embraces both neutral monism (1928a: §§162, 183), derived from Russell (1921), and parallelism (1928a: §§166–9, 183). No contradiction obtains because he distinguishes between the "problem of mind-body dualism" and the "psychophysical problem." The former is the question whether there are two separate types of things, be they substances, principles, or aspects. Carnap's answer is no. The conceptual system developed in the *Aufbau* recognizes only one type of object domain, and elements thereof are mental when categorized in a certain way and physical when categorized in another way: "the physical and the psychological must not be envisaged as two principles or aspects of the world. They are order forms of the one, unified domain of elements which are propertyless and merely connected through relations" (ibid.: §162). The psychophysical problem, on the other hand, is the question of how to interpret the parallelism of psychological and physiological events—or more accurately, how to interpret the fact that for every mental event, there is a corresponding physiological event—something which Carnap assumes to have been empirically established. It falls to the science of psychology to establish exactly which types of brain events are correlated with which types of mental events. These are mere empirical correlation problems solved or solvable by science. The philosophical psychophysical problem, which Carnap calls the "essence problem," is how to interpret the relation between psychological events and physiological events thus correlated. For Carnap, the parallelism can have no scientific explanation: "The quest for an explanation of these findings lies outside the range of science" (ibid.: §169) and so "the quest for an explanation of that parallelism belongs within metaphysics." Importantly, Carnap does not say at this point that this means the question is meaningless, only that it does not belong to science (ibid.: §182). Rather, since the constructional system of the *Aufbau* is concerned exclusively with the reduction or definition of scientific concepts, such metaphysical problems simply fall outside its purview. Carnap later stated that "regarding the criticism of traditional metaphysics, in the *Aufbau* I merely refrained from taking sides" (1963: 18–19, 44–46).

The positivism of Carnap, Schlick and Feigl

Very shortly thereafter, however, following his participation in the Vienna Circle discussions of Wittgenstein's *Tractatus* and its position that metaphysical statements are meaningless (*unsinnig*), Carnap explicitly commits himself to this view. In *Pseudoproblems of Philosophy*, written in 1927–8, he proposes a criterion of empirical meaningfulness or "factual content," according to which "Only statements with factual content are theoretically meaningful; (ostensible) statements which cannot, in principle, be supported by experience are meaningless" (1928b: §7). While not explicitly declaring the psycho-physical (or mind-body) problem to be a meaningless pseudo-problem, that was his new view.

Things get much more familiar to contemporary philosophers of mind with what is standardly but misleadingly called Carnap's "logical behaviorism" (a term he never used). He called his position "physicalism," which was an ambitious program to reduce or translate all scientific theories into a universal physical language that was supposed to serve as an intersubjective confirmation base. On this view, statements that cannot be so translated are unconfirmable and hence cognitively meaningless and the traditional metaphysical mind-body problem is deemed a pseudo-problem. Carnap's "logical behaviorism" (and that of the logical positivists generally) is simply part of their overall project of the physicalization of all of empirical science; that is, the translation of the sentences of all the special sciences into the "physicalist" language (either the language of physics or later the physical "thing language"), the only known "universal" language (the only one known to be intersubjective and intersensory). When applied to the science of psychology, physicalism becomes the "logical behaviorism" of Carnap (and Hempel 1935). It is in this light that one must view Carnap's later general thesis that "all statements of science can be translated into physical language" and the relevant sub-thesis that "all psychological statements can be translated into physical language." This sub-thesis is no different, in principle, from the sub-thesis for biology, that "every statement of biology can be translated into physical language" (1932a/1934: 70).

The crucial point to appreciate, however, which has been largely misunderstood by contemporary philosophers, especially contemporary philosophers of mind, is that when it comes to the empirical sciences, such as psychology, the translations in question are not analytic (and therefore not knowable a priori). Moreover, they are not even restricted to physical-*behavioral* translations but can also be *physiological* translations (unsurprisingly, given the influence of parallelism in the *Aufbau*). From the early 1930s onwards, Carnap invoked "rules of translation" (or transformation) of the physical language in which the translations are to be carried out. It is clear that not all of these rules are laws of logic and that some of them are intended to be laws of nature (Carnap 1935b). In *The Logical Syntax of Language* (1934) and "Testability and Meaning" (1936–7), Carnap explicitly distinguishes between the L-rules and the P-rules of a scientific language on the basis of which the transformations may be validly carried out: the former are logical laws and the latter physical laws. Both kinds of "translation rules" may be employed in physicalistic analysis or reduction. (Already earlier, he speaks of "the rules of transformation inside the physical language (*including the system of natural laws*)" (Carnap 1932a/1934: 88, emphasis added; compare 92 and 1932b/1959: 171).

In *Philosophy and Logical Syntax*, Carnap claims that "every sentence of any branch of scientific language is equipollent [equivalent] to some sentence of the physical language, and can therefore be translated into the physical language without changing its content" (1935a/1963: 455). Carnap is very clear there and in *The Logical Syntax of Language* (1934: §§51 and 82) that there can be two concepts of equivalence in a physical language: logical equivalence (L-equivalence) and physical equivalence (P-equivalence). Two sentences are L-equivalent when they are mutually derivable solely on the basis of logical laws; two sentences are P-equivalent when they are mutually derivable, in addition, on the basis of physical laws. Carnap explicitly allowed a psychological sentence, Q_1, and a physical translation of it, Q_2, to be P-equivalent, as Q_1 could be transformed into Q_2 on the basis of "a scientific law, that is, a universal sentence belonging to the valid sentences of the scientific language-system" (1935a/1963: 456; compare 1932b and 1935b). He made absolutely clear that this universal sentence "need not be analytic; the only assumption is that it is valid. It may be synthetic, in which case it is P-valid" (ibid.). Contrary to the received view fostered by Putnam (1965), Carnap never claimed to offer analytically true logical constructions of "mind talk" into either (overt or covert) "behavior talk" or "physical talk" (see Crawford 2013, 2014).

On the basis of his close association and conversations with Carnap and Wittgenstein, inside and outside the Vienna Circle, Schlick came to reject much of the double-access structuralist qualitative monism he earlier espoused in the *General Theory of Knowledge* and to embrace a form of the verifiability theory of meaning and adhere to the new translation physicalism. Although there are strong traces of his previous view lurking in his 1935 positivist paper "On the Relation between Psychological and Physical Concepts," he tries to strip these features of any metaphysical trappings, declaring his new adherence: "every psychological proposition can be translated into an expression in which physical concepts alone occur." With the term "physicalism" defined accordingly, he declared that "I therefore hold the thesis of physicalism to be correct" (1935/1949: 399). Contrasting this with his earlier, superficially similar position in *General Theory of Knowledge*, he remarks that "If, as a matter of fact, the physical language is characterized by complete universality, the setting down of this circumstance is in no way the assertion of a metaphysical monism" (ibid.: 407).

Like his teacher Schlick's 1935 paper, but even more so, Feigl's first paper on the mind-body problem, "Logical Analysis of the Psychophysical Problem: A Contribution to the New Positivism" presents a rather confusing hybrid of Schlick's old qualitative monism and the new positivist translation physicalism. But the main message, at least according to the later Feigl, was his version of translation physicalism: "the only non-metaphorical non-metaphysical formulation of the double aspect or double knowledge theory [is]: the formulation and detailed analysis of the mutual translatability of two universal languages. This, then, is the new positivist view of the psycho-physical relation" (1934: 437). Moreover, "we deem . . . Materialism strictly meaningless if [it] pretends to express [a] factual hypothesis" (ibid.: 443).

Unfortunately, unlike the ever-rigorous Carnap, neither Feigl nor Schlick were at all clear about what they meant by "translation." Schlick simply does not say. But his silence is less confusing than what Feigl does say. For Feigl talks about the "logical" nature of the psychophysical problem and its "logical" solution. At one point he says that "Logical mutual translatability, isomorphism, means simple *identity* of the two propositions" (ibid.: 436). He does not say what he means here by "logical" or by the "identity" of the "two" propositions. The natural interpretation would be that the "identity" of the two propositions arrived at by "logical mutual translatability" means that "they" are at least synonymous in the strong sense of analytically equivalent (and hence knowable a priori)—to use Carnap's terms, L-equivalent. But this is, of course, completely implausible and indeed absurd, as it would amount to an armchair natural science of psycho-physiology. On the other hand, "isomorphism" suggests, much more plausibly, a merely contingent or synthetic correlation between the two propositions.

The Aftermath: Feigl vs Carnap

By 1950, Feigl had clearly thought through the matter with greater care and now realized that there can be no "logical" (in the sense of analytic) connection between psychological statements and their physical translations, only a synthetic one. He must have actually realized this even before the publication, though presumably not the writing, of his 1934 paper. In a letter to Carnap in 1933, he suggested that the connection between psychological sentences and the physical sentences intended to give their "translations" will be synthetic, something Carnap in reply explicitly agrees with—unsurprisingly so, as this was his position all along (see Feigl 1963: 255 n. 28; Crawford 2013). Now Feigl returned to what he seems to have always felt was the right view to take of the mind-body problem all along, but for the positivists' peer pressure: that it was not a pseudo-problem (Feigl 1960) and that his teacher Schlick had basically got it right with his structuralist qualitative monism earlier in *General Theory of Knowledge*.

In "The Mind-Body Problem in the Development of Logical Empiricism," Feigl declared that "The precipitous assertion of logical equivalence [which he rejects] was of course based on the phenomenalistic claims of the explicit definability of the entities in one realm in terms of the entities of the corresponding realm" (1950/1953: 620). But who exactly made this "precipitous assertion of logical equivalence"? In his later classic long essay "The 'Mental' and the 'Physical,'" he repeated the point more clearly and explicitly:

> A most important *logical* requirement for the analysis of the mind-body problem is the recognition of the *synthetic* or *empirical* character of the statements regarding the correlation of psychological to neuro-physiological states. . . . I was tempted to identify, in the sense of *logical* identity, the mental with the neurophysiological. . .
>
> But if this theory is understood as holding a *logical translatability* (analytic transformability) of statements in the one language into statements in the other, this will certainly not do. . . .
>
> [T]he question which mental states correspond to which cerebral states is in *some* sense . . . an empirical question. If this were not so, the intriguing and very unfinished science of psychophysiology could be pursued and completed by purely a priori reasoning. . . .
>
> Subjective experience . . . cannot be *logically* identical with states of the organism; i.e., phenomenal terms could not explicitly be defined on the basis of physical_1 or physical_2 terms.
>
> (1958: 389–90, orig. emphasis; compare ibid.: 391).

According to Feigl, then, he himself did adhere to an analytic translatability thesis back in 1934 (or 1933), for it was he who made the "precipitous assertion of logical equivalence." Did anyone else? Feigl gives the impression that this was the general consensus during the early phase of logical positivism. But as we have seen, Carnap certainly did not hold any such view at any time. Whether anyone other than Feigl held this misbegotten view for more than a minute is not completely clear (on Hempel 1935 see Crawford 2014: 717 n. 8).

At any rate, Feigl and others influenced by him seem to have confused analytic translation with explicit definability. But for Carnap, explicit definitions need not be analytic. The constructional definitions of the *Aufbau*, for example, are extensional (Carnap 1928a: §§35, 43, 47–49). The point is that if a non-primitive expression, the *definiendum*, is explicitly definable in terms of primitive expressions, then it can be eliminated and replaced by its *definiens*, by the primitive expressions. Such explicit definitions may be extensional, offering merely materially necessary and sufficient conditions for the *definiendum*, that is, the construction of a material biconditional whose right-hand side, the *definiens*, contains only undefined primitive terms. The defined expressions will be so-called "theoretical" terms (e.g., mental ones) and the primitive expressions the "observation" terms (e.g., behavioral and physiological ones). To his credit, Carnap (1936–7) very early on saw that the search for explicit definitions of all empirical scientific terms in the extensional physical-thing language was badly misconceived, especially since many such definitions were supposed to be empirical laws, and he consequently (instead of invoking a stronger intensional language) weakened the project to one of providing (extensional) "reduction sentences," which linked the empirical term in question to observable physical conditions only under certain test conditions. Since these physical reduction sentences were not definitions of the terms they were reducing—they were only partial "conditional definitions"—they did not allow the terms to be eliminated and replaced, and hence they could not form the basis for translations. The crucial point to notice about this move from definition to reduction is that, with respect to the physicalization of psychology and other empirical sciences, it is not a shift from the category of analytic truths (knowable only a priori) to the category of synthetic truths (knowable only a posteriori). Rather, it is a shift *within* the single category of synthetic truths knowable only a posteriori from explicit (complete) definability, which permits elimination of the defined term, to conditional (incomplete) definability, which does not permit elimination of the partially defined term.

In "The Mental and the Physical," operating under the misapprehension (seemingly shared by Putnam [1957]) that he has discovered that the relation between mental statements and physical statements is not analytic as the early logical positivists allegedly thought, but synthetic, Feigl returns to what is in all essentials a version of Schlick's pre-positivist position (updated by replacing Schlick's original distinction between conceptual knowledge and intuitive acquaintance with an application of Russell's distinction between knowledge by description and knowledge by acquaintance):

> the physical sciences consist of knowledge-claims-*by-description*. That is to say that the objects (targets, referents) of such knowledge claims are "triangulated" on the basis of various areas of observational (sensory) evidence. What these objects are acquaintance-wise is left completely open as long as we remain within the frame of *physical* concept formation and theory construction. But, since in point of empirical fact, *I* am directly acquainted with the qualia of my own immediate experience, I happen to know (by acquaintance) what the neurophysiologist refers to when he talks about certain configurational aspects of my cerebral processes.
>
> (1958: 450, orig. emphasis)

Although he describes his own theory as an identity theory, he also, like Schlick, describes it as a kind monism, which seems much more appropriate given that they both view the referents of both mental and physiological terms—that is, the "realities in themselves"—as intrinsic qualities. Feigl, for instance, says that "The 'mental' states or events (in the sense of raw feels) are the referents (denotata) of both the phenomenal terms of the language of introspection, as well as of certain terms of the neurophysiological language" (ibid.: 447). And, in a later paper, he writes that "I take these referents [of mental and neurophysiological terms] to be the immediately experienced qualities" (1960: 38; compare 1958: 457, 474 and 1963: 262, 257). Strikingly, for Feigl, as for Schlick, "sentience (qualities experienced, and in human beings knowable by acquaintance) and other qualities (unexperienced and knowable only by description) [are] the basic reality" (1967a: 107). The concepts of theoretical physics "denote realities which are unknown by acquaintance, but which may in some way nevertheless be not entirely discontinuous with the qualities of direct experience" (ibid.: 40; compare 1971: 308).

The issue of whether this is a form of panpsychism arises, just as it did for Schlick. Feigl responds pretty much like Schlick: the world "as it is in itself" is a field of connected qualities some of which are mental and others which are not. But all these qualities can be known indirectly by description through physical science, hence, we again have a kind of structuralist physicalism: everything that exists is structurally describable physically and unfolds according to physical law. A small subset of the universe of qualities that is the content of this physical structure, and that the laws govern, is also known by acquaintance and therefore mental, which is to say that it, unlike the other qualities, can be referred to or described in psychological terms. While Feigl, like Schlick, explicitly rejects panpsychism, he is prepared to apply the term "pan-quality-ism" to his view that intrinsic qualities constitute all of nature—not just parts of our brains and nervous systems (1960: 39). (Compare Feigl's own exploration of the similarities between Russell [1927]—who also endorsed "pan-quality-ism"—and Schlick in his 1975.) It seems fair to say that neither Feigl nor Schlick satisfactorily managed to explain exactly what the difference between the mental and the non-mental qualities that constitute reality is, and so neither really managed successfully to shake off what seemed like the panpsychistic implications of their view. Feigl himself says that "one is tempted, with the panpsychists, to assume some unknown-by-acquaintance qualities quite cognate with those actually experienced" (1958: 485). But no clear account is ever given of what this "cognate" nature could be that falls short of being mental—calling forth Feigl's witty remark: "If this be metaphysics, make the least of it!"

It appears, then, that Schlick's and Feigl's (1967a: 142; 1975) structuralism about natural science combined with what Hatfield (2004) has called "respect for the phenomenal" ("raw feels") forms the basis for their monistic solution to the mind-body problem. Natural science, especially the foundational and comprehensive science of physics, is limited to providing indirect, quantitative and purely structural knowledge by description of the entire world in physical language. This structure has a content, however, that is purely qualitative, part of which can be accessed or known directly by acquaintance. Feigl, I think, is right, in holding that the name "double-language theory" (or "double-knowledge theory") is a better term than "identity theory" for this kind of monism, for neither he nor Schlick are really identifying what, pre-theoretically, seem like two different kinds of properties, qualitative raw feels and quantitively measurable physiological processes. Rather, when it comes to the mental part of intrinsically qualitative existence, the raw feels are the true existents—the things or realities "in themselves"—and they are referred to in two different ways.

What was Carnap's reaction to all this? Well, this is what he wrote in response to Feigl:

> The identity statement mentioned [that a certain psychological process *P* is identical with a certain neurophysiological process *N*] is a sentence of the object language; this fact may mislead the reader into believing that the controversy about the identity view concerns a question of fact. . . . It seems preferable to me to formulate the question in the metalanguage, not as a factual question about the world, but as a question concerning the choice of language. . . . Those facts Feigl proposes as evidence for the identity view are perhaps better regarded as reasons for preferring a monistic language . . . in this language the predicates "*P*" and "*N*," though not L-equivalent, are P-equivalent . . . I am willing to call my position an identity conception in the following sense: in agreement with Feigl I prefer the monistic language, and like him I believe that the evidence available today provides good reasons for the assumption that this language will also function well in the future.
>
> (1963: 885–6)

Unlike Feigl, Carnap continued to maintain that the mind-body problem, like all traditional metaphysical problems, was a pseudo-problem. In 1935 he had said that the "pseudo-object" identity-sentence of the material mode, "The evening-star and the morning-star are identical" is to be replaced by the syntactical formal-mode sentence "The words 'evening-star' and 'morning-star' are synonymous" (1935a/1963: 447); given the earlier more rigorous treatment (1934: §75), it is clear that he means P-synonymous. But even in his later semantic period, I think he would grant that his position on the mind-body problem could be expressed—albeit highly misleadingly—in the material mode as a synthetic identity theory. But that is as close to the Schlick-Feigl "identity theory" as he is willing to get.

Carnap made no remarks about Feigl's Schlick-inspired structuralist pan-quality monism. Obviously, he would flatly reject it as a metaphysical pseudo-doctrine. It is interesting to note, however, that around the time of his reply to Feigl, Carnap too, like Schlick and Feigl, explored a kind of structuralism, when he made essential use of the Ramsey sentence of a scientific theory—a key ingredient in some forms of structuralism—in his explication of the division between its analytic and synthetic components (1966: chs. 26–28). The difference is that Carnap's structuralism—unlike Schlick's and Feigl's—was, unsurprisingly, metaphysically neutral, eschewing any claims about the intrinsic ontological nature of the content of the structure as pseudo-questions. That said, despite their various changing and divergent attitudes towards metaphysics, in particular whether the mind-body problem is a pseudo-problem, all three of our scientific philosophers never wavered from some form of linguistic-conceptual physicalism.

References

Carnap, R. (1928a) *Der Logische Aufbau der Welt*, Berlin: Weltkreis-Verlag. Trans. *The Logical Structure of the World*, Berkeley: University of California Press, 1967. Repr. Chicago: Open Court, 2003.

——— (1928b) *Scheinprobleme in der Philosophie*, Berlin: Weltkreis-Verlag. Trans. "Pseudoproblems in Philosophy," in Carnap, *The Logical Structure of the World*, Berkeley: University of California Press, 1967, 301–43.

——— (1932a) "Die physikalische Sprache als Universalsprache der Wissenschaft," *Erkenntnis* 2: 432–65. Trans. *Unity of Science*, London: Kegan Paul, 1934.

——— (1932b) "'Psychologie in physikalischer Sprache," *Erkenntnis* 3: 102–42. Trans. "Psychology in Physical Language," in A. J. Ayer (ed.), *Logical Positivism*, New York: Free Press, 1959, pp. 165–98.

——— (1934) *Logische Syntax der Sprache*, Vienna: Springer. Rev. ed. trans. *The Logical Syntax of Language*, London: Kegan Paul, Trench, Trubner & Cie, 1937, repr. Chicago: Open Court, 2002.
——— (1935a) *Philosophy and Logical Syntax*. London: Routledge and Kegan Paul. Repr. in W. Alston and G. Nakhnikian (eds.), *Readings in Twentieth Century Philosophy*, New York: Free Press, 1963, pp. 424–59.
——— (1935b). "Les concepts psychologiques et les concepts physiques sont-ils foncièrement différent?" *Revue de Synthèse* 10: 43–53.
——— (1936–7) "Testability and Meaning," *Philosophy of Science* 3: 419–71 and 4: 1–40.
——— (1963) "Intellectual Autobiography" and "Comments and Replies," in Schilpp (1963), pp. 3–84, 859–1016.
——— (1966) *Philosophical Foundations of Physics*, New York: Basic Books. Rev. ed. *An Introduction to the Philosophy of Science*, 1974, repr. Mineola, NY: Dover, 1995.
Crawford, S. (2013) "The Myth of Logical Behaviorism and the Origins of the Identity Theory," in M. Beaney (ed.), *The Oxford Handbook of the History of Analytic Philosophy*, Oxford: Oxford University Press, pp. 621–55.
——— (2014) "On the Logical Positivists' Philosophy of Psychology: Laying a Legend to Rest," in M. C. Galavotti et al. (eds.), *New Directions in the Philosophy of Science*, Dordrecht: Springer, pp. 711–26.
Davidson, D. (1970) "Mental Events," in L. Foster and J. W. Swanson (eds.), *Experience and Theory*, Amherst: University of Massachusetts Press, pp. 79–101. Repr. in Davidson, *Essays on Actions and Events*, Oxford: Clarendon Press, 1980, pp. 207–24.
Feigl, H. (1934) "Logical Analysis of the Psychophysical Problem: A Contribution to the New Positivism," *Philosophy of Science* 1: 420–45.
——— (1950) "The Mind-Body Problem in the Development of Logical Empiricism," *Revue de Internationale de Philosophie* 4: 64–83. Repr. in H. Feigl and M. Brodbeck (eds.), *Readings in the Philosophy of Science*, New York: Appleton-Century Crofts, 1953, pp. 612–26.
——— (1958) "The 'Mental' and the 'Physical'," in H. Feigl, M. Scriven and G. Maxwell (eds.), *Concepts, Theories, and the Mind-Body Problem*, Minneapolis: University of Minnesota Press, 370–497. Repr. in Feigl (1967a), pp. 1–131.
——— (1960) "Mind-Body, *Not* a Pseudo-Problem," in S. Hook (ed.), *Dimensions of Mind*, New York: New York University Press, pp. 24–36.
——— (1963) "Physicalism, Unity of Science and the Foundations of Psychology," in Schilpp (1963), pp. 227–67.
——— (1967a) *The "Mental" and the "Physical"*, Minneapolis: University of Minnesota Press.
——— (1967b) "Postscript After Ten Years," in Feigl 1967a, pp. 133–60.
——— (1971) "Some Crucial Issues of Mind-Body Monism," *Synthese* 22: 295–312.
——— (1975) "Russell and Schlick: A Remarkable Agreement on a Monistic Solution of the Mind-Body Problem," *Erkenntnis* 9: 11–34.
Feigl, H. and Sellars, W. (eds.) (1949) *Readings in Philosophical Analysis*, New York: Appleton-Century Crofts.
Hatfield, G. (2004) "Sense Data and the Mind-Body Problem," in R. Schumacher (ed.), *Perception and Reality: From Descartes to the Present*, Paderborn: Mentis, pp. 305–31.
Hempel, C. G. (1935) "Analyse logique de la psychologie," *Revue de Synthèse* 10: 23–43. Trans. "The Logical Analysis of Psychology" (with a 1947 footnote addition) in Feigl and Sellars (1949), pp. 373–84.
Jackson, F. (1998) *From Metaphysics to Ethics: A Defense of Conceptual Analysis*, Oxford: Clarendon Press.
Kim, J. (2003) "Logical Positivism and the Mind-Body Problem," in Parrini, Salmon and Salmon 2003, pp. 263–78.
Parrini, P., Salmon, W. C. and Salmon, M. H. (eds.) (2003) *Logical Empiricism: Historical and Contemporary Perspectives*, Pittsburgh: University of Pittsburgh Press.
Putnam, H. (1957) "Psychological Concepts, Explication, and Ordinary Language," *Journal of Philosophy* 54: 94–100.
——— (1965) "Brains and Behavior," in R. Butler (ed.), *Analytical Philosophy. Second Series*. Oxford: Blackwell, pp. 1–19. Repr. in Putnam, *Mind, Language and Reality*, Cambridge: Cambridge University Press, 1975, pp. 325–41
Russell, B. (1921) *The Analysis of Mind*, London: G. Allen & Unwin.
——— (1927) *Analysis of Matter*, London: Routledge and Kegan Paul.
——— (1956) "Mind and Matter," in *Portraits from Memory*, London: George Allen and Unwin, pp. 145–65.
Schilpp, P. A. (ed.) (1963) *The Philosophy of Rudolf Carnap*, La Salle, IL: Open Court.

Schlick, M. (1918) *Allgemeine Erkenntnislehre*, Berlin: Springer, 1918, 2nd rev. ed. 1925. Trans. *General Theory of Knowledge*, Vienna: Springer, 1974, repr. Lasalle: Open Court, 1985.

——— (1935) "De la relation entre les notions psychologiques et les notions physiques," *Revue de Synthèse* 10: 1–20. Trans. "On the Relation Between Psychological and Physical Concepts," in Feigl and Sellars (1949), pp. 393–407.

26

HEMPEL AND CONFIRMATION THEORY

Jan Sprenger

Carl Gustav Hempel (1905–1997) was one of the primary exponents of logical empiricism. As a student and member of the *Gesellschaft für empirische Philosophie* in Berlin, alongside Hans Reichenbach and Kurt Grelling, he witnessed the emergence of logical empiricism as a philosophical program. From the mid-1930s onwards, his contributions shaped its development, too. Hempel studied primarily in Göttingen and Berlin, but in 1929–30, he also spent a semester in Vienna studying with Rudolf Carnap and participated in the activities of the Vienna Circle. Both societies joined forces for organizing scientific events, and founded the journal *Erkenntnis* in 1930, where many seminal papers of logical empiricism were published, with Carnap and Reichenbach as editors.

While the work of the Berlin philosophers is congenial to the project of the Vienna Circle, there are important differences, too. Neither Hempel nor his mentor Reichenbach identified "scientific philosophy" with the project of cleansing science of meaningless statements (e.g., Carnap 1930). Rather, Hempel extensively used a method that Carnap would apply in later works on probability and confirmation (Carnap 1950, 1952): *explication*, that is, the replacement of a vague and imprecise pre-theoretical concept (e.g., "confirmation") by a fruitful and precise concept (e.g., a formal confirmation criterion). Relying on the method of explication, Hempel developed adequacy conditions on a qualitative concept of confirmation (1943, 1945), a probabilistic measure of degree of confirmation (Hempel and Oppenheim 1945), and most famously, the DN model for explanation by means of natural laws (Hempel and Oppenheim 1948). Much of contemporary philosophy of science, including formally oriented literature (e.g., Sprenger and Hartmann 2019), is closer to Hempel's approach of explicating central concepts in ordinary scientific reasoning than it is to Carnap's reconstructive approach in the *Aufbau* (1928), or his later work on logical foundations of inductive inference (e.g., 1950). That said, Hempel and Carnap share the conviction that we must analyze the relationship between theory and evidence not only in terms of verifying the observable consequences of a theory, but also, and specifically, in terms of the *inductive* consequences of a given body of evidence for the assessment of a theory or hypothesis. More than Carnap, Hempel also worked on scientific reasoning in a broader context, especially in later years. Specifically, he engaged with Rudner's 1953 provocative thesis that the scientist's acceptance of a hypothesis always involves value judgments, and commented extensively on the most influential works from the next generation: Thomas S. Kuhn's *Structure* (1962) and Paul Feyerabend's *Against Method* (1975).

DOI: 10.4324/9781315650647-29

This chapter gives an overview of Hempel's work on confirmation and induction. We first explain Hempel's take on the problem of induction and his probabilistic explication of degree of confirmation. We then proceed to Hempel's explication of the classificatory or qualitative concept of confirmation, and the Satisfaction Criterion in particular. The subsequent section presents the famous paradox of the ravens, Hempel's analysis and its impact on later work. Finally, we briefly review Goodman's "new riddle on induction" that takes issue with Hempel's confirmation criteria, and Hempel's later work on values in inductive inference.

The modern problem of induction

Students of philosophy all learn about Hume's classical problem of induction: justifying beliefs and actions that are based on empirical, logically inconclusive evidence. According to Hempel, however, a second problem of induction is at least of equal importance: specifying the rules for a valid inference from empirical evidence (the premises) to a theoretical hypothesis (the conclusion); in other words, finding a *logic of inductive inference*. Such a logic would try to mirror the success of deductive logic for ampliative inferences and secure the objectivity of inductive inferences in science. Specifically, it would replace the subjective appraisal of a theory by objective, verifiable standards for confirmation (Hempel and Oppenheim 1945: 98–99) and contribute to the central aims of empiricist philosophy: to understand and to model the progress of science, the replacement of old by new theories, and the testability of abstract hypotheses by empirical observations.

Hempel (1965b: 30–34) is quick to point out that inductive inference cannot consist of an indiscriminate collection of facts, followed by their systematization and inductive generalizations. In fact, like Popper, Hempel stresses that the scientific process must be guided by tentative hypotheses which we later evaluate on the basis of empirical evidence. And like Popper, Hempel insists that a logic of scientific reasoning cannot cover this essentially creative and non-regulated process of inventing hypotheses. Neither can it prescribe the decision to accept or reject a hypothesis on the basis of evidence: for Hempel, this decision is entangled with pragmatic values (1960, 1983). "Rules of inductive inference will have to be conceived, not as canons of discovery, but as criteria of validation for proposed inductive arguments" (Hempel 1965b: 34): they do not *generate* a hypothesis from a given body of evidence, but *presuppose* that a hypothesis (or varying competing hypotheses) has been put forward independently, and evaluate that hypothesis against the available evidence. It is here, in the comparison of theoretical sentences that express a hypothesis, with observation reports expressing the evidence, that logical tools can make an important contribution.

In this context, Hempel stresses the importance of Carnap's (1947) Requirement of Total Evidence (RTE): the inductive support in favor of a hypothesis H should be calculated with respect to the total available evidence E. However, for Hempel the RTE is no rule of inductive inference (like, e.g., "observed deductive consequences of a theory confirm it"), but a rule that governs the *rational application* of inductive inferences (1965b: 43). As such, it is especially salient in contexts where we would like to base a decision on accepting or rejecting a theory on the quantitative degree of support $dc(\mathrm{H}, \mathrm{E})$ in favor of H.

Hempel and Oppenheim (1945) propose to measure inductive support based on the method of *maximum likelihood* that R. A. Fisher (1935) introduced into statistics a few years before. The idea is to find, for any evidence E, the "optimum distribution" Δ_E over the probability space, that is, the distribution that assigns maximal probability to E. Then H is assigned, as degree of confirmation with respect to E, the probability of H under Δ_E, that is, $dc(\mathrm{H},\mathrm{E}) = p_{\Delta_E}(\mathrm{H})$ (my notation, J.S.).

This procedure is, of course, well-known from maximum likelihood estimation: as a point estimate of an unknown parameter θ, one chooses the value $\breve{\theta}$ that assigns maximal probability to the observed data. And like maximum likelihood, Hempel and Oppenheim's method need not yield unique results, as the authors notice. For example, when the elements of the probability space are unrelated propositions of predicate logic, with H = *Fa* and E = *Gb*, then any degree of confirmation $0 \leq dc(\mathrm{H}, \mathrm{E}) \leq 1$ will be admissible. Hempel and Oppenheim's criterion will often just determine interval bounds for the inductive support of E for H. This lack of uniqueness is not necessarily a vice, however: Carnap (1952) defended in later work the multiplicity of inductive methods, and moreover, Hempel and Oppenheim show that the thus-defined degree of confirmation obeys several intuitive principles, such as:

- If H is a logical consequence of E (and E is consistent), then $dc(\mathrm{H}, \mathrm{E}) = 1$.
- For any optimum distribution Δ, $dc(\mathrm{H}, \mathrm{E}) + dc(\neg\mathrm{H}, \mathrm{E}) = 1$.

The function *dc*(H, E) acts in many respects like a probability function, and it can be connected to various plausible constraints on degree of confirmation. It differs from Carnapian confirmation functions (Carnap 1950) in various ways: it is not meant to explicate the pre-theoretical concept of "probability," it is strongly inspired by statistical reasoning, and it does not (unlike Carnap's confirmation functions) depend on the partitioning of the logical space.

Qualitative confirmation criteria

While Carnap quantified degree of confirmation without addressing the question of when a piece of evidence confirms a theory at all, Hempel thought that *qualitative adequacy conditions* were a necessary prolegomenon for a quantitative, probabilistic account of confirmation (1945: 30–33). Such adequacy conditions are supposed to capture the core elements of the concept of confirmation, and to constrain the quantitative analysis of confirmation in a successive stage. The first condition Hempel proposes is the

> *Entailment Condition (EnC)*: If hypothesis H logically follows from the observation report E, then E confirms H.

For example, if the hypothesis reads "there are black ravens," then the observation of a single black raven *proves* it and *a fortiori*, confirms it: logical implication is the strongest possible form of evidential support.

Then, in an inductive *logic*, confirmation should extend to the logical consequences of what is already confirmed. For instance, if we have evidence for Newton's law of gravitation, it must also be evidence for Kepler's laws, since the latter are a special case of the former. In other words, Hempel suggests the

> *Consequence Condition (CC)*: If an observation report E confirms every member of a set of sentences *S*, then it confirms every logical consequence of *S*, too (e.g. every sentence H for which $S \vDash \mathrm{H}$).

The Consequence Condition also implies the

> *Special Consequence Condition (SCC)*: If an observation report E confirms a hypothesis H, then it confirms every logical consequence of H, too.

The SCC clashes, however, with an important intuition about the link between prediction and confirmation: instead of testing theories directly, we often verify their observational consequences. For instance, the General Theory of Relativity was first tested by checking its predictions for the bending of light by massive bodies. This predictivist approach to confirmation motivates the

> *Converse Consequence Condition (CCC)*: If an observation report E confirms a hypothesis H, then it confirms every hypothesis H' that logically implies H (i.e., $H' \vDash H$).

Accepting both SCC and CCC, however, would trivialize the concept of confirmation. (By EnC, E confirms E; by CCC, E confirms $E \wedge H$ for any H; by SCC, E then confirms H—even when no actual link between E and H exists.) Faced with this choice, Hempel opts for SCC and dismisses CCC. This is mainly because CCC extends the confirmation relation too generously: it allows for the confirmation of mutually incompatible hypotheses (if E confirms H, then E confirms both $H \wedge X$ and $H \wedge \neg X$), and it licenses the tacking paradoxes for hypothetico-deductive confirmation (if E confirms H, then E confirms $H \wedge X$ for any X; see, e.g., Gemes 1998; Sprenger 2011). In fact, for an inductive *logic* it is a strange feature that incompatible conclusions follow from the same set of premises. In line with this reasoning, Hempel adopts the

> *Consistency Condition (CnC)*: If an observation report E confirms the hypotheses H and H', then H' must be logically consistent with H (i.e., there are models of H' that are also models of H).

The three cornerstones of Hempel's qualitative adequacy criteria are thus the Entailment Condition, the (Special) Consequence Condition and the Consistency Condition. Hempel then combines these formal criteria with a substantial confirmation criterion. Of course, logical entailment is usually too strong as a necessary criterion for confirmation: no finite set of observations will ever imply a universal statement of the form "all *F*s are *G*s." However, the hypothesis should agree with the evidence in the domain of the evidence. Specifically, Hempel suggests that if an observation report says something about a set of the singular terms (e.g., $S_E = \{a, b, c\}$), the evidence should provide a model of the restriction or development of the hypothesis to S_E (a precise definition is given in Hempel 1943). For instance, if $E = Fa \wedge Ga \wedge Fb$, then the development of $H = \forall x{:}\ Fx \rightarrow Gx$ to $S_E = \{a, b\}$ is $H_{|dom(E)} = (Fa \rightarrow Ga) \wedge (Fb \rightarrow Gb)$. This brings us to the

> *Satisfaction Criterion*: A piece of evidence E *directly Hempel-confirms* a hypothesis H if and only if E provides a model of the restriction of H to the domain of E. In other words, $E \vDash H_{|dom(E)}$, where $H_{|dom(E)}$ denotes the restriction of H to the singular terms that occur relevantly in E.

This criterion can be generalized as follows: anything that follows classically from a set of directly confirmed hypotheses counts as confirmed, in agreement with Hempel's Consequence Condition.

> *Hempel-Confirmation*: A piece of evidence E *Hempel-confirms* a hypothesis H if and only if H is entailed by a set of sentences Γ so that for all sentences $\varphi \in \Gamma$, φ is directly Hempel-confirmed by E.

It is easy to see that Hempel's account satisfies the Special Consequence and the Entailment Condition. It also improves upon several shortcomings of both the naïve account of confirmation by instances, and the hypothetico-deductive account. In fact, it also stands at the core of Glymour's (1980) account of bootstrap confirmation, and it can be connected to hypothetico-deductive confirmation too (Sprenger 2013). We will now move to the ravens' paradox as an important test case for formal theories of confirmation.

The ravens' paradox

Natural laws, and hypotheses about natural kinds, are often formulated in the form of universal conditionals, such as "all planets move in elliptical orbits," "all ravens are black," or "all lions are carnivores." According to a tradition in philosophy of science that goes back to Jean Nicod (1925/1961), hypotheses of the form "all *F*'s are *G*'s" are confirmed by their *instances*, that is, observations of an *F* that are also *G*s (e.g., E = $Fa \land Ga$). This suggests the following condition:

> *Nicod Condition (Confirmation by Instances)*: Universal conditionals such as H = $\forall x$: ($Fx \rightarrow Gx$) are confirmed by their instances, that is, propositions such as E = $Fa \land Ga$.

At the same time, as we have seen in the last section, theories of confirmation should respect certain logical principles. For example, if two hypotheses H and H' are logically equivalent, they should be equally confirmed by an observation E: inductive support should not depend on the chosen formulation of a hypothesis. This brings us to the

> *Equivalence Condition*: If observation E confirms hypothesis H, then it also confirms any hypothesis H' that is logically equivalent to H.

In fact, the Equivalence Condition is not only highly plausible; it also follows directly from Hempel's other criteria: if H is equivalent to H', H also implies H'. Thus, if E confirms H, E also confirms H' by SCC. However, Hempel (1945) observed that combining the Equivalence and the Nicod Conditions runs counter to established confirmatory intuitions. Take the hypothesis that no non-black object is a raven: H' = $\forall x$: $\neg Bx \rightarrow \neg Rx$. A white swan is an instance of that hypothesis. Thus, by the Nicod Condition, observing a white swan (E' = $\neg Ba \land \neg Ra$) confirms H'. By the Equivalence Condition, H' is equivalent to H = $\forall x$: $Rx \rightarrow Bx$ so that E' also confirms the hypothesis that all ravens are black. But obviously, observing a white swan should not affect our attitude toward the color of ravens.

> *Ravens Intuition*: Observations of a white swan or other non-black non-ravens do *not* confirm the hypothesis that all ravens are black.

Hence, we have three individually plausible, but incompatible claims—the Nicod Condition, the Equivalence Condition and the Ravens Intuition—at least one of which has to be discarded. Since this paradox of the ravens was first formulated by Hempel in his "Studies in the Logic of Confirmation" (1945), it is also known as *Hempel's paradox*. Even before, Hempel (1937: 221–2) proposed a structurally identical counterexample to a specific way of measuring the probability of universal conditionals (see also Hosiasson-Lindenbaum 1940).

Facing this trilemma, Hempel dismisses the Ravens Intuition and embraces the paradoxical conclusion: observing a white swan *confirms* the hypothesis that all ravens are black. To motivate that resolution, assume that we observe a gray bird resembling a raven. This bird may be a non-black raven and falsify our hypothesis. However, by conducting a genetic analysis, we learn that the bird is no raven, but a kind of crow. Here, it sounds correct to say that the results of the genetic analysis support the raven hypothesis—it was at risk of being falsified and has survived a test (= the genetic analysis). This Popperian line of response is also worked out by various papers in the 1950s and 1960s (e.g., Watkins 1957; Agassi 1958; Good 1966; Hempel 1967).

Hempel's analysis explains why white swans—or more generally, observations of the form $\neg Ra \wedge \neg Ba$—can confirm the raven hypothesis. But why did we have a different intuition in the first place? Hempel traces this back to an ambiguity in the paradox. In the crow/raven case, we did not yet know whether the newly observed bird was a raven or a crow. Therefore, its investigation has confirmatory (and falsificatory) potential. By contrast, in the white swan example, we know that *the object before us is no raven*: "this has the consequence that the outcome of the . . . test becomes entirely irrelevant for the confirmation of the hypothesis and thus can yield no new evidence for us" (Hempel 1945: 19). That is, the observation of a white swan should better be described as the observation of a non-black object (E' = $\neg Ba$) relative to the background knowledge that the object is not a raven (K' = $\neg Ra$) and such cases, that do not put the hypothesis at risk, should *not* count as confirming instances (cf. Popper 1934/1959). E = $\neg Ba \wedge \neg Ra$ and K = Ø, by contrast confirms the raven hypothesis. Accounting for background knowledge explains in particular why "indoor ornithology" cannot yield support for the raven hypothesis. Following this road, we have to rewrite the earlier confirmation criteria accounting for the role of background knowledge. For the Satisfaction Criterion, this can be done straightforwardly (i.e., the condition for direct Hempel-confirmation becomes $E \wedge K \vDash H_{|dom(E)}$).

Fitelson and Hawthorne (2011) pointed out that Hempel's Satisfaction Criterion does not square well with his analysis of the paradox. Since the Satisfaction Criterion is *monotonic* with regard to background knowledge, adding background knowledge cannot invalidate inductive support. In particular, even when we know that *a* is no raven (E = $\neg Ba$, K = $\neg Ra$), it will still be the case that

$$E \wedge K = \neg Ra \wedge \neg Ba \vDash (Ra \rightarrow Ba) = H_{|dom(E)}.$$

Thus, even for irrelevant evidence E, the raven hypothesis H is confirmed on Hempel's account. While Hempel spots correctly that the paradoxical conclusion of the raven example can be embraced by relegating the paradoxical aspect to implicit background knowledge, his own theory of confirmation does not implement that insight.

The raven paradox also anticipates Nelson Goodman's *new riddle of induction*. Goodman set up this problem in the third chapter of *Fact, Fiction and Forecast* (Goodman 1955/1983) as a challenge to the Satisfaction Criterion. If we observe only green emeralds up to time point $t = t_0$, this should, in any plausible logic of inductive inference, support the hypothesis that all emeralds are green. Hempel's Satisfaction Criterion agrees in fact if we formalize the hypothesis as the universal conditional H: $\forall x$: $Ex \rightarrow Gx$.

However, consider now the predicate "grue" which applies to an emerald e either (1) if e is green and has been observed up to time t_0; or (2) if e is blue and is observed for the first time after t_0. We can then redescribe our past observations as "emerald e_1 is grue," "emerald e_2 is grue," and so on. These observations support, according to the very same rules of inductive inference—the Satisfaction Criterion in particular—the hypothesis H' that all emeralds are grue. This conclusion violates the Consistency Condition: both H and H' are confirmed by the same observations. Moreover, they are not only incompatible with each other but also disagree on each single prediction for $t > t_0$. These are highly undesirable consequences. While

Hempel's paradox shows that observing instances of a hypothesis is no reliable guide to inductive inference, Goodman's new riddle demonstrates that for any universal generalization H that is confirmed according to the Satisfaction Criterion, we can construe a rival hypothesis H' such that the same observations confirm H', although H and H' make completely incompatible predictions. This casts doubt of the ability of a purely formal, syntactic account of confirmation to support rational expectations about the future.

In a postscript to the "Studies in the Logic of Confirmation" added in 1964, Hempel admits that the Consistency Condition may be too strong as an adequacy criterion for an inductive logic. In fact, it is no coincidence that in later works, Hempel did not work further on formal confirmation criteria. Specifically, in his article "Science and Human Values" (1960), Hempel took up arguments by Rudner (1953) and others that noncognitive, properly ethical values influence the decisions and inferences made by scientists. While Hempel stays faithful to his earlier views that ethical values do not have a logical relationship to theory and evidence, and therefore do not affect judgments of inductive support, he stresses that the *acceptance* or rejection of a hypothesis always carries a risk of error—the famous "inductive risk." Weighing these errors is not a purely logical process and needs to be done on the basis of definite utilities and losses assigned to correct and erroneous decisions (see also his 1981). This later work by Hempel has been very influential recently, for example in Heather Douglas's work (2000, 2009).

Finally, Hempel's late work shows a certain degree of skepticism toward the original logical empiricist project: the application of scientific theories for purposes of explanation and prediction depends on auxiliary assumptions, so-called "provisos"—for example, the absence of factors that could interfere with the forces postulated by the theory (Hempel 1988). In the light of this additional complexity, the task of formulating purely syntactic accounts of confirmation, explanation, and inductive inference becomes increasingly difficult.

Conclusion

Carl Gustav Hempel has been an ingenious researcher with manifold contributions to the development of twentieth-century philosophy of science (see the contributions in Fetzer 2000). The field of confirmation and induction is no exception. Some of his contributions in that area are nowadays mainly of historical interest: for example, the Satisfaction Criterion has mainly been superseded by hypothetico-deductive and Bayesian accounts of confirmation. Nonetheless, as one of the first systematic attempts to develop formal criteria for inductive inference, Hempel's work inspired important follow-up research, such as Goodman's "new riddle of induction," or Glymour's (1980) bootstrap confirmation. Hempel's work on the paradox of the ravens and the analysis he provides, by contrast, are seminal up to today and continue to generate numerous original research articles. From a methodological point of view, his insistence on developing adequacy criteria before moving to a quantitative analysis of confirmation has proven to be an extremely helpful strategy, and it is followed also in various parts of Bayesian philosophy of science (e.g., Fitelson 2001; Sprenger and Hartmann 2019). All in all, Hempel's contributions to the problem of induction and confirmation theory may not be as deep and detailed as Carnap's, but they equal them in terms of originality and interest, and they exceed them in terms of breadth of perspective.

Acknowledgments

Research on this article was supported through Starting Investigator Grant "Making Scientific Inferences More Objective" (grant No. 640638) by the European Research Council, and PRIN grant "From Models to Decisions" by the Italian Ministry for Universities and Research.

References

Agassi, J. (1958) "Corroboration Versus Induction," *British Journal for the Philosophy of Science* 9: 311–17.

Carnap, R. (1928) *Der logische Aufbau der Welt*, Berlin: Weltkreis-Verlag. Trans. *The Logical Structure of the World*, Berkeley: University of California Press, 1967, repr. Chicago: Open Court, 2003.

——— (1930) "Die alte und die neue Logik," *Erkenntnis* 1: 12–26. Trans. "The Old and the New Logic," in A. J. Ayer (ed.), *Logical Positivism*, New York: Free Press, 1959, pp. 60–81.

——— (1947) "On the Application of Inductive Logic," *Philosophy and Phenomenological Research* 8: 133–48.

——— (1950) *Logical Foundations of Probability*, Chicago: University of Chicago Press.

——— (1952) *The Continuum of Inductive Methods*, Chicago: University of Chicago Press.

Douglas, H. (2000) "Inductive Risk and Values in Science," *Philosophy of Science* 67: 559–79.

——— (2009) *Science, Policy, and the Value-Free Ideal*, Pittsburgh: Pittsburgh University Press.

Fetzer, J. H. (2000) *Science, Explanation, and Rationality: Aspects of the Philosophy of Carl G. Hempel*, Oxford: Oxford University Press.

Feyerabend, P. (1975) *Against Method*, London: Verso.

Fisher, R. A. (1935) *The Design of Experiments*, Edinburgh: Oliver & Boyd. Repr. New York: Hafner Press, 1974.

Fitelson, B. (2001) *Studies in Bayesian Confirmation Theory*, PhD thesis, University of Wisconsin, Madison.

Fitelson, B. and Hawthorne, J. (2011) "How Bayesian Confirmation Theory Handles the Paradox of the Ravens," in J. H. Fetzer and E. Eells (eds.), *The Place of Probability in Science*, New York: Springer, pp. 247–75.

Gemes, K. (1998) "Hypothetico-Deductivism: The Current State of Play; the Criterion of Empirical Significance: Endgame," *Erkenntnis* 49: 1–20.

Glymour, C. (1980) *Theory and Evidence*, Princeton, NJ: Princeton University Press.

Good, I. J. (1966) "The White Shoe Is a Red Herring," *British Journal for the Philosophy of Science* 17: 322.

Goodman, N. (1955) *Fact, Fiction and Forecast*, Cambridge, MA: Harvard University Press, 4th ed., 1983.

Hempel, C. G. (1937) "Le problème de la vérité," *Theoria* 3: 206–44. Trans. "The Problem of Truth," in Hempel (2000), pp. 35–75.

——— (1943) "A Purely Syntactical Definition of Confirmation," *Journal of Symbolic Logic* 8: 122–43.

——— (1945) "Studies in the Logic of Confirmation," *Mind* 5: 1–26, 97–121. Repr. in Hempel (1965a), pp. 3–52 (with a 1964 Postscript).

——— (1960) "Science and Human Values," in R. E. Spiller (ed.), *Social Control in a Free Society*, Philadelphia: University of Pennsylvania Press, pp. 39–64. Repr. in Hempel (1965a), pp. 81–96.

——— (1965a) *Aspects of Scientific Explanation and Other Essays in the Philosophy of Science*, New York: Free Press.

——— (1965b) "Recent Problems of Induction," in R. G. Colodny (ed.), *Mind and Cosmos*, Pittsburgh: Pittsburgh University Press, pp. 112–34. Repr. in Hempel (2001), pp. 29–48.

——— (1967) "The White Shoe: No Red Herring," *British Journal for the Philosophy of Science* 18: 239–40.

——— (1981) "Turns in the Evolution of the Problem of Induction," *Synthese* 46: 389–404. Repr. in Hempel 2001, pp. 344–56.

——— (1983) "Valuation and Objectivity in Science," in R. Cohen and L. Laudan (eds.), *Physics, Philosophy, and Psychoanalysis*, Dordrecht: Reidel, pp. 73–100. Repr. in Hempel (2001), pp. 372–95.

——— (1988) "Provisos: A Problem Concerning the Inferential Function of Scientific Theories," *Erkenntnis 28*, 147–64. Repr. in Hempel (2000), pp. 229–49.

——— (2000) *Selected Philosophical Essays* (ed. by R. Jeffrey), Cambridge: Cambridge University Press.

——— (2001) *The Philosophy of Carl G. Hempel. Studies in Science, Explanation and Rationality* (ed. by J. H. Fetzer), Oxford: Oxford University Press.

Hempel, C. G. and Oppenheim, P. (1945) "A Definition of 'Degree of Confirmation'," *Philosophy of Science* 12: 98–115. Repr. in Hempel (2000), pp. 135–61.

——— (1948) "Studies in the Logic of Explanation," *Philosophy of Science* 15: 135–75. Repr. in Hempel (1965a), pp. 245–96 (with a 1964 Postscript).

Hosiasson-Lindenbaum, J. (1940) "On Confirmation," *The Journal of Symbolic Logic* 5: 133–48.

Kuhn, T. S. (1962) *The Structure of Scientific Revolutions*, Chicago: University of Chicago Press, 2nd ed., 1970.

Nicod, J. (1925) *Le problème logique de l'induction*, Paris: Alcan. Repr. Paris: Presses Universitaires de France, 1961.

Popper, K. R. (1934) *Logik der Forschung*, Vienna: Springer. Rev. ed. trans. *The Logic of Scientific Discovery*, London: Routledge, 1959, repr. 2002.

Rudner, R. (1953) "The Scientist qua Scientist Makes Value Judgments," *Philosophy of Science* 20: 1–6.
Sprenger, J. (2011) "Hypothetico-Deductive Confirmation," *Philosophy Compass* 6: 497–508.
——— (2013) "A Synthesis of Hempelian and Hypothetico-Deductive Confirmation," *Erkenntnis* 78: 727–38.
Sprenger, J. and Hartmann, S. (2019) *Bayesian Philosophy of Science*, Oxford: Oxford University Press.
Watkins, J. W. N. (1957) "Between Analytic and Empirical," *Philosophy* 32: 112–31.

27
CARNAP AND ONTOLOGY

Gregory Lavers

On the one hand, there is much about Carnap's position on ontology that remains constant over the course of his career. This includes the views that ontological questions are closely tied to the choice of a language, and that they are generally not theoretical in character. On the other hand, there is much about his views on ontology that does change. One of the most significant changes involves his attitude toward *truth* and *reference*. At the time of *The Logical Syntax of Language* (1934, hereafter *Syntax*), Carnap recommended eliminating these concepts, but after Tarski's influence, he came to see these concepts as standing in need of clarification (or explication). To properly understand Carnap's mature position on ontology, it is essential to understand the position that it had emerged from.

Carnap's earlier views on ontology

Carnap describes himself as having realized, at the time of writing *Der Logische Aufbau der Welt* (1928a), that he had for a long time, according to whom he was speaking to, used languages that others might call *materialistic, idealistic, nominalistic*, or even *Platonic* (Carnap 1963: 17). Carnap says that others saw this as objectionable, but he gradually realized that he held a neutral position on ontological controversies. In fact, this attitude is built into this work itself by its being declared a matter of choice whether the system of concepts is constructed from a phenomenalistic or physicalistic basis. In §177, Carnap argues that his construction theory is not incompatible with realism, idealism, or phenomenalism, although these views presuppose a metaphysical concept of reality. In *Scheinprobleme* (1928b) as well, Carnap argues that the philosophical theses of realism, idealism, and phenomenalism employ a metaphysical concept of reality with no factual content. We can see, therefore, in this very early work, that Carnap already holds a version of the view that ontological positions do not concern anything factual but concern the manner in which we choose to describe the world.

In *Syntax*, Carnap wants to make this idea more precise—his goal is to generalize metamathematics in order to show which philosophical disputes are in reality concerned with the features of some language. Let us now examine the way in which Carnap deals with questions of ontology in *Syntax*. First, consider the division between the formal and material mode of speech. In order to understand what is truly being claimed by a philosophical assertion, we need to translate it from the material to the formal mode of speech. The central feature of the formal mode

 DOI: 10.4324/9781315650647-30

of speech is that it eliminates the concepts of meaning, truth, and reference. We also ought to eliminate ***universal words*** (words that apply to everything in their domain of applicability) such as *thing, number*, or *property*. Statements that do involve these notions are deemed in the material (*inhaltlich*) mode of speech. In the very opening of the book, Carnap discusses what it means to claim that logical syntax is formal:

> By the *logical syntax* of a language, we mean the formal theory of the linguistic forms of that language—the systematic statement of the formal rules which govern it together with the development of the consequences that follow from these rules.
>
> A theory, a rule, a definition, or the like is to be called formal when no reference is made in it either to the meaning [*Bedeutung*] of the symbols (for example, the words) or to the sense [*Sinn*] of the expressions (e.g. the sentences), but simply on the kind and serial order of the symbols from which the expression is constructed.
>
> (1934/1937: §1, orig. emphasis)

So we ought not ask or answer questions about the meaning of terms, since all properly posed philosophical questions must, for Carnap, be phrased in the formal mode of speech. Now, to understand Carnap's position on ontology at this time, we need to see how this stance relates to the existential assumptions made in the work. In Languages I and II, the statements that zero is not a successor and that distinct numbers have distinct successors are among the primitive sentences. So both Languages I and II assert the existence of infinitely many entities of type 0. Carnap, here, does not dispute the Wittgensteinean view that logic should not imply the existence of any objects:

> If logic is to be independent of empirical knowledge, then it must assume nothing concerning the *existence of objects*. For this reason Wittgenstein rejected the Axiom of Infinity, which asserts the existence of an infinite number of objects. And, for kindred reasons, Russell himself did not include this axiom amongst the primitive sentences of his logic.
>
> (ibid.: §38a, orig. emphasis)

Does this mean that Carnap takes these sentences to be a posteriori (i.e., not independent of empirical knowledge)? Of course not, but to avoid making such a claim, Carnap appeals to his distinction between ***name-languages*** and ***coordinate-languages***. A name-language picks out each entity of a domain by a name, whereas a coordinate-language picks out each element in a systematic way using a number. But worries concerning the existence of abstract objects are not worries about the manner in which the elements of the domain are picked out. So the appeal to this distinction does not alleviate any concerns about logic making existential presuppositions.

In fact, in terms of ontology, things get quite a bit worse in *Syntax*. When attempting to provide a definition of ***analyticity*** for Language II, Carnap had planned to give a substitutional interpretation of the higher-order quantifiers. This would allow him to claim that the definition is clearly syntactical. Gödel, however, showed that this would create a clear problem. Since there are only countably many predicate expressions in the language, but uncountably many subsets of elements of type 0, there would be certain mathematical falsehoods that come out as analytic. Carnap accepts Gödel's point:

> Thus the definition must not be limited to the syntactic properties which are definable in S, but must refer to all syntactic properties whatsoever. But do we not by this means

> arrive at a Platonic absolutism of ideas, which is non-denumerable and therefore can never be exhausted by definitions, is something that subsists in itself, independent of all construction and definition?
>
> (ibid.: §34d)

Carnap does not accept, of course, that this leads to Platonism. His reason for this is that we can define the valuations we require (the range over which we interpret the higher-order variables) either in the language itself or in some other language S_2. The reason he sees this as something of an answer to the charge of Platonism is that he takes an instrumentalist stance toward the logico-mathematical portion of the language.

> The S_l [logical sentences] (and with them all sentences of mathematics) are, from the point of view of material interpretation, expedients for the purpose of operating with S_d [descriptive sentences]. Thus, in laying down an S_l as a primitive sentence, only usefulness for this purpose is to be taken into consideration.
>
> (ibid.: §38a)

Carnap says that once a material interpretation of the descriptive portion of the language has been given, no further interpretation is required. Any sentence will have a truth-value once the descriptive vocabulary is interpreted, as all sentences of the logico-mathematical portion are either analytic or contradictory. But if this is the justification of his instrumentalist stance toward the logico-mathematical portion of the language, then it is clearly inappropriate to appeal to this instrumentalism while defending the definition of *analyticity* against the objection of Platonism.

Carnap would have likely realized the inadequacy of his response here if he did not see all questions about ontological assumptions as being improperly formed. Talk of meaning ought to be eliminated in order to attain a proper philosophical understanding of what is being claimed. For example, consider how Carnap translates a sentence about the content of arithmetical claims:

> 10a. The sentences of arithmetic *state* (or express) certain properties of numbers and certain relations between numbers.
>
> 10b. The statements of arithmetic are composed of numerical expressions and one- or many-termed numerical predicate in such and such a way.
>
> (ibid.: §75)

Likewise, we cannot ask if "5" refers to a number but must rest content with the translation of this into the formal mode as " '5' is a numeral." This is typical of ontological questions—they are pseudo-object questions of the material mode of speech. That is, they appear to concern objects but are, in fact, equivalent to sentences about terms, expressible in a syntax language. Notice that this translation from the material to the formal mode of speech involves eliminating the concept of reference as well as the universal word *number*.

We started discussing *Syntax* by saying that Carnap was trying, in this work, to make precise the view that ontological questions concern the features of some language. We are now in a position to see the flaw in Carnap's proposal. At the time of *Syntax*, Carnap was using what he called *syntax-languages* as his metalanguages. An object-language could include any collection of logical and descriptive vocabulary. For Carnap, however, metalanguages serve the sole purpose of discussing the properties of an object-language. Carnap saw it as following from this that there is no need for a significant amount of descriptive vocabulary in the metalanguage (only enough to describe which terms appear in which locations). This restriction, on the descriptive vocabulary in the metalanguages, allows Carnap to claim, without falling into triviality, that talk of reference to abstract objects is all done via pseudo-object sentences. Once semantic metalanguages are allowed, which contain a translation of the object-language and a truth predicate, all sentences become *pseudo-object sentences*, and this term would become trivial. "Rudolf Carnap was born in 1891" is equivalent to "A true sentence results from the substitution of 'Rudolf Carnap' for 'x' in the predicate 'x was born in 1891.'" But no one would take this as evidence that talk about people and their birth year is really just concerned with the features of some language. Carnap's metalanguages (syntax languages) contain as much logico-mathematical vocabulary as one could wish for, and are often rich enough to formulate a truth predicate (i.e., "analytic") for the logico-mathematical portion of the object-language. So Carnap's strategy for showing that talk of abstract objects is, properly considered, about terms could be extended in the context of semantic metalanguages to show that all sentences are pseudo-object sentences.

At best, *Syntax* dissolves rather than solves problems about ontology. Less charitably, Carnap simply ignores questions of ontology by dismissing them as improperly phrased. But this strategy is of course predicated on there being something inappropriate about talk of reference. Even in *Syntax*, Carnap considers the possibility of clarifying central concepts of the material mode of speech (like truth and reference):

> The material mode of speech is not itself erroneous it only readily leads itself to wrong use. *But if suitable definitions and rules are laid down and systematically applied, no obscurities or contradictions arise.* Since, however, the word-language is too irregular and too complicated to be actually comprehended in a system of rules, one must guard against the dangers of the material mode of speech as it is ordinarily used in the word-language by keeping in mind the peculiar character of its sentences.
>
> (ibid.: §81, emphasis added)

In the next section, we will see how Carnap's views change in response to such suitable definitions of central concepts of the material mode of speech like "truth" and "reference." He will continue to hold that, as they are used in ordinary language, they are likely to lead to confusion. But rather than needing to be eliminated, these notions stand in need of clarification.

Tarski's influence and the position of "Empiricism, Semantics, and Ontology"

In 1942, the Schilpp volume on G. E. Moore was published and contained an article by C. H. Langford called "The Concept of Analysis in Moore's Philosophy." Here, Langford argues that, in giving an analysis, the analysandum and analysans have to have the same meaning (in order for it to be correct). On the other hand, they must have different meanings in order to be informative and avoid complete triviality. This problem Langford describes as *the paradox of analysis*. Carnap read this work shortly after its publication, and it led to his formulating his account of *explication*. Carnap understood this paradox to rule out a certain conception of analysis. The goal cannot be to arrive at a concept identical to the one under analysis. If the analysans and analysandum are not identical, then they are distinct, and in giving an analysis we are introducing a new notion. Since correctness cannot be the goal of an analysis, Carnap suggests that explications (his term for philosophical analyses) must be governed by pragmatic norms like *simplicity*, *fruitfulness*, and *precision*. The explicatum has to be similar to the explicandum on Carnap's view, but only so much so that it could be used in its stead (see his 1950a: 1–18).

Of course, Carnap did not have his general account of explication until the mid-1940s, but we can see much of his earlier work as explicating various concepts. In fact, the quote that ended the previous section could be seen as a call for the explication of concepts of the material mode of speech. In his discussion of Tarski's work in his "Intellectual Autobiography" from the Schilpp volume, Carnap clearly states that he sees Tarski as having done just that. "I recognized that it provided for the first time the means for precisely explicating many concepts used in our philosophical discussions" (1963: 60–61). Once we accept semantic metalanguages that include the object language, not only can we define truth for formalized languages, but we can also define the concept of reference. We require of a reference predicate that it satisfy all instances of " 'a' refers to a." The concept of meaning, which was to be eliminated in order to achieve a proper philosophical understanding, can now be given a simple, precise, and fruitful definition.

In his 1939 monograph *Foundations of Logic and Mathematics*, a mere five years after *Syntax*, the change in the attitude toward reference is obvious. After defining the cardinal numbers and then laying out the Peano axioms, Carnap states:

> The *customary interpretation* of the Peano system may first be formulated this way: "b" designates the cardinal number 0; if ". . ." designates the cardinal number n then ". . .′" designates the next one, i.e., n + 1; "N" designates the class of finite cardinal numbers. Hence on this interpretation the system concerns the progression of finite cardinal numbers ordered according to magnitude.
>
> (1939/1955: 182, orig. emphasis)

Whereas the older position on terms for abstract objects was that they do not stand in need of interpretation and that all talk of interpretation was misguided, his new position abandons this instrumentalism completely. Talk of the reference to abstract objects, now sufficiently clarified, is seen by Carnap to be completely unproblematic. "Empiricism Semantics and Ontology" (1950b, hereafter ESO) identifies, as its main task, defending the use of abstract objects as the reference of terms in semantics. Let us turn now to the argument presented there.

Although the term *explication* is absent from ESO, ESO is all about explication. When he discusses the system of numbers, he is considering a Frege-style definition of number. Frege's definition of number, followed closely by Tarski's definition of truth, is Carnap's favorite example of a successful explication. The argument concerning the central issue of the paper (abstract objects in semantics) is all about explications as well. His argument, on this point, is given extremely briefly in the penultimate section. Consider the claim " '5' refers to a number." Before we can assess such a claim, we first require an explication of both the arithmetical and the semantic vocabulary used. Carnap has in mind Fregean definitions of not just the individual numbers, but also the general concept of number. Relative to this explication, it holds that 5 is a number. Also, we require of a reference (or designation) predicate that it satisfies all instances of " 'a' refers to a." So, in particular, " '5' refers to 5." But then clearly relative to both of these explications, it holds that "5" refers to a number. Since at every step here we are giving explications, and correctness is not a norm of explications (as shown by the paradox of analysis), there is no question remaining as to whether or not these explications "get things right." We could do something similar for any other system of abstract objects which we wish to introduce. For example, Carnap, in *Meaning and Necessity*, explicitly describes himself as giving an explication of proposition (1947: §2). A semantic metalanguage for such a system would see it as containing terms that refer to propositions. All of the *linguistic frameworks* considered in ESO are to be understood as explications of a certain range of vocabulary. Given a suitable explication, we can say that various terms for abstract objects refer. These, of course, are answers to internal questions, being relative to explications of the relevant vocabulary. The person who asks the *external question* of whether the terms really refer, is not satisfied with such relativized answers and assumes that such questions have definitive answers independent of what Carnap sees as much-needed clarifications of the relevant vocabulary (Carnap's relativism on ontology has inspired a number of contemporary metaphysicians who call themselves neo-Carnapians: for discussions, see Thomasson 2015; Hirsch 2010; Eklund 2016).

In considering the sentence "numerical expressions refer" we see that what Carnap is relativist about is how the relevant vocabulary is explicated. Questions about whether the sentences of such systems are true or whether the terms refer are to be answered affirmatively according to the most straightforward explication of the semantic vocabulary (as it applies to such systems). "Thus the question of the admissibility of entities of a certain type or of abstract entities in general as designata is reduced to the question of the acceptability of the linguistic framework for those entities" (1950b/1956: 217). Given the clarity with which semantic concepts could be defined in semantic metalanguages, Carnap saw no motivation for treating any of the object-languages for abstract objects as empty formalisms. All of this, of course, is not to say that Carnap's final position is a form of metaphysical realism or Platonism. Throughout his career, Carnap rejected the doctrine of Platonism as meaningless metaphysics. That said, Carnap is prepared to accept sufficiently clearly explicated versions of many statements we would take to characterize Platonism (for considerations of space, I have focused on Carnap's views on abstract objects. For a discussion of Carnap's views on theoretical entities in science: see Lavers 2016a).

Relation to Frege, Russell, and Quine

As Michael Beaney has argued (1996, 2004), already Frege was concerned with the problem that we call the paradox of analysis. In fact, in his lecturer notes for his 1914 course "Logic in Mathematics" (published as Frege 1969), Frege puts forward a view on explication very similar to Carnap's. Here Frege talks of analyzing terms to gain a clearer understanding of their meanings. But then, in the construction of the system, the same term can be used, but its meaning is strictly what is conferred on it by the constructive definitions. Carnap was one of the few students in this class and took detailed notes (now published in Reck and Awodey 2004). Carnap's notes, however, do not include the discussion of definition and analysis just mentioned. Furthermore, Carnap did not come up with his own solution to the paradox of analysis until 1945 after reading Langford's article. So, the extent to which Carnap's views on explication have their origin in Frege can only be speculated.

Despite these similarities, one may want to claim that, for Frege, mathematical propositions are objective, mind-independent truths, and for Carnap they are human-made conventions. This would involve a severe distortion of at least the later Carnap's position. It is true that Carnap's early view was a type of conventionalism. The principle of tolerance, after all, speaks of establishing conventions. However, in his later period he saw talk of conventions as unhelpful for explaining logico-mathematical truth.

> Once the meanings of the individual words in a sentence [of the form "All black dogs are dogs"] are given (which may be regarded as a matter of convention), then it is no longer a matter of convention or of arbitrary choice whether to regard the sentence as true; the truth of such a sentence is determined by the logical relations holding between the given meanings.
>
> (1963: 916)

So, at least according to his later philosophical view, Carnap is a conventionalist about logico-mathematical truth in only a trivial sense. The meaning of basic vocabulary (individual words) is conventionally determined, but this has to be admitted in all domains of knowledge whatsoever. Carnap is, however, a pluralist in that he holds there may be more than one way to satisfactorily explicate the logical vocabulary. It is important to distinguish between pluralism and conventionalism, and to recognize that only the latter implies that truths are somehow ultimately dependent on human beings.

In fact, Carnap, sounds positively Fregean when discussing propositions in *Meaning and Necessity*:

> We take as the extension of the sentence its truth-value, and as its intension the proposition expressed by it. This is in accord with the identity conditions for extensions and for intensions stated in the preceding section. Propositions are here regarded as objective, nonmental, extra-linguistic entities. It is shown that this conception is applicable also in the case of false sentences.
>
> (1947/1956: 25)

Carnap's justification for this is that neither people nor languages, nor anything subjective, enter into the truth conditions for statements about propositions. The same could be said relative to an explication of our mathematical vocabulary. Therefore, Carnap—at least the later Carnap—would maintain that arithmetical statements are objective, mind-independent truths, and furthermore, are not *true by convention* but true in virtue of the logical relations that hold between the meanings of the basic elements that they contain.

Frege, in *Grundgesetze* (1893–1903), recognizes that in constructing his system he had to make several pragmatic choices (e.g., treating the truth-values as objects), but he shows a certain confidence that anyone trying to attain the same goal would have to make those same choices. This optimism on Frege's part, regarding his chosen principles working better than any alternatives, as compared to Carnap's attitude of tolerance, is the main difference between their philosophical views (see Lavers 2013, 2016b for further comparisons).

Russell, in his discussion of ontological questions, displays an attitude at greater odds with Carnap's views on ontology. Russell thought that logical analysis, together with a sense for what is real, will reveal the basic inventory of the world. In *The Philosophy of Logical Atomism*, Russell describes his changing attitude toward the belief in propositions, saying that the belief in false propositions "is more than one can manage to believe, and I do think no person with a vivid sense of reality can imagine it" (1918/2009: 55). We saw earlier that Carnap introduces propositions in *Meaning and Necessity*. Here, he shows that he is facing the problem that Russell was facing in the previous quote. Carnap says, "[t]he greatest difficulty in the task of explicating the concept of proposition is involved in the case of the false proposition" (1947/1956: 29). That said, there is no theoretical question of the existence of propositions that is to be decided by our sense of reality. A theoretical question would be one that is sufficiently clear to admit of a definite answer. But in the case of the existence of propositions, this is not the case. Here, there are only practical questions of how they can be defined, and the practical utility of doing such.

Quine described one of his principal differences with Carnap as concerned with ontology. For many years after the debate between Carnap and Quine, Quine was widely seen as the clear winner. However, due to a growing interest in Carnap, and the history of analytic philosophy generally, this standard interpretation has been called into question (see, e.g., Coffa 1991; Creath 2007; Friedman 1999). Quine famously argued that even if ontological questions tend to be relative to a language, there is a language-independent formulation of the questions of ontology.

Can we find a language, sufficient for the purposes of science, that avoids quantifying over a certain kind of entity? This question is not tied to the features of any particular language. Quine saw Carnap, through even the later years of Carnap's life, as trying to hold on to the ontological position of *Syntax*. There, Carnap maintained an instrumentalist attitude toward the logico-mathematical portion of the language. Carnap also, at the time, saw *universal words* as part of the misleading material mode of speech. Universal words (inspired by Wittgenstein's notion of a *formal concept*) include *thing*, *number*, or *property*. They are, as previously mentioned, terms that apply to everything in their respective domains. Even in 1968, Quine sees Carnap as wanting to hold on to the thesis of the illegitimacy of universal words and sees Carnap's position in ESO as a minor reformulation of this doctrine. "In his later writing this doctrine of universal words takes the form of a distinction between internal and external questions, in which people come to grips with the relative merits of theories" (1969: 52). Carnap's published version of *Meaning and Necessity* contains a remark made in response to a comment from Quine on an earlier draft. Here Carnap clearly disavows the view that there is anything illegitimate about universal words. "It is important to emphasize the point just made that, once you admit certain variables, you are bound to admit the corresponding universal concept" (1947/1956: 44). Quine also believes that Carnap is still trying to defend some kind of instrumentalism towards the logico-mathematical vocabulary:

> Carnap . . . has recognized that he is able to preserve a double standard for ontological questions and scientific hypotheses only by assuming an absolute distinction between the analytic and the synthetic; and I need not say again that this is a distinction which I reject. The issue over there being classes seems more a question of convenient conceptual scheme; the issue over there being centaurs, or brick houses on Elm Street, seems more a question of fact. But I have been urging that this difference is only one of degree.
>
> (1951/1963: 45–46)

It is the *Syntax* view that contains a double standard toward ontology. There, we need only interpret the descriptive portion of a language. Carnap in his later period, however, has abandoned his instrumentalist stance toward abstract ontology (see Lavers 2015). In fact, after all, it was a stated goal of ESO to show the acceptability of abstract objects as designata in semantics, and also to help empiricists overcome their nominalistic scruples.

References

Beaney, M. (1996) *Frege: Making Sense*, London: Duckworth.

——— (2004) "Carnap's Conception of Explication: From Frege to Husserl?" in S. Awodey and C. Klein (eds.), *Carnap Brought Home: The View from Jena*, Chicago: Open Court, pp. 117–50.

Blatti, S. and Lapointe, S. (2016) *Ontology after Carnap*, Oxford: Oxford University Press.

Carnap, R. (1928a) *Der Logische Aufbau der Welt*, Berlin: Weltkreis-Verlag. Trans. *The Logical Structure of the World*, Berkeley: University of California Press, 1967, repr. Chicago: Open Court, 2003.

——— (1928b) *Scheinprobleme in der Philosophie*, Berlin: Weltkreis-Verlag. Trans. "Pseudoproblems in Philosophy," in Carnap (1928a/1967), pp. 301–43.

——— (1934) *Logische Syntax der Sprache*, Vienna: Springer. Rev. ed. trans. *The Logical Syntax of Language*, London: Kegan, Paul, Trench Teubner & Cie, 1937, repr. Chicago: Open Court, 2002.

——— (1939) *Foundations of Logic and Mathematics*, Chicago: University of Chicago Press. Repr. in O. Neurath, R. Carnap and C. Morris (eds.), *International Encyclopaedia of Unified Science*, Chicago: University of Chicago Press, 1955.

——— (1947) *Meaning and Necessity: A Study in Semantics and Modal Logic*, Chicago: University of Chicago Press, 2nd ed., 1956.

——— (1950a) *Logical Foundations of Probability*, Chicago: University of Chicago Press.

——— (1950b) "Empiricism, Semantics and Ontology," *Revue International de Philosophie* 4: 20–40. Repr. in Carnap, *Meaning and Necessity*, 2nd ed., 1956, pp. 205–21.

——— (1963) "Intellectual Autobiography" and "Comments and Replies," in P. A. Schilpp (ed.), *The Philosophy of Rudolf Carnap*, La Salle, IL: Open Court, pp. 3–84, 859–1016.

Coffa, A. (1991) *The Semantic Tradition From Kant to Carnap: To the Vienna Station*, Cambridge: Cambridge University Press.

Creath, R. (2007) "Quine's Challenge to Carnap," in M. Friedman and R. Creath (eds.), *The Cambridge Companion to Carnap*, Cambridge: Cambridge University Press, pp. 316–35.

Eklund, M. (2016) "Carnap's Legacy for the Contemporary Metaontological Debate," in Blatti and Lapointe (2016), pp. 165–89.

Frege, G. (1893–1903) *Grundgesetze der Arithmetik*, Jena: Verlag Hermann Pohle, vol. 2. Part. trans. *The Basic Laws of Arithmetic: Exposition of the System*, Berkeley: University of California Press, 1964, complete trans. *The Basic Laws of Arithmetic*, Oxford: Oxford University Press, 2013.

——— (1969) "Logik in der Mathematik" (orig. 1914, unpublished), in H. Hermes, F. Kambartel and F. Kaulbach (eds.), *Nachgelassene Schriften*, Hamburg: Meiner, pp. 219–70. Trans. "Logic in mathematics," in H. Hermes, F. Kambartel and F. Kaulbach (eds.), *Posthumous Writings*, Oxford: Basil Blackwell, 1979, pp. 203–50.

Friedman, M. (1999) *Reconsidering Logical Positivism*, Cambridge: Cambridge University Press.

Hirsch, E. (2010) *Quantifier Variance and Realism: Essays in Metaontology*, Oxford: Oxford University Press.

Langford, C. H. (1942) "The Notion of Analysis in Moore's Philosophy," in P. A. Schilpp (ed.), *The Philosophy of G. E. Moore*, La Salle, IL: Open Court, pp. 321–42. Repr. 1968.

Lavers, G. (2013) "Frege, Carnap, and Explication: 'Our Concern Here Is to Arrive at a Concept of Number Usable for the Purpose of Science,'" *History and Philosophy of Logic* 34: 225–41.

——— (2015) "Carnap, Quine, Quantification and Ontology," in A. Torza (ed.), *Quantifiers, Quantifiers, and Quantifiers: Themes in Logic, Metaphysics, and Language*, Dordrecht: Springer, pp. 271–99.

——— (2016a) "Carnap on Abstract and Theoretical Entities," in Blatti and Lapointe (2015), pp. 200–19.

——— (2016b) "Frege the Carnapian and Carnap the Fregean," in S. Costrie (ed.), *Early Analytic Philosophy—New Perspectives on the Tradition*, Dordrecht: Springer, pp. 353–73.

Quine, W. V. O. (1951) "Two Dogmas of Empiricism," *Philosophical Review* 60: 20–43. Repr. in Quine, *From a Logical Point of View*, New York: Harper & Row, 1933, pp. 20–46.

——— (1969) "Ontological Relativity," in Quine, *Ontological Relativity and other Essays*, New York: Columbia University Press, pp. 26–68.

Reck, E. and Awodey, S. (eds.) (2004) *Frege's Lectures on Logic: Carnap's Student Notes, 1910–1914*, Chicago: Open Court.

Russell, B. (1918) "The Philosophy of Logical Atomism," *The Monist* 28–29. Repr. in Russell, *The Philosophy of Logical Atomism*, London: Routledge, 2009, pp. 1–124.

Thomasson, A. L. (2015) *Ontology Made Easy*, Oxford: Oxford University Press.

28
NEURATH ON POLITICAL ECONOMY

John O'Neill

While Otto Neurath's contributions to the philosophy of the Vienna Circle continue to be influential, his contributions to political economy, the discipline in which he trained, are less well known. This is the case despite their theoretical and practical significance. Theoretically, his proposals for socialization formulated after the First World War formed a starting point for the socialist calculation debates. Ludwig von Mises's contribution starts from criticism of Neurath, as does Max Weber's contribution on the subject in *Economy and Society*. Neurath's criticisms of monetary valuation, his theory of well-being, and his physicalist understanding of the economy contributed to the development of ecological economics. His account of well-being is a precursor to more recent multidimensional theories of well-being. Neurath's work also had a practical significance. His socialization proposals were developed partly in his role as director of socialization in the Bavarian revolution. He was deeply engaged with what he characterized as the "communal economy" of the Viennese settler movement which led to further work in urban planning and housing in Red Vienna. He was involved in experiments in urban planning in the UK after the Second World War. He created the Social and Economic Museum in Vienna. His work on isotypes, the visual presentation of social facts, had a major impact in the presentation of social statistics. Given this significance and influence, why has his work in political economy not been more widely recognized?

Part of the answer to that question lies in the reception of logical empiricism which created influential misunderstandings of Neurath's political economy. From very different political positions, both the Austrian School of economics and the Frankfurt School take logical empiricism generally, and Neurath in particular, to be committed to a technocratic politics, grounded in a scientistic understanding of the social sciences. The lens of the Frankfurt School has had a particular influence on accounts of the place of the Vienna Circle in histories of Western Marxism. It appears as a foil to mainstream Western Marxism that is taken to be marked by the influence of philosophical idealism, a shift from concern with economy to the superstructure, in particular culture, and critical distance from emphasis on the scientific nature of socialism. The commitment to empiricism and scientism is claimed to lead to a social science that is incapable of criticism of the existing social order and a political practice that is conservative and technocratic. The actual history of Western Marxism before the Second World War is more complex than this standard history allows. Central figures in the Western Marxist tradition were closer to the left Vienna Circle than the account suggests. Karl Korsch was associated with the Berlin

DOI: 10.4324/9781315650647-31

Group of logical empiricists, and Bertolt Brecht's notebooks show the influence of Neurath's social behaviorism.

More recently, logical empiricism's contribution to economics has been criticized from another direction for being responsible for the conceptual and normative impoverishment of modern neo-classical economics. Amartya Sen's revival of a critical and descriptively rich welfare economics is taken by Hilary Putnam and Vivian Walsh to involve excising the influence of logical positivism on neo-classical economics (e.g., Putnam 2002; Putnam and Walsh 2014). The claims are similarly problematic. Neurath's political economy was critical of neo-classical approaches. His multidimensional account of well-being, like that of Sen, employs vocabulary that is rich in its characterization of the different dimensions of well-being and the social, biophysical and environmental conditions required for their realization (Lessman 2007; O'Neill and Uebel 2008). A prevalent founding myth of recent political philosophy is that Rawls revived political philosophy that the influence of logical empiricism had silenced. The result is the loss from view of political and social thought of the left Vienna Circle which is marked by an economic radicalism and an understanding of the environmental dimensions of economic and political institutions that was absent from much liberal political philosophy after Rawls.

A central aim of this chapter is to correct some of these mischaracterizations of Neurath's political economy and its relationship to logical empiricism. To do so is not to say that his position is without problems. The intent is rather to reveal something of the richness of Neurath's political economy that shows that more detailed critical scrutiny is warranted. The source of many of the misconceptions about the political economy of Neurath stem from the very different traditions of Austrian economics and the Frankfurt School. Correspondingly, this chapter focuses on Neurath's debates with Mises and Hayek in the Austrian tradition and with Horkheimer in the Frankfurt tradition.

Socialist calculation and the limits of monetary valuation

Central to Neurath's political economy are arguments about the limits of monetary measures and a defense of measures in kind—*in natura*—that employ physical and social measures of the conditions for and dimensions of human well-being. Neurath makes a number of claims that need to be distinguished. The first concerns economic theory. Against standard economic theory which characterizes economic activity through "monetary and credit relations" (1916/2004: 301), Neurath defends the tradition of political economy, typified by work of Aristotle through to Smith, and later of Marx and of Popper-Lynkeus and Ballod-Atlanticus, in which the concept of real wealth and the analysis of social relations are central. The second claim concerns the rationality of decisions in kind and calculations in kind. Decisions in kind using a multiplicity of measures of value are contrasted with decisions with some single measure of value being money or any other single unit. The third claim is about economies as institutional orders. An economy in kind is an "institutional order of a society" (ibid.: 304) with a distinctive mode of resource allocation to be contrasted with the institutional order of market economies: "We should suggest looking at markets and finance and at the whole reckoning in money as an institution like any other" (Neurath 1944: 39).

Economies in kind feature in two distinct ways in Neurath's work. First, they are an object of study. Neurath's early empirical work in economics examined the functioning of various non-monetary economies in the ancient world and during periods of war (1909, 1910). Second, Neurath's contributions to the socialization debates advocated a radical in-kind economy of associations in which in-kind measures replaced monetary measures. While physical and social statistics would be required to make choices of resource allocation, no single unit of comparison would be adequate for decision making: "There are no units that can be used as the basis of a decision, neither units of money nor hours of work. One must directly judge the desirability of the two possibilities" (1919/1973: 145). Neurath's arguments turn on claims about commensurability. Choices between alternatives require a multidimensional understanding both of productive inputs and of well-being. In making this claim, Neurath was critical not just of monetary valuation, but of any alternatives that used some single unit of comparison, such as labor time or energy units.

Neurath's arguments were the occasion for Ludwig von Mises's contribution to the socialist calculation debates. Mises argues that in the absence of a single measure of value, rational choice about the use of higher-order production goods is impossible (1922/1981: 13). Without a single cardinal measure of value for production goods, no rational choice is possible between the alternative uses of "the bewildering mass of intermediate products and potentialities of production" (1920/1935: 103). Monetary exchange value provides a common unit of measurement of the relative worth of different productive factors for comparing their employment: "calculations based upon exchange values enable us to reduce values to a common unit" (1922/1981: 99). Market exchange makes possible the imputation of the relative worth of productive factors on the basis of consumer valuations. Market prices in the factors of production in turn require the private ownership of the means of production. Hence, rational choices between the alternative uses of productive resources are not possible in a socialist economy (1922/1981: 15).

The debate between Neurath and Mises raised questions about the intergenerational valuation of goods, the measurement of human well-being, and the nature of practical rationality. Consider the question of intergenerational valuation of goods. A problem with monetary valuation that Neurath's arguments highlighted is that market choices by current consumers fail to capture the relative values of different uses of productive factors for future generations. The value of productive resources for future generations cannot be directly captured in current market exchange. Intergenerational comparisons also raise problems for socialist alternatives to the market that employ single units in making decisions, such as labor time or energy units. Using labor time alone allows for no consideration for the effects of the use of energy and resources for future generations (Neurath 1925/2004: 468), and the use of energy units alone could not capture the impact on the quality and quantity of labor time undertaken within current generations (1928/1973: 263). Intergenerational choice in the use of productive resources requires multi-criteria decision procedures and judgments.

A second problem that Neurath raises against single-unit measures of value is their failure to capture changes in well-being. The rejection of the existence of a cardinal measure of welfare is already to be found in Neurath's early work (1912). In his contributions to the socialist calculation debate, the argument turns on the "multidimensional" nature of welfare concepts (1937a/2004: 520). While Neurath's account of the "quality of life" is hedonic, it is measured indirectly though objective "conditions of life, i.e. housing, food, clothing, working hours, etc." (1920–1/2004: 356). The conditions of life include social goods such as quality of personal and institutional relationships. Calculation in kind links these plural conditions of life with the various external conditions required for their realization, including different uses of productive resources. Single measures of value cannot capture either the plural dimensions of well-being or their external conditions (1925/2004: 426–7).

Thirdly, the argument between Neurath and Mises turns on differences about the nature of practical rationality. Mises's assumption that rational choice in the use of productive goods requires a single cardinal measure exhibits what Neurath characterized as "pseudorationalism." Reflective rationalism recognizes the boundaries of reason in decision making: "Rationalism sees its chief triumph in the clear recognition of the limits of actual insight" (1913/1983: 8).

In particular, one cannot capture the different value dimensions of options by a single measure and reduce choices to a matter of calculation. To employ an environmental example he uses, consider alternative sources of energy such as coal, hydraulic power, and solar energy: a variety of ethical and political judgments come into play, for example, around intergenerational impacts and the distribution of risks. It is not possible to arrive at a single optimal outcome through some computational procedure using a single cardinal metric, either monetary or non-monetary.

Mises's arguments against Neurath can be contrasted with those of Max Weber which are more careful in their claims about the rationality of choices in the absence of a monetary metric. Weber distinguishes two concepts of rationality, formal and substantive. The "formal rationality of economic action" refers to "the extent of quantitative calculation or accounting which is technically possible and which is actually applied" (1921–2/1978: 85). In contrast, "substantive rationality" refers to "the degree to which the provisioning of a given group of persons . . . with goods is shaped by economically orientated social action under some criterion . . . of ultimate values" (ibid.). Weber argues that formal rationality is best realized through monetary calculations based on exchange values and that economies in kind lack in this type of rationality. Thus far, his position parallels Mises's. However, Weber does not identify formal rationality with rationality as such. Economic systems can still be judged in terms of their substantive rationality with respect to some ends where " 'purely formal' rationality of calculation in monetary terms is of quite secondary importance or even is fundamentally inimical to their respective ultimate ends" (ibid.: 86). Weber's contrast recognizes different dimensions of rational choice that are absent in the work of Mises.

These early exchanges between Neurath, Mises, and Weber involved dimensions of argument that were lost in mainstream socialist calculation debates. The contributions of Lange and Taylor shifted the debate. Lange's neo-classical model of socialism accepts Mises's argument that rational economic action requires prices and rejected Neurath's proposals along with those of Marx (1936–7/1964: 135). The debate narrows on whether prices on productive resources should be determined by actual market transactions, as Mises claims, or whether shadow accounting prices could be employed to determine their use. The earlier arguments about the limits of monetary valuation disappeared in the subsequent exchanges in the socialist calculation debates. The arguments about the incommensurability of values, the intergenerational impacts of economic decisions, and the nature of rational economic decision making were largely lost. Where they had a continuing influence is in the tradition of ecological economics, particularly through the work of K. William Kapp (1974: 38) who drew on the debate initiated by Neurath (Martinez-Alier 1987; Uebel 2005, 2018; O'Neill and Uebel 2015; O'Neill 2019). With the failures of market-based approaches to environmental policy making, and to public policy making generally, Neurath's work has taken on new significance (O'Neill, 1998, 2016; Martinez-Alier, Munda and O'Neill 1998).

Science, knowledge, and planning

Arguments about rationality are also evident in Neurath's exchanges with Fredrich Hayek, whose contributions to the socialist calculation debates shifted the ground to epistemic questions. Hayek's arguments against Lange and other defenders of socialist planning turned on the limits of knowledge available to any central planning board (1937/1948, 1942–4, 1945/1948). The source of these limits lies in "the division of knowledge" in society. The argument appeals not just to the dispersal of knowledge, but to the nature of knowledge dispersed. Hayek contrasts the universal, generic, explicit, and propositional knowledge of the scientist with the particular, local, tacit, and practical knowledge of social actors. The latter forms of knowledge cannot be articulated in a form that could be passed on to a central planning body. The project of socialist planning is founded on the scientistic illusion that identifies knowledge only with the generic, explicit propositional knowledge of the sciences. It fails to recognize the particular, practical knowledge of actors and hence the limits in the knowledge available to any planning board. The rationalist illusion of the omniscient social planner fails to recognize the limits of reason: "it may . . . prove to be far the most difficult and not the least important task for human reason rationally to comprehend its own limitations" (1942–4: 162).

For Hayek, Neurath's work exhibits these scientistic illusions: "The most persistent advocate of. . . *in natura* calculation is, significantly, Dr. Otto Neurath, the protagonist of modern 'physicalism' and 'objectivism'" (ibid.: 170). Neurath's commitment to replacing calculation through prices with *in natura* calculation is taken to have its foundations in his "physicalism," understood as a program to eliminate all concepts from the social sciences that cannot be characterized in physical terms: given this elimination, economic decision making can take place solely in terms of physical inputs and outputs. Against this physicalist "objectivism," Hayek defends a form of "subjectivism" that denies the possibility of characterizing the objects of social science in purely physical terms without reference to a mental vocabulary (ibid.: 53).

Neurath responded to Hayek's criticisms in unpublished notes (1945a) and correspondence with Hayek. He affirmed his commitments to physicalism and *in natura* calculation but disputed their characterization by Hayek. Physicalism is not a form of eliminativism of mental and intentional vocabulary from the social science, but the claim that the statements of the social science are controllable by statements containing "spatio-temporal expressions" or "when, where, how terms": "what Professor von Hayek and others call 'mental' appears manifestly in my language as 'speech behaviour' or 'arguing' etc., i.e., a good where, when, how item" (ibid.). Likewise, Neurath is not committed to planning that uses only physical units. The social and institutional conditions of human welfare also matter: "often a change in a man's food and shelter is of less importance than a change in his state of being bullied or humiliated by certain institutions" (1942/1973: 425).

Hayek's criticisms miss their target. Neurath's work is concerned with the institutional conditions of well-being. Indeed, as Neurath stresses in his reply, he shares Hayek's skepticism of planners with complete knowledge able to arrive at some optimal outcome. Hayek's comment about the need for "human reason rationally to comprehend its own limitations" (1942–4: 162) parallels his own early rejection of "pseudorationalism" (1913/1981: 8). Neurath invoked his version of empiricism about the sciences to question the claim that there exist rules or methods able to determine a single optimal decision: the holism of theory and its underdetermination by observational evidence; the provisionality and uncertainty of empirical evidence given the revisability of observation statements; the principle of methodological pluralism and tolerance against "the absolutism of falsificationism . . . and the absolutism of verificationism" (1935/1983: 131). Neurath takes this pluralist understanding of science to have implications for decision making, since they undermine the assumptions underlying technocratic planning. The model of the social engineer offered by Hayek, an agent with complete knowledge aiming at some "technical optimum" (1942–4: 170), is rejected for this reason. Hence, Neurath's criticism of "the 'technocratic' movement" which assumes there exists "one best solution with its 'optimum happiness,' with its 'optimum population,' with its 'optimum health,' with its 'optimum working week,' with its 'optimum productivity' or something else of this kind" and which "asks for a particular authority which should be exercized by technicians and other experts in selecting 'big plans'" (1942: 426–7). While there are parallels between Neurath's and Hayek's criticisms of the assumptions underpinning technocratic planning, there are clearly important differences (O'Neill 2006). Hayek's argument starts from the contrast of generic, explicit, scientific knowledge with local, practical, and tacit knowledge. Neurath grants the significance of local and practical knowledge, but he argues that unpredictability and incompleteness are features of scientific knowledge itself.

Hayek's arguments for markets and against planning appealed to two contrasts. First, the possibility of planning within enterprises is contrasted with its impossibility between them in an economy as a whole: the epistemic problems of dispersed knowledge require markets to solve them (1945/1948: 77–78, 1976: 107–8). Second, organization based on a common "hierarchy of ends" (1944/2014: 101) is contrasted with the pluralism of ends of agents fostered within markets. The market is a condition for people with different ends to coordinate action and live together:

> The discovery that by substituting abstract rules of conduct for obligatory concrete ends made it possible to extend the order of peace beyond the small groups pursuing the same ends, because it enabled each individual to gain from the skill and knowledge of others whom he need not even know and whose aims could be wholly different from his own.
>
> (1976: 109; see also 1944/2014: 100–6, 125–33)

Neurath's response was to turn Hayek's position back on itself (1945a, 1945c/2004). First, the epistemic problems that Hayek raises against socialist planning already arise within the spheres of planning Hayek grants (1945a). Since the epistemic problems are ubiquitous, appeals to the necessity of markets as a solution to them fail (O'Neill 2007). Second, markets undermine pluralism, while planning can foster it. Neurath was committed to pluralism about the variety of ways in which a good life could be led. However, he had deep differences with Hayek about the conditions in which such a pluralism could be realized (O'Neill 2007; Whyte 2020). Already in his early work, Neurath argued that, far from encouraging a diversity of different ways of life, market economies were inimical to such diversity: "it was the tendency to organise the economy in all civilisations after the same pattern which made the free market society so much hated" (1920/2004: 402). The aim of planning in contrast should be to acknowledge and foster variety: "within a socialized economy a far greater multiplicity of ways of life can be made possible than in a free trade economy" (1919/1973: 145). This defense of planning that fosters pluralism and variety in ways of life is developed in his later work on "planning for freedom" (1942). In his review of Hayek's *The Road to Serfdom*, Neurath thus argues against the picture of planning as necessarily leading to totalitarianism, that it is possible for planning to be consistent with social plurality and freedom. Representative bodies could distribute goods "based on an orchestration of the various wishes of its members," with "safeguards of the rights of smaller groups in matters which vitally affect their happiness" where groups could develop their own "types of settlement or even types of work" (1945c/2004: 546). He concludes by rejecting the choice of the "painful market society of the past" and "dictatorial planning."

Debates with the Frankfurt School

The standard mischaracterization of logical empiricism as necessarily committed to a technocratic form of politics has its origins not just in Hayek's work, but also, from a different political direction, in that of the Frankfurt School (O'Neill and Uebel 2004). As with Hayek, this commitment to a technocratic politics is taken to follow from a commitment to scientism. However, where for Hayek scientism is associated with post-capitalist socialization and planning, for Max Horkheimer, in virtue of the identification of knowledge with the sciences, logical positivism cannot criticize the existing capitalist order. In particular, it cannot capture the role the sciences play in legitimizing the existing social order: "In view of the fact that the ruling economic powers use science as well as the whole of society for their special ends, this ideology, this identification of thought with the special sciences, must lead to the perpetuation of the status quo" (1937a/1972: 179). Subsequently, Horkheimer argued that scientific knowledge is constituted by an interest in "the manipulation of physical nature" which "at least in . . . the current period" serves the capitalist order: "The technological advances of the bourgeois period are inseparably linked to this function of the pursuit of science" (1937b/1972: 194). A mark of "traditional" as against "critical" theory is the absence of self-reflection about this role of science. It takes a particular form of scientific knowledge as given and thus serves "the conservation and continuous renewal of the existing state of affairs" (ibid., 196). Logical positivism is a traditional theory and a conservative ideology.

In his response to Horkheimer, Neurath rejects the claim that his account of the sciences ruled out the possibility of critical reflection on the social role of the sciences. While expressing skepticism about the possibility of philosophical reflection beyond empirical control, rational reflection on the individual sciences is still possible from within a naturalistic perspective: "Whatever is claimed with one scientific discipline can be criticized by a more comprehensive scientific standpoint, without regard to any divisions between the disciplines, but *we know of no court of appeal beyond the science that judges science and investigates its foundations*" (1937b/2011: 20–21, orig. emphasis, trans. revised). Neurath's perspective on the sciences contrasted with the more formal perspectives of logical empiricism associated with its post-World War II orthodoxy in drawing on the history and sociology of science. This allowed for a sociologically informed conception of scientific self-reflection which also acknowledges a role for the social determination of scientific belief.

> Historical changes do not only alter that which we call "theoretical formulations" or "constructions" but also the stock of protocol sentences. . . . Some of our observations prove themselves to be very stable, but in principle nothing is certain—everything is flux. It is plain that a consistent thinker will seek to apply these considerations, which are based on experience, to his own life and will ask himself how he would act, how he would argue if he would be positioned differently. He will realise that decisive changes in the pursuit of science are not only determined by intensive reflexions of a generation of scholars, but also what happens in social life generally, which the scholars are part of.
>
> (ibid.: 16, trans. revised)

Naturalistic reflection on the special sciences appeals not just to wider empirical inquiry in the history and sociology of science, but also to wider everyday empirical knowledge. Neurath's naturalistic perspective on the sciences, rather than grounding a technocratic politics, underpinned a participatory model of planning. This recognized both the dependence on science that is a feature of modern decision making, and the need for decisions to answer to the voice of ordinary citizens: "Our life is connected more and more with experts, but on the other hand, we are less prepared to accept other people's judgements, when making decisions." Democracy is "the continual struggle between the expert . . . and the common man" (1945b/1996: 251).

Neurath's responses to Hayek and Horkheimer reveal the relationships between the theoretical and practical dimensions of Neurath's work. His arguments for participatory forms of planning that recognized pluralism in the ways a good life could be lived informed, and were informed by, his own involvement in urban planning and housing. From his work in Vienna through to his engagement with urban planning for postwar Britain, he defended a pluralist perspective that allowed for different kinds of housing and settlement that recognized local habits, traditions, and ways of life, against "a totalitarian undercurrent, pressing forward some way of life" (1945c/2004: 247) which he detected in much city planning and modernist architecture. The role of the architect, planner, and "social engineer" was not to offer some optimal single design or state of affairs, but to offer alternatives open to democratic deliberation of affected citizens (Blau 2006; Hochhäusl 2011). His work on visual education aimed to humanize knowledge, to foster public participation in which informed citizens could engage in the planning process (Nemeth 2019). Far from fostering a form of technocratic planning, Neurath's empiricism grounded arguments for planning shaped through the participation and deliberation of ordinary people.

References

Blau, E. (2006) "Isotype and Architecture in Red Vienna: The Modern Projects of Otto Neurath and Josef Frank," *Austrian Studies* 14: 227–59.

Cat, J. and Tuboly, A. (eds.) (2019) *Neurath Reconsidered: New Sources and Perspectives*, Cham: Springer.

Hayek, F. (1937) "Economics and Knowledge," *Economica* 4: 33–54. Repr. in Hayek (1948), pp. 33–56.

——— (1942–4) "Scientism and the Study of Society," *Economica* 9–11. Repr. in Hayek, *The Counter-Revolution of Science. Studies in the Abuse of Reason*, Indianapolis: Liberty Fund, Indianapolis, 1979, pp. 17–182.

——— (1944) *The Road to Serfdom*, Chicago: University of Chicago Press.

——— (1945) "The Uses of Knowledge in Society," *American Economic Review* 35. Repr. in Hayek (1948), pp. 77–91.

——— (1948) *Individualism and Economic Order*, Chicago: University of Chicago Press.

——— (1976) *Law, Legislation and Liberty*, London: Routledge and Kegan Paul, vol. 2.

Hochhäusl, S. (2011) *Otto Neurath—City Planning: Proposing a Socio-Political Map for Modern Urbanism*, Innsbruck: Innsbruck University Press.

Horkheimer, M. (1937a) "Der neueste Angriff auf die Metaphysik," *Zeitschrift für Sozialforschung* 6: 4–53. Trans. "The Latest Attack on Metaphysics," in Horkheimer (1972), pp. 132–87.

——— (1937b) "Traditionelle und Kritische Theorie," *Zeitschrift für Sozialforschung* 6: 245–94. Trans. "Traditional and Critical Theory," in Horkheimer (1972), pp. 188–243.

——— (1972) *Critical Theory: Selected Essays*, New York: Seabury Press.

Kapp, K. W. (1974) *Environmental Policies and Development Planning in Contemporary China and Other Essays*, Paris: Mouton.

Lange, O. (1936–7) "On the Economic Theory of Socialism," *Review of Economic Studies* 4. Repr. in B. Lippincott (ed.), *On the Economic Theory of Socialism*, New York: McGraw-Hill, 1964, pp. 55–143.

Lessmann, O. (2007) "A Similar Line of Thought in Neurath and Sen. Interpersonal Comparability," in Nemeth et al. (2007), pp. 115–30.

Martinez-Alier, J. (1987) *Ecological Economics*, Oxford: Blackwell, 2nd ed., 1990.

Martinez-Alier, J., Munda, G. and O'Neill, J. (1998) "Weak Comparability of Values as a Foundation for Ecological Economics," *Ecological Economics* 26: 277–86.

Mises, L. V. (1920) "Die Wirtschaftsrechnung im sozialistischen Gemeinwesen," *Archiv für Sozialwissenschaft* 47. Trans. "Economic Calculation in the Socialist Commonwealth," in F. A. Hayek (ed.), *Collectivist Economic Planning*, London: Routledge and Sons, 1935, pp. 89–130.

——— (1922) *Die Gemeinwirtschaft*, Jena Fischer. Trans. *Socialism*, Indianapolis: Liberty Press, 1981.

Nemeth, E. (2019) "Visualizing Relations in Society and Economics: Otto Neurath's Isotype-Method Against the Background of His Economic Thought," in Cat and Tuboly (2019), pp. 117–40.

Nemeth, E., Schmitz, S. and Uebel, T. (eds.) (2007) *Neurath's Economics in Context*, Dordrecht: Kluwer.

Neurath, O. (1909) *Antike Wirtschaftsgeschichte*, Leipzig: Teubner. Excerpts trans. "Economic History of Antiquity," in Neurath (2004), pp. 120–52.

——— (1910) "Die Kriegswirtschaft," *Jahresbericht der Neuen Wiener Handelsakademie* 16: 5–54. Trans. "War Economy," in Neurath (2004), pp. 153–99.

——— (1912) "Das Problem des Lustmaximums," *Jahrbuch der Philosophischen Gesellschaft an der Universität Wien 1912*. Trans. "The Problem of the Pleasure Maximum," in Neurath (1973), pp. 113–22.

——— (1913) "Die Verirrten des Cartesius und das Auxiliarmotiv," *Jahrbuch der Philosophischen Gesellschaft an der Universität zu Wien 1913*, pp. 45–59. Trans. "The Lost Wanderers of Descartes and the Auxiliary Motive," in Neurath (1983), pp. 1–12.

——— (1916) "Die Naturalwirtschaftslehre und der Naturalkalkül in ihren Beziehungen zur Kriegswirtschaftslehre," *Weltwirtschaftliches Archiv* 8: 245–58. Trans. "Economics in Kind, Calculation in Kind and War Economics," in Neurath (2004), pp. 299–313.

——— (1919) *Wesen und Weg der Sozialisierung*, Munich: Callwey. Trans. "Character and Course of Socialisation," in Neurath (1973), pp. 135–50.

——— (1920) *Vollsozialisierung*, Jena: Diederichs. Trans. "Total Socialisation," in Neurath (2004), pp. 371–404.

——— (1920–21) "Ein System der Sozialisierung," *Archiv für Sozialwissenschaft und Sozialpolitik* 48: 44–73. Trans. "A System of Socialisation," in Neurath (2004), pp. 345–71.

——— (1925) *Wirtschaftsplan und Naturalrechnung*, Berlin: Laub. Trans. "Economic Plan and Calculation in Kind," in Neurath (2004), pp. 405–65.

——— (1928) *Lebensgestaltung und Klassenkampf*, Berlin: Laub. Trans. "Personal Life and Class Struggle," in Neurath (1973), pp. 249–98.

——— (1935) "Pseudorationalismus der Falsifikation," *Erkenntnis* 5: 353–65. Trans. "Pseudorationalism of Falsification," in Neurath (1983), pp. 121–31.

——— (1937a) "Inventory of the Standard of Living," *Zeitschrift für Sozialforschung* 6: 140–51. Repr. in Neurath (2004), pp. 513–26.

——— (1937b) *Einheitswissenschaft und logischer Empirismus. Eine Erwiderung*, ms., Otto Neurath Nachlass, Wiener Kreis Archiv, Rijksarchif Noord-Holland, Haarlem. Trans. "Unified Science and Logical Empiricism," in J. Symons, O. Pombo and J. M. Manuel (eds.), *Otto Neurath and the Unity of Science*, Dordrecht: Springer, 2011, pp. 15–30.

——— (1942) "International Planning for Freedom," *New Commonwealth Quarterly*. Repr. in Neurath (1973), pp. 422–40.

——— (1944) *Foundations of the Social Sciences*, Chicago: University of Chicago Press.

——— (1945a) "Physicalism, Planning and the Social Sciences: Bricks Prepared for a Discussion v. Hayek—Neurath, 26th July, 1945," ms. 202 K.56, Otto Neurath Nachlass, Wiener Kreis Archiv, Rijksarchif Noord-Holland, Haarlem.

——— (1945b) "Visual Education: Humanisation versus Popularisation," in E. Nemeth and F. Stadler (eds.), *Encyclopedia and Utopia*, Dordrecht: Kluwer, 1996, pp. 245–335.

——— (1945c) "Alternatives to Market Competition," *The London Quarterly of World Affairs*, January issue: 121–2. Repr. in Neurath (2004), pp. 546–8.

——— (1973) *Empiricism and Sociology* (ed. by M. Neurath and R. S. Cohen), Dordrecht: Reidel.

——— (1983) *Philosophical Papers* (ed. by R. S. Cohen and M. Neurath), Dordrecht: Reidel.

——— (2004) *Economic Writings* (ed. by T. Uebel and R. S. Cohen), Dordrecht: Kluwer.

O'Neill, J. (1998) *The Market: Ethics, Knowledge and Politics*, London: Routledge.

——— (2006) "Knowledge, Planning and Markets: A Missing Chapter in the Socialist Calculation Debates," *Economics and Philosophy* 22: 55–78.

——— (2007) "Pluralism and Economic Institutions," in Nemeth et al. (2007), pp. 77–100.

——— (2016) "Markets, Ethics and Environment," in S. Gardiner and A. Thompson (eds.), *Oxford Handbook of Environmental Ethics*, Oxford: Oxford University Press, pp. 40–50.

——— (2019) "From Socialist Calculation to Political Ecology," in J. McDonnell, V. Prashad and M. Davis (eds.), *Marx 200. The Significance of Marxism in the 21st Century*, London: Praxis Press, pp. 104–18.

O'Neill, J. and Uebel, T. (2004) "Horkheimer and Neurath: Restarting a Disrupted Debate," *The European Journal of Philosophy* 12: 75–101.

——— (2008) "Logical Empiricism as Critical Theory? The Debate Continues," *Analyse & Kritik* 30: 373–98.

——— (2015) "Analytical Philosophy and Ecological Economics," in J. Martinez-Alier and R. Muradian (eds.), *Handbook of Ecological Economics*, Cheltenham: Edward Elgar, pp. 48–73.

Putnam, H. (2002) *The Collapse of the Fact/Value Dichotomy and Other Essays*, Cambridge, MA: Harvard University Press.

Putnam, H. and Walsh, V. (eds.) (2014) *The End of Value-Free Economics*, Abingdon: Routledge.

Uebel, T. (2005) "Incommensurability, Ecology, and Planning: Neurath in the Socialist Calculation Debate, 1919–1928," *History of Political Economy* 37: 309–42.

——— (2018) "Calculation in Kind and Substantive Rationality: Neurath, Weber, Kapp," *History of Political Economy* 50: 289–320.
Weber, M. (1921–2) *Wirtschaft und Gesellschaft*, Tübingen: Mohr (Siebeck). Trans. *Economy and Society*, Berkeley: University of California Press, 1978.
Whyte, J. (2020) "Calculation and Conflict," *South Atlantic Quarterly* 119: 31–35.

PART III

Relations to philosophical contemporaries

29

THE VIENNA CIRCLE'S RELATIONSHIP WITH WITTGENSTEIN

Johannes Friedl

Ludwig Wittgenstein's *Tractatus Logico-Philosophicus* appeared in a bilingual book edition in 1922 (after its first troubled German publication in the last issue of *Annalen der Naturphilosophie*), the same year Moritz Schlick was appointed at the University of Vienna. Schlick began studying the *Tractatus* no later than the summer of 1924, while Rudolf Carnap had read excerpts even earlier (see Schlick 1924; Carnap 1963: 24). Otto Neurath reported that Hans Hahn had been a driving force in drawing the attention of the just emerging Circle to the book (1936/1981: 697). Another initial push had been a talk of the mathematician Kurt Reidemeister in the fall of that year, which, as Schlick wrote in his first letter to Wittgenstein, "made a great impression on us all" (McGuinness 1967/1979: 13). For the whole academic year of 1925–6, and at least the beginning of the next academic year, the Circle devoted its meetings to a line-by-line-reading. In February 1927 Schlick succeeded in contacting Wittgenstein, who had returned to Vienna to do architectural work after his career as a primary school teacher ended.

Personal relations

Wittgenstein refused to take part in the Circle's meetings, but he agreed to meet with Friedrich Waismann, Carnap, Herbert Feigl, and Maria Kasper (later Feigl's wife). At these informal meetings, not only were philosophical issues discussed; sometimes Wittgenstein preferred reading poetry to the group. Especially Schlick and Waismann were deeply impressed by Wittgenstein's personality. Schlick expressed his admiration in numerous letters, while Waismann soon began to imitate Wittgenstein's gestures and speech (Feigl 1969b/1981: 63). At the beginning of 1929, Wittgenstein relocated to Cambridge but regularly visited Vienna in university vacations during the following years. From then on, he only met with Schlick and Waismann; most of these meetings (from 1929 to the end of 1932) were recorded in shorthand by Waismann and later published (McGuinness 1967).

In the Circle's manifesto, Wittgenstein (along with Einstein and Russell) is celebrated as one of the "leading thinkers of the present day who represent the scientific world-conception in the public eye most effectively and who also exert the greatest influence on the Vienna Circle" (Verein Ernst Mach 1929/2012: 108). This is partly misleading, because, outside of Cambridge, Wittgenstein had been made known only by the Circle. With a few exceptions, his book had not yet been noticed. At the conference in Prague in 1929—where the Circle presented itself

DOI: 10.4324/9781315650647-33

for the first time as a unified group—the question came up as to whether Wittgenstein was a real person or just a "collective synonym" for the Circle's views (Neider in Haller and Rutte 1977: 35).

Within the Circle, Waismann took over the role of Wittgenstein's spokesman, regularly reporting on the latest developments. Especially noteworthy are Waismann's "Theses" (1967), an account of central themes of the *Tractatus* together with some of Wittgenstein's new ideas that were discussed intensely at meetings in spring 1931 (see the minutes in Stadler 1997/2015: 97–107). Waismann also promoted Wittgenstein's views outside of the Circle, most notably at the conferences in Prague in 1929 and Königsberg in 1930 (Waismann 1930, 1982).

From 1933 Wittgenstein met separately with Schlick and Waismann, but both had access to some important papers of Wittgenstein of that time, including at least a significant portion of the *Big Typescript* and the *Blue Book* (see Uebel 2017: 704). Additional dictations from Wittgenstein to Waismann and numerous notes and excerpts by Waismann were published in Baker (2003). In the period of 1934–6, small portions of Wittgenstein's writings were discussed in the Circle, which at that time consisted mostly of younger participants, but Schlick's murder in 1936 marked the end of any direct connection.

The impact of the *Tractatus*

Wittgenstein's conception of philosophy was welcomed without reservations: the task of philosophy is not to gain insight to truths beyond the reach of science, nor is it a study of the cognitive capacities of the human mind—philosophy is analysis of language. Clarification of language is the only way to overcome once and for all the bewilderment and endless controversies of traditional philosophy, demarcating what can be said from what cannot be expressed in meaningful propositions:

> No special "philosophical propositions" are propounded, but propositions are merely clarified. . . *there is no such thing as philosophy as a basic or universal science alongside or above the various fields of the one empirical science;* there is no path to genuine knowledge other than the path of experience; there is no realm of ideas that stands over or beyond experience.
>
> (Verein Ernst Mach 1929/2012: 89, orig. emphasis; see also *Tractatus* 4.112).

In particular, Wittgenstein's conception of logic was regarded as the decisive milestone. The propositions of logic are tautological and say nothing about the world; likewise, logical constants are not names for peculiar logical objects (this was Wittgenstein's "fundamental idea": 4.0312). On this understanding, the validity of logic no longer creates an obstacle to empiricism. Hahn's statement that "to me, the *Tractatus* has explained the role of logic" (Menger 1980: xii) was representative for the whole Circle. The thesis that all knowledge a priori is analytic and says nothing about the world was accepted unanimously. (This was already expressed in Schlick's *General Theory of Knowledge* of 1918, but due to deficiences in logic and semantics all too unclearly, as its author confessed in the preface to the second edition in 1925.)

An essential aspect of the *Tractatus*'s picture theory is Wittgenstein's atomism, his demand for elementary propositions. Notoriously, Wittgenstein gave no examples for them, but their existence was guaranteed as long as there are propositions with a determined sense at all. With elementary propositions taken as reports on the immediate data of experience, the *Tractatus* fits well with empiricism, a reading that was backed by Wittgenstein later on, when he spoke of the *Tractatus*'s simple objects as "such things as a colour, a point in visual space etc." (1980: 120).

The view that all complex propositions are built up from elementary propositions using truth-functional operations paved the way for the adoption of verificationism. While it is controversial whether verificationism is implicit in the *Tractatus* (see Hacker 1996: 50–59), Wittgenstein held a strong form of it when he returned to philosophy. The slogan "The sense of a proposition is its verification" is found in Waismann's notes from December 1929 (McGuinness 1967/1979: 47), and in notes from the same year, Wittgenstein spoke of verificationism as "our old principle" (Wittgenstein 1994: 157). Verificationist ideas were in the air at that time—most notably from Mach and from Einstein's treatment of the conception of simultaneity (see Friedman 1983/1999: 42)—but all members of the Circle unanimously attributed the principle of verification to Wittgenstein (Kraft 1950/1953: 31 and 197, n. 29; Carnap 1963: 27).

Wittgenstein's ideas were not accepted by all members in their entirety, important qualifications were made. Clarifying the concept of "verification" raised a number of issues, including questions about Wittgenstein's elementary propositions and the relationship between propositions in the process of verification. Even though all accepted that logic was tautologous, not everyone was convinced by the picture theory. Neurath rejected the radical separation of philosophy and science; even Schlick had previously suggested that philosophical theses were to be tested against scientific results (Schlick 1929 [written 1925]). The rejection of a particular domain of philosophy left its nature underdetermined. Did philosophy become a theory of language or stop being a theory at all? The latter view is Wittgenstein's, for whom philosophy consists only of acts of clarification. It is impossible to state the conditions of representation; they can only be shown, not specified in language. This rendered the propositions of the *Tractatus* themselves meaningless. Schlick fully embraced this view of philosophy as an activity (*Tractatus* 4.112 is the only passage ever explicitly referred to in his writings). By contrast, Carnap did not agree with this "anti-theoretical" conception of philosophy, and Neurath, resisting the separation between philosophy and science, refused to follow even Wittgenstein's first step (Neurath 1931/1983: 59).

Waismann and Schlick: admiration and tension

Waismann's plan to write a book on Wittgenstein was formed even before the Circle made personal contact, spurred on by the Circle's close reading. Schlick wrote that Waismann's original intention was to compose "essentially an exposition and elucidation of the ideas of Ludwig Wittgenstein, as set forth by him in his *Tractatus Logico-Philosophicus*" (Schlick 1976: 136). The rapid development of Wittgenstein's views in the years after his return to philosophy foiled this original plan. Wittgenstein then agreed to cooperate for a book written by Waismann that presented his present views and made use of the discussions and papers Wittgenstein provided. Under the title *Logik, Sprache, Philosophie*, the book was announced to appear as volume one of the series *Wissenschaftliche Weltauffassung*. This form of their collaboration ended in the fall of 1932, possibly because of Wittgenstein's dissatisfaction with Waismann's "Theses." At the next stage, the book was conceived to be co-authored by Wittgenstein and Waismann, and drafts written by Waismann were discussed whenever Wittgenstein was in Vienna. Nevertheless, this new form of collaboration was not successful either. In the summer of 1934, Waismann expressed his frustration in a letter to Schlick, stating that Wittgenstein "has the great gift of always seeing things as if for the first time . . . it doesn't matter at all how the thoughts are put together since in the end nothing is left as it was" (quoted in Baker 2003: xxvii).

In the end, Wittgenstein declined co-authorship and Waismann continued the work on his own (for the exact, quite dramatic course of the events of 1934, see Manninen 2011: 262–5). Later on, the publication of Waismann's book was delayed by the plan to publish an English translation simultaneously, and eventually it fell victim to the outbreak of war. What Waismann wrote to Wittgenstein in response to Wittgenstein's complaints about insufficient acknowledgement in Waismann's paper on identity (1936)—"The fundamental idea of this investigation is from you, the execution is mine" (quoted in Iven 2015: 169, trans. JF)—may also serve to characterize his book. Nevertheless, one must not underestimate Waismann's achievement in providing a systematic exposition of the rapidly developing views of the "middle" Wittgenstein, including both criticism of the *Tractatus* and commentary on other philosophers (Waismann 1965, 1976).

From 1937 to 1940, Waismann lectured in Cambridge. Although there were some meetings at that time, the relationship did not flourish, with Wittgenstein making dismissive remarks (McGuinness 2011a: 12–14). In personal communication after the war, Waismann called Wittgenstein "the great disappointment of my life," who "wholly entered the camp of obscurantists" (Neider in Haller and Rutte 1977: 33, trans. in Uebel 2017: 704). Philosophizing in a way still deeply influenced by Wittgenstein, Waismann developed ideas like the conception of "open texture" (1945), based on some earlier ideas by Wittgenstein; in other work, he explicitly argued against the later Wittgenstein (see Hacker 1996: 163–7).

The relationship between Schlick and Wittgenstein was of a different nature; at the personal level, there was no irritation on either side. Their meetings continued until Schlick's death, including a common sojourn in Istria in September 1933 (see Iven 2009). Unlike Waismann, none of Schlick's papers were merely expository. To be sure, in programmatic papers from 1928 to 1931, as well as in several letters from that time, Schlick's praise was enthusiastic, and he undoubtedly saw a decisive breakthrough in the *Tractatus*. To what extent its achievements were anticipated in Schlick's early work proved a contentious issue already among his contemporaries (see Feigl 1969a/1981: 21 and Menger 1982: 84). A special case is Schlick's use of the saying/showing-dichotomy in his posthumously published "Form and Content" (1938). Obviously inspired by Wittgenstein, in Schlick's hands this dichotomy is close to his older distinction between experience and cognition ("*Erleben*" and "*Erkennen*"), but the old thesis of the indefinability of experienced content is changed to the one of its incommunicability; thus the Tractatus' doctrine of the incommunicable (i.e., of logical form) is turned upside down.

Not only did Schlick never utter any criticism of Wittgenstein, but he was also keen to incorporate certain of Wittgenstein's latest achievements, including the new "meaning as use" doctrine. Schlick attempted to combine this thesis with verificationism, still believing to be in basic agreement at a time when Wittgenstein already had left verificationism behind (Schlick 1936: 458, written in 1934). A more important feature of their philosophical relationship is how Schlick made use of Wittgensteinian ideas and style of thinking in developing further his own views; this holds, e.g., for his treatments of the problem of the self and the first person, the question of privacy, or the psycho-physical problem (see Friedl 2013: ch. 6). The same holds true of Schlick's most basic agenda, which he described (in a self-penned dictionary article about himself) as the elaboration of a "consistent and thoroughly pure empiricism" (1950: 462, trans. JF). For him it was essential to connect all genuine knowledge claims to experience, by which he always meant sensation. In Schlick's final years the basic conception to secure this empiricism was the theory of affirmations ("*Konstatierungen*"). Again, in elaborating this conception, Schlick made use of a Wittgenstein-style argumentation, holding that it is the grammar of such peculiar sentences, demanding ostensive gestures, which connect them immediately with reality and secure their character as immune to doubt (1934).

It was not only with regard to ostensive definition—"the only explanation which can work without any previous knowledge" (1936/1979: 458)—that the use of Wittgensteinian ideas to develop his own older views resulted in disharmony. More importantly, there was disagreement on the nature of propositions like the affirmations. Already in 1932, Wittgenstein questioned whether propositions referring to one's own present experiences were genuine knowledge claims (see Glock 1996: 50–54). Another difference emerged when Schlick dealt with the doctrine of physicalism: by mixing Wittgensteinian fragments with his own older position, Schlick ended up defending the possibility of a private language (1935/1979: 432–4).

Although Schlick refrained from pointing out his differences from Wittgenstein's views, it is highly implausible to assume that he was unaware of them. Even if we account for the incompleteness of Schlick's work due to his sudden death, it is difficult to see how his own commitment to empiricism can be brought into full harmony with the thought of the philosopher he could "hardly exaggerate [his] indebtedness to" (1936/1979: 458).

Neurath and Carnap: aversion and overcoming

Neurath was unique in taking a negative stance towards Wittgenstein from the start. The talk about "what cannot be said but only shown" especially provoked his displeasure. Like Ramsey's well-known criticism, Neurath held that one should simply be silent instead of being silent about something (1931/1983: 60). This was, of course, not the only point on which Neurath disagreed. During the Circle's reading of the *Tractatus*, he thought he detected "metaphysics" so many times that he shortened his objections to simply uttering "M"; after a while, he suggested that it would minimize interruptions to utter "non-M" when the participants were *not* indulging in metaphysics (reported in Hempel 1969/2001: 256). Even later on, Neurath was not able to read the cardinal propositions of the *Tractatus* "without being shocked" (McGuinness 1991/2002: 193).

Neurath acknowledged some of Wittgenstein's discoveries, of course, in particular the tautologous nature of logic, but whenever he touched upon Wittgenstein in his papers, he added a warning about the metaphysical commitments. Especially telling is the quite modest role Neurath was willing to grant Wittgenstein in his account of the development of the logical empiricist's movement (1936/1981: 697). There is, moreover, the huge difference between the *Tractatus*'s claim to offer the ultimate solution to all philosophical problems and Neurath's emphasis on cooperation, gradual progress, and joint development of thought. In addition, Neurath held Wittgenstein (due to his influence on Schlick and Waismann) responsible for the increasing split within the Circle in the 1930s, to which he was highly sensitive. Arne Naess remembered that

> it was touching but also somewhat alarming to watch Otto Neurath embrace aloof and aristocratic Polish logicians of various philosophical affiliations and proclaim, "We agree! You are one of us!" If Neurath sensed that one was *somehow* on the right side, one was identified as a sort of logical positivist. Protestations were of little use and disagreements were conceived as due only to "unhappy formulations" (*unglückliche Formulierungen*), and there was always a remedy for that.
>
> (1972: 135, orig. emphasis)

From the beginning, Neurath strongly sensed that Wittgenstein *somehow* was on the wrong side (later, he sensed this about Karl Popper too). Carnap, on the contrary, unambiguously acknowledged the significance of Wittgenstein for his own work: "For me personally, Wittgenstein was perhaps the philosopher who, besides Russell and Frege, had the greatest influence on my thinking" (1963: 25). Although Carnap had read parts of the *Tractatus* before he arrived in Vienna, his early chief work, *The Logical Structure of the World*, written mainly in 1925, does not show marks of Wittgenstein's influence. The few references to Wittgenstein were presumably added only in 1927, when Carnap reworked and shortened the book for publication.

In the spring of 1932, Wittgenstein accused Carnap of plagiarism (the relevant correspondence is now published in Iven 2015; for translations of key passages see Hintikka 1989). The accusations concern "Die physikalische Sprache als Universalsprache der Wissenschaft" (1932) and list a number of complaints, including the startling claim that physicalism is already dealt with in the *Tractatus*. It is beyond my scope here to enter the highly contentious debate whether Wittgenstein's priority claims are legitimate (see Hintikka 1989; Stern 2007). Although Wittgenstein's reaction to Carnap's work raises highly interesting questions, it may be even more important to see that major tenets of Carnap's *Logical Syntax of Language* (1934) can be characterized by his attempt to retain central insights of the *Tractatus* but to shed views deemed unacceptable. While holding on to the basic ideas that all knowledge is language-dependent and philosophy is only concerned with an analysis of language, Carnap was not willing to accept Wittgenstein's conclusion that this analysis required the employment of nonsensical propositions ("elucidations"). Carnap sought to overcome this obstacle by introducing the formal mode of speech, allowing language to be studied independently of its representational function while focusing exclusively on purely syntactical features. That Wittgenstein included the formal mode of speech in his accusations of plagiarism is bewildering, but maybe due in part to the fact that Carnap's ideas concerning it became fully intelligible only in *Logical Syntax*.

At least three points must be noticed here. First, the formal mode of speech offered a way to demarcate philosophical investigations from those in the natural sciences, the latter proceeding in the material mode. While philosophy is sharply distinguished from natural sciences in the *Tractatus* as well, Carnap's way of drawing the line offers a new approach to characterizing (and ultimately to rejecting) metaphysics (1934/1937: Part V). Second, the syntactical structure of a language can be stated in meaningful propositions; the divergence between this point and the conception of the *Tractatus* is emphasized at length by Carnap himself (ibid.: § 73). Philosophy was no longer doomed to be (important) nonsense; as Neurath put it, this move provided the final building block for logical empiricism as a self-contained and comprehensive conception (1936/1981: 697). Third, the syntactical structure of a language is no longer restricted by the ontological structure of the state of affairs it depicts. There is not one and only one form of meaningful language that shared its structure with what is represented (a view that was discarded as "absolutism"). Ultimately, the structure of language is determined by rules that are not discovered but stipulated, resulting in Carnap's famous principle of tolerance (1934/1937: §17). In this Carnap was supported or even anticipated by Hahn, who rejected any ontological commitment (Hahn 1933/1987: 29–34; see Uebel 2009: 71–75).

With just these three points, Carnap saw his work take a major step beyond the *Tractatus* (see Awodey and Carus 2009). Nevertheless, the relationship between the *Logical Syntax of Language* and Wittgenstein's thought is a complicated matter (for a discussion that stresses similarities, see Kuusela 2012). The plagiarism charges found an echo in 1934 when Schlick urged Carnap to include a remark on Wittgenstein's development beyond the *Tractatus* (1934/2002: xvi). Carnap's turn to semantics in 1935 cannot be treated here, but it must not be considered as a "relapse" into some Tractarian conception, because it did not mean a return to an *explanatory* conception of meaning (see Leitgeb and Carus 2020).

Conclusion

The Circle's relation to Wittgenstein cannot be captured in a simple formula; complexity arises on both sides. On Wittgenstein's side, there is both the enigmatic character of the *Tractatus* and the exceedingly rapid development of his thought after his return to philosophy. Only a retrospective perspective reveals that something new was in the making, something that went far beyond corrections and amendments of the only philosophical book ever published in his lifetime. On the Circle's side, individual members reacted to Wittgenstein in quite different ways. Beside Neurath, members or allies like Frank or Reichenbach remained unaffected by Wittgenstein at large (see Haller 1984/1988: 38–39). Furthermore, no one in the Circle (except Waismann, for a certain time) ever intended to follow Wittgenstein in the manner of faithful students. For this reason, the question of whether the Circle misinterpreted Wittgenstein is an odd one, since faithful interpretation was never intended. Looking back on the Circle's line-by-line reading of the *Tractatus*, Carnap recalled: "Often long reflections were necessary in order to find out what was meant. And sometimes we did not find any clear interpretation. . . . We learned much by our discussions of the book, and accepted many views as far as we could assimilate them to our basic conceptions" (1963: 24).

The entanglement of Wittgenstein's solutions with ontological commitments and the "mysticism" of the saying/showing-dichotomy met with wide resistance. Without question, then, there were deep differences in fundamental attitude between Wittgenstein and most of the members of the Circle. Perhaps these differences are best exhibited by the contrast between the foreword of Wittgenstein's *Philosophical Remarks* (written in 1930) and the programmatic declarations in the Circle's manifesto, which stresses the scientific attitude and cooperation and embeds the scientific world-conception in the life of the present and its strife for progress (Verein Ernst Mach 1929/2012: 90). Wittgenstein expressed his hostility to the "vast stream of European and American civilization" and contrasted it with his own attempt to grasp the world "at its center—in its essence" (Wittgenstein 1964/1975: 7).

References

Achinstein, P. and Barker, S. (eds.) (1969) *The Legacy of Logical Positivism*, Baltimore: John Hopkins Press.

Awodey, S. and Carus, A. W. (2009) "From Wittgenstein's Prison to the Boundless Ocean: Carnap's Dream of Logical Syntax," in Wagner 2009, pp. 79–106.

Baker, G. (ed.) (2003) *The Voices of Wittgenstein: Ludwig Wittgenstein and Friedrich Waismann*, London: Routledge.

Carnap, R. (1932) "Die physikalische Sprache als Unversalsprache der Wissenschaft," *Erkenntnis* 2: 432–65. Trans. *The Unity of Science*, London: Kegan, Paul, Trench Teubner & Co., 1934.

——— (1934) *Logische Syntax der Sprache*, Vienna: Springer. Rev. ed. trans. *The Logical Syntax of Language*, London: Kegan Paul, Trench, Trubner & Co., 1937, repr. Chicago: Open Court, 2002.

——— (1963) "Intellectual Autobiography," in P. A. Schilpp (ed.), *The Philosophy of Rudolf Carnap*, La Salle, IL: Open Court, pp. 3–84.

Feigl, H. (1969a) "The Origin and Spirit of Logical Positivism," in Achinstein and Barker 1969, pp. 3–24. Repr. in Feigl 1981, pp. 21–37.

——— (1969b) "The Wiener Kreis in America," in D. Fleming and B. Baylin (eds.), *The Intellectual Migration*, Cambridge, MA: Harvard University Press, pp. 630–74. Repr. in Feigl (1981), pp. 57–93.

——— (1981) *Inquiries and Provocations. Selected Writings, 1929–1974* (ed. by R. Cohen), Dordrecht: Reidel.

Friedl, J. (2013) *Die Entwicklung von Moritz Schlicks Erkenntnistheorie im Wiener Kreis*, Vienna: Springer.

Friedman, M. (1983) "Moritz Schlick's *Philosophical Papers*," *Philosophy of Science* 50: 498–514. Repr. in Friedman, *Reconsidering Logical Positivism*, Cambridge: Cambridge University Press, pp. 17–33.

Glock, H. J. (1996) *A Wittgenstein Dictionary*, Oxford: Blackwell.

Hacker, P. (1996) *Wittgenstein's Place in Twentieth Century Analytic Philosophy*, Cambridge: Blackwell.

Hahn, H. (1933) *Logik, Mathematik und Naturerkennen*, Vienna: Gerold. Trans. "Logics, Mathematics, and Knowledge of Nature," in B. McGuinness (ed.), *Unified Science*, Dordrecht: Reidel, 1987, pp. 24–45.

Haller, R. (1984) "Was Wittgenstein a Neopositivist?" *Fundamenta Scientae* 5: 271–84. Repr. in Haller, *Questions on Wittgenstein*, London: Routledge, 1988, pp. 27–43.

Haller, R. and Rutte, H. (1977) "Gespräch mit Heinrich Neider," *Conceptus* 28: 21–42.

Hempel, C. G. (1969) "Logical Positivism and the Social Sciences," in Achinstein and Barker 1969, pp. 163–94. Repr. in Hempel, *The Philosophy of Carl G. Hempel: Studies in Science, Explanation, and Rationality* (ed. by J. H. Fetzer), Oxford: Oxford University Press, pp. 253–75.

Hintikka, J. (1989) "Ludwig's Apple Tree: Evidence concerning the Philosophical Relations between Wittgenstein and the Vienna Circle," in W. L. Gombocz et al. (eds.), *Traditionen und Perspektiven der analytischen Philosophie*, Vienna: Hölder-Pichler-Tempsky, pp. 187–202. Repr. in Hintikka, *Ludwig Wittgenstein: Half-Truths and One-and-a-Half Truths*, Dordrecht: Kluwer, 1996, pp. 125–44.

Iven, M. (2009) "Wittgenstein und Schlick. Zur Geschichte eines Diktats," in F. Stadler and H. J. Wendel (eds.), *Stationen. Dem Philosophen und Physiker Moritz Schlick zum 125. Geburtstag*, Wien/New York: Springer, pp. 63–80.

——— (ed.) (2015) "Er 'ist eine Künstlernatur von hinreissender Genialität.' Die Korrespondenz zwischen Ludwig Wittgenstein und Moritz Schlick sowie ausgewählte Briefe von und an Friedrich Waismann, Rudolf Carnap, Frank P. Ramsey, Ludwig Hänsel und Margaret Stonborough," *Wittgenstein-Studien* 6: 83–174.

Kraft, V. (1950) *Der Wiener Kreis*, Wien: Springer, 2nd ed., 1968. Trans. *The Vienna Circle: The Origin of Neopositivism*, New York: Philosophical Library, 1953.

Kuusela, O. (2012) "Carnap and the *Tractatus'* Philosophy of Logic," *Journal for the History of Analytical Philosophy* 1 (3): 1–25.

Leitgeb, H. and A. Carus (2020) "Rudolf Carnap," in E. N. Zalta (ed.), *The Stanford Encyclopedia of Philosophy*, Summer 2020 ed., https://plato.stanford.edu/archives/sum2020/entries/carnap/.

Manninen, J. (2011) "Waismann's Testimony of Wittgenstein's Fresh Starts in 1931–35," in McGuinness (2011b), pp. 243–65.

McGuinness, B. (ed.) (1967) *Wittgenstein und der Wiener Kreis: Gespräche aufgezeichnet von Friedrich Waismann*, Oxford: Blackwell. Trans. *Wittgenstein and the Vienna Circle: Conversations Recorded by Friedrich Waismann*, Oxford: Blackwell, 1979.

——— (1991) "Wittgensteins Beziehungen zum Wiener Kreis," in W. Hochkeppel (ed.), *Jour Fixe der Vernunft. Der Wiener Kreis und die Folgen*, Vienna: Hölder-Pichler-Tempsky, pp. 108–26. Trans. "Relations with and within the Vienna Circle," in McGuinness, *Approaches to Wittgenstein. Collected Papers*, London: Routledge, 2002, pp. 184–200.

——— (2011a) "Waismann: The Wandering Scholar," in McGuinness (2011b), pp. 9–16.

——— (ed.) (2011b) *Friedrich Waismann: Causality and Logical Positivism*, Dordrecht: Springer.

Menger, K. (1980) "Introduction," in Hahn, *Empiricism, Logic, and Mathematics* (ed. by B. McGuinness), Dordrecht: Reidel, pp. ix–xviii.

——— (1982) "Memories of Moritz Schlick," in E. T. Gadol (ed.), *Rationality and Science. A Memorial Volume for Moritz Schlick in Celebration of the Centennial of His Birth*, Wien/New York: Springer, pp. 83–103.

Næss, A. (1972) "Historical Note on Possibilistic Pluralism," in A. Næss, *The Pluralist and Possibilist Aspect of the Scientific Enterprise*, Oslo: Universitetsforlaget, pp. 134–7.

Neurath, O. (1931) "Soziologie im Physikalismus," *Erkenntnis* 2: 393–431. Trans. "Sociology in the Framework of Physicalism," in Neurath, *Philosophical Papers 1913–1946* (ed. by R. S. Cohen and M. Neurath), Dordrecht: Reidel, 1983, pp. 58–90.

——— (1936) *Le développement du Cercle de Vienne et l'avenir de l'empirisme logique*, Paris: Hermann & Cie. Trans. "Die Entwicklung des Wiener Kreises und die Zukunft des Logischen Empirismus," in Neurath, *Gesammelte philosophische und methodologische Schriften* (ed. by R. Haller and H. Rutte), Vienna: Hölder-Pichler-Tempsky, 1981, pp. 673–702.

Schlick, M. (1918) *Allgemeine Erkenntnislehre*, Berlin: Springer, 1918, 2nd rev. ed. 1925. Trans. *General Theory of Knowledge*, Vienna: Springer, 1974, repr. LaSalle, IL: Open Court, 1985.

——— (1924) Letter to Hans Reichenbach, 5 August 1924, Archives of Scientific Philosophy, Pittsburgh, Collection Hans Reichenbach, ASP-HR 016-42-16.

——— (1929) "Erkenntnistheorie und modern Physik," *Scientia* 45: 307–86. Trans. "Epistemology and Modern Physics," in Schlick (1979), pp. 91–98.

——— (1934) "Über das Fundament der Erkenntnis," *Erkenntnis* 4: 79–99. Trans. "On the Foundation of Knowledge," in Schlick (1979), pp. 370–87.

——— (1935) "De la relation entre les notions psychologiques et les notions physiques," *Revue de Synthèse* 10: 1–20. Trans. "On the Relation Between Psychological and Physical Concepts," in Schlick (1979), pp. 420–36.

——— (1936) "Meaning and Verification," *Philosophical Review* 45: 339–69. Repr. in Schlick (1979), pp. 456–81.

——— (1938) "Form and Content," in Schlick, *Gesammelte Aufsätze 1926–1936*, Vienna: Gerold, pp. 151–250. Repr. in Schlick (1979), pp. 285–369.

——— (1950) "Schlick, Moritz," in W. Ziegenfuß and G. Jung (eds.), *Philosophen-Lexikon. Handwörterbuch der Philosophie nach Personen*, Berlin: de Gruyter, vol. II, pp. 462–4.

——— (1976) "Vorrede," in Waismann (1976), pp. 11–23 [orig. 1928]. Trans. "Preface to: Friedrich Waismann, *Logik, Sprache, Philosophie*," in Schlick (1979), pp. 130–8.

——— (1979) *Philosophical Papers, vol. 2: 1925–1936* (ed. by H. Mulder and F. B. van de Velde-Schlick), Dordrecht: Reidel.

Stadler, F. (1997) *Studien zum Wiener Kreis. Ursprung, Entwicklung und Wirkung des Logischen Empirismus im Kontext*, Frankfurt a. M.: Suhrkamp. Trans. of 2nd rev. ed. *The Vienna Circle: Studies in the Origins, Development, and Influence of Logical Empiricism*, Dordrecht: Springer, 2015.

Stern, D. (2007) "Wittgenstein, the Vienna Circle, and Physicalism: A Reassessment," in A. Richardson and T. Uebel (eds.), *The Cambridge Companion to Logical Empiricism*, Cambridge: Cambridge University Press, pp. 305–31.

Uebel, T. (2009) "Carnap's *Logical Syntax* in the Context of the Vienna Circle," in Wagner (2009), pp. 53–78.

——— (2017) "Wittgenstein and the Vienna Circle," in H. J. Glock and J. Hyman (eds.), *A Companion to Wittgenstein*, Oxford: Wiley, pp. 699–717.

Verein Ernst Mach (1929) *Wissenschaftliche Weltauffassung. Der Wiener Kreis*, Vienna: Wolf. Trans. "The Scientific Conception of the World. The Vienna Circle," in O. Neurath, *Empiricism and Sociology* (ed. by R. S. Cohen and M. Neurath), Dordrecht: Reidel, 1973, pp. 299–318; rev. trans. (with orig. annotated bibliography) "The Scientific World-Conception. The Vienna Circle," in F. Stadler and T. Uebel (eds.), *Wissenschaftliche Weltauffassung. Der Wiener Kreis. Hrsg. vom Verein Ernst Mach (1929)*, Vienna: Springer, 2012, pp. 75–116.

Wagner, P. (ed.) (2009) *Carnap's Logical Syntax of Language*, Basingstoke: Palgrave Macmillan.

Waismann, F. (1930) "Logische Analyse des Wahrscheinlichkeitsbegriffs," *Erkenntnis* 1: 228–48. Trans. "A Logical Analysis of the Concept of Probability," in Waismann (1977), pp. 4–21.

——— (1936) "Über den Begriff der Identität," *Erkenntnis* 6: 56–64. Trans. "The Concept of Identity," in Waismann (1977), pp. 22–29.

——— (1945) "Verifiability," *Proceedings of the Aristotelian Society*, Suppl. vol. 19: 119–50. Repr. in Waismann, *How I See Philosophy* (ed. by R. Harré), London: Palgrave Macmillan, 1968, pp. 39–66.

——— (1965) *Principles of Linguistic Philosophy* (ed. by R. Harré), Basingstoke: Macmillan, 2nd ed., 1997.

——— (1967) "Thesen," in McGuinness (1967), pp. 233–61 [orig. 1930–1]. Trans. "Theses," in McGuinness (1967/1979), pp. 233–61.

——— (1976) *Logik, Sprache, Philosophie* (ed. by G. Baker and B. McGuinness), Stuttgart: Reclam.

——— (1977) *Philosophical Papers* (ed. by B. McGuinness), Dordrecht: Reidel.

——— (1982) "Über das Wesen der Mathematik: Der Standpunkt Wittgensteins," in Waismann, *Lectures on the Philosophy of Mathematics* (ed. by W. Grassl), Amsterdam: Rodopi, pp. 157–67 [orig. 1930].

Wittgenstein, L. (1922) "Logisch-Philosophische Abhandlung," *Annalen der Naturphilosophie* 14: 185–262. Bilingual ed. trans. by F. Ramsey and C. K. Ogden *Tractatus Logico-Philosophicus*, London: Kegan Paul, Trench Trubner & Co., 1922, rev. ed. 1933, repr. London: Routledge & Kegan Paul, 1983; trans. by D. F. Pears and B. F. McGuinness, London: Routledge & Kegan Paul, 1961, repr. 1974.

——— (1964) *Philosophische Bemerkungen* (ed. by R. Rhees), Oxford: Blackwell. Trans. *Philosophical Remarks* (ed. by R. Rhees), Oxford: Blackwell, 1975.

——— (1980) *Wittgenstein's Lectures, Cambridge 1930–1932* (ed. by D. Lee), Oxford: Blackwell.

——— (1994) *Philosophische Betrachtungen/Philosophische Bemerkungen* (*Wiener Ausgabe* vol. 2) (ed. by M. Nedo), Vienna: Springer.

30
CASSIRER AND THE LOGICAL EMPIRICISTS

Matthias Neuber

Logical idealism, as it was defended by Ernst Cassirer, is obviously something different than logical empiricism, as it was defended by the members of the Vienna Circle and the Berlin Group. However, both philosophical traditions were closer to each other than the mere contraposition of two "isms" might suggest. This pertains primarily to the rejection of any form of metaphysics and to the common aim of establishing a "scientific" philosophy. Further commonalities can be identified in relation to certain Kantian elements in the respective analytical methods, to the privileged status of the logic of relations, and to the rather marginal treatment of ethical issues. In short, Cassirer and the logical empiricists shared a lot of ground.

To be sure, Cassirer from the very beginning criticized empiricism in its *radical* version. That is, he distanced himself from the attempt to reduce all of knowledge to a purely sensory basis. But much the same holds true of the logical empiricists. Furthermore, it has for a long time been neglected that many (though not all) of the logical empiricists began their philosophical careers within the Kantian tradition. More recently, it was especially Michael Friedman who disclosed a whole range of neo-Kantian elements in the early works of Moritz Schlick, Hans Reichenbach, and Rudolf Carnap (see his 1999). Cassirer who, as is well known, belonged to the Marburg School of neo-Kantianism, was entirely aware of this fact. It is therefore not all too astonishing that, in a document to be found in his *Nachlass*, Cassirer declares that there is no other "school" in contemporary philosophy that is closer to his own approach than the Vienna Circle (2011: 206). As John Michael Krois has pointed out, there were also many *divergences*, especially concerning the issue of physicalism and the role of the cultural sciences (see his 2000: 110–18). But it is more than obvious that both sides thought of themselves as taking part in the same sort of rational, enlightenment-based discourse.

In what follows, the attempt will be made to shed some light on the various interrelationships between Cassirer and the logical empiricists. I will confine myself to the most outstanding points and protagonists and leave room for subsequent more fine-grained reconstructions.

Cassirer and Schlick

As is well known, Schlick's *Allgemeine Erkenntnislehre*, first published in 1918, contained an extended critique of the "philosophy of immanence," which came pretty close to a rejection of radical empiricism (1918/1985: §26). On the other hand, Schlick argued in this early *magnum*

DOI: 10.4324/9781315650647-34

opus for a standpoint best characterized as "critical realism." This view was *critical* insofar as it implied an epistemological reflection on the preconditions of scientific knowledge. It was *realistic* insofar as it was centered around the claim that Kantian things-in-themselves are knowable. Thus, Schlick at one place declared that "the only natural continuation of Kant's theory of knowledge, to which his system points from various angles, lies not in the idealist but the realist direction, and we arrive at it by a revision of Kant's utterances about the so-called thing-in-itself and its knowability" (1919/1979: 282). In order to drive this point home, Schlick severely downgraded the role of *intuition*, declaring that "intuition . . . provides, to be sure, an *acquaintance* with objects, but never a knowledge of them" (ibid.). In consequence, he arrived at a conception of *purely conceptual* knowledge according to which "the import of the conceptual function consists precisely in *signifying* or *designating*. Here, however, to signify means nothing more than to *coordinate* or *associate* (*zuordnen*), that is, to place in a one-one or at most a many-one correspondence" (1918/1985: 23). Since Schlick categorically rejected the (Aristotelian) account of objects as substances and instead conceived of (scientific) objects as clusters of lawful relations (ibid.: §40), he ended up with a thoroughly *relationalist* (or "structuralist") account of science and scientific knowledge, all the more since the coordinative conceptual function itself was relational (rather than "monadic").

Cassirer had put forward a very similar approach as early as 1910, in his seminal monograph *Substanzbegriff und Funktionsbegriff*. There, one can already find both a relationalist view of scientific knowledge and a coordinative account of the "conceptual function" (see Ryckman 1991; Gower 2000; Neuber 2013). It is therefore somewhat surprising that Schlick, in a long footnote to §5 of his *Allgemeine Erkenntnislehre*, severely criticized Cassirer's entire approach, but in so doing relied on the same Aristotelian logic that had been the main target of Cassirer's own relationalist account. Although Schlick had deleted the footnote from the second edition of 1925, Cassirer replied in his 1927 essay "Erkenntnistheorie nebst Grenzfragen der Logik und Denkpsychologie" by criticizing Schlick's conception of scientific knowledge as a form of conventionalism and fictionalism (1927: 69; see Friedman 2000: 117–19; Neuber 2012a: 214–17; Ferrari 2016: 98–101). Siding with Schlick's rejection of the philosophy of immanence, Cassirer confronted the latter's view of concepts as mere signs with the corresponding "realist" account of scientific objects. According to Cassirer, Schlick's commitment to the knowability of Kantian things-in-themselves was, in fact, illusory (1927: 70–71). In his view, Schlick had merely *renamed* the Kantian object of experience under the label "thing-in-itself." The Kantian things-in-themselves, as Schlick conceived of them, were *nothing but* appearances in the original Kantian sense. It was for this reason that Cassirer thought of Schlick as being actually indebted to his own 1910 relationalist account in *Substanzbegriff und Funktionsbegriff*, the only difference being that Schlick, due to his alleged epistemic "conventionalism," denied the existence of synthetic a priori preconditions within science and scientific theory construction (ibid.: 75–77).

It was exactly this issue of the synthetic a priori that had already been the focus of Schlick's 1921 review of Cassirer's book *Zur Einsteinschen Relativitätstheorie* (which too had appeared in 1921). In that review, Schlick challenged Cassirer's neo-Kantian reading of Einstein's theory of relativity with the following diagnosis: "Every attempt to reconcile Einstein with Kant must discover synthetic *a priori* principles in the theory of relativity; otherwise it must be regarded from the outset as a failure, since it has not even got to the point of stating the problem correctly" (1921/1979: 325). As is well known, Schlick himself favored an empiricist interpretation of Einstein's theory and therefore refused any attempt to reconcile that theory with the Kantian assumption of synthetic a priori preconditions. However, as Thomas Ryckman has pointed out, Schlick *misinterpreted* Cassirer's approach toward the a priori. More exactly, he overlooked both Cassirer's "dynamical" reading of the a priori and its reconceptualization in terms of Kantian *regulative* (rather than "constitutive") principles. "Schlick completely ignored the genetic character and historical evolution of the 'regulative principles' and 'rules of understanding' that comprised the core of Cassirer's account of the development of the concept of physical objectivity culminating in general covariance" (2005: 51). Cassirer himself had hinted at this very point in a letter to Schlick from 23 October 1920 (see Neuber 2012a: 160–2; Ferrari 2016: 88–89). Yet Schlick did not see any reason to remove his initial diagnosis and instead argued for the *replacement* of the synthetic a priori by *conventions* in a Poincaréan sense (see Friedman 1994).

According to Alberto Coffa, with Schlick we are "back in the world of Kantian questions and semi-Kantian answers" (1991: 171). This is exaggerated, given Schlick's rather vehement opposition to Cassirer's neo-Kantian approach. Nevertheless, it can hardly be denied that the exchange between Cassirer and Schlick was embedded in Kantian presuppositions. It is therefore regrettable that Cassirer never realized his plan to publish an extended critique of Schlick's later, Viennese, views for the journal *Erkenntnis*, a project proposed to him by the journal's co-editor Hans Reichenbach. But, as Cassirer wrote in a letter to Reichenbach from 1 September 1936, Schlick's assassination (in June 1936) had deprived him of the respective "inner impulse" (2009: 151).

Cassirer and Reichenbach

As regards Cassirer's relation to the members of the Berlin Group, it is primarily Reichenbach who comes into focus. To be sure, Cassirer was criticized as early as 1908 by a later member of the Berlin Group, Kurt Grelling (see Grelling 1908; Cassirer 1906). But this episode can be neglected here. In the case of Reichenbach, however, a closer look is in order. Not only did Reichenbach study with Cassirer in Berlin (before completing his dissertation in Erlangen), but it must also be noted that Reichenbach, in his 1920 monograph *Relativitätstheorie und Erkenntnis Apriori*, argued in a significantly similar way to Cassirer in his *Zur Einsteinschen Relativitätstheorie* of 1921. Reichenbach himself was aware of this fact. In a footnote of his book, he explicitly mentioned Cassirer as the first "outstanding representative" of neo-Kantianism who attempted at a philosophical reconstruction of the theory of relativity (1920: 108–9 n. 20). Thus *both* Cassirer and Reichenbach contributed to a *revisionist* Kantian reading of Einstein's theory.

To begin with, it was the idea of the "relativized" a priori that set the stage for Reichenbach's Kantian-inspired reconstruction (see CH. 21). Distinguishing two meanings of the a priori, he rejected the first which meant "necessarily and unrevisably true," but endorsed the second according to which a priori meant "constitutive of the object of knowledge" (ibid.: 46). The resulting conception was relativized insofar as it claimed apriority for certain core principles of *historically evolving* theories. Reichenbach called these principles "axioms of coordination" and demarcated them from what he called "axioms of connection," which in turn he equated with empirical laws. Thus, for example, Newtonian physics presupposed, according to Reichenbach, Euclidean spatial geometry and Galilean kinematics, which he thought were a priori in the constitutive sense but only *relative to* Newtonian physics, because in, say, Einsteinian physics these principles were supplanted by other (likewise constitutive) axioms of coordination. As Thomas Oberdan noted, "axioms of coordination function in Reichenbach's epistemology in much the same way as the synthetic apriori in Kant's; they serve to explicate the insight that the object of knowledge is constructed from conceptual elements which are not given in experience" (2009: 195).

Now, in Cassirer's account, the relativization of the a priori also played a crucial role. Relying on related ideas by his teacher Hermann Cohen (see Cohen 1885), he developed the kind of "dynamical" reading of the a priori that Schlick in his critique had obviously ignored. So

Ryckman is right to state that "[b]oth Cassirer and Reichenbach drew upon a revisionist conception of the role of *a priori* elements in physical theory as not fixed for all time but "relative," changing with the advance of science" (2005: 14). However, it is also important to note that Cassirer's relativization of the a priori was completely divorced from the assumption of constitutive a priori principles. In his view, those principles were merely *regulative* in the original Kantian sense, the main reason being that he took over from Cohen the view that Kant's principled distinction between the two independent "faculties" of sensibility and understanding is misguided, since it is the logic of the concept *alone* by which the epistemic work in science is done (see Cohen 1902; Cassirer 1910: ch. 1). Kant's conception of the "schematism of pure concepts" became redundant and with it the whole account of principles in the a priori constitutive sense. As Friedman pointed out:

> Constitutive principles, for Kant, arise from applying the intellectual faculties (understanding and reason) to the distinct faculty of sensibility, whereas regulative principles arise from the intellectual faculties independently of such application. By rejecting Kant's original account of the transcendental schematism of the understanding with respect to a distinct faculty of sensibility in favor of a teleologically oriented "genetic" conception of knowledge, Cassirer (and the Marburg School more generally) has thereby replaced Kant's constitutive a priori with a purely regulative ideal.
>
> (2000: 117)

So even though they agreed about the need of a relativized understanding of the Kantian a priori, Cassirer and the early Reichenbach arrived at significantly different conceptions of what is *implied* by such a relativization. Furthermore, it must be remembered that Reichenbach eventually converted to the sort of "conventionalized empiricism" that was defended by Schlick (see CH. 21).

Yet, Reichenbach and Cassirer stayed in contact. As late as April 1945, Reichenbach invited Cassirer to teach for one semester at the University of California at Los Angeles (see Cassirer 2009: 237–8); unfortunately, Cassirer died one and a half weeks later.

Cassirer and Carnap

While Schlick and Reichenbach rather directly referred to Cassirer's views, Carnap's approach toward Cassirer (and the Marburg School in general) was more indirect. To be sure, Carnap, like Schlick and Reichenbach, was philosophically "socialized" within the neo-Kantian tradition (see Carus 2007: ch. 4). But his attitude toward the diverse contemporary philosophical currents was from the very beginning rather eclectic. That is, he combined elements from both the Marburg School and the Southwest School of neo-Kantianism with elements from positivism, from conventionalism, and even from phenomenology. This can be seen as early as in his dissertation *Der Raum* from 1922 where Carnap developed a kind of "hybrid" conception of space. Interestingly enough, Cassirer was one of the few contemporary philosophers who took notice of Carnap's dissertation. Thus, in the third volume of his *Philosophy of Symbolic Forms*, he explicitly appreciated Carnap's distinction between "formal," "intuitive," and "physical" space, especially because of its affinity with the sort of "pluralism" he himself attempted to establish with regard to the diverse symbolic forms of spatial representation (1929: 493 n. 1).

Carnap, on the other hand, repeatedly mentioned the importance of certain insights from the neo-Kantian tradition for his own "constitutional project" in *Der logische Aufbau der Welt.*

Cassirer, in particular, was appreciated for his theory of relational concepts in *Substanzbegriff und Funktionsbegriff* which fit well with Carnap's general claim that "*each scientific statement can in principle be so transformed that it is nothing but a structure statement*" (1928/1967, §16, orig. emphasis). Carnap specifies his agreement with Cassirer in this respect as follows:

> Cassirer . . . has shown that a science which has the aim of characterizing unique entities through contexts of laws without loss of individuality must utilize, not class ("generic") concepts, but *relational concepts*, since these can lead to the formation of series and thus to the establishment of systems of ordering. Since one can easily make the transition from relations to classes, and since the opposite is possible only very rarely, it follows that it is relations extensions which must initially be posited.
>
> (ibid.: §75, orig. emphasis)

Nevertheless, as Friedman has made plain, Carnap had a "neutral and distant attitude toward the neo-Kantian tradition" (1992/1999: 134). On the one hand, he explicitly opposed the commitment to the synthetic a priori that was still at work within that tradition (1928/1967: §179) while, on the other, he attempted to remain neutral in the controversy between (transcendental) idealism and (metaphysical) realism.

> Are the constituted structures "generated in thought," as the Marburg School teaches, or "only recognized" by thought, as realism asserts? Constitutional theory employs a neutral language; according to it the structures are neither "generated" nor "recognized" but rather "*constituted*"; and it is already here to be expressly emphasized that this word "constitution" is always meant completely *neutrally*. From the point of view of constitutional theory the dispute involving "generation" versus "recognition" is therefore an idle linguistic dispute.
>
> (ibid.: §5, orig. emphasis)

It might be questioned whether Carnap's "neutralism" was indeed a tenable position (see Neuber 2013), but it is important to note that the principal aim of his constitutional project was to integrate elements from different philosophical traditions within his own conception. Thus, he claimed that "[*c*]*onstruction theory and idealism . . . do not contradict one another in any point*" (1928/1967, §177, orig. emphasis). The very same conclusion holds, he claimed, with regard to realism and positivism (resp. phenomenalism). Thus, he wrote:

> [T]wo entirely different and frequently hostile philosophical positions have the merit of both having discovered the necessary basis of the constructional system. Positivism has emphasized that the only material of cognition consists in the undigested experientially given. It is here that we have to look for the basic elements of the constructional system. Transcendental idealism, especially the Neo-Kantian school (Rickert, Cassirer, Bauch), has justly emphasized that these elements do not suffice. Order concepts, our basic relations, must be added.
>
> (ibid.: §75)

It was the accentuation of the importance of the logic of relations that, for Carnap, marked the significance of Cassirer's contribution. His own constitutional project agreed with Cassirer's logical (transcendental) idealism in the view that "the constructed objects are objects of conceptual knowledge only *qua* logical forms which are generated in a certain way" (ibid.: §177). Cassirer had discussed the logic of relations especially in its application to the concept of number (1910: ch. 2), but Carnap's ambition was to expand the logic of relations even to the basic elements, the so-called *Elementarerlebnisse*, of his own system.

> Even though these basic elements are initially introduced as unanalyzable units, eventually, in the progress of construction, different properties are attributed to them, and they are analyzed into (quasi) constituents . . . It is only through this procedure, that is, only as constructed objects, that they become objects of cognition in the proper sense of the word, in particular, objects of psychology.
>
> (1928/1967: §177)

Given this expansion of the logic of relations to the basic elements of the constitutional system, the view of Carnap's *Aufbau* as a variant of radical empiricism, or even sensualism, is wholly inadequate. Carnap's indebtedness to the neo-Kantian tradition, and especially to Cassirer's relationalist approach, precludes such an interpretation (see Friedman 1992).

However, there were also points of severe disagreement between Carnap and Cassirer. As both Massimo Ferrari (2016) and Thomas Mormann (2016) have pointed out, they disagreed over the issue of physicalism. In an unpublished document titled "Ausdrucksfunktion und 'Wiener Kreis'" (written presumably in 1935–6), Cassirer explicitly opposed Carnap's characterization of the problem of other minds as a metaphysical "pseudo-problem" (2011: 176–8). Carnap's attempt at a physicalistic reconstruction of the language of science (see his 1932) was deemed a failure because it was held to be unable to account for the fact of the cultural sciences. Physics *itself*, Cassirer maintained, had to be regarded as a cultural fact and therefore as one of the diverse *symbolic forms* which, he thought, were not reducible to something physical (2011: 212). On the whole, many of the problems that the members of the Vienna Circle disqualified as metaphysical ("meta-*physisch*") were in fact merely *meta*-physical ("nur meta-*physikalisch*") and hence to be regarded as serious philosophical problems (ibid.: 210). More precisely, the meta-physical perspective was supposed by Cassirer to provide insights, not into the nature of "mind-independent'" reality (as with traditional metaphysics), but rather into the structure of the diverse symbolic forms: it served to illuminate the *modus operandi* of the *animal symbolicum*.

While Cassirer's general attitude toward the philosophy of the Vienna Circle was positive and favorable, his critical discussion of Carnap's (and other logical empiricists') plea for physicalism provides material for further research.

Cassirer and Frank

Among the founding members of the so-called First Vienna Circle, it was primarily Philipp Frank who commented on Cassirer's work. To be sure, Otto Neurath invited Cassirer to participate at the 'Second International Congress for the Unity of Science' in 1936 in Copenhagen. In the case of Frank, however, the relation was more substantial (though not intense). In his book on the law of causality, Frank characterized Cassirer's *Zur Einsteinschen Relativitätstheorie* as an example of the "process of self-destruction" of "school philosophy" (1932: 283), a trend that Frank appreciated as progressive. He repeated the compliment in a 1937 letter to Cassirer (2009: 176) and in his review of Cassirer's *Determinismus und Indeterminismus in der modernen Physik* (1936). There Frank defined his aim as judging "Cassirer's exposition from the standpoint of logical empiricism, according to which only those statements may occur in science that can be justified through logical derivation or empirical tests" (1938: 179). His conclusion was affirmative:

> Cassirer's book is to be welcomed from the standpoint of logical empiricism as a highly successful attempt to continue the adjustment of traditional idealist philosophy to the progress of science, which in my opinion can end only with the complete disintegration of the traditional philosophy.
>
> (ibid.: 184–5)

It is rather doubtful whether Cassirer agreed with this characterization (see Mormann 2016: 164), yet clearly the logical empiricists recognized Cassirer as a competent analyst of modern physics, including quantum mechanics.

Cassirer and Kaila

As is well known, Cassirer in the mid-1930s had to emigrate—via Oxford—to Sweden, where he obtained a professorship at the University of Gothenburg. In 1939 he became a Swedish citizen. In Finland, another Nordic country, Eino Kaila promoted the logical empiricist movement (see CH. 33). Already in his 1930 monograph *Der logistische Neupositivismus*, Kaila referred to Cassirer as an interpreter of Kant within the paradigm of a "scientific philosophy" (1930/1979: 3). The closeness of Kaila's own philosophical conception to that of Cassirer is remarkable. Though arguing in terms of a "realist" interpretation of the logical empiricist program (see Niiniluoto 1992; Neuber 2012b), Kaila was obviously inspired by certain core elements of Cassirer's logical idealism.

The most noteworthy point in this regard is the connection between the concepts of invariance and objectivity. In his 1936 *Über das System der Wirklichkeitsbegriffe* Kaila relied on Cassirer's *Substanzbegriff und Funktionsbegriff* in order to critically review the "logical peculiarity" (1936/1979: 102) of ancient Greek—especially Aristotelian—thought. This peculiarity consisted, according to Kaila, in "the *preference of predicative invariances of a logically lower order over relational invariances of a logically higher order*" (ibid., orig. emphasis). Kaila's principal claim here was the same as Cassirer's: in modern scientific thought the hierarchy became reversed, such that relational (functional) concepts formed the medium of establishing *invariant systems of laws*. In his 1941 *Über den physikalischen Realitätsbegriff*, Kaila elaborated this account by introducing a "principle of invariance." According to this principle, a description of reality is the more objective the more invariant it is. Kaila declared categorically:

> The physical reality of everyday is a system of invariances of experience, in which a large part of the phenomena is adjudged as "illusion" and eliminated. Physico-scientific reality is the system of higher invariances of everyday reality, in which again a large part of the latter reality is adjudged as "illusion" and eliminated.
>
> (1941/1979: 185)

From this, Kaila arrived at the conclusion that

> physico-scientific reality . . . is in logical respects the highest reality we can attain. The disclosure and representation of this reality—and not, say, the "analysis of

> sensations"—is the aim of physical research which determines the formation of its concepts and theories.
>
> (ibid.)

Kaila's account of objectivity was based on the assumption of a hierarchical system of invariant systems of relations.

It cannot be denied that Cassirer had set the stage for such an account, especially in his *Zur Einsteinschen Relativitätstheorie* (see Ihmig 1997). Furthermore, Cassirer himself was aware of Kaila's work, but it was the psychological rather than the philosophical writings of Kaila that engaged his interest. For example, in his essay on "The Concept of Group and the Theory of Perception," he quoted from an article by Kaila that dealt with issues of visual perception (1944: 13 on Kaila 1923). Whether Cassirer was aware of how close Kaila's *invariantist account of objectivity* was to his own must be left as an open question.

Concluding remark

The relation between Cassirer and the logical empiricists is worth considering in greater detail. Perhaps the most striking feature of their various existing interactions is that they show them to be more "allied" in certain respects than one might have expected. But it must be kept in mind that this makes logical empiricism and logical idealism by no means identical.

References

Carnap, R. (1922) *Der Raum. Ein Beitrag zur Wissenschaftslehre*, Kant Studien Ergänzungshefte 56. Trans. "Space. A Contribution to the Theory of Science," in Carnap, *Collected Works* (ed. by A. W. Carus et al.), Oxford: Oxford University Press, 2019, pp. 22–208.

——— (1928) *Der logische Aufbau der Welt*, Berlin: Weltkreis-Verlag. Trans. *The Logical Structure of the World*, Berkeley: University of California Press, 1967, repr. Chicago: Open Court, 2003.

——— (1932) "Die physikalische Sprache als Universalsprache der Wissenschaft," *Erkenntnis* 2: 432–65. Trans. *The Unity of Science*, London: Kegan Paul, Trench Teubner & Co., 1934.

Carus, A. (2007) *Carnap and Twentieth-Century Thought: Explication as Enlightenment*, Cambridge: Cambridge University Press.

Cassirer, E. (1906) *Der kritische Idealismus und die Philosophie des "gesunden Menschenverstandes"*, Gießen: Töpelmann.

——— (1910) *Substanzbegriff und Funktionsbegriff: Untersuchungen über die Grundfragen der Erkenntniskritik*, Berlin: Bruno Cassirer. Trans. "Substance and Function," in Cassirer (1923), pp. 1–350.

——— (1921) *Zur Einsteinschen Relativitatstheorie: Erkenntnistheoretische Betrachtungen*, Berlin: Bruno Cassirer. Trans. "Einstein's Theory of Relativity," in Cassirer (1923), pp. 351–456.

——— (1923) *Substance and Function and Einstein's Theory of Relativity*, Chicago: Open Court.

——— (1927) "Erkenntnistheorie nebst Grenzfragen der Logik und Denkpsychologie," *Jahrbücher der Philosophie* 3: 31–92.

——— (1929) *Philosophie der symbolischen Formen. Dritter Teil: Phänomenologie der Erkenntnis*, Berlin: Bruno Cassirer. Trans. *The Philosophy of Symbolic Forms. Part Three: The Phenomenology of Knowledge*, New Haven: Yale University Press, 1957.

——— (1936) *Determinismus und Indeterminismus in der modernen Physik: Historische und systematische Studien zum Kausalproblem*, Göteborg: Wettergren & Kerber.

——— (1944) "The Concept of Group and the Theory of Perception," *Philosophy and Phenomenological Research* 5: 1–36.

——— (2009) *Nachgelassene Manuskripte und Texte. Band 18: Ausgewählter wissenschaftlicher Briefwechsel* (ed. by J. M. Krois), Hamburg: Meiner.

——— (2011) *Nachgelassene Manuskripte und Texte. Band 4: Symbolische Prägnanz, Ausdrucksphänomen und "Wiener Kreis"* (ed. by C. Möckel), Hamburg: Meiner.

Coffa, A. (1991) *The Semantic Tradition from Kant to Carnap: To the Vienna Station*, Cambridge: Cambridge University Press.

Cohen, H. (1885) *Kants Theorie der Erfahrung*, Berlin: Dümmler.

——— (1902) *Logik der reinen Erkenntnis*, Berlin: Bruno Cassirer.

Frank, P. (1932) *Das Kausalgesetz und seine Grenzen*, Wien: Springer. Trans. *The Causal Law and its Limits*, Dordrecht: Kluwer, 1998.

——— (1938) "Bemerkungen zu E. Cassirer: Determinismus und Indeterminismus in der modernen Physik," *Theoria* 4: 70–80.

Ferrari, M. (2016) "Cassirer, Schlick und der Neukantianismus—Philosophische Streitfragen im Kontext," in M. Neuber (ed.), *Husserl, Cassirer, Schlick: 'Wissenschaftliche Philosophie' im Spannungsfeld von Phänomenologie, Neukantianismus und logischem Empirismus*, Cham: Springer, pp. 85–106.

Friedman, M. (1992) "Epistemology in the *Aufbau*," *Synthese* 93: 15–57. Repr. with a Postscript in Friedman (1999), pp. 114–62.

——— (1994) "Geometry, Convention and the Relativized A Priori," in W. Salmon and G. Wolters (eds.), *Logic, Language, and the Structure of Scientific Theories*, Pittsburgh: University of Pittsburgh Press, pp. 21–34. Repr. in Friedman (1999), pp. 59–70.

——— (1999) *Reconsidering Logical Positivism*, Cambridge: Cambridge University Press.

——— (2000) *A Parting of the Ways: Carnap, Cassirer, and Heidegger*, Chicago: Open Court.

Gower, B. (2000) "Cassirer, Schlick and 'Structural' Realism: The Philosophy of the Exact Sciences in the Background to Early Logical Empiricism," *British Journal for the History of Philosophy* 8: 71–106.

Grelling, K. (1908) "Das gute, klare Recht der Freunde der anthropologischen Vernunftkritik, verteidigt gegen Ernst Cassirer," *Abhandlungen der Friesschen Schule* 2: 153–90.

Ihmig, K. N. (1997) *Cassirers Invariantentheorie der Erfahrung und seine Rezeption des "Erlanger Programms,"* Hamburg: Meiner.

Kaila, E. (1923) "Gegenstandsfarbe und Beleuchtung," *Psychologische Forschung* 3: 18–59.

——— (1930) *Der logistische Neupositivismus—Eine kritische Studie*, Turku: Annales Universitatis Fennicae Aboensis BXIII. Trans. "Logistic Neopositivism. A Critical Study," in Kaila (1979), pp. 1–58.

——— (1936) *Über das System der Wirklichkeitsbegriffe: Ein Beitrag zum Logischen Empirismus, Acta Philosophica Fennica Facs. II*, Helsinki. Trans. "On the System of the Concepts of Reality. A Contribution to Logical Empiricism," in Kaila (1979), pp. 59–125.

——— (1941) *Über den physikalischen Realitätsbegriff: Zweiter Beitrag zum Logischen Empirismus, Acta Philosophica Fennica Facs. IV*, Helsinki. Trans. "On the Concept of Reality in Physical Science. Second Contribution to Logical Empiricism," in Kaila (1979), pp. 126–258.

——— (1979) *Reality and Experience: Four Philosophical Essay* (ed. by R. S. Cohen), Dordrecht: Reidel.

Krois, J. M. (2000) "Ernst Cassirer und der Wiener Kreis," in F. Stadler (ed.), *Elemente moderner Wissenschaftstheorie*, Vienna: Springer, pp. 105–22.

Mormann, T. (2016) "Wissenschaftliche Philosophie im Exil: Cassirer und der Wiener Kreis nach 1933," in M. Neuber (ed.), *Husserl, Cassirer, Schlick: 'Wissenschaftliche Philosophie' im Spannungsfeld von Phänomenologie, Neukantianismus und logischem Empirismus*, Cham: Springer, pp. 159–79.

Neuber, M. (2012a) *Die Grenzen des Revisionismus: Schlick, Cassirer und das "Raumproblem"*, Vienna: Springer.

——— (2012b) "Invariance, Structure, Measurement: Eino Kaila and the History of Logical Empiricism," *Theoria* 78: 358–83.

——— (2013) "Treffpunkt Struktur—Cassirer, Schlick und Carnap," *Archiv für Geschichte der Philosophie* 95: 206–33.

Niiniluoto, I. (1992) "Eino Kaila and Scientific Realism," in I. Niiniluoto, M. Sintonen and G. H. von Wright (eds.), *Eino Kaila and Logical Empiricism*, Helsinki: Societas Philosophica Fennica, pp. 102–16.

Oberdan, T. (2009) "Geometry, Convention, and the Relativized Apriori: The Schlick-Reichenbach Correspondence," in F. Stadler and H. J. Wendel (eds.), *Stationen. Dem Philosophen und Physiker Moritz Schlick zum 125. Geburtstag*, Vienna: Springer, pp. 186–211.

Reichenbach, H. (1920) *Relativitätstheorie und Erkenntnis Apriori*, Berlin: Springer. Trans. *The Theory of Relativity and A Priori Knowledge*, Berkeley: University of California Press, 1965.

Ryckman, T. (1991) "*Conditio sine qua non? Zuordnung* in the Early Epistemologies of Cassirer and Schlick," *Synthese* 88: 57–95.

——— (2005) *The Reign of Relativity: Philosophy in Physics 1915–1925*, Oxford: Oxford University Press.

Schlick, M. (1918) *Allgemeine Erkenntnislehre*, Berlin: Springer, 1918, 2nd rev. ed. 1925. Trans. *General Theory of Knowledge*, Vienna: Springer, 1974, repr. Lasalle, IL: Open Court, 1985.

——— (1919) "Erscheinung und Wesen," *Kant-Studien* 23: 188–208. Trans. "Appearance and Essence," in Schlick (1979), pp. 11–28.

——— (1921) "Kritizistische oder empiristische Deutung der neuen Physik?" *Kant Studien* 26: 96–111. Trans. "Critical or Empiricist Interpretation of Modern Physics?" in Schlick (1979), pp. 322–35.

——— (1979) *Philosophical Papers, vol. 1 (1909–1922)* (ed. by H. Mulder and B. van de Velde-Schlick), Dordrecht: Reidel.

31

CRITICAL RATIONALISM, THE VIENNA CIRCLE, AND THE EMPIRICAL BASIS PROBLEM

Artur Koterski

The beginnings of critical rationalism are embedded in the history of the Vienna Circle on many levels. Its originator, Karl Popper (1902–1994), was born and educated in Vienna, where he studied and attended lectures by Hans Hahn and Moritz Schlick, the founders of the Vienna Circle; later on he took part in Rudolf Carnap's and Karl Menger's seminars. Popper was also acquainted with other Circle members. In 1929 he was introduced to Victor Kraft, who considerably influenced his views on anti-inductivism and the hypothetico-deductive model of scientific theories (see Kraft 1925/1973: 53 and Radler 2013). A year later Popper met Herbert Feigl; with his support, Popper joined Feigl and Carnap on their holiday in Tyrol in August 1932, which gave him the opportunity to learn about the recent developments of physicalism in the Vienna Circle and discuss at length the manuscript of his own book, *Die beiden Grundprobleme der Erkenntnistheorie* (1978; see Popper 1974a/2002: 87–97; Hacohen 2001: 94, 103–5, 172, 184–5). This encounter was momentous. On Carnap's recommendation, a version of the book was published in 1934 in the series edited by Schlick and Philipp Frank as *Logik der Forschung* (hereafter *LdF*), not translated until 1959 with more recent additions as *The Logic of Discovery*. Furthermore, Otto Neurath's invitations gave Popper the opportunity to present himself at important international conferences, the Congresses for the Unity of Science in Paris in 1935, and in Copenhagen in 1936. While the Vienna Circle philosophers in general regarded Popper's work as an important contribution to the scientific philosophy that they were trying to create, he was never asked by Schlick—despite Carnap's and Feigl's efforts—to participate in their Thursday meetings. Popper himself deprecated (nearly all) affinities between his views and logical empiricism and stressed the differences; unfortunately, his attacks repeatedly were directed against straw men (see, e.g., 1934/2002: 10–5, 76–9, 1963: 253–92). This is perhaps most evident in the case of Neurath's conception of protocol sentences and the alleged inductivism of the Vienna Circle (see Koterski 2012).

Critical rationalism and the empirical basis problem

Critical rationalism is a theory of scientific rationality that originated in strict opposition to justificationism and inductivism, verificationism, psychologism, historicism, and conventionalism, as well as some other traits of empirical positivism—all *as interpreted* by Popper. In his understanding, rationality is best conceived as a critical attitude, i.e., "a readiness to accept criticism"

DOI: 10.4324/9781315650647-35

(1974a/2002: 118). To prove that science is rational in this sense and to show how to maximize its rationality, Popper presented a version of empiricism for which falsificationism, indeterminism, objectivism, scientific realism, and, later on, verisimilitude of truth are fundamental. These aspects were critically assessed and developed further by his students and collaborators; in not just a few cases this led eventually to the abandonment of basic precepts defended by their teacher. Furthermore, this development resulted in a multiplicity of conceptions that call themselves "critical rationalism" but are in conflict with each other. If nevertheless something does unite them, it is the highly critical attitude towards logical empiricism that was retained by virtually all major critical rationalists.

The critical rationalist critique of justificationism alleged its comprehensive failure both in the descriptive and normative dimensions. Since inductive procedures were deemed illegitimate, justificationism could not give a historically acceptable account of the growth of science (Popper 1934/2002: 18, 1963: 185–90); furthermore, the prescriptions it offers promote uncritical attitudes and scientific dishonesty (Popper 1983: 163; Lakatos 1978: 103). Popper's own conception of arguing and learning from experience requires an entirely different framework.

The new platform was founded on falsificationism. Its central claim concerns the question of scientificity: it demands that hypotheses accepted in science should satisfy the requirement of falsifiability: "Those theories are scientific which are capable of being tested experimentally, where tests of a hypothesis are attempts to refute it" (Agassi 1975: 40–1). According to Popper, falsification has a hypothetico-deductive character and is based on *modus tollens*, $((h \rightarrow p) \land {\sim}p) \rightarrow {\sim}h$, where h is a hypothesis under the test, p is a prediction inferred from h, and ${\sim}p$ is an accepted falsifier, i.e., a singular, observational statement about the negative outcome of an experiment. For falsification to serve as an instrument of criticism in science, the problem of the empirical basis needs to be solved. This problem, or "Fries's trilemma" (see Wettersten 1992: 159, n. 26), concerns the nature of basic sentences (initial conditions, test-statements) and acceptance rules for them. Is the acceptance of basic sentences to be justified by other, already accepted sentences, or by perceptions, or by convention?

To achieve objectivity (understood as intersubjectivity) Popper chose a physicalistic language: the empirical basis of science consists of existential statements about determinate spatiotemporal events. Popper emphatically ruled out all competing theories that employed a phenomenal language and, rejecting altogether their aim of justification, he opted for a conventionalist strategy for the acceptance of basic statements. It consisted of two steps. The first one was to stipulate a distinction between "observational" and "theoretical" statements. The second one was to decide which part of the class of observational sentences temporarily rises above doubt and should be included into the so-called background knowledge, i.e., the body of the empirical knowledge unquestioned at the moment (Popper 1978/2009: 475).

The solution to accept the empirical basis on conventionalist grounds, i.e., without justification (his official opposition to conventionalism notwithstanding), remains in perfect agreement with the principles of critical rationalism which recognizes its own limits. These principles would be contradicted by a comprehensive rationalism which demands that "any assumption which cannot be supported either by argument or by experience is to be discarded" (such a version of rationalism is uncritical and self-contradictory) (Popper 1945: 217–8).

Basic statements are accepted as a result of a decision, and *in this sense* they are conventional; however, such decisions are motivated by experiences, so this solution complies with empiricism. It also follows Popper's anti-inductivist policy, because basic statements are accepted in the course of testing of a theory and are not simply collected for the sake of future generalizations. Moreover, since they can be subjected to further tests and their acceptance can be revoked, they do not form a solid base: thus the critical rationalist anti-foundationalism.

In principle Popper never revised the answer he gave in the 1930s to the problem of the empirical basis (see 1934/2002: 74–94). The only concession he made was to admit that sensations were not only "motives" but rather "inconclusive reasons" for accepting basic statements (1974b: 1114). Some other critical rationalists, however, called for further changes, e.g., to ascribe to perceptual reports not only causal but also an epistemological role in the solution of this problem.

According to John Watkins, the conventionalist strategy adopted by Popper breaks the asymmetry between falsifiability and verifiability and makes falsifications impossible because it does not really allow the testing of basic statements. To make that possible, a compromise with phenomenalism was necessary: while any empirical basis consists of physicalistic or "level-1" statements, they can be tested only with help of "level-0" first-person perceptual reports (of the "here-and-now so-and-so" kind) that themselves do not belong to science (see Watkins 1984: 249–54, 264–5, 273, 1992: 150–3).

Elie Zahar likewise thinks that the conventionalist solution of Fries's trilemma rescinds the asymmetry; his proposal to save it deviates from Popperian orthodoxy even further. He believes, unlike Watkins, that level-0 reports can directly test scientific theories. Sentences like "I seem to be perceiving *x*" are infallible in the moment they are uttered because our consciousness has a direct access to its own states. It is a "myth" that justification is an inter-sentences relation (Zahar 1989: 325–6, 332–5). Intersubjectivity is not a "desirable feature" of basic statements anymore and, moreover, they do not have to refer to reproducible states of affairs (ibid.: 112 and 1983: 158). The conclusion is that "a phenomenological account of the empirical basis forms a necessary complement to Popper's falsificationism" (Zahar 1995: 45; see also 1997: 82–7, 2001: 22–4, 26, 30–3).

A still different form of infallibility was invoked by John Worrall. In case of a conflict of a theoretical system with experience, there are always some "crude data" involved that are either "entirely unquestionable" or are at least never questioned. In the examples that he gives, they are the "records of the angles of inclination of certain telescopes at certain times (that is, when certain clocks showed particular readings)," or "the needle in this meter pointed to (or close to) the mark '10'"; therefore, they qualify as Popperian basic statements. Thus, he also disagrees with "the allegedly inevitable fallibility of basic statements" (1995, 83–5; see also 1991: 331–6, 351–2, n. 10).

Some of Popper's followers rejected his theory of rationality and defended a more comprehensive critical rationalism. They disagreed with "Popper's unfortunate tendency to demand convention or irrational decision whenever some point is reached which cannot be justified" (Bartley 1984: 215; see also 1982: 166–7). These critics did not consent to any "phenomenological complements" either. According to Bartley, Fries's problem arises only when falsifications are intended to prove the falsity of theories—or when they are meant to justify the claims about falsity of theories. However, there is not even a need for the acceptance of an empirical basis: in case of a clash of a theory with a test-statement, none of them proves that the other is false, and both can be false. Rational criticism rather consists in establishing that the theory is problematic with regard to some basic statements and *vice versa*; this is followed by making a (testable) conjecture as to their accuracy and provisional rejection of one of them (see Miller 1994: 92–3).

Others considered the conditions imposed by Popper on test-statements as too narrow. The proposed changes differed: minimalists modify merely the formal condition and include into the empirical basis "all types of singular statements about observable events" (Andersson 1994: 71 and 1998: 151; see also Carnap 1932a/1987: 465), whereas maximalists break all possible constraints and deprive the empirical basis problem of its significance.

The relative unimportance of the empirical basis for modern falsificationism was emphasized by Lakatos. The objects of methodological appraisals in his methodology of scientific research programs are series of theories *plus* their heuristics. Single basic statements are unable to falsify and reject a scientific program; this can be done only with a competing program that exhibits specified methodological advantages over its rival. Non-falsifiability is not tied to the absence of potential falsifiers of the empirical basis and can be imposed by a methodological *fiat* as in the case of a program's "hard core." On the other hand, hypotheses without potential falsifiers can be falsified if they conflict with a well-corroborated research theory, or rather with a progressive research program. However, falsifications are not even necessary for scientific progress; what matters is the growth of science. Therefore "[t]he sophisticated falsificationist allows any part of the body of science to be replaced but only on the condition that it is replaced in a 'progressive' way, so that the replacement successfully anticipates novel facts" (Lakatos 1978: 99). While the importance of Fries's trilemma is significantly lowered in his methodology, Lakatos in principle retains Popper's solution, but enhanced it with an "appeal procedure" to deal with faulty interpretations of empirical data (ibid.: 42–5; see also Kreuzer 1985: 59).

Some Popperians went still further. Agassi—together with Bartley and other comprehensively critical rationalists—holds that Popper's solution to the empirical basis problem and any acceptance rules for test-statements are "superfluous." He maintains that observational reports should be understood as explananda; when a report of some type keeps on being repeated, the task consists in finding highly testable hypotheses that can explain it and, thereafter, in picking the best of them, *regardless of* whether the explanation given assumed the truth or *falsity* of those reports. Additional rules for rejecting a basic statement are likewise redundant. In case of an empirical conflict, the basic statement should be rejected if the rejection "is the more testable option" than the rejection of the hypothesis (1975: 110–5, 148–9; also Berkson 1987: 29). It is very important for Agassi that finally there exists a way to deal with false observational reports.

The gamut of views about the empirical basis problem shows that critical rationalists either accept the conventionalist solution adopted by Popper, or its phenomenalistic counterpart, or they reject the very question of the basis: all possibilities were explored. Nowadays, many philosophers sympathize with this last opinion and say that the protocol sentences debate is a neopositivistic relic, solely of interest to historians, but this is not true of critical rationalists. When Popper formulated his principle of critical rationalism which demands "that our adoption and our rejection of scientific theories should depend upon our critical reasoning," he indicated that the latter has to be "combined with the results of observation and experiment," and that therefore, "our adoption and our rejection of scientific theories should depend . . . upon singular observation statements" (1983: 32). The empirical basis problem is inevitably involved in the critical rationalists' theories of scientific rationality; even if some of them think it is not a problem anymore, they feel obligated to explain why it is so.

Moreover, one's attitude toward it serves as a litmus paper test of one's conception of rationality which can bring surprising results. Popper himself was classified as an irrationalist: "No doubt, Popper comes close to being a rationalist. . . . His logical rationality is limited by his claim that the acceptance or rejection of protocol statements is largely a matter of convention" (Bunge 1987: 11; see also Lakatos 1978: 165–6; Bartley 1993: 211). Accordingly, Bartley, Agassi, or Andersson insist that critical rationalists should remove this conventional trait.

As noticed earlier, some critical rationalists weakened the conventionalism of Popper's original conception by introducing an appeal procedure and admitting that sensations can be reasons for accepting protocols. However, the second of these amendments was not received by all with equal enthusiasm. Lakatos never mentioned it, and David Miller reproached Watkins that he forgot that "experience can only motivate the decision to accept a test statement." He agreed with Lakatos that without some conventional steps, criticism would be impossible (Miller 1994: 30, 93; see also his 1999: 65 and 2006: 127; Lakatos 1978: 42, and 1999: 90).

This met with protests. Alan Musgrave—together with Watkins and Zahar—claimed that perceptual experiences are good reasons for the acceptance of perceptual judgments. They hold that there are infallible statements that can be used in testing, so for them "in Miller's hands so-called critical rationalism has turned into wholesale irrationalism" (Musgrave 1989: 335; see also Watkins 1995: 614–5). For others, however, it is the position of Musgrave, Watkins, and Zahar that does not classify as critical rationalism, because "the view that test statements are fallible is basic for critical rationalism" (Andersson 1991: 284; see also Albert 1988: 2).

Critical rationalism and the protocol sentences debate

The discussion among critical rationalists resembles the Vienna Circle's protocol sentences debate from the 1930s and is linked to it by Popper's contribution, mainly in *LdF*. How similar are these two disputes, and which one was more fruitful in terms of yielding better understanding of testing procedures in science?

As noted, the critical rationalists presented principally three approaches to the empirical basis problem. The first one is represented by Popper, his disciples who accept his solution as it is (Miller and Shearmur [Shearmur 2004: 105, and 2006: 197]), and those who introduced some minor changes (Lakatos and Radnitzky [Radnitzky 1993: 291, 1987: 168–9, 1989: 387–8]). In essence, it is a physicalist solution of the problem. The second approach is chosen by those who oppose the claim that no statement can ever be established with certainty (Musgrave, Watkins, and Zahar). In essence it is a phenomenalist solution of the problem. The third approach is advanced by radicals like Agassi. They maintain that the problem is superfluous. Since they insist that no statement is more basic than any other, their view amounts essentially to saying that statements of all kinds can be used as the test-statements.

These positions have close counterparts in the theories held in the Vienna Circle during the protocol sentences debate. *The physicalist conception* of protocol sentences was formulated by Otto Neurath and developed since early 1931 (see, e.g., Neurath 1932) so it predates that of Popper. As shown elsewhere, Popper's theory is as an impoverished version of Neurath's (see Koterski 2006). Admittedly, already in 1932, Popper disagreed with the opinion that their views converged (see Popper 1963: 267–8). However, his dissent was accompanied through years by distortions of Neurath's views. Whether it was a failure to understand him, or a willful misinterpretation, it is hard to take Popper's protestations wholly seriously.

The critical rationalists who complained about the conventionality of Popper's solution to the problem of empirical basis would not be satisfied with Neurath's theory of test-statements either. However, they could see that it copes better with the problems they raised in their criticism of Popper. First and foremost, Neurathian protocols explicitly indicate the link with experience, including the sense modality involved. Perceptions do not prove them; however, they are not classified in true/false categories.

Popper objected that while Neurath's protocols are not irrevocable, he did not give cancellation rules for them and took this to show that Neurath inadvertently abandoned empiricism (1934/2002: 78–9). However, Popper was wrong; it has been demonstrated that a consistent naturalistic reading of Neurath's theory is possible where all the supposedly missing rules, including those for cancellation, are indicated (see Uebel 2007: 381–9, 393–4). Ironically, some influential Popperians—Agassi, Bartley, and also Watkins (and no one seems to claim otherwise)—agree that Popper did not state such rules either (see Agassi 1975: 144–6; Bartley 1982: 214, n. 66).

Agassi suggested that his account of observation reports is a significant improvement over that of Popper, and eliminates the problem of the empirical basis because his theory allows accepting such reports as false (see his 1975: 115). Since Neurath's theory explains how to deal with protocols obtained as a result of an observational error or a lie (see Uebel 2007: 384–5, 391–3), Agassi should also admit that Neurath's conception superseded that of Popper as well.

Once the empirical basis problem is disposed of, any statement can be used as a test-statement. This disagrees with Neurath's conception. It may be noted, however, that such a view was supported by Carnap already earlier (and before he met Popper on 2 August 1932). On 25 May 1932, Carnap noted in his diary a conversation with Philipp Frank: "I hold the conventionalist view that the protocol sentences may look as they like, and that any [stipulation of a] condition etc. can be taken as a protocol sentence of another when the adequate rules are established" (1932b).

The phenomenalist conception of test-statements was developed in the Vienna Circle by Moritz Schlick and Béla Juhos as a critique of conventionalism and radical fallibilism of the physicalist theories as defended by Neurath, Carnap, and also Popper (Schlick 1934/1979: 374, n. 2). Schlick argued that the justificatory chain for any statement has to end up not with a convention, but with a statement *and* an extra-linguistic, deictic indication of the fact which constitutes its meaning. Such statements, "affirmations," give an account of the most fundamental cognitive acts and grasp their content. They have quite peculiar characteristics.

Affirmations are of the form "here-and-now so-and-so," so they are very close counterparts of the level-0 statements—all the more so since Schlick placed them outside of the system-language of science like Watkins. In the moment of being uttered, they are absolutely certain to the speaker, but after that, they become hypothetical. Zahar and Watkins agree with Schlick "that we can infallibly know the truth of our autopsychological statements—at least while asserting them" (Zahar 1989: 332). Musgrave, who in principle endorses a physicalistic theory of observation (see his 1999: 343), permits a whiff of foundationalism: "Of course, my reasonable perceptual belief might turn out to be false. If evidence comes in . . . of perceptual error, I may concede that my perceptual belief was wrong—but that does not mean that I was wrong to have believed it" (2011: 232; see also his 2009: 5–13, 1993: 174, 281–2).

One of the most troublesome questions for the phenomenalists is whether perceptual judgments can be infallible despite the possibility of unintentionally wrong choice of words, as in the case of *lapsus memoriae*, when someone utters "here yellow" upon seeing a blue spot. Schlick answered that in "affirmations" the terms acquire the meaning intended by the speaker and it does not matter how they are employed by others (1935/1979: 412). While the argument given by Schlick, and especially by Juhos, that the truth of affirmations is guaranteed by their "grammar," does not feature as such in considerations by Popperians, a very similar line of reasoning is presented by Zahar:

> Doubts have been raised about the meaning of the universals occurring in level-0 reports: how do we know that they subsume what we actually perceive? But these meanings are *as intended* by, and during, the conscious activity and do not extend beyond it. In other words: there is no guarantee that other people, or I at a later date, should attach the same meaning to 'redness' as I do now.
>
> (Zahar 1989: 336)

Zahar, Watkins, and Musgrave, agreeing that infallible judgments are possible, reject the "logomaniac" assumption from *LdF* that statements can be justified only by statements. The closest critical rationalist counterpart of Schlick's conception is, however, Watkins's theory which does better than its neopositivist forerunner at least in one respect. His level-0 reports can be written down since they do not have to be made together with gestures (they are already absent in Juhos's theory which, however, relies on questionable arguments to justify infallibility of his version of affirmations).

Conclusion

John Wettersten presents Popper's ideas as a breakthrough and is persuaded that they led to considerable progress, but he worries that "many of his views will be, in many cases have been, surreptitiously adopted, perhaps ascribed to others" (Wettersten 2005: 120). His words sound strangely ironic in the context of the empirical basis problem. The conflict which disunites critical rationalists brings to mind the protocol sentences debate known from *Erkenntnis* and *Analysis* and not only for its substance. Knowingly or not, the critical rationalists' dispute followed very closely the tracks left behind by logical empiricists, also acquired personal overtones along the way, and ended up with no common agreement that would corroborate Wettersten's claim of progress.

In comparison with Schlick's version, the phenomenalist view obtained a more satisfactory formulation from Watkins and was further broadened by Zahar. For some this may count as progress, but not for Popper, who strongly opposed this view in *LdF*. On the other hand, the position Popper defended there can hardly be regarded as progressive because a better theory already existed then, as we have seen. Neither Popper nor other critical rationalists have ever admitted this fact. By contrast, an important aspect of Popper's thought was recognized quite early on by Neurath in a letter to Carnap:

> Someone from Vienna [and] raised in the atmosphere that I and also [Philipp] Frank know very well, and you too to a large degree, wants to appear, so to speak, as an original genius. This will not work. If one has the misfortune to grow up in such a highly developed environment, then to a significant extent one is always *a continuator*.
> (25 June, 1935, trans. AK)

References

Agassi, J. (1975) *Science in Flux*, Dordrecht: D. Reidel.

Agassi, J. and Jarvie, I. C. (eds.) (1987) *Rationality: The Critical View*, Dordrecht: Martinus Nijhoff.

Albert, H. (1988) "Critical Rationalism: The Problem of Method in Social Sciences and Law," *Ratio Juris* 1: 1–19.

Andersson, G. (1991) "Feyerabend on Falsifications, Galileo, and Lady Reason," in Munévar (1991), pp. 281–95.

——— (1994) *Criticism and the History of Science: Kuhn's, Lakatos's, and Feyerabend's Criticism of Critical Rationalism*, Leiden: E.J. Brill.

——— (1998) "Basisprobleme," in H. Keuth (ed.), *Logik der Forschung*, Berlin: Akademie Verlag, pp. 143–64.

Bartley, W. W. (1982) "Critical Study: The Philosophy of Karl Popper. Part III: Rationality, Criticism, and Logic," *Philosophia* 11: 121–221.

——— (1984) *The Retreat to Commitment*, La Salle: Open Court, 2nd ed.

——— (1993) "Theories of Rationality," in Radnitzky and Bartley (1993), pp. 205–14.

Berkson, W. (1987) "Skeptical Rationalism," in Agassi and Jarvie (1987), pp. 21–43.

Bunge, M. (1987) "Seven Desiderata for Rationality," in Agassi and Jarvie (1987), pp. 1–15.

Carnap, R. (1932a) "Über Protokollsätze," *Erkenntnis* 3: 215–28. Trans. "On Protocol Sentences," *Noûs* 21 (1987): 457–70.

——— (1932b) Entry 25 May, Carnap's *Tagebücher*, Sig. RC 025-75-10, Archives of Scientific Philosophy (University of Pittsburgh). Quoted by permission of the University of Pittsburgh. All rights reserved.

Hacohen, M. H. (2001) *Karl Popper: The Formative Years 1902–1945*, Cambridge: Cambridge University Press.

Jarvie, I. C., Milford, K. and Miller, D. (eds.) (2006) *Karl Popper: A Centenary Assessment*, Aldershot: Ashgate, vol. 2.

Koterski, A. (2006) "Basic Statements versus Protocols," in Jarvie, Milford, and Miller (2006), pp. 186–96.

——— (2012) "The Backbone of the Straw Man: Popper's Critique of the Vienna Circle's Inductivism," in D. Dieks et al. (eds.), *Probabilities, Laws, and Structures*, Dordrecht: Springer, pp. 457–68.

Kraft, V. (1925) *Die Grundformen der wissenschaftlichen Methoden*, Wien: Verlag der österreichischen Akademie der Wissenschaften, 2nd rev. ed., 1973.

Kreuzer, F. (ed.) (1985) *Die Zukunft ist offen. Das Altenberger Gespräch*, Munich: Piper.

Lakatos, I. (1978) *The Methodology of Scientific Research Programmes. Philosophical Papers* (ed. by J. Worrall and G. Currie), Cambridge: Cambridge University Press, vol. 1.

——— (1999) "Lectures on Scientific Method," in I. Lakatos and P. K. Feyerabend, *For and Against Method* (ed. by M. Motterlini), Chicago: University of Chicago Press, pp. 19–109.

Miller, D. W. (1994) *Critical Rationalism: A Restatement and Defence*, Peru: Open Court.

——— (1999) "Popper and Tarski," in I. C. Jarvie and S. Pralong (eds.), *Popper's Open Society After Fifty Years*, London: Routledge, pp. 57–71.

——— (2006) *Out of Error: Further Essays on Critical Rationalism*, Aldershot: Ashgate.

Munévar, G. (ed.) (1991) *Beyond Reason: Essays on the Philosophy of Paul Feyerabend*, Dordrecht: Kluwer.

Musgrave, A. (1989) "Deduction versus Psychologism," in M. A. Notturno (ed.), *Perspectives on Psychologism*, Leiden: E.J. Brill, pp. 315–40.

——— (1993) *Common Sense, Science and Scepticism: A Historical Introduction to the Theory of Knowledge*, Cambridge: Cambridge University Press.

——— (1999) *Essays on Realism and Rationalism*, Amsterdam: Rodopi.

——— (2009) "Experience and Perceptual Belief," in Z. Parusniková and R. S. Cohen (eds.), *Rethinking Popper*, Dordrecht: Springer, pp. 5–19.

——— (2011) "Popper and Hypothetico-Deductivism," in D. M. Gabbay, S. Hartmann and J. Woods (eds.), *Handbook of the History of Logic*, Amsterdam: North Holland, vol. 10, pp. 205–34.

Neurath, O. (1932) "Protokollsätze," *Erkenntnis* 3: 204–14. Trans. "Protocol Sentences," in A. J. Ayer (ed.), *Logical Positivism*, New York: Free Press, 1959, pp. 199–208, and "Protocol Statements," in Neurath, *Philosophical Paper 1913–1946* (ed. by R. S. Cohen and M. Neurath), Dordrecht: Reidel, 1983, pp. 91–99.

——— (1935) Letter to Carnap, 25 June 1935. Neurath Nachlass. Wiener Kreis Stichting, Rijksarchief in Noord-Holland, Haarlem, The Netherlands.

Popper, K. R. (1934) *Die Logic der Forschung*, Vienna: Springer. Rev. ed. trans. *Logic of Scientific Discovery*, London: Huchinson, 1959, repr. London: Routledge, 2002.

(1945) *The Open Society and Its Enemies*, London: Routledge & Sons, repr. 2011, vol. 2.

——— (1963) *Conjectures and Refutations: The Growth of Scientific Knowledge*, London: Routledge and Kegan Paul.

——— (1974a) "Intellectual Autobiography," in Schilpp 1974, pp. 3–182. Repr. as *Unended Quest. An Intellectual Autobiography*, LaSalle, IL: Open Court, 1976, repr. London: Routledge, 1992.

——— (1974b) "The Philosopher Replies," in Schilpp 1974, pp. 961–1200.

——— (1978) *Die beiden Grundprobleme der Erkenntnistheorie. Aufgrund von Manuskripten aus den Jahren 1930–1933* (ed. by T. E. Hansen), Tübingen: Mohr. Trans. *The Two Fundamental Problems of the Theory of Knowledge* (ed. by T. E. Hansen), London: Routledge, 2009.

——— (1983) *Realism and the Aim of Science* (ed. by W. W. Bartley), London: Routledge.

Radler, J. (2013) "Victor Kraft und Karl Popper—ein Verhältnis gekennzeichnet von freundschaftlicher Wertschätzung und kritischer Distanz," in R. Neck and H. Stelzer (eds.), *Kritischer Rationalismus heute. Zur Aktualität der Philosophie Karl Poppers*, Frankfurt a M: Peter Lang, pp. 282–95.

Radnitzky, G. (1987) "The 'Economic' Approach to the Philosophy of Science," *The British Journal of the Philosophy of Science* 38: 159–79.

——— (1989) "Falsification Looked at from an 'Economic' View," in K. Gavroglu, Y. Goudaroulis and P. Nicolacopoulos (eds.), *Imre Lakatos and Theories of Scientific Change*, Dordrecht: Kluwer, pp. 383–95.

——— (1993) "In Defense of Self-Applicable Critical Rationalism," in Radnitzky and Bartley (1993), pp. 279–312.

Radnitzky, G. and Bartley, W. W. (eds.) (1993) *Evolutionary Epistemology, Rationality, and the Sociology of Knowledge*, La Salle, IL: Open Court.

Schilpp, P. A. (ed.), (1974) *The Philosophy of Karl Popper*, La Salle, IL: Open Court.

Schlick, M. (1934) "Über das Fundament der Erkenntnis," *Erkenntnis* 4: 79–99. Trans. "The Foundation of Knowledge," in Schlick (1979), pp. 370–87.

——— (1935) "Sur les 'constatations'," in *Sur le Fondement de la Conaissance*, Paris: Hermann & Cie. Trans. "On Affirmations," in Schlick (1979), pp. 408–13.

——— (1979) *Philosophical Papers* (ed. by H. L. Mulder and B. F. B. van de Velde-Schlick), Dordrecht: D. Reidel, vol. 2.

Shearmur, J. (2004) "Popper versus Analytical Philosophy?" in P. Catton and G. Macdonald (eds.), *Karl Popper: Critical Appraisals*, London: Routledge, pp. 99–113.

——— (2006) "Karl Popper and the Empirical Basis," in Jarvie, Milford, and Miller (2006), pp. 197–210.

Uebel, T. (2007) *Empiricism at the Crossroads: The Vienna Circle's Protocol-Sentence Debate*, Chicago: Open Court.

Watkins, J. W. N. (1984) *Science and Scepticism*, Princeton: Princeton University Press.

——— (1992) "A Neo-Popperian Advance on Popper?" in W. Newton-Smith, T. Chiang and E. James (eds.), *Popper in China*, London: Routledge, pp. 138–57.

——— (1995) "Critical Rationalism: A Restatement and Defence by David Miller," *The British Journal for the Philosophy of Science* 46: 610–6.

Wettersten, J. R. (1992) *The Roots of Critical Rationalism*, Amsterdam: Rodopi.

——— (2005) "Popper's Historical Role: Innovative Dissident," *Journal for General Philosophy of Science* 36: 119–33.

Worrall, J. (1991) "Feyerabend and the Facts," in Munévar (1991), pp. 329–53.

——— (1995) "'Revolution in Permanence:' Popper on Theory-Change in Science," in A. O'Hear (ed.), *Karl Popper: Philosophy and Problems*, Cambridge: Cambridge University Press, pp. 75–102.

Zahar, E. G. (1983) "The Popper—Lakatos Controversy in the Light of 'Die beiden Grundprobleme der Erkenntnistheorie'," *The British Journal for the Philosophy of Science* 34: 149–71.

——— (1989) "John Watkins on the Empirical Basis and the Corroboration of Scientific Theories," in F. D'Agostino and I. C. Jarvie (eds.), *Freedom and Rationality: Essays in Honor of John Watkins*, Dordrecht: Kluwer, 325–41.

——— (1995) "The Problem of the Empirical Basis," in A. O'Hear (ed.), *Karl Popper: Philosophy and Problems*, Cambridge: Cambridge University Press, 45–74.

——— (1997) *Leçons d'épistémologie*, Paris: Imprimerie de l'École polytechnique.

——— (2001) *Poincaré's Philosophy: From Conventionalism to Phenomenology*, Chicago: Open Court.

32
THE LVOV-WARSAW SCHOOL AND LOGICAL EMPIRICISM

Jan Woleński

The Lvov-Warsaw School (hereafter LWS), a significant Polish analytic movement, had several connections with logical empiricism (hereafter LE). I will outline the historical development of LWS and selected views characteristic for this group and its particular representatives. Here I will concentrate on personal relations between LWS and LE and related philosophical views. LWS was complex and numerous (with about 80 members) and oriented toward many problems from all branches of philosophy, from formal logic to normative axiology (see Woleński 2015). My account will focus mainly on logic and its applications to philosophy, meaning semantics (the philosophy of language), formal logic, and the philosophy of science.

The origin and development of the Lvov-Warsaw School

LWS was established by Kazimierz Twardowski (1866–1938) at the end of the nineteenth century in Lvov, a city belonging to the Austro-Hungarian Empire at that time. Twardowski was a student of Franz Brentano and shared his view that the method of philosophy is the same as that of science. Guided by this principle, Twardowski intended to introduce scientific philosophy to the Polish philosophical community.

Two stages in the history of LWS can be distinguished. The first begins in 1895, when Twardowski became professor in Lvov. He was a charismatic teacher and very soon taught several young philosophers, including (in alphabetical order) Kazimierz Ajdukiewicz (1890–1963), Tadeusz Czeżowski (1889–1981), Tadeusz Kotarbiński (1886–1981), Stanisław Leśniewski (1886–1939), Jan Łukasiewicz (1878–1956) and Zygmunt Zawirski (1882–1948), all of whom were deeply interested in logic. This stage of the history of LWS ends in the years 1915–18. Warsaw, formerly in the Russian Empire, was occupied by German troops just at the beginning of World War I. The new authorities allowed the re-opening of Warsaw University, closed in 1831, and Kotarbiński, Leśniewski, and Łukasiewicz were appointed as professors.

After 1918, the date of renewed Polish independence, Warsaw became the second center of LWS. The formation of the Warsaw Logical School was the most significant fact in the entire history of LWS. This school became a common creation of philosophers and mathematicians, particularly of those interested in the foundations of mathematics, set theory, and topology. This project afforded very good opportunities for serious logical research. Leśniewski, and Łukasiewicz, who were appointed as professors of logic, very soon found gifted students,

 DOI: 10.4324/9781315650647-36

like Alfred Tarski (1901–1983), Adolf Lindenbaum (1904–1941), Mordechaj Wajsberg (1902–1943?), Stanisław Jaśkowski (1906–1965), Jerzy Słupecki (1904–1987), Andrzej Mostowski (1916–1975), Bolesław Sobociński (1906–1980) and Czesław Lejewski (1913–2001). Kotarbiński was another important teacher in Warsaw. His students include Janina Hosiasson-Lindenbaum (1899–1941), Edward Poznański (1901–1976), Dina Sztejnbarg (later Janina Kotarbińska; 1901–1997) and Aleksander Wundheiler (1902–1957). Twardowski and Ajdukiewicz remained in Lvov as professors of philosophy. They taught, among others, Izydora Dąmbska (1904–1983), Maria Kokoszyńska (1905–1980) and Henryk Mehlberg (1904–1978). To complete the picture, Czeżowski taught in Vilna, Zawirski in Poznań and later in Cracow. LWS also influenced catholic philosophy. Fathers Józef M. Bocheński (1902–1995) and Jan Salamucha (1903–1944) tried to modernize neo-scholasticism by using contemporary logical tools.

World War II ended the history of LWS as a separate philosophical group. Twardowski and Leśniewski died before 1939; the Lindenbaums, Wajsberg (who were Jewish), and Salamucha were murdered by the Nazis; Zawirski died in the first years after the war; Lejewski, Łukasiewicz, Mehlberg, Sobociński, and Tarski left Poland. Although many members of LWS became active in Poland as professors and found many students after 1945, LWS was not revived and existed only in individual activities of its former members. The new political environment in Poland did not favor the resurrection of analytic philosophy. Yet one should note that the high standards for doing philosophy that had been established by LWS contributed much to preserving academic philosophy in Poland in 1945–89, perhaps more successfully than in other communist countries.

Some historical facts about contacts between LWS and the Vienna Circle

The first information about the Vienna Circle appeared in Poland in a journal *Ruch Filozoficzny* (Philosophical Movement) in 1929. A short note characterized it as an informal group intending to promote a scientific understanding of the world and looking for contacts with philosophers sharing such an approach. Yet, the manifesto of the Vienna Circle did not list any Polish names. Personal as well as scientific contacts between representatives of LWS and the Circle appears to have begun only in 1929, when Karl Menger visited Warsaw and Carnap did so in November 1930. Both were very impressed by the situation of philosophy in Poland, particularly by works of Polish philosophers in logic (see Carnap 1963: 31). They also noted several affinities in views of both groups. Menger reported:

> It was not only formal logical work of the Polish logicians, however, that impressed me. I also noted that they were interested in philosophical problems similar to some discussed in the Vienna Circle, but they attacked them in connection with and partly on the basis of, their, exact logical studies. They always confined themselves to *concrete* questions and completely eschewed those vague generalities which seemed to me becloud [*sic*] some of the Vienna discussions in the late 1920s discussions. So I decided to familiarize the Vienna Circle as well as the members of my Mathematical Colloquium with logico-philosophical work of the Warsaw School and invited Tarski to deliver three lectures before the Colloquium, to two of which I planned to invite the entire Circle.
>
> (1994: 146–7, orig. emphasis)

Tarski visited Vienna in 1930 and had important discussions with Carnap and Kurt Gödel on semantics and metamathematics. (Tarski's contacts with Popper can also be mentioned, although the latter was not a member of the Vienna Circle.) Dąmbska, Kokoszyńska and Mehlberg received grants for study in Vienna in the early 1930s. Ajdukiewicz, Hosiasson-Lindenbaum, Kokoszyńska, Łukasiewicz, Mehlberg, and Zawirski participated in the Prague Conference in 1934. Sizeable groups of members of LWS participated in the congresses for the unity of science (Paris, 1935; Copenhagen, 1936; Paris, 1937; Cambridge, England, 1938); the war prevented Poles from participation in Cambridge, Massachusetts, in 1939 except for Tarski. (Kotarbiński and Łukasiewicz belonged to the international committee organizing these meetings.) A particularly significant event were the discussions about semantics at the Paris Congress in 1935 between Tarski, Kokoszyńska, Carnap, and Otto Neurath. Carnap and Popper helped with the German translation of Tarski's famous work on truth (first published in Polish in 1933, in German in 1936, in English 1956).

Some main logical results and ideas in the Warsaw school of logic

Łukasiewicz and his students intensively worked on propositional calculus and its metalogic. They defined many important metalogical and metamathematical concepts, for example, the logical matrix, the Lindenbaum algebra (algebra of formulas), logical consequence (defined and axiomatized), and deductive system (defined and axiomatized). Lindenbaum proved that every consistent formal system can be extended to a maximally consistent one. These investigations were considered paradigms of logical exactness also by many representatives of LE.

Tarski's semantic theory of truth (1933) was perhaps the most important and influential idea developed in LWS. This theory has two aspects: philosophical and metamathematical. Tarski himself stressed that his truth-theory supports the Aristotelian idea that a sentence *A* is true if and only if it says how things are. More formally: "*A*" is true if and only if *A*; for instance, "Snow is white" is true if and only if snow is white. There has been disagreement ever since whether the truth schema commits one to a correspondence conception of truth: Popper thought so, but many—for example, Neurath—disagreed with him. Tarski also solved the so-called semantic paradoxes, in particular the Liar paradox, by distinguishing object language and metalanguage; his definition of truth is formulated in the latter. Metamathematically speaking, truth definitions belong to model theory, one of the main parts of contemporary mathematical logic, but model theory did not exist prior to Tarski's truth definition. Tarski's undefinability theorem is the central result about truth as formally defined: it says that if *S* is a formal system sufficient for the formalization of Peano arithmetic, then the set of its truths is not defined in it. It is one of so-called limitative theorems showing the limits of the strength of formal systems. One can summarize Tarski's statement as showing that semantics exceeds syntax in its expressive power. Tarski's theorem was inspired by Gödel's incompleteness result (there exist undecidable mathematical sentences). As regards the definability of truth, although Gödel argued that truth is not definable he meant that it is not treatable in mathematics at all. On the other hand, Tarski showed that a mathematical definition of the concept of truth is generally available, but noted that if **L** is a language sufficient for the Peano arithmetic, the set of **L**-truths is not definable in **L**-itself. An important difference between both theorems is that Gödel's theorem is constructively provable, but Tarski's result is not.

Another of the major achievements of Łukasiewicz was the discovery of many-valued logic. In this he was guided by philosophical considerations about future contingencies and the concept of possibility. In order to solve these problems, Łukasiewicz admitted propositions which are neither true nor false. The sentences about future contingent states of affairs are natural candidates for having the third value (½). In particular, if p has the value ½, its negation has it as well. This idea led to three-valued logic in which the laws of the excluded middle and contradiction do not hold. Later, Łukasiewicz generalized three-valued logic to systems with an arbitrary (also infinite) number of values. According to Łukasiewicz, the foundation of two- or many-valued logic lies not in this or that logical theorem, but in metalogic; in particular, it is related to by accepting or rejecting the principle of bivalence. Many-valued logic became for Łukasiewicz the basis for modal logic in which all modal functors are extensional. At first, Łukasiewicz intended to defend indeterminism via three-valued logic, but later sharply distinguished logic and philosophy.

Łukasiewicz also revolutionized the history of logic by considering the history of logical ideas from the vantage point of mathematical logic. Convinced of the continuity of formal logic from Aristotle to modern mathematical logic, Łukasiewicz showed that the Stoics invented propositional calculus. Another of Łukasiewicz's historical discoveries consisted in the rehabilitation of medieval logic, commonly neglected as fruitless scholasticism. Łukasiewicz also elaborated a modern interpretation of Aristotle's syllogistic.

Leśniewski constructed a comprehensive system for the foundations of mathematics. It consisted of three parts: protothetic (generalized propositional calculus), ontology (a theory of a form '*a* is *b*') and mereology (a theory of the parthood relation; it was intended to be a substitute for set theory). All of Leśniewski's systems are axiomatic. According to his nominalistic preferences, formulas are regarded as sequences of concrete inscriptions. One of Lesniewski's influences should be especially noted: it concerns the theory of syntactic categories developed by Ajdukiewicz in the early 1930s which became a prototype for categorial grammar.

In general, most Polish logicians treated logical studies as independent of philosophical commitments; similarly, they thought that mathematics should not be limited by philosophical assumptions. Only Leśniewski had explicit nominalistic commitments which influenced the shape of his systems. On the other hand, several specific projects were influenced by philosophical ideas (recall Łukasiewicz's many-valued logic and Tarski's truth-definition). And although they rejected psychologism, Polish logicians also were strongly inclined towards empiricism as a general epistemological attitude. This attitude often resulted, ontologically speaking, in sympathies for nominalism or constructivism or expressed itself in skepticism concerning the sharp distinction between logical and extralogical truth. Leśniewski maintained that his ontology formulates general principles of being. Łukasiewicz at first believed that experience could decide which logical system, two-valued or many-valued, might be satisfied in the world, but his later view was more in the spirit of conventionalism. Anti-psychologism was a common view in the philosophy of logic in LWS.

Metaphilosophy

In principle, LWS shared Brentano's view that philosophy can be a science. On the other hand, representatives of this school were skeptical about so-called great philosophical problems as too speculative and related to subjective worldviews. Hence, philosophy should concentrate on concrete issues because they can be clearly formulated and eventually solved. Logic was considered an indispensable tool for scientific philosophy as was careful linguistic analysis. LWS strongly rejected irrationalism and claimed that every accepted proposition should be intersubjectively communicable and testable (the principle of anti-irrationalism). Beyond that, no general definition of science was assumed. Yet, in later years the idea of philosophy as a science was gradually replaced in LWS by a more analytical approach according to which philosophical work is perhaps something pre-scientific but still (anti)-irrational.

Philosophy of science

Philosophy of science was intensively pursued by the LWS. Science qua science was considered rational and unified by its logical structure and tools used in scientific justifications. Inductivism was a prevailing view about justification in empirical science. Some philosophers (e.g., Hosiasson) formulated an axiomatic system of inductive logic, while others (Czeżowski, Zawirski) combined many-valued logic and probability. Łukasiewicz worked on induction in his early period but later favored deductivism, anticipating some of Popper's views. Most philosophers accepted scientific realism, but some elements of an anti-realistic interpretation of science appeared in the work of Ajdukiewicz, Łukasiewicz, and Poznański and Wundheiler. Of the many investigations concerning special problems, let me mention only Mehlberg's version of the causal theory of time, some works on the causality problem in quantum mechanics, and his and Zawirski's defense of causality in quantum mechanics. Particularly, Zawirski argued that the unpredictability of the future (Heisenberg) does not entail that the principle of causality fails.

Kotarbiński's Reism

Reism is a general doctrine having two aspects, ontological and semantic. In general, reism goes against the acceptance of the existence of general (abstract) objects, that is, facts, properties, states of affairs, relations, etc. More precisely, reism clams that any object is a material, spatiotemporal, concrete thing, while states of affairs, properties, or relations do not count as objects. These ontological theses are supplemented by semantic conceptions. The key idea consists in the distinction between genuine names and apparent names (onomatoids). A name is genuine if and only if refers to things. On the other hand, onomatoids are words that allegedly refer to abstract entities, that is, non-existing referents. Apparent names are not empty terms, because the latter are genuine names and can always be decomposed into non-empty genuine names (e.g., "round square"). Now, a sentence is meaningful if and only if it consists (except for logical constants) only of genuine names or is reducible to such sentences. For example, the sentence "all lions are animals" is reistically meaningful, but "relations are abstract objects" is not. Furthermore, "a square triangle is rectangular" ("square triangle" is a genuine name, but empty) is acceptable, but "sets exist outside of time and space" (the name "set" is abstract, unless we identify sets with mereological classes) is not. The sentence "whiteness is a property of snow" can be reduced to "snow is white." This example shows how to translate some sentences with apparent names into purely reistic statements. In some respects Kotarbiński's view is very similar to physicalism, for instance, in that the principle of semantic reism offers a criterion of meaningfulness of sentences. In the Vienna Circle, Rose Rand reported and wrote her PhD about Kotarbiński's work (see her 1938 and CH. 13).

Radical conventionalism and semantic epistemology

Ajdukiewicz developed an epistemological doctrine called radical conventionalism. It is based on a conception of language and meaning where the latter category serves as a primitive idea (thus, we assume that linguistic expressions have meanings). Furthermore, the meaning of expressions induces the rules (meaning-rules) for accepting its sentences. In other words, if we know the meanings of sentence-constituents, we can formulate directives of accepting sentences in questions). Ajdukiewicz distinguishes three kinds of meaning-rules: (a) axiomatic (they demand the unconditional acceptance of sentences, for example, "*A* is *A*"; (b) deductive (they demand the acceptance of a sentence relatively to the prior acceptance of other sentences, for example, $\neg A$ follows from $A \to B$ and $\neg B$), (c) empirical (they demand the acceptance of a sentence in a definite empirical situation, for example "it is raining" when it rains). Omitting some technical details, the acceptance or rejection of a sentence depends on the meaning-rules used. In the empirical sciences, directives of the type (c) play a crucial role. If we have a language **L**, the set of meanings of its expressions constitutes its conceptual apparatus. Across scientific languages, conceptual apparatuses are either identical or incommensurable (non-translatable). Thus, empirical data do not force us to accept or reject any sentence, because there always remains the possibility of changing a given conceptual apparatus. By contrast, with Poincaré, according to whom we can freely only modify theoretical principles because they are conventional, unlike experiential reports which are stable, Ajdukiewicz extended conventionalism to all sentences, because any sentence, no matter whether experiential or theoretical, depends on a conceptual apparatus.

Ajdukiewicz later rejected radical conventionalism. Inspired by Tarski's semantics, he developed the program of semantic epistemology, mainly directed towards the defense of realism against various forms of idealism. Typically, Ajdukiewicz paraphrased idealism (for instance, that of Rickert or Berkeley) in such a way that it presupposed the reduction of syntax to semantics and then showed that, due to the incompleteness theorems of Gödel and Tarski, this reduction cannot be effected. This leads to the refutation of Rickert's claim that reality is a correlate of the transcendental subject or of the principle that *esse* = *percipi* (*percipi* can be interpreted as a syntactic category, but *esse* belongs to semantics).

A philosophical comparison of LWS and LE

Since LWS arose earlier than the Vienna Circle, the formation of the former was, of course, independent of the latter, but we may speak about changes in the views in the LWS that came about under the influence of the Circle and vice versa. Some members of the LWS provided explicit statements about the relations between both schools:

> There are in Poland no absolute adherents of the Vienna Circle. I do not know any Polish philosopher who would have assimilated and accepted the material theses of the Vienna Circle. The affinity between some Polish philosophers and the Vienna Circle consists in the affinity between some Polish philosophers and the fundamental methodological attitude and the affinity of problems analyzed.
>
> (Adjukiewicz 1935: 151)

> Professor Ajdukiewicz was right when he wrote about logistic anti-irrationalism in Poland that he did not know any Polish philosopher who would accept the material theses of the Vienna Circle as his own. We are, it seems, too sober to do so.
>
> (Łukasiewicz 1936/1970: 233)

> Polish philosophers . . . were treated by representatives of logical positivism as coming closely to their standpoint. That was right to some extent but not very much, because Polish scientific philosophy did not share the most important point of old and new positivism. A radically anti-metaphysical attitude is the essence of positivism. Yet Polish scientific philosophy did not preclude the possibility of that at least some issues of traditional metaphysics . . . should be treated in a scientific manner.
>
> (Zawirski 1948: 6–7, trans. JW)

Yet such declarations should be taken *cum grano salis*. Hence, a closer examination of the attitude of LWS toward the main theses of LE is required in order to see the relations between both schools (see also Woleński 1989). It is unproblematic that both schools accepted anti-irrationalism (see earlier in this chapter) and the separation of science from worldviews.

The verifiability theory of meaning. No philosopher of LWS accepted a definition of meaning according to which it consisted in how a statement was verified (or tested). Meaning was considered as a primitive category, as it was for Ajdukiewicz or Tarski. Some younger philosophers, like Mehlberg or Kokoszyńska, considered verifiability (falsifiability, testability) to be providing criteria for understanding the meanings of expressions. According to Ajdukiewicz, meanings can be fixed either by phenomenological insight in Husserl's sense or by proposing meaning postulates. Some younger philosophers, like Mehlberg and Kokoszyńska, considered verifiability (falsifiability, testability) to be providing (partial) criteria for proving expressions meaningful according to empirical data. (A similar sentiment animates reism, as noted.) Although one could say that some theses of some of the members of LWS about meaning are similar to those of LE, there is no identity.

Metaphysics is meaningless. The main thesis of LE was: metaphysical utterances are meaningless and neither true nor false. LWS was also anti-metaphysical, but for a different reason, if the LE doctrine is interpreted crudely. Already Twardowski criticized so-called metaphysicism as admitting sheer speculation, for instance, in the mind-body problem. He and other philosophers of LWS considered speculative metaphysics not as meaningless *tout court*, but rather as a body of statements which cannot be decided as true or false for their vagueness, unclarity, etc. LE would agree that such statements are cognitively meaningless. On the other hand, LWS held that metaphysical problems (in the traditional sense) should be carefully investigated. Hence, several problems banished by LE as metaphysical were investigated by LWS; for instance, Kotarbiński did not abstain from developing a theory of ontological categories.

Philosophy is a logical analysis of the language of science. Carnap started out with a syntactic understanding of logical analysis and, accordingly, philosophy became the logical syntax of language (of science), and he only later on added semantics. LWS also considered the analysis of language to of fundamental importance for philosophy, but the attitude of Twardowski and his students was attentive to matters of meaning from the beginning and later led to Tarski's semantic ideas. This difference is important, because semantics allows one to speak about the relations between expressions and their referents.

Physicalism, naturalism, and the unity of science. Kotarbiński's reism is close to physicalism and naturalism, but it was not a view commonly shared in LWS. However, it is perhaps significant that the later Kotarbiński's preference for semantic reism was influenced by Carnap's distinction of the formal and material mode in the discussion of philosophical issues. Most philosophers belonging to LWS considered science to be unified by its logical structure, not by the type of language it used. Physicalism and naturalism were considered in Poland as inconsistent with the nature of the humanities, ethics, and aesthetics. Yet some members of LWS were very sensitive to physicalism in the case of particular problems. For instance, Tarski argued that the axiomatic approach to the concept of truth would be at odds with claims of physicalism and the unity of science.

Relativism. Ajdukiewicz maintained that conceptual apparatuses are relative to languages, and this and his conventionalism are close to some ideas of LE, particularly those developed by Carnap and Neurath. On the other hand, LWS entirely rejected relativism in the understanding of truth and values.

Inductivism. The views of LWS and LE on induction were similar, In particular, Hosiasson elaborated an axiomatic system of inductive confirmation (similar to that of Carnap), and

Zawirski (like Hans Reichenbach) tried to combine many-valued logic with probability as a way towards developing inductive logic. But not all representatives of LWS were inductivists (for instance, Łukasiewicz rejected induction as a method of science). On the other hand, it would be difficult to say that inductivism was a typical mark of LE.

The analyticity of logic and mathematics. LWS considered logic as true of the empirical world. The same view was taken concerning the propositions of mathematics. In particular, Tarski saw no essential difference between mathematical and empirical truth. On this point, the empiricism of LWS differed from that of LE. Both schools considered experience as the only source of reliable knowledge of the world, but LE also distinguished the truths of logic and mathematics as analytic and a priori from the rest as synthetic and a posteriori.

Emotivism. Emotivism in ethics and aesthetics was entirely rejected in LWS.

Conclusion

A general comparison of LWS and LE summarily reads as follows. The views of the former school were even more diversified and less compact than those of the latter. This was the result of various circumstances, in particular, different traditions and a longer period of formation of the Polish school. The Vienna Circle from the beginning was a very radical and revolutionary movement with controversial claims and distinctions, and it attempted to produce a coherent philosophical anti-metaphysical doctrine (I remain non-committal about its success). By contrast, several views of LWS were moderate and expressed individual views without any intention to fit into a uniform overall scheme. This resulted in some gaps (e.g., the lack of a fully developed theory of meaning) and even some incoherence (e.g., the status of the humanities in the light of the evident scientism of LWS). On the other hand, LE gradually liberalized its extremely radical initial views and later became quite similar in certain respects to LWS, for instance, in its relation to existence questions as admissible in concrete situations (Carnap's internal questions). Investigations concerning concrete influences are difficult, but it seems that the second generation of LWS, the younger members who graduated in the interwar period, were more inclined to adopt views of LE than its senior members who had developed their views before 1918. In particular, topics in the philosophy of science considered by younger members of LWS were very similar to those investigated by VC in their content as well as the method of research.

One point is evident, however: the influence of semantics on the development of LE. Tarski convinced Carnap to take semantics seriously as a fruitful philosophical device, but there was also opposition in the Vienna Circle. Here is Carnap's report:

> When I met Tarski again in Vienna in the spring of 1935, I urged him to deliver a paper on semantics and on his definition of truth at the International Congress for Scientific Philosophy to be held in Paris in September. I told him that all those interested in scientific philosophy and the analysis of language would welcome this new instrument with enthusiasm and would be eager to apply it in their own philosophical work. But Tarski was very skeptical. He thought that most philosophers even those working in modern logic would be not only indifferent, but hostile to the explication of the concept of truth. I promised to emphasize the importance of semantics.
>
> (Carnap 1963: 61)

Tarski's worries were justified. His papers at the Paris Congress in 1935 were discussed by Neurath, Reichenbach, Carl Gustav Hempel, and Arne Naess, who were strongly opposed to semantics as useful for philosophy. Neurath was the main opponent and argued that semantics introduces metaphysics into philosophy. Later, Kokoszyńska joined the debate, and ultimately semantics won (see Mancosu 2008; Woleński 2018).

The reorientation of LE toward semantics coincided with essential changes in some views of this movement, particularly of Carnap. For instance, he abandoned his earlier treatment of theoretical terms as fully defined by observational terms and passed to a model-theoretic (semantic) treatment on which theories are only partially interpreted by experiential data. Insofar as the encounter of LWS and LE resulted in the latter's acceptance of semantics, it must also be seen as an episode of the utmost importance for the development of contemporary analytic philosophy, whatever its specialization.

References

Ajdukiewicz, K. (1935) "Der logistische Antiirationalismus in Polen," *Erkenntnis* 5: 151–64.

Carnap, R. (1963) "Intellectual Autobiography," in P. Schilpp (ed.), *The Philosophy of Rudolf Carnap*, LaSalle: Open Court, pp. 3–84.

Łukasiewicz, J. (1980) "Logistic and Philosophy," in J. Łukasiewicz (ed.), *Selected Works*, Amsterdam: North-Holland (orig. 1936 in Polish.).

Mancosu, P. (2008) "Tarski, Neurath and Kokoszyńska on the Semantic Conception of Truth," in D. Patterson (ed.), *New Essays on Tarski and Philosophy*, Oxford: Oxford University Press, pp. 192–224.

Menger, K. (1994) *Reminiscences of the Vienna Circle and the Mathematical Colloquium*, Dordrecht: Kluwer.

Rand, R. (1938) "T. Kotarbińskis Philosophie auf Grund seines Hauptwerkes: 'Elemente der Erkenntnistheorie, der Logik und der Methodologie der Wissenschaften'," *Erkenntnis* 7: 92–120.

Tarski, A. (1956) "The Concept of Truth in Formal Languages," in A. Tarski (ed.), *Logic, Semantics, Metamathematics Papers from 1923–1939*, Oxford: Clarendon Press 1956, pp. 152–278, 2nd ed., Indianapolis: Hackett, 1984 (orig, 1933 in Polish, trans. 1935 in German.).

Woleński, J. (1989) "Lvov-Warsaw School and the Vienna Circle," in K. Szaniawski (ed.), *The Vienna Circle and the Lvov-Warsaw School*, Dordrecht: Nijhoff, pp. 443–53.

——— (2015) "The Lvov-Warsaw School," in E. N. Zalta (ed.), *The Stanford Encyclopedia of Philosophy*, https://plato.stanford.edu/entries/lvov-warsaw/.

——— (2018) "The Semantics Controversy at the Paris Congress 1935," *Philosophia Scientae* 22: 199–211.

Zawirski, Z. (1948) *O współczesnych kierunkach filozofii* (*On Contemporary Philosophical Currents*), Kraków: Wiedza-Zawód Kultura.

33

LOGICAL EMPIRICISM IN NORTHERN EUROPE

Ilkka Niiniluoto

Logic has a long academic tradition in the Nordic countries as a subfield of theoretical philosophy. Logicians have always had an interest in the theory of knowledge and the methods of science, but the study of the philosophy of science in the modern sense was started in the 1920s by Eino Kaila in Finland and Jørgen Jørgensen in Denmark, and in the 1930s by Åke Petzäll in Sweden and Arne Naess in Norway. Logical empiricism made a decisive impact in Northern Europe with these pioneering authors, whose work has been continued in the style of analytic philosophy by their successful colleagues and students already for several generations.

Historical background

The first universities in Northern Europe were at Uppsala in Sweden in 1477 and at Copenhagen in Denmark in 1479. Both countries were converted to Lutheranism in the early sixteenth century, and new universities in the Swedish kingdom were established in Dorpat (Tartu) in Estonia in 1632, Åbo (Turku) in Finland in 1640, and Lund in Skåne in 1666. In most of these universities, the Faculty of Arts had two chairs in philosophy: theoretical (logic and metaphysics) and practical (ethics and politics).

In 1814 Norway was separated from Denmark and joined Sweden in a union until its independence in 1905. The University of Oslo was established in 1813. In 1809 Finland was separated from Sweden to become a Grand Duchy of the Russian empire, and in 1828 the Academy of Turku was moved to the new capital, Helsinki. When Finland became an independent republic in 1917, its name was changed to the University of Helsinki.

In the nineteenth century, philosophy in all Nordic countries was dominated by German idealism. The influence of Kant, Schelling, and Fichte was soon changed to Hegel, taught in Oslo by Marcus Jacob Mortrand and in Helsinki by Johan Vilhelm Snellman. Snellman's successor Thiodolf Rein, who founded the Philosophical Society of Finland in 1873, supported Lotze's philosophy and his successor Arvi Grotenfelt neo-Kantian philosophy of history. In Copenhagen, Søren Kierkegaard protested against the Hegelian school, led by Hans Lassen Martensen. In Uppsala the main idealist figure was Christopher Jacob Boström.

New naturalist trends in the Nordic philosophy were represented by Harald Höffding (1843–1931) in Denmark and Edward Westermarck (1862–1939) in Finland. Höffding, whose works were read throughout all Scandinavian countries, was inspired by the positivism of Comte, Mill,

DOI: 10.4324/9781315650647-37

and Spencer, and (like Westermarck) defended the parallel or double-aspect theory of mind and body. Westermarck, who was at the same time professor of practical philosophy in Helsinki and professor of sociology at the London School of Economics, studied the origins and evolution of moral ideas along the lines of British empiricism. With his student and friend Rolf Lagerborg (1874–1959), Westermarck was an ethical relativist and a sharp critic of religion, but his school had no interest in logic. In 1919–32 he became the first professor of philosophy in the Swedish-language Åbo Academy University, followed by Lagerborg. In Sweden a parallel development of emotivist noncognitivist ethics, but with an emphasis on epistemological realism, can be found in the anti-metaphysical Uppsala School led by Axel Hägerström (1868–1939) and Adolf Phalén (1884–1931).

Empiricist tendencies within philosophy were also supported by experimental psychology, first developed in Copenhagen by Alfred Lehmann (1858–1921). From Helsinki Grotenfelt and Hjalmar Neiglick (1860–1889) went to study in Wundt's laboratory at Leipzig in the mid-1880s, but psychology was separated from theoretical philosophy as late as 1948. In Oslo the second professor of philosophy, Harald Schjelderup (1896–1974), changed his chair to psychology in 1928.

Finland: Eino Kaila and his school

Eino Kaila (1890–1958), who wrote his doctoral dissertation in experimental psychology in 1916, had broad interests in philosophy. He came from a clerical family of priests (his uncle and father were archbishops of Finland), but was himself always skeptical about organized religions. Kaila's early essays are critical reviews of Haeckel, Bergson, and James. In 1920 he argued, against vitalism, that mental life is a biological phenomenon. As an anti-reductionist monist, Kaila was attracted by Mach's ideas but concluded that the phenomenalist position had to be replaced by critical realism which accepts the reality of both ordinary physical objects and atoms. In Finland a radical defense of Mach came from Rolf Lagerborg.

Kaila was appointed the first professor of philosophy at the new Finnish University of Turku in 1922. In the mid-1920s he sought contacts by correspondence with Hans Reichenbach, Moritz Schlick, and Rudolf Carnap. He started a series of monographs in German on causality, probability logic, deduction, and philosophy of nature. Already in his *Die Prinzipien der Wahrscheinlichkeitslogik* in 1926, Kaila characterized his position as "ein logischer Empirismus," in contrast to psychological empiricism, so that he coined the term "logical empiricism" for the new movement of exact philosophy. When the Vienna Circle announced in 1929 its manifesto on the "scientific conception of the world," Kaila was mentioned as one of the thinkers close to the Circle. Kaila made his first visit to Vienna in May 1929 on the invitation of Schlick, defending there his theory that *die Aussenwelthypothese* (the hypothesis about the external world) has a very high probability given empirical evidence.

In 1930 Kaila published in Turku the first detailed analysis and sharp critique of Carnap's phenomenalistic constitution system in *Der logischen Aufbau der Welt* (see Kaila 1930). In spite of his admiration of the new logical method of philosophy, Kaila found Carnap's conclusions to be "catastrophic," since they deprive empirical research of the "realist" language of science. Kaila argued that without the concept of probability, Carnap's constitution by means of explicit definitions is unable to leave the realm of the given. Carnap gave his response in *Erkenntnis* in 1931.

Kaila's cooperation with the new radical movement was well known and debated by conservative humanists in the faculty meeting, when Kaila in the summer of 1930 was appointed professor of theoretical philosophy at the University of Helsinki. His main rival was Grotenfelt's student J. E. Salomaa (1891–1960) who represented the traditional historical approach to philosophy. While the German neo-Kantian referees Bruno Bauch and Ernst Cassirer did not express a preference over either of the candidates, the decisive statement was given by Grotenfelt, who concluded that Kaila is "a more original thinker" than Salomaa.

Kaila's victory in Helsinki was a major change in Finnish philosophy. As a charismatic lecturer and professor in Helsinki in 1930–48, Kaila introduced to his academic audience symbolic logic, modern epistemology, and philosophy of science, continuing also his work in Gestalt psychology. His 1934 book on the psychology of personality, *Persoonallisuus*, was soon translated into Swedish, Danish, and Estonian.

Kaila returned to Vienna in the spring terms of 1932 and 1934, corresponded with Carnap, contributed articles and reviews to the journals *Erkenntnis* and *Theoria*, and participated in international conferences in Europe. However, the politically conservative Kaila disliked the socialist agenda of Otto Neurath and Philipp Frank. Like Reichenbach, he separated his logical empiricism from the narrowly positivist views of some members of the Vienna Circle. He distinguished science from metaphysics by the condition of testability, but he did not accept the verification criterion of meaning that Schlick and Waismann had adopted from Wittgenstein, since it excludes "all-sentences" or general laws of nature.

Kaila thought that philosophy and science together can solve the riddle of reality. In his review of Carnap's *Logische Syntax* in *Theoria* in 1936, he argued that philosophy cannot be restricted to the formal mode of speech. In his German monographs in 1936 and 1941, which he characterized as "contributions to logical empiricism," Kaila attempted to analyze the concept of reality by means of the concept of invariance. Reality includes perceptual p-objects, physical f-objects and scientific s-objects in the order of increasing invariance, and increasing "degree of reality," and the higher objects are definable as invariances of lower-level objects. In his widely read Finnish textbook *Inhimillinen tieto* (1939; translated into Swedish in the same year, and into English as *Human Knowledge*; see Kaila 2014), Kaila characterized logical empiricism by four theses: (1) A statement is analytic if and only it is a priori (denial of synthetic a priori); (2) every statement concerning reality has to have real content (principle of testability); (3) every theory concerning reality has to be translatable into the language of experience (principle of translatability); and (4) sentences about subjective experiences are equivalent to statements about the body (principle of logical behaviorism). Value statements were analyzed as expressions of emotions and prescriptions.

Kaila was elected as one of the twelve members of the Academy of Finland in 1948. He continued to his death in 1958 the "synthetic" task of philosophy in formulating a comprehensive world-outlook on the basis of best scientific theories, especially holistic "field theories" like quantum physics, biological systems theory, and Gestalt psychology. In the 1950s he gave up the principle of translatability, coming back to the critical realism of his youth. But Kaila's later work was isolated from the new Anglo-Saxon school of philosophy of science: he did not publish in English, and the first English translations of his studies appeared only in 1979 (see Kaila 1979).

Kaila was not a formal logician, but he had broad knowledge about the new approaches and results, so that he could encourage his students to work in this field. The first doctoral dissertation on logic was Uuno Saarnio's (1913–1977) study of symbol systems in 1935, which opened the new monograph series *Acta Philosophica Fennica* of the Philosophical Society of Finland. Saarnio, who became the chief librarian of the City of Helsinki, published later on set theory and transfinite numbers.

Georg Henrik von Wright (1916–2003) started his studies with Kaila in 1934. He wrote his doctoral dissertation *The Logical Problem of Induction* in 1941 and continued with important contributions to probability and eliminative induction. In 1946 he published a book-length survey, *Den logiska empirismen*, on logical empiricism. The young von Wright's period as Wittgenstein's successor at the University of Cambridge in 1948–51 was the real international breakthrough of Finnish philosophers. After returning to Finland in 1952, von Wright was elected in 1961 as a member of the Academy of Finland. He wrote studies in philosophical logic, modal logic, and action theory and became the founder of modern deontic logic. In 1971 he published his *Explanation and Understanding*, which departs from the logical empiricist thesis of the unity of science by its claim that the explanation of action cannot be given by the deductive-nomological model but that human actions have to be intentionally understood by practical reasoning. He also defended a manipulation notion of causality. His work on analytic hermeneutics forged connections between analytic philosophy and Continental trends.

The two "father figures" of von Wright were Kaila and Wittgenstein, and the latter distanced him from the scientific optimism of logical empiricism. With Wittgenstein he shared an interest in Spengler, and in writings published in the 1980s he developed a cool cultural pessimism about modernity, the myth of progress, and the harmful social and environmental consequences of the scientific-technological civilization. G. H. von Wright (with Elizabeth Anscombe and Rush Rhees) was one of the executors of Wittgenstein's literary estate, and he devoted 50 years of his life to careful editorial projects of Wittgenstein's works. The University of Helsinki continues to support The von Wright and Wittgenstein Archives (WWA).

Oiva Ketonen (1913–2000) moved to philosophy from mathematics. He studied proof theory in 1938–9 in Göttingen with Gerhard Gentzen. His doctoral dissertation in 1944 contained an invertible system of sequence calculus, which was a remarkable improvement of Gentzen's results. Ketonen became Kaila's successor on the chair of theoretical philosophy in Helsinki from 1951 to 1977. His main rival was Sven Krohn (1903–1999), Salomaa's student, with interests in parapsychology and phenomenology, who had written a critical dissertation on logical empiricism in 1948. During his visit in 1949–50 to the United States, Ketonen met Carl G. Hempel at Yale and Ernest Nagel at Columbia University (New York). With these influences he later lectured mainly on the philosophy of science. Ketonen's and Saarnio's student Raili Kauppi (1920–1995) wrote her doctoral dissertation in 1960 on Leibniz's logic and was appointed at the University of Tampere in 1966 as the first female professor of philosophy in the Nordic countries.

Erik Stenius (1911–1990), who also studied with Paul Bernays in Zürich, made his main contributions in logic and the philosophy of language, but his careful "critical essays" included topics related to the philosophy of mathematics and physics. His doctoral dissertation in 1949 dealt with logical paradoxes, and his book *Wittgenstein's 'Tractatus'* in 1960 gave a precise explication of the picture theory of language by means of the concept of isomorphism. Stenius was first appointed at Åbo Academy University and later became the Swedish-language professor of philosophy in Helsinki after von Wright in 1963–74. The emphasis on logic in Åbo was continued by the Swedish professors Stig Kanger in 1963–8 and Krister Segerberg in 1968–80. Ingmar Pörn (1935–2014), a student of Stenius and Kanger, studied the logic of power and became the successor of Stenius in Helsinki.

G. H. von Wright's student Jaakko Hintikka (1929–2015), who was also influenced by Kaila, wrote his dissertation in 1953 on distributive normal forms in first-order logic. In 1957 he discovered (independently of Stig Kanger and the later work of Saul Kripke) the possible worlds semantics for modal logic, and in 1962 he published his pioneering study of epistemic and doxastic logic, *Knowledge and Belief*. After a fellowship at Harvard University in 1956–9, Hintikka was appointed professor of practical philosophy at Helsinki 1959 (with Krohn again as the main rival), and in the 1960s he divided his time between Helsinki and Stanford University. In 1964 Hintikka published an improvement of Carnap's system of inductive logic by showing how universal generalizations may receive non-zero probabilities in infinite universes. He also applied this system to study semantic information and the explanatory power of theories. In the 1970s Hintikka worked in Helsinki as research professor at the Academy of Finland, but after 1980 he was mostly active as professor in the United States (Tallahassee and Boston). For several decades Hintikka was one of the most successful editors of philosophical journals (*Synthese*) and monograph series (*Synthese Library*). Besides his studies in the philosophy of mathematics, philosophy of language, game-theoretical semantics, and independence-friendly IF-logic, he developed an interrogative model of scientific inquiry, based on information-seeking questions. Hintikka is also known for his bold interpretations of classical logicians and philosophers like Aristotle, Descartes, Leibniz, Kant, Peirce, Hilbert, Russell, Gödel, and Wittgenstein. Among Finnish philosophers, one can say that Hintikka remained most faithful to logical empiricism. In 2001 he argued, against the criticism of "new philosophers of science" like Thomas Kuhn, that the mistake of logical empiricists was not the use of formal logic as a tool of solving analytic problems, but only the use of too-weak logical theories.

Hintikka was extremely active in stimulating and supervising research work in logic and philosophy, and many of his students have become university professors in Finland. Risto Hilpinen (born 1943, Krohn's successor at Turku, later in Miami), Juhani Pietarinen (1938–2018, professor at Turku), and Ilkka Niiniluoto (born 1946, Ketonen's successor at Helsinki) wrote their doctoral dissertations on inductive logic, thus continuing "the Finnish school of induction" started already by Kaila and von Wright. Later, Niiniluoto investigated scientific realism and the concept of truthlikeness by using Hintikka's normal forms. Raimo Tuomela (1940–2020, professor of the methodology of the social sciences in Helsinki) applied distributive normal forms to study the methodological gains due to theoretical concepts. Later, he concentrated in philosophical issues of social action. Veikko Rantala (born 1933, professor at Tampere after Kauppi) wrote important studies on theories and theory-change with David Pearce. Juha Manninen (born 1945, professor of general history of ideas at Oulu) has explored Kaila's relations to the Vienna Circle. Matti Sintonen (born 1951, professor at Helsinki) has worked on scientific explanation and the interrogative model of inquiry. Esa Saarinen (born 1953, professor of applied philosophy at Aalto University) dedicated his early work to intensional logic and philosophy of language. Gabriel Sandu (born 1954 in Rumania, professor at Helsinki) has contributed to game-theoretical semantics and the IF-logic.

Jan von Plato (born 1951, Swedish-language professor of philosophy at Helsinki) worked on the history and foundations of probability theory before turning to structural proof theory with his Italian wife Sara Negri (born 1967, Professor at Helsinki). Together they have given a reappraisal of Ketonen's logical works, and von Plato has continued with historical studies in Gentzen's and Gödel's logic. In the Department of Mathematics in Helsinki, Jouko Väänänen (born 1950) has built up a successful program on the foundations of mathematics, set theory, model theory, and generalized quantifiers. Tuomela's student and successor Uskali Mäki (born 1951), the director of a center of excellence, TINT, in the philosophy of the social sciences, has defended a realist interpretation of economic theories.

The Philosophical Society of Finland, chaired in 1975–2015 by Niiniluoto after Kaila, Ketonen, and von Wright, has actively promoted logic and philosophy of science in many conferences and publications. With a series of nine Soviet–Finnish Logic Colloquia in 1976–97, the influence of logical empiricism and analytic philosophy was extended to the Northern region in Russia.

Denmark: Jørgen Jørgensen and his students

After Höffding, the leading philosopher in Denmark was Jørgen Jørgensen (1894–1969). Like Kaila, he turned his Christian upbringing into a critical view of religion. As a student he supported the anti-metaphysical critical idealism of the neo-Kantian Marburg School, but after completing his thesis on Herman Cohen in 1918 he was attracted by Mach and Pearson via the radical empiricism of Herbert Iversen (1890–1920). In 1924 Jørgensen started to write a long thesis of 1000 pages supporting Russell's logicism. In 1926 he was awarded a gold medal for this thesis and was appointed professor of philosophy at the University of Copenhagen. The thesis was published in English in 1931 in three volumes with the title *A Treatise of Formal Logic* (see Jorgensen 1931). (It was reviewed by Kaila in *Erkenntnis*.) In his attempt to avoid Platonist metaphysics, Jørgensen linked logic to empirical matters from psychology and linguistics, but later he adopted the syntactical approach under Carnap's influence.

Jørgensen met Schlick in 1930 in Oxford, became an active member of the unified science movement and also shared Neurath's socialist views. He arranged for Carnap and Neurath to visit Copenhagen, the hometown of the physicist Niels Bohr (1885–1962). Bohr, whose interest in philosophy was initiated already by Höffding, was (with Werner Heisenberg) the main advocate of "the Copenhagen interpretation" of quantum mechanics. During this time, Jørgensen defended Bohr against Einstein in the name of "neopositivism," but he later adopted a more realist view. At the 1935 international congress in Paris, Jørgensen was elected to the organizing committee of the *International Encyclopedia of Unified Science*, and in June 1936 he organized in Copenhagen the Second International Congress for the Unity of Science, which was the most important meeting on logical empiricism in the Nordic countries. Since 1927 Jørgensen had himself developed multidisciplinary themes in his extensive philosophical lectures, and his book on psychology in 1942–5 based this field upon biology (this work was reviewed by Kaila in *Theoria*). His survey *The Development of Logical Empiricism* was published in 1951 as volume II, number 9 of the series *Foundations of the Unity of Science*. Jørgensen served as professor until his retirement in 1964.

Jørgensen accepted the dichotomy of facts and values by defending an emotivist version of noncognitivism. His approach in ethics was continued by Mogens Blegvad (1919–2001). Close to Jørgensen's view were also the studies on imperatives by Alf Ross (1899–1979), professor of law at Copenhagen in 1938–69, who had studied with Hägerström in Stockholm, and Hans Kelsen in Vienna. Ross was a key figure of Scandinavian realism which interpreted laws as predictions about a judge's behavior in court.

Jørgensen's heritage can be seen in Danish philosophers of science, among them Johannes Witt-Hansen (1908–1986, professor in Copenhagen), David Favrholdt (1931–2012, professor in Odense), and Jan Faye (born 1947, professor in Copenhagen). Witt-Hansen has written on the philosophy of physics. Favrholdt's dissertation dealt with Wittgenstein's *Tractatus*. Favrholdt and Faye are both experts on Bohr. Justus Hartnack (1912–2005, professor in Aarhus) investigated the later Wittgenstein in the spirit of ordinary language philosophy. A new wave of philosophical logic, formal epistemology, and information studies has been started in Copenhagen by Vincent Hendricks (born 1970).

Philosophy in Iceland, which gained its independence from Denmark in 1944, has a broad international scope, but no specific influence of logical empiricism.

Sweden: Åke Petzäll and the Uppsala tradition

In Sweden, the movement of logical empiricism was introduced by Åke Petzäll (1901–1957), who wrote his doctoral thesis on Locke's critique of innate ideas at the University of Gothenburg (Göteborg) in 1928. In the summer periods of 1930 and 1931 he was able to follow the meetings of the Vienna Circle, with Friedrich Waismann as his main contact. Waismann's notes of discussions with Wittgenstein were available to Petzäll. Petzäll's essay *Logistischer Positivismus* in 1931 gave, in 36 pages, a critical treatment of Schlick and Waismann's doctrine of verification as a criterion of empirical meaningfulness. He argued that terms and statements always refer to something beyond the immediately given sensory experience. He also questioned the idea that reality itself has a logical form. He received positive feedback from Carnap and Viktor Kraft. After discussions with Neurath and Carnap on the physicalist approach, Petzäll published in 1935 his second work *Zum Methodenproblem der Erkenntnisforschung*, which led to a lively debate with Neurath on protocol sentences, the coherence theory of truth, and the possibility of comparing language and reality. Petzäll was sensitive to the split within the Vienna Circle about the possibility of epistemology, and suggested that the study of "the meaning of validity claims" should be the starting point in the investigation of knowledge, so that the roles of logic and psychology (or normative and causal inquiries) can be separated (Petzäll 1936). Neurath answered in the journal *Theoria* in 1936 by appealing to a purely historical criterion of validity within an "encyclopedia." From 1939 to his death in 1957, Petzäll served as professor of practical philosophy at the University of Lund.

Petzäll helped the neo-Kantian philosopher Ernst Cassirer receive a visiting professorship in Gothenburg in 1935–41 as a German refugee. In 1935 Petzäll founded a new journal, *Theoria: Tidskrift för filosofi och psykologi*, which published studies in philosophy and psychology both in the Scandinavian languages and in German, English, and French. Meanwhile, *Acta Philosophica Fennica*, which was established in the same year, published monographs by Finnish authors (like Kaila, von Wright, and Hintikka), *Theoria* under Petzäll's editorship became an important international forum of logical empiricism, with contributions from among others by Carnap, Neurath, Frank, Kraft, Hempel, Cassirer, Jørgensen, Kaila, von Wright, Naess, and Petzäll himself. In 1937 Petzäll established in Paris an internal philosophical institute, Institut International de Philosophie, which invited prominent philosophers as its members and published the *Bibliographie de la Philosophie*.

Petzäll never received a similar leading position in Sweden as Kaila in Finland and Jørgensen in Denmark. His international program of supporting logical empiricism was met with suspicion by the followers of the Uppsala School, most notably by Konrad Marc-Wogau (1902–1991), professor of theoretical philosophy at Uppsala in 1946–68, and Anders Wedberg (1913–1978), professor of theoretical philosophy in Stockholm from 1949 to 1976.

Marc-Wogau published in 1936 the treatise *Inhalt und Umfang des Begriffs*, where he discussed Schlick's and Cassirer's theories of concepts, which led to an exchange with Cassirer in *Theoria* in 1936–40 (see Cassirer 1936), while Cassirer also wrote a monograph on Hägerström in 1939. Marc-Wogau wrote on classical philosophers like Kant, on epistemology, and a textbook in logic. His most important work, *Die Theorie der Sinnesdaten* (1945), examined British sense datum theories. He eventually succeeded Petzäll as the editor of *Theoria*. The controversy between rival philosophical schools was largely dissolved in Sweden when the young Ingemar Hedenius (1908–1982), philosopher of ethics and religion and later Hägerström's successor at Uppsala, presented logical empiricism (and A. J. Ayer's related work in England) as a natural continuation of the "value nihilism" of the Uppsala School. Marc-Wogau's student Sören Halldén (1923–2010), professor at Lund, made the first Swedish contributions in philosophical logic with his studies in preference logic.

Wedberg published highly systematic studies in the history of philosophy. His students were linked to the tradition of logical empiricism with their works in philosophical logic. Wedberg's student Stig Kanger (1924–1988), who succeeded Marc-Wogau in Uppsala in 1968, was a pioneer of possible worlds semantics for modal logic with his dissertation from 1957. Kanger was in close cooperation with the Finnish logicians von Wright and Hintikka, and their influence in intensional logic and decision theory can be seen in the studies in Uppsala by Lennart

Åqvist (1932–2019) and Krister Segerberg (born 1936) and in Lund by Bengt Hansson (born 1943), Peter Gärdenfors (born 1949), Wlodek Rabinowicz (born 1947), and Nils-Erik Sahlin (born 1954). Åqvist has developed the logic of questions, Segerberg dynamic modal logic, while Hansson and Rabinowicz have investigated preference logic and decision theory. Gärdenfors is famous for models of belief revision, and Sahlin is known for his studies on the Cambridge philosopher Frank Ramsey. The first woman to defend a doctoral thesis in philosophy in Sweden was the Finnish logician Ghita Holmström-Hintikka (born 1935). Wedberg's student Dag Prawitz (born 1936), professor in Stockholm, is a leading expert in Gentzen-type proof theory. Per Martin-Löf (born 1942) has worked on probability and intuitionistic type theory. The chair of theory of science in Gothenburg was held by Håkan Törnebohm (1919–2016), who worked on relativity theory and confirmation.

Norway: Arne Naess and his colleagues

A young philosopher from Norway, Arne Naess (1912–2009) joined the meetings of the Vienna Circle in 1934–6 and while there wrote his 1936 doctoral thesis *Erkenntnis und wissenschaftliches Verhalten* submitted at the University of Oslo. He approached logic and epistemology from an anti-foundational perspective including behavioral, psychological, and sociological aspects. At the Third International Congress for the Unity of Science in Paris in 1937 he had intense discussions on truth with Neurath and Carnap, which led to his monograph on empirical semantics *Truth as Conceived by Those Who Are Not Professional Philosophers* in 1938. Naess combined semantics with empirical interviews of non-philosophers about concepts like truth and logical inference. In Vienna he also worked with the psychologist Egon Brunswik. Supported by Kaila and Jørgensen, Naess was appointed to the chair of philosophy in Oslo in 1939 at the age of 27. Later, he investigated communication theory, and a critical manuscript on the Vienna Circle, showing many affinities with Neurath's anti-foundationalism, was published in 1956 (Naess 1956). The students of Naess in empirical semantics include Harald Ofstad (1920–1994, professor in Stockholm) and Ingemund Gullvåg (1925–1998, professor in Trondheim).

During the war Naess was active in the anti-Nazi resistance movement. Later, he became known in the international peace movement and among the adherents of Mahatma Gandhi. In 1958 he founded the journal *Inquiry*. Naess retired from his chair in 1970 but continued his work as an influential environmental thinker, defending the intrinsic value of nature in his "ecosophy" (Naess 1989). His *Selected Works* (2005) include ten volumes. Naess died in 2009 at the age of 96.

Another notable trend in Norway is its strong tradition in mathematical logic. Thoralf Skolem (1887–1963) made pioneering studies in model theory and set theory already in the 1920s. Jens Erik Fenstad (1935–2020) has taught both mathematicians and philosophers in Oslo. Dagfinn Føllesdal (born 1932) started in mathematics as Skolem's student. In his master's thesis in 1956 he related Frege's notion of *Sinn* to Husserl's *noema*, which started new comparisons between analytic philosophy and phenomenology. Føllesdal took his PhD in philosophy at Harvard with Quine on reference and quantified modal logic, and later shared his time between Oslo and Stanford (as a long-time friend and colleague of Hintikka). In Oslo, philosophy of language is studied also by Andrew Jones (born 1947), epistemology and philosophy of mind by Olav Gjelsvik (born 1956), and philosophy and history of science by Nils Roll-Hansen (born 1938).

Norwegian scholars have a special interest in Wittgenstein who had a hut near Bergen. Knut Erik Tranøy (1918–2012), who studied with Wittgenstein at Cambridge and was a good friend of von Wright, worked on research ethics and medical ethics in Bergen and Oslo. An electronic edition of Wittgenstein's *Nachlass* has been prepared in The Wittgenstein Archives at the University of Bergen (WAB) under the directorship of Alois Pichler (born 1966).

References

Carnap, R. (1931) "[Review of Kaila 1930]," *Erkenntnis* 2: 75–78.

Cassirer, E. (1936) "*Inhalt und Umfang des Begriffs*," *Theoria* 2: 207–32. Repr. in Cassirer, *Erkenntnis, Begriff, Kultur* (ed. by R. A. Bast), Hamburg: Meiner, 1993, pp. 165–97.

Jørgensen, J. (1931) *A Treatise on Formal Logic: Its Evolution and Main Branches, with Its Relation to Mathematics and Philosophy*, Copenhagen: Levin & Munksgaard.

——— (1951) *The Development of Logical Empiricism*, Chicago: University of Chicago Press.

Kaila, E. (1930) *Der logistische Neupositivismus: Eine kritische Studie*, Turku: Annales Universitatis Fennicae Aboensis BXIII. Trans. "Logistic Neopositivism: A Critical Study Positivism," in Kaila (1979), pp. 1–58.

——— (1936) *Über das System der Wirklichkeitsbegriffe: Ein Beitrag zum Logischen Empirismus, Acta Philosophica Fennica Facs. II*, Helsinki. Trans. "On the System of Concepts of Reality: A Contribution to Logical Empiricism," in Kaila (1979), pp. 59–125.

——— (1941) *Über den physikalischen Realitätsbegriff: Zweiter Beitrag zum Logischen Empirismus, Acta Philosophica Fennica Facs. IV*, Helsinki. Trans. "On the Concept of Reality in Physical Science: Second Contribution to Logical Empiricism," in Kaila (1979), pp. 126–258.

——— (1979) *Reality and Experience. Four Philosophical Essays* (ed. by R. S. Cohen), Dordrecht: Reidel.

——— (2014) *Human Knowledge: A Classic Statement of Logical Empiricism* (trans. by A. Korhonen), Chicago: Open Court.

Marc-Wogau, K. (1936) *Inhalt und Umfang des Begriffs. Beitrag zur Theorie des Begriffs*, Uppsala: Almqvist & Wiksell.

——— (1945) *Die Theorie der Sinnesdaten. Probleme der neueren Erkenntnistheorie in England*, Uppsala: A.-b. Lundequitska Bokhandeln.

Naess, A. (1936) *Erkenntnis und wissenschaftliches Verhalten*, Oslo: Det Norske Videnskaps-Akademi.

——— (1938) *Truth as Conceived by Those Who Are Not Professional Philosophers*, Oslo: Det Norske Videnskaps-Akademi.

——— (1956) *Wie fördert man heute die empirische Bewegung? Eine Auseinandersetzung mit dem Empirismus von Otto Neurath und Rudolph Carnap*, Oslo: Universitetsforlaget. English trans. in Naess (2005), vol. 8, pp. 163–216.

——— (1989) *Ecology, Community and Lifestyle*, Cambridge: Cambridge University Press.

——— (2005) *Selected Works of Arne Naess* (ed. by H. Glasser), Dordrecht: Springer, vol. 10.

Neurath, O. (1936) "Physikalismus und Erkenntnisforschung," *Theoria* 2: 97–105, 234–7. Trans. "Physicalism and the Investigation of Knowledge," in Neurath (1983), pp. 159–66, 168–71.

——— (1983) *Philosophical Papers 1913–1946* (ed. by R. S. Cohen and M. Neurath), Dordrecht: Reidel.

Petzäll, Å. (1931) *Logistischer Positivismus—Versuch einer Darstellung und Würdigung der philosophischen Grundanschaungen des sog. Wiener Kreises der wissenschaflichen Weltauffassung*, Göteborg: Wettergren & Kerbers.

——— (1935) *Zum Methodenproblem der Erkenntnisforschung* (Göteborgs Högskolas Arsskrift 41), Göteborg: Wettergren & Kerbers.

——— (1936) "Replik zu Neurath," *Theoria* 2: 231–3. Trans. in Neurath (1983), pp. 166–8.

34

LOGICAL EMPIRICISM IN THE ANGLOPHONE WORLD: EARLY RECEPTIONS

Christopher Pincock

This chapter considers how four representative Anglophone philosophers interpreted logical empiricism in the 1930s. These philosophers are L. Susan Stebbing (1885–1943), A. J. Ayer (1910–1989), Charles Morris (1901–1979), and Ernest Nagel (1901–1985). I argue that Stebbing and Ayer took logical empiricists to be practicing a certain philosophical method, the method of analysis. By contrast, Morris and Nagel considered logical empiricism in light of the defense of a viable form of naturalism. This historical reconstruction clarifies the limited and distorted ways that logical empiricism was received in the Anglophone world. Note that I use "logical empiricism" and "logical positivism" interchangeably. (For additional discussion and references, see Hardcastle and Richardson 2003; Reisch 2005; Stadler 2007; Uebel 2013; Richardson 2017; Tuboly 2021.)

The United Kingdom: Stebbing and Ayer

Stebbing's most sustained engagement with logical empiricism was her 1933 British Academy lecture "Logical Positivism and Analysis" (1933a), although a number of other publications in this period help to clarify her criticisms (see also Chapman 2013: chs. 4–5; Beaney 2016; and CH. 13). Stebbing took logical positivism to be "in no small measure due to the inspiration of Wittgenstein." She mentions the Vienna Circle and "especially" Schlick, Carnap, Waismann and Neurath. All are said to embrace "the repudiation of metaphysics" in Comte's sense along with the "new logic" of "Frege, Peano, Whitehead, and Russell, as developed by Wittgenstein." This approach to philosophy is throughout contrasted with "the philosophical practice of Moore" (1933a: 2–3). By her own reports, after 1917 Stebbing was very much influenced by Moore. Along with many other British philosophers in the 1920s and 1930s, Stebbing aimed to make sense of Moore's new way of philosophizing as instantiating some specific method dubbed "analysis" (Baldwin 2013). In essence, Stebbing's criticism of logical positivism is that it has not learned the lessons found in Moore and so is confused in its arguments, methods, and conclusions.

For Stebbing, Moorean analysis is properly thought of as metaphysical analysis. This is distinct from traditional "deductive" metaphysics, as it gives up the quest for certain, systematic principles. Analysis starts with a proposition of the sort that is known to be true such as that I am now seeing a pen (1933b: 71). Everyone can easily understand this proposition, and

 DOI: 10.4324/9781315650647-38

this understanding consists in knowing the proposition's "immediate reference," but beyond a proposition's immediate reference is "everything that it refers to, however indirectly" (1933b: 78–79). A metaphysical, so-called directional analysis of a proposition would reveal the basic facts that must obtain for that proposition to be true. With Moore, Stebbing insists that we are ignorant of the analysis of many of the propositions that we know to be true. Analysis extends our knowledge by indicating what the analysis of known propositions is. Thus, "the chief task of philosophy is to discover the correct analysis of expressions which every one would agree are sometimes used to say what is true" (1933a: 8). Analysis of this sort must be kept separate from justification. The propositions that are analyzed may be known non-inferentially, and so whatever directional analysis they have will not reveal our justification for our knowledge of that proposition. This is especially important for Stebbing in connection with perceptual propositions such as that I am now seeing a pen. She admits that sense-data appear in the analysis of perceptual propositions. Sense-data are the objects that we see directly, while objects like pens are seen only indirectly. But for Stebbing, this is consistent with maintaining that I can know non-inferentially that I am now seeing a pen: "the perceptual object [the pen] is not inferred at all; it is discerned within a given perceptual situation. The problem is not one of justifying an inference; it is a problem of analysis" (1933b: 72).

Stebbing argues that logical positivists have confused analysis with justification in their discussions of verificationism. Here, she draws on Schlick, Waismann, and Carnap. There is an "innocuous" form of verificationism that merely insists that the analysis of a proposition proceeds via "facts of experience," which Stebbing takes to involve sense data. But in deploying "the new logic" in their analysis, they have made two errors. First, they have substituted a kind of linguistic analysis for what was a genuine analysis of facts. Second, they have confused Stebbing's preferred directional analysis with "constructed deductive systems" (1933a: 13–14). Linguistic analysis considers the meanings of expressions like sentences, how they arise out of the meanings of words, and how words get their meanings. Stebbing complains that Wittgenstein and those influenced by him have offered a theory of meaning that entails some form of solipsism. She objects that any form of solipsism undercuts the known propositions that are the starting point of analysis. She is thus confident that this species of linguistic analysis is flawed (ibid.: 26).

The second, and more significant, mistake is to conflate the development of a constructed deductive system with a genuine metaphysical, directional analysis. The former "postulational" analysis can indeed derive sentences from other sentences that are deemed to be axioms or postulates of that system. It is this sort of derivation that is central to Russell's logical constructions of material objects, points, and instants. Stebbing insists, though, that this sort of postulational analysis is completely different from her own directional analysis: "they [the logical positivists] regard tables, for instance, as constructs of the given. But a table is not a construct" (1933a: 32). The same point is clearer in the 1934 Aristotelian Society Presidential Address "Constructions." There she rejects Carnap's *Aufbau* as a logical construction of ordinary objects like tables. A construction is something that is made, but tables are not built in the manner sketched in Carnap's constitution systems: "It does not make sense to say that a logical construction can be substituted for a persistent substantival object, although . . . every statement about this table can be finally translated into a set of sentences in which the word 'table' does not occur" (1934: 23). A logical construction arises as part of a postulational system, and although it can be interpreted or coordinated with known objects, a logical construction cannot replace such an object. This is consistent with various conclusions concerning the results of linguistic analysis. The key point is that "Analysis is entirely different from construction" (ibid.: 25): the former is directional and univocal, while the latter is flexible and abstract. There are many postulational systems, but only one full directional analysis of a known proposition.

For Stebbing, then, logical positivism should adapt itself to a genuine, directional analysis whose starting points had been identified by Moore, but whose presuppositions remained unexamined. Ayer took a very different attitude towards logical positivism. Although Ayer agreed with Stebbing that the key question for philosophers was the nature of analysis, he endorsed a much more linguistic conception of analysis and rejected Stebbing's directional analysis as a holdover from an outdated and discredited metaphysics. A concise statement of Ayer's attitude towards logical positivism comes in the first paragraph of his 1935 lecture "The Analytic Movement in Contemporary British Philosophy":

> Logical positivists['] . . . conception of philosophy as an activity of analysis, so far from being a novelty in England, is implicit in the work of all the great British empiricists from Locke to Mill. But in the latter half of the nineteenth century British philosophy, succumbing to the influence of Hegelian idealism, became for the first time predominantly metaphysical; and from this lapse it has not yet altogether recovered. There has, however, during the last thirty years been a growing tendency among British philosophers to revert to empiricist principles.
>
> (Ayer 1936a: 53)

Genuine analysis is opposed to metaphysics. While Ayer gives Russell, Moore, and Wittgenstein considerable credit for initiating this return to empiricist analysis, each is also criticized for their failure to take the final step and drop metaphysics of any sort. Stebbing is singled out for criticism for retaining a commitment to atomic facts as the ultimate residue of her directional analysis. This is labeled "a relic of metaphysical realism which is inconsistent with the general tenor" of Wittgenstein's *Tractatus* (ibid.: 58).

Ayer famously visited Vienna at the urging of his tutor Ryle, arriving in late 1932 and attending the meetings of Schlick's circle through the spring of 1933, partly overlapping with Quine's visit (see Ayer 1992: 15–17). Ayer notes the strong impression that Carnap's *Aufbau* made on him. Carnap's "Overcoming of Metaphysics" clearly resonated with Ayer and solidified his conviction that the core of logical positivism was an opposition to metaphysics. However, when it came to the protocol sentence debates then dominating the group, Ayer reports that he sided with Schlick (against Carnap and Neurath), and maintained that observation statements refer to sense-data. This early combination of views is prominent in Ayer's "Demonstration of the Impossibility of Metaphysics." Wittgenstein, Schlick's "Positivism and Realism," and Carnap's "Overcoming" are mentioned in a foreword to the paper. Ayer says that they have already presented the arguments of Ayer's paper, but that these views have "so far been ignored or misunderstood" (1934: 335).

The argument against metaphysics is a simple one: metaphysics considers "the nature of reality underlying or transcending the phenomena which the special sciences are content to study," but "to indicate the situation which verifies a proposition is to indicate what the proposition means" (ibid.: 335–7). As "any proposition which would be verified by an empirical observation is ipso facto not metaphysical" (ibid.: 339), it follows that there are no meaningful metaphysical propositions. On this approach, philosophy consists in an analysis of propositions that traces each proposition back to its verification base. This is a kind of clarification that conforms to the Tractarian dictum that philosophy is an activity that avoids any distinctive philosophical propositions.

Ayer was aware that his argument turns in large part on the character of the basic propositions that mark the end points of analysis. He initially sided with Schlick in calling these propositions "not descriptive at all but ostensive" (ibid.: 337). In a note appended to the published paper, however, Ayer notes that he has shifted his view on elementary propositions since 1933 when the paper was written. Ayer's new position is defended in the paper "Atomic Propositions," which was the first article in *Analysis*. All genuine propositions are now deemed hypotheses that cannot be conclusively verified.

Crucially, the new attitude towards atomic propositions results from a more thorough elimination of metaphysics: the most common attempt to tie such propositions to atomic facts is flawed as the "conception of atomic fact is a senseless metaphysical conception." The sensible alternative is to "define the atomic propositions as those propositions which describe the observations by which any given proposition is verified" (1933: 2–3). These atomic propositions are not certain, for they may be revised just like any other description.

Given this base of atomic propositions, a pressing issue for Ayer is the character of his verification principle. In the 1934 paper he offers an interesting defense of this account of meaning:

> For the business of philosophy is to give definitions. And in setting out to define meaning or any other concept we must adopt some rule according to which we conduct our enquiry, and by reference to which we determine whether its conclusions are correct. In formulating our criterion we are attempting to show what this rule should be.
>
> (1934: 343)

Using his criterion for meaning, Ayer claims that we can make genuine empirical claims, and we can compare the dictates of the criterion with "some prima facie propositions which by universal agreement are given as significant and some expressions which are agreed to be meaningless" (ibid.: 345). Ayer thus defends verifiability on largely pragmatic grounds. He tries to avoid taking a stand on the nature of meaning or propositions, as this would presumably reintroduce dreaded metaphysical assumptions.

When Ayer came to write his influential *Language, Truth and Logic*, the argument against metaphysics from "Impossibility" was expanded as the book's first chapter "The Elimination of Metaphysics." Now Ayer maintains that only a weak sort of verifiability is necessary for a proposition to be meaningful. But when it comes to the justification of the criterion itself, the earlier motivation is replaced by the following: "it will be shown that all propositions which have factual content are empirical hypotheses; and that the function of an empirical hypothesis is to provide a rule for the anticipation of experience" (1936b/1946: 41).

More generally, philosophy is concerned with definitions of a certain sort, usually tied to the conventions of some group (ibid.: 70). Ayer offers a convention for "factual content" that he argues to be a match to our practice. By these means, he hopes to convince us that metaphysics is without factual content.

Unsurprisingly, Stebbing was quite hostile to Ayer's book. Stebbing notes that Ayer has a thoroughly linguistic approach to the analysis of sentences, where it ultimately turns on "the conventions of the language to which the sentence belongs." Against this, Stebbing first insists that the necessity of logic cannot be interpreted in the terms of linguistic conventions. Her second concern focuses on the need for basic propositions which do stand in some special relation to the facts. Reverting to Moorean perceptual examples, Stebbing claims that "an empirical view of science" requires "some propositions that are incorrigible." Otherwise, one "is bound to accept Carnap's principle of tolerance, and thus to make science itself wholly conventional"

(1936: 360–1). Without the known propositions of Moorean analysis, no genuine science is possible. Stebbing does not here insist that these propositions conform to common sense. But without some such incorrigible base, no analysis of knowledge is possible, and analysis devolves into merely linguistic system-building.

Stebbing and Ayer thus explored competing aspects of logical empiricism. They both sought to recruit logical empiricism to clarify the nature of analysis. Both philosophers, in very different ways, selectively adapted logical empiricist claims and methods to their native philosophical context.

The United States: Morris and Nagel

Morris's pamphlet *Logical Positivism, Pragmatism and Scientific Empiricism* reprints five articles from the mid-1930s that argue that logical positivism's future lies in a fruitful integration with American pragmatism. The resulting "scientific empiricism" makes space for three aspects of the meanings of signs: "syntactics, semantics, and pragmatics" (1937: 4). Carnap's focus on the relations between signs in *Logical Syntax* instantiates a narrow "formalism," while semantics is pursued by Wittgenstein, Schlick, and Waismann when they restrict philosophy to the activity of clarifying meanings (ibid.: 11). Neither project can be sustained in isolation from the investigation of how people use signs in their biological and social activities. Dewey is praised for being "peculiarly sensitive to the instrumental relation of symbols to the life of the individual and the community" (ibid.: 14). This sweeping vision of a "wider positivism" (ibid.: 54) marked one of the most nuanced Anglophone receptions of logical positivism.

Morris agrees with Stebbing when he argues that the strands of logical positivism developed by Carnap and Schlick face the twin challenges of avoiding linguistic formalism and any form of solipsism. The difference between Morris and Stebbing can be tied to the stark contrast between Moore's conception of analysis and Dewey's broad form of naturalism. Moore's starting points are some known propositions, which include knowledge that other minds exist and that these other minds know much of what I know. On Morris's reconstruction, the starting point is instead humans as they are in nature and in communities, using signs to make sense of their world and to obtain their goals. Dewey's emphasis on the social character of meaning is the ground for an "emergent naturalism or realism" that would integrate human biology into a naturalistic cosmology: "science needs both speculative boldness and the greatest fidelity to fact. Metaphysics has given speculative boldness and vision; positivism has given intellectual asceticism and technique" (ibid.: 45). This vision would generate a kind of "empirical cosmology" of the sort developed by Whitehead. It is a necessary ingredient in the broader study of humans and the meanings they ascribe to their signs. With this element in place, the scientific study of humans and the world can address "two of the deepest needs and impulses which man has shown": the need to test all speculation and "the demand for an object of devotion and of knowledge in some sense independent of and more stable than oneself as knower and actor" (ibid.: 31).

One major benefit of this kind of comprehensive project is that it allows logical positivists and their allies to recognize what is special about science and how to defend science from its enemies (ibid.: 55). Neurath was thus a natural ally for Morris, as both sought to develop the ambitious Unity of Science Encyclopedia project. However, one can already detect in Morris's 1937 pamphlet a tension between Neurath's and Morris's aspirations for logical or scientific empiricism. What Morris championed as a "demand for an object of devotion" should have struck Neurath as an endorsement of a tacit metaphysics that Neurath argued is best left behind. Morris sought an intellectual and theoretical unification through sweeping principles. Neurath, by contrast, aimed to create a unified science without any relic of the realism or cosmology that Whitehead was so eager to offer.

These divisions became explicit when Morris published his book *Paths of Life: Preface to a World Religion* (1942). This book endorsed the psychologist William Sheldon's speculative theory of personality types. Sheldon posited a link between physical characteristics ("somatotype") and personality type. As Reisch has summarized Morris's project, the conclusion was that "human beings consciously cultivate a personality type or temperament that synthesized and balanced the main features of all dominant temperaments. Thus Morris outlined nothing less than a new world religion." Neurath was "shocked" at this development, and Carnap refused to engage with this aspect of Morris's work (2005: 46–47). At the same time, Morris was roundly criticized for his scientific work on language. Dewey and his collaborator Bentley, in particular, contested Morris's claims to be extending Peirce's original semiotic project (ibid.: 334–8). Morris's aspirations to unify logical empiricism and pragmatism were thus rejected on both sides.

The broad naturalism that Morris deployed to ground his scientific empiricism can be fruitfully compared to the more modest naturalism that Ernest Nagel extracted from his sustained engagement with logical empiricism. The central issue between Morris and Nagel is the proper way to conduct a scientific study of human values. Morris recommended an all-encompassing cosmology precisely so that human values and ways of life could be studied and structured in scientifically respectable ways. Nagel adopted instead a more modest piecemeal approach that took a certain form of naturalism for granted without supporting any global restructuring of human values.

Nagel called his own position "contextualistic naturalism" (1956: xviii), and he was remarkably consistent in tying a viable naturalism to limited, "contextual" investigations. This position is clear even in Nagel's dissertation. Naturalists share the "common conviction that reflective inquiry may discover the order of the birth and decay of things; that thought and ideal forms have physical efficacy only in virtue of the operations of matter of which they are expressions; and the ideals of life must be based upon the satisfactions achievable in a material world" (1930: ii). The "ideals of life" are thus constrained by the fruits of "reflective inquiry" into the material order of nature. Nagel took the character of this inquiry to be his focus, especially scientific inquiry. Following his mentor Morris R. Cohen, he saw logic as the study of the ways that these inquiries should be conducted. The first chapter of the dissertation was published as "Measurement" in the 1931 volume of *Erkenntnis*. This early article makes clear how modest the results of scientific inquiry must be. For example, Nagel affirms a "relative view of quantity" that eschews abstract magnitudes of the sort that Russell had defended in the *Principles of Mathematics* (1931: 324). When two weights are said to be equal, this statement is tied to the operational context in which the measurement is made. There is no need to posit an additional abstract magnitude that the two objects have in common. Among other worries, this would violate the contextual restrictions on the meaningfulness of the original measurement result. More generally, all scientific claims should be given a contextual interpretation. The philosopher can get clear on their meaning by paying attention to the operations and broader assumptions that go into the formulation and justification of the statement.

Nagel traveled to Europe in 1934 in time to attend the August "Vorkonferenz" in Prague organized by logical empiricists to meet in advance of the Prague Congress of Philosophy (see his reports in 1934b, 1936b). Prior to this trip Nagel expressed some skepticism about logical empiricism. Of special interest is the 1933 APA address "Verifiability, Truth, and Verification." Many confuse these three notions, including "American as well as foreign representatives of modern positivism." The crucial route to avoiding confusion is to recognize that "no proposition can be established as true without a fund of pre-existing knowledge of subject-matter; and hence neither the proposition under consideration, nor this pre-existing knowledge can be established as true by any *one* appeal to the immediately given" (1934a: 141–2). A scientist thus selects out some aspect of nature for investigation, but for the investigation to succeed, some background assumptions must be made. Again, the operation of measurement is central to Nagel's case. The truth of the statement must be tied to this subject matter, and it is not exhausted by the process of verification. One upshot of these distinctions that Nagel emphasizes is that this approach to meaning need not assume any kind of metaphysical atomism: "the operational theory need not sell its valuable birthright, for the mess of pottage of atomic propositions and immediate knowledge" (ibid.: 147).

A pressing question for Nagel's contextual naturalism is how one should integrate the results of all these restricted scientific investigations. From the start, Nagel affirmed some kind of anti-reductionist metaphysics that he traced to Peirce and Dewey. For his Prague Congress talk, Nagel took up one of the topics that he would pursue for the rest of his career: scientific reduction. "Reduction and Autonomy in the Sciences" distinguishes six senses of reduction and considers their applicability to the social sciences. All inquiry begins with a "selective reduction" of some complex subject that involves picking out certain entities and "phases" of their behavior. This step is "essential to all discursive thinking" (1936a: 182). In certain cases, a selected entity E may be "constitutively reduced" by decomposing it into parts that are the subject matter of some other investigation. However, this does not guarantee the more demanding form of "characteristic reduction" which involves theoretically inferring the phase of the entity E from the phases of its constituents. An even more demanding "complete reduction" requires linking phase φ_1 of E_1 to phase φ_2 of E_2 using a function *f* that is theoretically derived from laws that apply to the constituents of these entities.

Using these distinctions, Nagel then argues that the constitutive reduction of social entities is perfectly consistent with the autonomy of the social sciences. Characteristic reduction fails: "entities behaving physically or biologically simply exhibit phases of behavior in social contexts for which physical and biological laws are inadequate" (ibid.: 185). In the discussion following this paper, Charles Morris made what might seem to be an obvious connection to the physicalism of the logical empiricists: Nagel shows how to avoid "the metaphysical materialistic reduction which the critics of physicalism falsely attribute to it, misled by its defenders' failure to distinguish clearly various types of reduction" (Schiller et al. 1936: 199).

On Morris's reading, Nagel's distinctions allow a viable physicalism that dropped the metaphysics of materialism. But, as we have seen, Morris aspired to a kind of metaphysical completion of the sciences that would bring together in some systematic form the results of various scientific inquiries. On this point, Nagel moved in a decidedly different anti-systematic and anti-metaphysical direction. In the version of his Prague paper published in *Erkenntnis*, Nagel added a short preface to clarify the link to logical positivism: "it is fundamental to logical positivism to hold that the meaning of terms is to be ascertained by examining them in the contexts wherein they are used, and in no other fashion; and to maintain that in the clarification of meanings by such procedural analysis lies the task of philosophy" (1935: 46). Nagel here recasts logical positivism to be essentially the contextual naturalism that he had pursued in his own work. For Nagel, there was no way to consistently extend a piecemeal investigation of nature into some system. The metaphysical urge that Morris embraced was eventually dismissed by Nagel, who argued that a "logic without metaphysics" was in fact the key to avoiding various charges against naturalism.

Reception or transformation?

Stebbing, Ayer, Morris, and Nagel represent four of the most sustained Anglophone engagements with logical empiricism in the 1930s (apart from W. V. O. Quine and Max Black). However, our discussion has shown four pairwise contrary interpretations of the significance of logical empiricism. One way to view this disarray is the unfortunate result of philosophers cynically using logical empiricism for their own polemical purposes. However, historians of philosophy have found again and again that new movements in philosophy are often absorbed in a selective and distorted way. Perhaps, then, there is nothing pernicious about these Anglophone receptions of logical empiricism. In general, philosophical innovations are difficult to understand, and we might expect that outsiders will interpret these innovations using concepts, methods, and programs that they are already familiar with. An American pragmatist like Morris will then naturally understand the logical empiricist writings they encounter in connection to American pragmatism. A further point to consider is the motivation that an outsider will have to study a new movement in philosophy like logical empiricism. Only a few outsiders will take the trouble to immerse themselves in the new movement, through time spent reading, traveling, and corresponding. These few will be motivated by problems that they detect in their own setting, and so will be more likely to continue to engage with the new movement if this movement seems to hold out some solutions to these problems.

As we have seen, Stebbing and Ayer had a problem with the character of analysis and its presuppositions. Morris and Nagel sought to clarify a viable form of naturalism. It was through these filters that logical empiricism was received in the United Kingdom and the United States, respectively. Successful logical empiricists like Carnap and Reichenbach adapted themselves to these new philosophical situations, while others were unable or unwilling to adapt. In this way, the Anglophone reception and transformation of logical empiricism may exhibit a similar structure to the transmission of other philosophical movements.

References

Ayer, A. J. (1933) "Atomic Propositions," *Analysis* 1: 2–6.

——— (1934) "Demonstration of the Impossibility of Metaphysics," *Mind* 43: 335–45.

——— (1936a) "The Analytic Movement in Contemporary British Philosophy," in *Actes du Congrès International de Philosophie Scientifique, Sorbonne, Paris, 1935. VIII: Histoire de la Logique et de la Philosophie Scientifique*, Paris: Hermann & Cie, pp. 53–59.

——— (1936b/1946) *Language, Truth and Logic*, New York: Dover, 2nd ed.

——— (1992) "My Mental Development," in P. A. Schilpp (ed.), *The Philosophy of A. J. Ayer*, La Salle, IL: Open Court, pp. 3–40.

Baldwin, T. (2013) "G. E. Moore and the Cambridge School of Analysis," in Beaney (2013), pp. 430–50.

Beaney, M. (2016) "Susan Stebbing and the Early Reception of Logical Empiricism in Britain," in C. Damböck (ed.), *Influences on the Aufbau*, Dordrecht: Springer, pp. 233–56.

——— (ed.) (2013) *Oxford Handbook of the History of Analytic Philosophy*, Oxford: Oxford University Press.

Chapman, S. (2013) *Susan Stebbing and the Language of Common Sense*, Basingstoke: Palgrave-Macmillan.

Hardcastle, G. and Richardson, A. (eds.) (2003) *Logical Empiricism in North America*, Minneapolis: University of Minnesota Press.

Morris, C. W. (1937) *Logical Positivism, Pragmatism and Scientific Empiricism*, Paris: Hermann & Cie.

——— (1942) *Paths of Life: Preface to a World Religion*, New York: Harper.

Nagel, E. (1930) *On the Logic of Measurement*, New York: Privately Printed.

——— (1931) "Measurement," *Erkenntnis* 2: 313–35.

——— (1934a) "Verifiability, Truth, and Verification," *Journal of Philosophy* 31: 141–8.

——— (1934b) "The Eighth International Congress of Philosophy," *Journal of Philosophy* 31: 589–601.

——— (1935) "The Logic of Reduction in the Sciences," *Erkenntnis* 5: 46–52.

——— (1936a) "Reduction and Autonomy in the Sciences," in *Actes du Huitième Congrès International de Philosophie à Prague 2–7 Septembre 1934*, Nedeln/Liechtenstein: Kraus Reprint, n.d., pp. 181–6.

——— (1936b) "Impressions and Appraisals of Analytic Philosophy in Europe," *Journal of Philosophy* 33: 5–24, 29–53.

——— (1956) *Logic Without Metaphysics*, Glencoe, IL: The Free Press.

Reisch, G. (2005) *How the Cold War Transformed Philosophy of Science*, New York: Cambridge University Press.

Richardson, A. (2017) "From Scientific to Analytic: Remarks on How Logical Positivism Became a Chapter of Analytic Philosophy," in A. Preston (ed.), *Analytic Philosophy: An Interpretive History*, London: Routledge, pp. 146–59.

Schiller, F. C. S. et al. (1936) "Discussion," in *Actes du Huitième Congrès International de Philosophie à Prague 2–7 Septembre 1934*, Nedeln/Liechtenstein: Kraus Reprint, n.d., pp. 197–200.

Stadler, F. (2007) "History of the Philosophy of Science: From *Wissenschaftslogik* (Logic of Science) to Philosophy of Science: Europe and America, 1930–1960," in T. A. F. Kuipers (ed.), *General Philosophy of Science: Focal Issues*, Amsterdam: Elsevier, 577–658.

Stebbing, S. (1933a) "Logical Positivism and Analysis," *Proceedings of the British Academy* 19: 53–87. Repr. *Logical Positivism and Analysis*, London: H. Milford.

——— (1933b) "The Method of Analysis in Metaphysics," *Proceedings of the Aristotelian Society* 33: 65–94.

——— (1934) "Constructions," *Proceedings of the Aristotelian Society* 34: 1–30.

——— (1936) "[Review of *Language, Truth and Logic* by Alfred J. Ayer]," *Mind* 45: 355–64.

Uebel, T. (2013) "Early Logical Empiricism and its Reception: The Case of the Vienna Circle," in Beaney (2013), pp. 518–45.

Tuboly, A. T. (ed.) (2021) *The Historical and Philosophical Significance of Ayer's Language, Truth and Logic*, Cham: Palgrave-Macmillan.

35
PRAGMATISM AND LOGICAL EMPIRICISM

Massimo Ferrari

At the turn of the twentieth century, American pragmatism arrived on the European philosophical scene. Its reception represents a major theme for scholars aiming to explore the intellectual climate of that time and, in particular, its relationship with logical empiricism. Yet the story is more complicated than the dominant narrative has for a long time suggested. The earliest encounter of future logical empiricists with pragmatism—largely neglected until recent scholarship—focused mainly on William James's views rather than on Charles S. Peirce's or John Dewey's. The later convergence between the Viennese supporters of the "scientific world-conception" and pragmatist philosophers (such as Charles Morris), which has been usually depicted by the protagonists themselves as a decisive event for logical empiricism (Morris 1963: 87), represents only one part of the history of the relations between the two movements. The story rather begins in Vienna, long before the intellectual migration of European philosophy of science in the 1930s.

Peirce's verificationism

Charles S. Peirce is not mentioned in the manifesto of 1929 promoting the "scientific world-conception." Its album of ancestors includes many European philosophers and scientists endorsing the new spirit of scientific philosophy (Mach, Russell, Poincaré, Duhem, among others), but some of the leading figures of American pragmatism—especially Peirce and John Dewey—are omitted. Only William James is mentioned, albeit with some precautions (Verein Ernst Mach 1929/2012: 78). This is not surprising. The leading members of the Vienna Circle were only vaguely acquainted with Peirce's views, as one can see from Schlick's brief reference in *General Theory of Knowledge* (1918/1985: 165) and the absence of any in Rudolf Carnap's *The Logical Structure of the World* (1928), though he praised Peirce's contribution to the logic of relations elsewhere (1930/1959: 139). Neurath himself, whose fallibilism, holism, and criticism of Cartesian accounts of knowledge have been deemed to be quite close to Peirce (Mormann 1999), will refer to Peirce and Dewey only in later years.

To be sure, Peirce received less attention because most of his work became available only in the 1920s when Morris R. Cohen edited a first collection of his essays (Peirce 1923), followed by the edition of the scattered or in part unpublished writings of Peirce in the *Collected Papers*

DOI: 10.4324/9781315650647-39

in the early 1930s. Until then, Peirce was the great forgotten master of American philosophy. Nevertheless, Willard V.O. Quine suggested that Peirce could be considered the first to elaborate a verifiability theory of meaning similar to that one shared later by the Vienna Circle (1981: 30). Peirce's famous pragmatic maxim in "How to Make Our Ideas Clear" was as short as it was full of promise: "Consider what effects, which might conceivably have practical bearings, we conceive the object of our conception to have. Then, our conception of those effects is the whole of our conception of the object" (1878/1923: 45). According to Quine, this criterion was even more fruitful than Mach's one based on sense-data, because Peirce focused on the crucial role of sentences instead of single terms as Mach did (1981: 30). Moreover, Peirce's further development offered a sophisticated account of what verification means, insofar as he never believed in the "myth" of definite testability of empirical sentences (Chauviré 2001). Peirce anticipated the "liberalization" of empiricism later pursued by Carnap, though his conception of experience was at any rate "much richer" than the Viennese verificationism allowed (Hookway 2013: 33).

In any case, Peirce remained unknown in Vienna for a long time. It was in Florence that Peirce became an influential figure at the beginning of the twentieth century. While James was a great inspiration for the "magic" pragmatism professed by Giovanni Papini and Giuseppe Prezzolini, the young founders in Florence of the review *Leonardo*, both of the other supporters of this combative journal, Giovanni Vailati and Mario Calderoni, endorsed Peirce's verificationism by framing a "logical" version of pragmatism (Ferrari 2014). In 1905 Calderoni suggested that only the pragmatism of Peirce (understood as a sequel to the genuine positivism of John Stuart Mill) succeeded in criticizing traditional philosophy, rejecting metaphysics, and applying a kind of experimental philosophy to the various fields of human reasoning. Calderoni was convinced that pragmatism could not be identified with the Jamesian "Gospel of the will to believe" professed by Papini and Prezzolini; he stressed that, once beliefs were distinguished from the will as rules for the verification of propositions, they had simply to be understood as rules formulating forecasts about what *could* be, starting from some premises expressed by linguistic propositions or practical intentions (Calderoni 1924, vol. 1: 329–58).

For his part, Vailati was essentially in agreement with Calderoni since he believed that pragmatism was able to account for scientific thought. The main idea Vailati attempted to make plausible was that pragmatism—whose "initiator" Peirce promoted "an original trend in logical-mathematical studies"—was based on the convergence with mathematics and mathematical logic "in their tendency to eliminate all superfluity and redundancy of wording and concept" (Vailati 2010: 169). For Vailati, Peirce was central, and he regarded methodological rules of significance as fundamental for the new perspective. Thus, in an article written in 1909 together with his friend Calderoni, Vailati defended a kind of verificationism à la Peirce, claiming that to clarify the meaning of an assertion means to indicate "which particular experiences, according to such an assertion, are going to take place, or would take place under specific given circumstances" (ibid.: 234).

The short adventure of "logical" pragmatism is worth considering because both Vailati and Calderoni were moving in the direction of the later developments of scientific philosophy in Vienna. In particular, given Vailati's own work in the field of logic as collaborator of Giuseppe Peano's *Formulario* as well as his acquaintance with contemporary philosophy of mathematics, it is not surprising that he was mentioned in the manifesto of the scientific world-conception (Verein Ernst Mach 1929/2012: 80) This posthumous acknowledgement testifies to a neglected history that must be discussed on another occasion.

James and the First Vienna Circle

William James is generally recognized as the most prominent philosopher involved in the European discussions of pragmatism at the beginning of the twentieth century. However, and differently from France and Italy, in the German-speaking countries James was suspected as the prophet of a utilitarian conception of truth, or even of a typical "Yankee" way of thinking. Nevertheless, there is at least one exception of some importance for the history of the so-called first Vienna Circle between ca. 1907 and 1912 (Uebel 2000; see CH. 10). The figure at issue here is Wilhelm Jerusalem, an almost unknown Privatdozent at the Vienna University who passionately but solitarily defended James's pragmatism during the Third International Congress of Philosophy in Heidelberg in September 1908. Jerusalem enthusiastically embraced James's philosophical standpoint, and his German translation of *Pragmatism* of 1907 was published already in 1908. Jerusalem added a preface in which he urged readers to abandon contemporary German idealism (i.e., phenomenology and neo-Kantianism) and to welcome the new philosophy. Elsewhere in the same year, Jerusalem presented James as an ally in the effort to replace Kant's theory of knowledge and he stressed James's and Mach's kinship in their battles against the pseudo-problems of metaphysics, drawing in this way an ideal axis between Vienna and the United States (Jerusalem 1908).

Contrary to a widespread view, James was deeply interested in the rapid changes in scientific thought since the second half of the nineteenth century. James was not only the advocate of the "will to believe" or the philosopher devoted to religious experience and faith, but he also urged abandoning the "bad *a priori* reasons, . . . fixed principles, closed systems, and pretended absolutes and origins" (James 1907/1975: 31). James was well acquainted with contemporary philosophy of science. Mach, Duhem, Poincaré, Jevons, Stallo, Pearson, and even Russell's *Principles of Mathematics* were discussed and quoted by him in the first decade of the new century (e.g., ibid.: 93 and 1911/1979: 91–93). James's familiarity with the main philosophers of science of his time emerges very clearly from his general perspective that is best regarded as a kind of *holism* both concerning the nature of knowledge and the "plastic . . . process of truth's growth" that "preserves the older stock of truths with a minimum of modification" (1907/1975: 35). So, James's very metaphorical way of expressing philosophical ideas should not divert the attention of the reader of *Pragmatism* from the epistemological core he sketched in a manner quite similar to Duhem (or, later, to Neurath and Quine).

To be sure, for Neurath, Frank, and the other members of the First Vienna Circle, the pragmatist method heralded by James had a great significance in the struggle against metaphysics. As James said in his book, echoing Peirce's pragmatic maxim, we have "a method of settling metaphysical disputes that otherwise might be interminable" (ibid.: 28). Discussing the matters in a coffee house, the future adherents of the scientific world-conception appreciated the originality of James's philosophical insights, influenced by Mach's heritage that had "paved the way" to their agreement with Peirce's and James's maxim (Uebel 2015: 11). Since 1907 Hans Hahn, Neurath, Richard von Mises, and Frank had lively discussions of the recent innovations in logic, mathematics, and the natural sciences, with Poincaré's conventionalism a favored philosophical topic. As Frank stated, it seemed highly plausible that "one has to distinguish between what is logically possible and what is helpful in empirical science. In other words, logic needs a drop of pragmatic oil" (Frank 1949: 11). Frank's recollections suggested, therefore, that in the early days of the Vienna Circle, the "pragmatic oil" was a central element in framing a new conception of science able to provide a logical and epistemological account of scientific theories. It was against this background that, years later, some members of the Circle became aware of the great affinities with the pragmatist movement flourishing "beyond Atlantic in the United States" (ibid.:

33). In Frank's opinion, both pragmatism and logical empiricism were rooted in a "natural common ground" (ibid.: 48).

This "common ground" was cultivated early on by Neurath too. Neurath's criticism of foundationalism in 1913 represents the first attempt to refute "pseudo-rationalism" based, to some extent, on a pragmatist point of view. The leading insight was the essentially provisional and unavoidable fallible nature of our knowledge. Like James, Neurath was convinced that the "bad *a priori* reasons" were to be dismissed along with the unfounded pretension of rationalism to reliance on "the method" (1913/1983: 10–11). Neurath also relied systematically on Duhem's holism, maintaining that "the correctness of each statement" concerning reality is related "to that of all others" and hence holding theories in general—and this long before Quine—to be empirically underdetermined (ibid.: 3). This was exactly the topic James had dealt with in his *Pragmatism* and that Neurath seemed to have in mind in emphasizing how "an infinite number of systems can be indicated which are applicable in the part that can be interpreted empirically but exclude each other in the part that so far eludes empirical interpretations" (1916/1983: 25). Hence, it is significant that in the later description of his early intellectual development, Neurath referred not only to Mach, Duhem, and Poincaré, but also to pragmatism, and in particular to James, as inspiration for the history of the Vienna Circle (1936/1981: 697; 1938/1981: 842; 1941/1983, 213, 217). Neurath's anti-foundationalism, his fallibilism, his criticism toward Descartes's rationalism, and, finally, his historical view of the growth of knowledge remain at least partly unexplained unless the Jamesian pragmatist blood flowing through his epistemological veins is also considered.

Pragmatism and the Vienna Circle

In 1933 Hahn stated that the "great problem" of truth required a new solution. According to him, the concept of truth has nothing to do with the metaphysical view of correspondence between mind and reality, but it is only comprehensible as the "confirmation" (*Bewährung*) of propositions. "We side with the pragmatist conception [of truth]," Hahn stated by referring explicitly to William James's *Pragmatism* and to John Dewey's *Studies in Logical Theories*. Truth thus has no longer an absolute character, and it becomes rather a relativized, human truth. Only in this way truth can be really "applicable" and—Hahn believed—useful for human knowledge (Hahn 1933/1987: 44).

In the late 1920s, Frank also emphasized connections with James's pragmatism. First of all, Frank had no doubt that Carnap's *The Logical Structure of the World* showed an unexpected affinity with James as regards the criterion of empirical significance. For him, Carnap had believed that "the meaning of any statement is given by its 'cash value,' that is by what it means as a direction to human behavior" (Frank 1949: 33). Although is not clear whether in 1928 Carnap was willing to acknowledge a kind of sympathy for pragmatism and Frank's remarks on its "kindred spirits" working in the United States, it is undeniable that Frank pointed out an important aspect. In 1929 Frank praised the "attack against the truth concept of the school philosophy" made by James, who didn't conceive truth as "a faithful copy of reality" but rather as a system of principles allowing "to change our experiences according to our wishes" (1930/1949: 101). The physicist tests the truth of a theory through "agreements," that is, by comparing experiences with other experiences following a procedure which "has been made by Mach and James into a general conception of the criteria of truth" (ibid.: 102).

The "transatlantic truth" (to use Bertrand Russell's provocative phrase) clearly influenced the view of some logical empiricists, leaving a notable mark on Hahn's, Neurath's, and Frank's intellectual development up to the 1930s. As Frank later summarized their agreement, what is important in science is not its alleged metaphysical interpretation, but, "to repeat the words of William James, . . . the 'cash-value' of a set of principles" (1951: 102; see also his 1932: ch. 1). These hidden convergences once more document the extent to which, long before the emigration of logical empiricism to the United States, American pragmatism was already at home in Vienna. Yet, until recently this fact has been obscured by the official history, which stressed instead the role of "natural supplement" played by both pragmatists and operationalists like Percy W. Bridgman for European logical empiricism (Joergensen 1951: 55).

In the United States during the 1920s Clarence I. Lewis developed a pragmatist conception of the a priori that constitutes a long-neglected original contribution to the "relativization of the a priori" much debated in the early logical empiricism (Mormann 2012: 113–14; see CH. 21). Lewis's idea of the "pragmatic a priori" and of "conceptual pragmatism" rested upon the articulation of the theory of knowledge in three respects. Firstly, only logical-mathematical concepts established through definitions are a priori; second, the given of experience is an element totally independent from thought; third, the interpretation of experience rests upon the activity of mind and is "more" than experience itself: "truth is made by mind" (Lewis 1970: 240, 248–9). However, for the pragmatist Lewis, the active interpretation provided by mind is not rooted in a fixed system of categories as conceived by Kant.

Independently of this unrealized convergence, Lewis contacted Schlick and Carnap in the early 1930s to bridge the gap between pragmatism and logical empiricism. The first important result of this intellectual exchange was his "Experience and Meaning" which called attention to the danger of solipsism in the criterion of verification through direct experience. For Lewis, cognition is guided by "an element of anticipation" foreshadowing the datum, and he stressed the conditions under which the verification can be projected: "to know (empirically) is to be able to anticipate correctly further possible experience" (1934/1970: 268; compare Lewis 1929: 29–31).

In his response, Schlick sought to establish broad agreement with the concept of verifiability stressed by Lewis. Already in 1910 Schlick had criticized James's concept of truth as "unscientific" for confusing the nature of truth with the criteria of its verification and its practical consequences and claimed, by contrast, truth to be the "one-to-one coordination" of judgments with "states-of-affairs" (1910/1979: 67 and 88). This point he repeated in the *General Theory of Knowledge*, where, however, he also lauded, as he had done in 1910, "the great merit" of pragmatism in considering "the process of verification" as the only way "to *establish* truth" (1918/1985: 165, orig. emphasis). Now Schlick declared Lewis to be in "perfect agreement" with the Viennese empiricism: verifiability means solely "possibility of verification," not at all "verifiable here now" (1936/1979: 460–1). Schlick's late convergence with Lewis, together with his interest (1929) in Percy W. Bridgman's *The Logic of Modern Physics* (1927), testifies to the increasing dialogue between logical empiricism and American pragmatism before the dissolution of the European scientific community.

A sequel: logical empiricism in the land of pragmatism

Logical empiricism in Europe came to an end with the emigration of its major exponents to the United States (Hardcastle and Richardson 2003). Carnap's and Neurath's collaboration with John Dewey and Charles Morris on the ambitious project of the *International Encyclopedia of Unified Science* represented a new trend in the relationship between the "very fertile American manner of thinking" and the "scientific philosophy" escaped from German-speaking countries (Neurath 1937/1983: 190). Later on, Frank noted that the "classical source" of their conception of meaning went back to Peirce's maxim and had been further developed by the "most prominent" pragmatist, William James (Frank 1951: 33–35). For Frank, the great achievement of logical empiricism was to have formulated "the pragmatist criteria of meaning in a strictly logical way which satisfied the most rigid requirements of formal logic" (ibid.: 39).

In this context, Charles Morris played a crucial role. Morris got in touch with the Vienna Circle in the late 1920s, and their cooperation soon developed, notably on the occasion of the Congrès International de Philosophie Scientifique in Paris in September 1935 (Mormann 2016). For Morris, pragmatism offered a detailed account of the functioning of mind by stressing, thanks to Peirce, the central role of symbolism (1932: 282–3). The project of semiotics, left incomplete by Peirce, represented his own most important legacy, even that of pragmatism in general (Morris 1970: 27, 144; see also Frank 1957: 349). While Morris declared that the time had come for close cooperation between their "essentially complementary" movements (1937: 23), he did not shrink from criticizing Carnap's view of philosophy as logical analysis of science for its neglect of essential non-formal features of language (1938a: 21). It was precisely these features that pragmatism sought to develop an enlarged theory of meaning called "semiotic" which incorporated pragmatics, the study of the uses made of language (ibid.: 58–59). Morris stressed that "all empirical science involves experimentation, and experimentation is an activity, a practice," which is in turn connected to a psychological and sociological backdrop (1938b: 72).

Lewis also stressed that thinking of experience as a kind of "activity" constituted the main difference between pragmatism and logical empiricism. "The pragmatic emphasis upon relevance to some active intent is largely or wholly omitted in logical positivism" (1941/1970: 94). Lewis meant here exclusively Carnap, who, in his eyes, excluded what ought to be at the core of a pragmatic conception of meaning, the "content of experience" and that "which concerns the relation of expressions to what may be given in experience" (ibid.: 96–97).

One can now ask whether, and to what extent, the difference from pragmatism influenced significant developments of logical empiricism. To be sure, for Neurath and the collaborators of the *Encyclopedia*, the pragmatist milieu was an exciting opportunity, though he disagreed with Morris's semiotic and stressed more than his American colleagues the political implications of the unified science. But it was Carnap who was the most prominent figure in this new intellectual age, though he became the target of Quine's attack against the "two dogmas of empiricism." One may ask whether Carnap's encounter with pragmatism entailed decisive changes in his thought. Ever since the *Logical Syntax of Language* and its "principle of tolerance" (1934: §17), Carnap regarded the choice between different formal languages as a practical question (Richardson 2007: 300) and emphasized that linguistic frameworks are to be accepted or refused solely by a decision (1936/1937: 2). This "practical question" regarding the "efficiency as instruments" of linguistic constructions (1950/1956: 207 and 221) represents plainly a kind of pragmatism, but these developments originated within the protocol sentences debate in Vienna rather than under the influence of American pragmatists (Limbeck-Lilienau 2012: 93). Carnap also agreed with Lewis on the impossibility of absolute verification (1936/1937: 426) and was able to share Lewis's conception of the a priori or Morris's idea of a variable a priori (Morris 1937: 51; Richardson 2003: 14). However, one may question whether these convergences are enough to speak of a pragmatist turn in Carnap's intellectual career.

Likewise, it should be noted that Hans Reichenbach's alleged pragmatism too must not be taken for granted as sometimes has been done. In 1938 Reichenbach presented himself, albeit with some caution, as the heir of pragmatism in modifying an unduly strict empiricist theory of meaning, centering it instead on the notions of probability and prediction (1938: 68). "The pragmatic idea that the definition of meaning is to be chosen in adaptation to the system of human actions, that it is to be determined by the postulate of utilizability, decides, therefore, against the strictly positivistic language" (ibid.: 150). Yet, probability and the inductive method had been "inextricably intertwined" since the very beginning of Reichenbach's logical empiricism (Galavotti 2011: 100). Hence, as with Carnap, it must be stated that Reichenbach was not a follower of pragmatism, but a partner in scientific philosophy inspired by pragmatic motives.

Finally, the history of logical empiricism and pragmatism in the United States is marked by a crucial difference concerning the distinction between cognitive statements of science and ethical discourse or values in general. The American pragmatists, Morris once said, are all "value-oriented philosophers"; in particular, Dewey is "the axiologist of the pragmatic movement" (1970: 81). What Dewey had once and for all clarified, Morris firmly believed, was the social and institutional relevance of science as "social factor" in human life. Moreover, science permits to extend its method of "freed intelligence" into ethical and social questions, whereas philosophy has to mediate between facts and values (Morris 1937: 14–15). As a matter of fact, Dewey's *Theory of Valuation*, published in 1939 as volume 2 of the *International Encyclopedia of Unified Science*, prompted behind-the-scenes criticism by Carnap, signaling thereby a notable difference between the pragmatist account of the cognitive status of values and the noncognitivist theory of ethics endorsed by Carnap and logical empiricism (Reisch 2005: 83–95). Lewis once remarked that even truth implies an "imperative significance for conduct"—in other words, that to determine truth means to determine that which is correct to believe. In refusing any division between normative and descriptive, the pragmatist conceives knowledge as "for the sake of action," and action is conversely directed to realize "what is valuable" (1941/1970: 109–12).

The consequences of this philosophical conflict can hardly be underestimated. Nonetheless, what was common to pragmatism and logical empiricism was the commitment to scientific knowledge considered as a wall against alleged "absolute values" as well as a defense of "humanitarian democracy" (Frank 1951: 109). Some may wonder whether the noncognitivism vis-à-vis value statements represents Carnap's and other logical empiricists' "third dogma". But whether it did or not, this did not exclude—as Carnap said later—"an intense interest in moral problems" as well as the engagement for a "scientific humanism" (1963: 82–83). At least in this latter sense too, pragmatism and logical empiricism shared a common goal.

References

Bridgman, P. W. (1927) *The Logic of Modern Physics*, New York: The Macmillan Company.

Calderoni, M. (1924) *Scritti*, Firenze: La Voce, vol. 2.

Carnap, R. (1928) *Der logische Aufbau der Welt*, Berlin: Weltkreis-Verlag. Trans. *The Logical Structure of the World*, Berkeley: University of California Press, 1967, repr. Chicago: Open Court, 2003.

——— (1930) "Die alte und die neue Logik," *Erkenntnis* 1: 12–26. Trans. "The Old and the New Logic," in A. J. Ayer (ed.), *Logical Positivism*, New York: Free Press, 1959, pp. 60–81.

——— (1934) *Logische Syntax der Sprache*, Vienna: Springer. Rev. ed. trans. *The Logical Syntax of Language*, London: Kegan, Paul, Trench Teubner & Cie, 1937, repr. Chicago: Open Court, 2002.

——— (1936/37) "Testability and Meaning," *Philosophy of Science* 3: 419–71, 4: 1–40.

——— (1950) "Empiricism, Semantics and Ontology," *Revue Internationale de Philosophie* 4: 20–40. Repr. in Carnap, *Meaning and Necessity. A Study in Semantics and Modal Logic*, Chicago: The University of Chicago Press, 2nd ed., 1956, pp. 205–20.

——— (1963) "Intellectual Autobiography," in Schilpp (1963), pp. 3–84.

Chauviré, Ch. (2001) "De Cambridge à Vienne: La maxime pragmatiste et sa lecture vérificationniste," in J. Sebestik and A. Soulez (eds.), *Le Cercle de Vienne. Doctrines et controverses*, Paris: L'Harmattan, pp. 43–58.

Ferrari, M. (2014) "Pragmatism and European Philosophy: William James and the French-Italian Connection," in M. C. Galavotti et al. (eds.), *New Directions in the Philosophy of Science*, Dordrecht: Springer, pp. 609–25.

Frank, P. (1930) "Was bedeuten die gegenwärtigen physikalischen Theorien für die allgemeine Erkenntnislehre?" *Erkenntnis* 1: 126–57. Trans. "Physical Theories of the Twentieth Century and School Philosophy," in Frank (1949), pp. 90–121.

——— (1932) *Das Kausalgesetz und seine Grenzen*, Vienna: Springer. Trans. *The Law of Causality and Its Limits*, Dordrecht: Kluwer, 1998.

——— (1949) "Historical Introduction," in P. Frank (ed.), *Modern Science and Its Philosophy*. Cambridge: Harvard University Press, pp. 1–51.

——— (1951) *Relativity—A Richer Truth*, London: Cape.

——— (1957) *Philosophy of Science: The Link between Science and Philosophy*, Englewood Cliffs, NJ: Prentice-Hall.

Galavotti, M. C. (2011) "On Hans Reichenbach's Inductivism," *Synthese* 181: 95–111.

Hahn, H. (1933) *Logik, Mathematik und Naturerkennen*, Vienna: Gerold. Trans. "Logic, Mathematics, and Knowledge of Nature," in B. McGuinness (ed.), *Unified Science*, Dordrecht: Reidel, 1987, pp. 24–45.

Hardcastle, G. L. and Richardson, A. (eds.) (2003) *Logical Empiricism in North America*, Minneapolis: University of Minnesota Press.

Hookway, C. (2013) "'The Principle of Peirce' and the Origins of Pragmatism," in A. Malachowski (ed.), *The Cambridge Companion to Pragmatism*, Cambridge: Cambridge University Press, pp. 17–35.

James, W. (1907) *Pragmatism: A New Name for Some Old Ways of Thinking*, London: Longmans, Green & Co. Repr. Cambridge, MA: Harvard University Press, 1975.

——— (1911) *Some Problems of Philosophy*, New York: Longmans, Green and Co. Repr. Cambridge, MA: Harvard University Press, 1979.

Jerusalem, W. (1908) "Der Pragmatismus. Eine neue philosophische Methode," *Deutsche Literaturzeitung* 29: 197–206.

Joergensen, J. (1951) *The Development of Logical Empiricism*, Chicago: University of Chicago Press.

Lewis, C. I. (1929) *Mind and World Order: Outline of a Theory of Knowledge*, London: Charles Scribner's Sons.

——— (1934) "Experience and Meaning," *Philosophical Review* 43: 121–47. Repr. in Lewis 1970, pp. 258–76.

——— (1941) "Logical Positivism and Pragmatism," unpublished issue of *Revue Internationale de Philosophie*, in Lewis 1970, pp. 92–112.

——— (1970) *Collected Papers* (ed. by J. D. Goheen and J. L. Mothershead, Jr.), Stanford: Stanford University Press.

Limbeck-Lilienau, Ch. (2012) "Carnap's Encounter with Pragmatism," in *Rudolf Carnap and the Legacy of Logical Empiricism* (ed. by R. Creath), Dordrecht: Springer, pp. 89–111.

Mormann, T. (1999) "Neuraths anticartesische Konzeption von Sprache und Wissenschaft," in E. Nemeth and R. Heinrich (eds.), *Otto Neurath: Rationalität, Planung, Vielfalt*, Vienna: Oldenbourg, pp. 32–61.

——— (2012) "Toward a Theory of the Pragmatic A Priori: From Carnap to Lewis and Beyond," in R. Creath (ed.), *Rudolf Carnap and the Legacy of Logical Empiricism*, Dordrecht: Springer, pp. 113–32.

——— (2016) "Morris' Pariser Programm einer wissenschaftlichen Philosophie," in C. Bonnet and E. Nemeth (eds.), *Wissenschaft und Praxis. Zur Wissenschaftsphilosophie in Frankreich und Österreich in der ersten Hälfte des 20. Jahrhunderts*, Vienna: Springer, pp. 73–88.

Morris, C. (1932) *Six Theories of Mind*, Chicago: University of Chicago Press.

——— (1937) *Logical Positivism, Pragmatism and Scientific Empiricism*, Paris: Hermann.

——— (1938a) *Foundations of the Theory of Signs*, Chicago: The University of Chicago Press.

——— (1938b) "Scientific Empiricism," in O. Neurath et al. (eds.), *Encyclopedia and Unified Science*, Chicago: The University of Chicago Press, pp. 63–75.

——— (1963) "Pragmatism and Logical Empiricism," in Schilpp (1963), pp. 87–98.

——— (1970) *The Pragmatic Movement in American Philosophy*, New York: Braziller.

Neurath, O. (1913) "Die Verirrten des Cartesius und das Auxiliarmotiv (Zur Psychologie des Entschlusses)," *Jahrbuch der Philosophischen Gesellschaft an der Universität zu Wien 1913*, pp. 45–59. Trans. "The Lost Wanderers and the Auxiliary Motive (On the Psychology of Decision)," in Neurath (1983), pp. 1–12.

——— (1916) "Zur Klassifikation von Hypothesensystemen," *Jahrbuch der Philosophischen Gesellschaft an der Universität Wien 1914 und 1915*, pp. 39–63. Trans. "On the Classification of Systems of Hypotheses," in Neurath (1983), pp. 13–31.

——— (1936) *Le développement du Cercle de Vienne et l'avenir de l'empirisme logique*, Paris: Hermann & Cie. Trans. "Die Entwicklung des Wiener Kreises und die Zukunft des Logischen Empirismus," in Neurath (1981), pp. 673–703.

——— (1937) "Die neue Enzyklopädie des wissenschaftlichen Empirismus," *Scientia* 62: 309–20. Trans. "The New Encyclopedia of Scientific Empiricism," in Neurath (1983), pp. 189–99.
——— (1938) "Weren en Zijn," *Algemeen Nederlands Tijdschrift voor Wijsbegeerte en Psychologie* 31: 223–39. Trans. "Wissen und Sein," in Neurath (1981), pp. 833–43.
——— (1941) "Universal Jargon and Terminology," *Proceedings of the Aristotelian Society* New Series 41: 127–48. Repr. in Neurath (1983), pp. 213–29.
——— (1981) *Gesammelte philosophische und methodologische Schriften* (ed. by R. Haller and H. Rutte), Vienna: Hölder-Pichler-Tempsky.
——— (1983) *Philosophical Papers 1913–1946* (ed. by R. S. Cohen and M. Neurath), Dordrecht: Reidel.
Peirce, C. S. (1878) "How to Make Our Ideas Clear," *Popular Science Monthly* 12: 286–303. Repr. in Peirce (1923), pp. 32–60.
——— (1923) *Chance, Love and Logic, Philosophical Essays* (ed. by M. R. Cohen), New York: Harcourt, Brace and Co.
Quine, W. V. O. (1981) "The Pragmatists' Place in Empiricism," in R. J. Mulvaney and Ph. M. Zeltner (eds.), *Pragmatism: Its Sources and Prospects*, Columbia: University of South Carolina Press, pp. 21–39.
Reichenbach, H. (1938) *Experience and Prediction*, Chicago: University of Chicago Press.
Reisch, G. (2005) *How the Cold War Transformed Philosophy of Science: To the Icy Slopes of Logic*, New York: Cambridge University Press.
Richardson, A. (2003) "Logical Empiricism, American Pragmatism, and the Fate of Scientific Philosophy in North America," in G. Hardcastle and A. Richardson (eds.), *Logical Empiricism in North America*, Minneapolis: University of Minnesota Press, pp. 1–24.
——— (2007) "Carnapian Pragmatism," in M. Friedman and R. Creath (eds.), *The Cambridge Companion to Carnap*, Cambridge: Cambridge University Press, pp. 295–315.
Schilpp, P. A. (ed.) (1963) *The Philosophy of Rudolf Carnap*, La Salle, IL: Open Court.
Schlick, M. (1910) "Das Wesen der Wahrheit nach der modernen Logik," *Vierteljahresschrift für wissenschaftliche Philosophie und Soziologie* 34: 386–477. Trans. "The Nature of Truth According to Modern Logic," in Schlick (1979), vol. 1, pp. 41–103.
——— (1918) *Allgemeine Erkenntnislehre*, Berlin: Springer, 1918, 2nd rev. ed. 1925. Trans. *General Theory of Knowledge*, Vienna: Springer, 1974, repr. LaSalle: Open Court, 1985.
——— (1929) "[Review of *The Logic of Modern Physics* by P. W. Bridgman]," *Die Naturwissenschaften* 17: 549–50. Repr. in Schlick, *Kritische Gesamtausgabe, Section I, vol. 6, Die Wiener Zeit. Aufsätze, Beiträge, Rezensionen 1926–1936* (ed. by J. Friedl and H. Rutte), Vienna: Springer, 2008, pp. 190–1.
——— (1936) "Meaning and Verification," *Philosophical Review* 45: 339–69. Repr. in Schlick (1979), vol. 2, pp. 456–81.
——— (1979) *Philosophical Papers* (ed. by H. L. Mulder and B. F. B van de Velde-Schlick), Dordrecht: Reidel, vol. 2.
Uebel, T. (2000) *Vernunftkritik und Wissenschaft: Otto Neurath und der erste Wiener Kreis*, Vienna: Springer.
——— (2015) "American Pragmatism and the Vienna Circle: The Early Years," *Journal for the History of Analytical Philosophy* 3 (3): 1–35.
Vailati, G. (2010) *Logic and Pragmatism. Selected Essays* (ed. by G. Arrighi, P. Cantù, M. De Zan, P. Suppes), Stanford: CSLI Publications.
Verein Ernst Mach (1929) *Wissenschaftliche Weltauffassung. Der Wiener Kreis*, Vienna: Wolf. Trans. "The Scientific Conception of the World. The Vienna Circle," in O. Neurath, *Empiricism and Sociology* (ed. by R. S. Cohen and M. Neurath), Dordrecht: Reidel, 1973, pp. 299–318; rev. trans. (with orig. annotated bibliography) "The Scientific World-Conception. The Vienna Circle," in F. Stadler and T. Uebel (eds.), *Wissenschaftliche Weltauffassung. Der Wiener Kreis. Hrsg. vom Verein Ernst Mach (1929)*, Vienna: Springer, 2012, pp. 75–116.

PART IV

Leading post-positivist criticisms and legacy

36

QUINE AND POST-POSITIVISM

Richard Creath

In 1967 John Passmore said "Logical positivism, then, is dead, or as dead as a philosophical movement ever becomes" (1967: 57). Such a death would hardly have been instantaneous. But by the late 1960s there was a widely accepted narrative according to which logical empiricism (equated here and in Passmore with logical positivism) was a combination of highly suspect doctrines that had been decisively refuted by W. V. O. Quine, among others, thus initiating an era, sometimes known as "post-positivism." The latter period was characterized by such views as holism, naturalism, the rejection of both foundationalism and certainty, and the resurgence of metaphysics, especially modal metaphysics. Recent historical scholarship has shown that this narrative needs serious revision.

The present three-part essay identifies the most characteristic positions of the primary philosophers involved. The first section deals with logical positivism. That movement was sufficiently varied, often with sharply contrasting views, and showed sufficient development over time that this chapter can touch on only a few particularly relevant themes and matters where recent scholarship has changed our understanding. The second section concerns Quine and addresses those aspects of his work most relevant to the narrative concerning the shift from logical empiricism to post-positivism. The third and final section deals with post-positivism, outlining the extent of the changes from the preceding period and the extent of Quine's influence in bringing them about. None of the three sections attempts to assess the merits of the views or arguments of the positivists or post-positivists themselves.

Logical empiricism

Often thought to be a specific set of related doctrines, logical empiricism or logical positivism is better understood as a diverse movement of thinkers with a common concern for the use of logical methods in understanding the methodology of the empirical sciences and for the use of science thus understood in reforming and improving the customs and institutions that make meaningful social life possible. These thinkers also wanted to find a proper home within their understanding of scientific methods, not only for logic and mathematics, but also for philosophy itself. Though held together by these common concerns, there was probably no doctrine that they all held in common, not even empiricism. And those who were empiricists held often sharply divergent views on what version of empiricism to adopt.

 DOI: 10.4324/9781315650647-41

Until the mid-1930s the two main centers of logical empiricism were Vienna and Berlin. In both places there was strong but not universal support for both holism and naturalism of various forms, especially by Rudolf Carnap and Otto Neurath. Consider this brief quotation from Carnap's *The Logical Syntax of Language*:

> Further, it is, in general, impossible to test even a single hypothetical sentence. In the case of a single sentence of this kind, there are in general no suitable L-consequences of the form of protocol-sentences; hence for the deduction of sentences having the form of protocol-sentences the remaining hypotheses must also be used. Thus *the test applies, at bottom, not to a single hypothesis but to the whole system of physics as a system of hypotheses.*
>
> (Carnap 1934/1937: 318, orig. emphasis)

On the same page Carnap endorses the revisability of protocol (that is, observation) sentences and even logic and mathematics. He in effect rejects foundationalism and claims to certainty. This view was, or at least became, the prevailing view of the so-called left wing of the Vienna Circle (e.g., Neurath, Carnap, Philip Frank, and Hans Hahn), but also of Hans Reichenbach and most of the Berlin Group. Foundationalism and limited claims of certainty, however, were defended by others, most notably by Moritz Schlick.

A number of themes are widely and rightly associated with the movement. Among these are: (1) verificationism or an anti-metaphysical stance, often treated separately, though they are two sides of the same coin; (2) the rejection of the synthetic a priori; and (3) the unity of science.

(1) Verificationism was most prominently an account of intelligibility according to which a claim is intelligible (meaningful) if and only if it is either analytic or testable, at least in principle, by comparison with some set of observations sentences (see CH. 16). The complaint against this was, and occasionally still is, that the verificationist principle itself is neither analytic nor testable and hence is self-undercutting. This objection was well known to the positivists and was specifically addressed by Carnap. The objection is readily countered by treating the principle as a proposal for structuring the language of science. As a proposal, it would be neither true nor false. But if adopted, the language of science would be such that a sentence that asserted that all sentences of that language were either analytic or testable would be not only true but analytic.

The anti-metaphysical stance is closely connected to verificationism and can likewise be thought of as a proposal. This stance is not, as is often assumed, the rejection of an entire branch of philosophy. Rather, it rejects only that specific kind of philosophizing that assumes that philosophers have some a priori access to independent and objective facts that are somehow deeper than or behind those to which empirical science has access and to which that science must conform. Such an assumption of special a priori access that outweighs scientific findings was felt by nearly all the logical empiricists, and especially by Carnap and Neurath, to be harmful to the development of science. It was thought to be harmful politically and socially as well, by reducing its ability to inform the reasoned consideration of new and perhaps beneficial forms of social engagement and community life. Instead of rejecting an entire branch of philosophy, Carnap seeks to transform and reconceive it as making proposals for structuring the language of science. Different such proposals could coexist and scientists would be free to adopt whatever system of concepts they find fruitful.

(2) Many of the logical empiricists had started out as neo-Kantians, but all or nearly all came to reject the idea that there is a domain of substantive truths about the world that can be known a priori. Kant had taken this synthetic a priori domain to include geometry and mathematics,

and he rightly saw that such truths could not be derived from the Aristotelian logic then available. But by the early twentieth century, a new and more powerful logic was available within which mathematics and (pure) geometry was constructible. Thus, all this could be seen as analytic, that is, true in virtue of meaning or, alternatively, logical truths in a broad sense or definitional abbreviations thereof. Given Ludwig Wittgenstein's "no-content" theory of logic, the way seemed open to a thoroughgoing empiricism in which all of the truths that had content were synthetic and fully empirical (see CH. 14). It was Carnap who most forcefully defended the analytic-synthetic distinction and the idea that the best (clearest and most defensible) way of treating those sentences heretofore thought to be a priori is to treat them as analytic.

(3) The idea that the sciences are unified was most fully embraced by the left wing of the Vienna Circle and most of the Berlin group as well (see CH. 18). Neurath and Carnap were its most vocal champions. Their views overlap almost completely, though there is some difference in formulation. For Neurath, we need to be able to use the results of both the natural and human sciences together in addressing urgent social and political problems. Thus, the various sciences must be inferentially connected to rather than isolated from one another. Together, they must all be empirically testable and applicable. As a result, Neurath's rejection of metaphysics and his call for the unity of science are largely interchangeable.

For Carnap, the unity of science divides into two parts: the unity of the language of science, and the unity of the laws of the various sciences. As to the former, Carnap thought that there must be sufficient logical connections between what we say in science (in whatever field) and what is publicly observable so as to allow our scientific claims to be tested. This, he thought, would both guarantee the inferential connections and practical applicability that Neurath sought and overcome metaphysics. Carnap was prepared to argue for the unity of the language of science, but on the unity of laws he was concerned chiefly with framing the question as one of whether the laws of one science that were established at a given time imply those of other sciences. This, Carnap thought, should depend on what the laws were and hence was an empirical question. On this empirical question Carnap took no official position, whatever his private opinions. And he thought that philosophers generally, insofar as they were not empirical scientists, should take no official view.

W. V. O. Quine

W. V. O. Quine was one of the most important and provocative philosophers of the twentieth century. Whatever the ultimate fate of his views and arguments, and we are not evaluating them here, they make a permanent contribution to our philosophical understanding by raising questions and opening possibilities not previously seen.

Quine once said (1971: xxiii) that at the start of his own career he was Carnap's disciple for six years. They began to diverge in 1940, as Quine developed a reference-centered account of language in which ontological questions and commitment were central and modality was considered suspect and even unintelligible. By the publication of "Two Dogmas of Empiricism" (1951), he publicly rejected the analytic-synthetic distinction that was central to Carnap's work. But the argument that Quine used was a demand for "behavioral criteria," and this is a version of the verificationist demands on meaningfulness. At the end of that paper, he outlined a thoroughly holist epistemology in terms not unlike those quoted from Carnap's *Logical Syntax*. The previous year Quine had used a shorter but similar epistemic sketch to suggest how to account for the apparent apriority of logic and mathematics without actually granting them that status (1950). Thus, Quine's view was an empiricism no less thoroughgoing than that of the logical empiricists.

Over the next nearly 50 years, Quine went on to champion a number of distinctive themes. Among the most noteworthy is the indeterminacy of translation (1960: ch. 2), i.e., the idea that there is no fact of the matter about which, if any, of the many behaviorally adequate translations from one language into another (or even into itself) preserves either the referents of expressions or their meanings. The argument for this was based on an underdetermination thesis that springs directly from his holism. A second theme was ontological relativity, the idea that the ontological commitment of what we say is by no means obvious, and any determination of it will have to be relativized to a manual of translation (Quine 1969a). A third theme was naturalized epistemology, though the exact and proper formulation of this was not always clear to Quine's readers. Some took the initial formulation (1969b: 75) that we should "settle for psychology" at face value as meaning that a purely descriptive account of how we do in fact process data and fix on theory is all we need or should want in epistemology. What Quine actually meant was more complicated, for he said, "Naturalization of epistemology does not jettison the normative and settle for the indiscriminate description of ongoing procedures" (Quine 1986a/1998: 664). Quine (1969b) did present his epistemological naturalism as though it were in opposition to Carnap. But it is difficult to discern that there is a real difference of opinion, or if there is, whether it is larger than the differences of opinion between one logical empiricist and another. Finally, and most enduringly, Quine continued to defend an extensionalist approach to language.

How close, then, Quine was to the logical empiricists on the three themes (1–3) discussed in the previous section? Regarding (1), there is certainly some prima facie reason to think that Quine is a defender of metaphysics (see Rosen 2014: 552) In "Two Dogmas," Quine said that his arguments there have the effect of "blurring of the supposed boundary between speculative metaphysics and natural science" (1951/1953: 20). Moreover, there and elsewhere ontology is central to Quine's concerns, and ontology is part of the sub-discipline we call metaphysics. Finally, in the same article, he rejects the verificationist theory of meaning, though later (Quine 1986b/1998: 155) he said he finds the view attractive, put off only by our inability to find, for sentences taken one by one, sentences at the observation level that would confirm or disconfirm them.

But prima facie reasons cannot be the last word. Carnap never considered Quine to be a metaphysician in his sense, saying rightly that Quine was only trying to turn ontological questions into empirical ones. Quine was therefore not trying to go beyond science by philosophic means. Moreover, Quine himself generally saw his own outlook as broadly anti-metaphysical. The verificationist theory of *meaningfulness*, and not the verificationist theory of *meaning*, was the logical empiricists' tool of choice in their battle against metaphysics. Quine accepted the former, without endorsing any particular criterion of significance, and used it in his argument against the analytic-synthetic distinction. Finally, even if the distinction between speculative metaphysics and natural science is blurred, this would not show that Quine engaged in speculative metaphysics or that he wanted to.

Regarding (2) and (3), Quine was no friend of the synthetic a priori. In fact, he rejected the a priori altogether. As we have seen, while he challenged the analytic-synthetic distinction, the argument he used was a verificationist one. Neither Quine nor Carnap argued that all of science is implied by the truths of a single science. Rather, they argued that all claims and concepts of whatever science must be appropriately tied back to a public observation language. That is what Carnap meant by the unity of the language of science. Both Carnap and Neurath had argued that the various sciences needed to be sufficiently integrated inferentially so as to apply the science to urgent political and social problems. Quine does not explicitly state such a motive, but his holism guarantees this result.

Quine's views, or most of them, such as holism, naturalism, and empiricism, not to mention the revisability of our observational judgments and even of logic, the rejection of foundationalism and certainty, are well within the range of views commonly accepted within the logical empiricist movement. Beyond the details of doctrine, Quine likewise shared the aims of the movement, that is, of using the methods of logic to understand empirical science and the place of logic and mathematics within that overall science. It is small wonder, then, that Hilary Putnam called Quine "the greatest logical positivist" (1988/1990).

Post-positivism

Post-positivism is by no means a well-defined movement or set of doctrines. Here, the term refers to philosophical work done in the English-speaking world during the 1960s, 70s, and beyond that conceives of itself as done in opposition to logical positivism or logical empiricism. Doctrinally, as said earlier, the new era was characterized by such views as holism, naturalism, and the rejection of both foundationalism and certainty. Attitudinally, there was the perception of having turned a page. If there is anything that holds post-positivism together, it is this narrative that the logical empiricists had held a great many views that had now been decisively discredited by Quine among others. Now at last we could move confidently into the future without having to worry about the logical empiricist critique. Nowhere was this more visible than in the resurgence of metaphysics, and especially of modal metaphysics.

We saw that many of the central doctrines that prevailed in the post-positivist decades were not all that different from those (holism, naturalism, and views on foundationalism and certainty) that prevailed among the logical empiricists. Quine inspired the widespread use of both the word "holism" in the context of confirmation and meaning and the expression "naturalized epistemology." And he undoubtedly persuaded many to adopt these various doctrines as well. But the narrative that there was vast change on these topics from the logical empiricists is not supported by the historical record.

There were genuine changes on some topics, and the task now is to survey some of the most salient in order to gauge the nature of the change and to estimate Quine's influence thereon. Quine had challenged the analytic-synthetic distinction. It did not disappear, but it was rarely invoked for Carnap's purposes. The distinction was now considered to be controversial, and that was clearly due to Quine. Post-positivist philosophers did largely reject the verificationist theory of meaningfulness that positivists had used to attack philosophical attempts to go beyond or behind empirical science, i.e., to attack what the logical empiricists called metaphysics. Perhaps it would be more accurate to say that post-positivist philosophers stopped worrying about the issue altogether. Quine had not himself rejected this form of verificationism but had used it in his challenge to analyticity. Quine had argued against the other form of verificationism as a theory of meaning. It is possible that a confusion of the two forms operated to undermine the former among the post-positivists. Quine was not confused, and a version of the verificationist theory of meaning survived in the writings of Wilfrid Sellars and of Robert Brandom (1994), and in various versions of conceptual role semantics.

As indicated, there was in the 1960s and 70s a dramatic renewal of interest in metaphysics. But largely it was not the metaphysics that Carnap, Neurath, and others had attacked. It was much more logic and science friendly. Undoubtedly, there was much philosophical work that Carnap and probably Quine would have found objectionably metaphysical. Quine himself, as indicated earlier, was generally anti-metaphysical, but some may have seen in his writings all the encouragement they needed. Among the most salient of developments in metaphysics broadly understood was a significant increase in work on modal logic and using modal concepts to understand a wide range of philosophical issues. It is hard to see that as due to Quine's influence, since he had spent much of his career campaigning against modality. Indeed, the new concern for modality is in some ways closer to Carnap than to Quine.

What, then, shall we say of the standard narrative about Quine and post-positivism? As usual, the facts are more complicated than any single narrative would have it. There was indeed, in the 1960s and 70s, a strong reaction against what philosophers thought the logical positivists/logical empiricists had said. But in so reacting, many philosophers adopted views that were identical, or at least close, to those actually promoted by the logical empiricists. Of course, there were differences of detail. But such differences as occurred were often no greater than those exhibited in the natural evolution of logical empiricism itself. And not infrequently, those changes were prompted by arguments pioneered by the logical empiricists.

Quine's criticisms of the analytic-synthetic distinction and of Carnap's early work was undoubtedly influential in getting philosophers to think that they were engaged in a wholesale rejection of logical positivism. Quine was not the only such influence, though certainly a major one. But Quine did more than that. He was an enormously creative and persuasive philosopher who shaped how philosophers framed their questions, what vocabulary they used, and what conclusions they drew. While Quine was closer to the logical empiricists than philosophers knew, and while the conclusions these later philosophers drew were closer to those of the logical empiricists than they generally recognized, that is certainly no argument against those views or against Quine.

References

Brandom, R. (1994) *Making It Explicit*, Cambridge, MA: Harvard University Press.

Carnap, R. (1934) *Logische Syntax der Sprache*, Vienna: Springer. Rev. ed. trans. *The Logical Syntax of Language*, London: Kegan Paul, Trench, Trubner, & Co., 1937. Repr. Chicago: Open Court, 2002.

Hahn, L. E. and Schilpp, P. A. (eds.) (1986) *The Philosophy of W.V. Quine*, Chicago: Open Court, 2nd expanded ed., 1998.

Passmore, J. (1967) "Logical Positivism," in P. Edwards et al. (eds.), *Encyclopedia of Philosophy*, New York: Macmillan and The Free Press, 1967, vol. 5, pp. 52–57.

Putnam, H. (1988) "[Review of W. V. O. Quine, *Quiddities: An Intermittently Philosophical Dictionary*]," *London Review of Books* 10.8 (21 April 1988): 11–13. Repr. as "The Greatest Logical Positivist," in H. Putnam (ed.), *Realism with a Human Face*, Cambridge, MA: Harvard University Press, 1990, pp. 268–77.

Quine, W. V. O. (1950) *Methods of Logic*, New York: Holt.

——— (1951) "Two Dogmas of Empiricism," *Philosophical Review* 60: 20–43. Repr. in Quine, *From a Logical Point of View*, Cambridge, MA: Harvard University Press, 1953, pp. 20–46.

——— (1960) *Word and Object*, Cambridge, MA: MIT Press.

——— (1969a) "Ontological Relativity," in Quine (1969c), pp. 26–68.

——— (1969b) "Epistemology Naturalized," in Quine (1969c), pp. 69–90.

——— (1969c) *Ontological Relativity and Other Essays*, New York: Columbia University Press.

——— (1971) "Homage to Carnap," in *PSA 1070: In Memory of Rudolf Carnap*, Dordrecht: Reidel, pp. xxii–xxv.

——— (1986a) "Reply to Morton White," in Hahn and Schilpp (1986/1998), pp. 663–5.

——— (1986b) "Reply to Roger F. Gibson, Jr.," in Hahn and Schilpp (1986/1998), pp. 155–7.

Rosen, G. (2014) "Quine and the Revival of Metaphysics," in G. Harman and E. Lapore (eds.), *A Companion to W.V.O. Quine*, Chichester: Wiley-Blackwell, pp. 552–70.

37

KUHN, CARNAP, AND LOGICAL EMPIRICISM

Gürol Irzik

According to the conventional wisdom, Thomas Kuhn's *The Structure of Scientific Revolutions* (1962, 2nd ed. 1970; henceforth SSR) (1) played a major role in the demise of logical empiricism (henceforth LE) by (2) demolishing its key tenets and replacing them with an alternative picture of science that has virtually nothing in common with them. However, the relationship between Kuhn's views and LE is not at all as straightforward as this claim suggests. While (1) is undoubtedly correct, the revisionist historiography of LE in the last two and a half decades documented that (2) is far from true. In this chapter, I highlight the major findings of this historiography, focusing on "the Carnap-Kuhn connection."

Kuhn's critique of LE and his alternative image of science

In SSR Kuhn criticizes only three tenets that he explicitly attributes to the logical empiricists: (a) the accumulationist view, according to which science progresses by incremental accumulation of individual discoveries of facts; (b) the theory-observation dichotomy, according to which observations are independent of theories and provide secure foundations for them; and (c) "the probabilistic theories of verification" (as Kuhn calls them), which take the process of confirmation to be a matter of degree of support bestowed on a scientific theory by a given body of evidence expressed in a neutral observation language. Kuhn argues that all three tenets are closely bound up with a dysfunctional traditional foundationalist epistemology that treats what is "immediately given" in experience as the incorrigible building blocks out of which theories are constructed. He also targets the context of discovery –the context of justification distinction, though only implicitly, by saying early on that his book may have violated it. According to him, these faulty tenets are derived mainly from an analysis of the logical structure of finished scientific achievements, as these are recorded in the classics, which are written from the perspective of normal science and therefore make scientific revolutions invisible. So he sets himself the task of drawing an entirely novel picture of science based on historical case studies of actual scientific practice.

Astonishingly for a book that has the lion's share of responsibility in the dethronement of LE, SSR refers critically to only a single one of all the logical empiricists' works, namely, Ernest Nagel's brief book on probabilistic theories of confirmation (1962/1970: 145 n. 1). Kuhn's attack on the accumulationism of LE is also weak because it confines itself to arguing against the

 DOI: 10.4324/9781315650647-42

construction of theory-neutral observation languages. Nagel's views on inter-theoretic reduction and Carl Hempel's deductive-nomological model of explanation, both of which imply accumulation at the theoretical level, are conspicuously absent in SSR (see Irzik 2012).

In its barest outline, Kuhn's alternative picture of science goes as follows. Development of mature sciences is necessarily discontinuous, ruptured by revolutions that follow periods of normal science. Normal science consists of the activity of puzzle-solving governed by a paradigm. A paradigm has two components: the constellation of community commitments, and shared exemplars. The latter refers to concrete model solutions to typical problems and can guide research even in the absence of explicit rules. The former, on the other hand, consists of symbolic generalizations (such as $F = ma$), metaphysical and methodological commitments, and values like quantitative accuracy, consistency, simplicity, broad scope and fruitfulness (later dubbed "the five values" in Kuhn 1977).

During normal science the paradigm is never questioned, and puzzle-solving can be understood in terms neither of confirmation nor of falsification since it is not an attempt to test the paradigm. Failure to achieve a solution to a puzzle discredits only the scientist and not the paradigm. Yet, those puzzles that resist solution, despite persistent attempts, turn into anomalies and eventually might trigger a crisis. The eventual loss of faith in the existing paradigm and the search for an alternative are typical signs of the crisis. When one of the alternatives succeeds in solving some of the outstanding and generally recognized problems and thereby gains the confidence of most members of the relevant scientific community, it is embraced as the new paradigm. In short, a scientific revolution has occurred. "Scientific revolutions," then, "are those non-cumulative developmental episodes in which an older paradigm is replaced in whole or in part by an incompatible new one" (1962/1970: 92).

Kuhn likens the psycho-social experiences of the scientists going through a paradigm shift to a gestalt switch and political or religious conversion. For that reason, paradigm choice cannot be forced by logic and experience alone. This gives the impression that the paradigm choice is subjective and lacks rational grounds, an impression that is buttressed by his famous thesis that the new paradigm is not only incompatible, but also incommensurable with the old one. There is no common measure to compare them point by point, since the paradigm shift involves a change in scientific puzzles and standards for solving them, a meaning change, and finally a change in the world itself. What is at stake, therefore, is no less than the rationality of science, along with scientific progress which can no longer be seen directed towards truth in the sense of correspondence with a mind-independent world. Different paradigms constitute different phenomenal worlds (ontological relativism), and all standards of puzzle-solving are relative to the paradigm (methodological relativism).

In his "Postscript" of the second edition, Kuhn responds to the charge that his views deprive science of its progress and rationality. Scientific progress, he says, is not toward some fixed goal like "objective truth" as scientific realists claim; rather, it consists of increased puzzle-solving power, analogous to biological evolution, i.e., "the selection by conflict within the scientific community of the fittest way to practice future science" (ibid.: 172). He then hints that philosophers' usual criteria like quantitative accuracy, simplicity, and the like are often shared across paradigms but notes that they function as values. This means that two scientists who share the same criteria may yet interpret or apply them differently. For example, they may attach different weights to them because of their personal preferences and thus end up choosing rival paradigms without being irrational at all. For Kuhn, then, there is no algorithm for paradigm choice even in the presence of shared values, so the parties have no recourse other than appealing to techniques of persuasion.

Let us now look at what Carnap thinks about the three tenets that Kuhn has attributed to the logical empiricists and criticized in SSR.

Anti-foundationalism and the theory-observation distinction

Logical empiricists discussed the topic of foundationalism intensely during the protocol-sentence debate from the late 1920s to the mid-1930s. The central issue was the empirical basis of science in light of the failure of the Kantian project as well as the utter inadequacy of classical empiricism. In the end almost all logical empiricists adopted an explicitly anti-foundationalist epistemology, a form of physicalism, according to which the empirical basis of science consisted of protocols that refer to fallible and hence revisable statements about intersubjectively observable states of affairs (see Uebel 2007).

To be sure, Carnap never gave up the idea of constructing an observation language distinct from the theoretical language of science. Yet, this does not mean that he believed in a sharp theory-observation separation. The protocols debate already made clear that observation statements cannot provide secure foundations for theories. Moreover, Carnap was well aware that any distinction between observational and non-observational predicates and entities has to be arbitrary since they form a "field of continuous degrees of observability" (1936–7: 455; 1966: 226). Yet, some kind of relative and pragmatic distinction was required, he thought, for the testing of theories.

Development as non-accumulation

That the non-accumulational character of scientific development was not foreign to Carnap's mature philosophy of science was first argued by Reisch (1991). In that article, Reisch drew attention to the similarities between Kuhn's paradigm change and Carnap's theoretical language change and also published two hitherto-unknown letters Carnap wrote to Kuhn in his capacity as an editor of the series *International Encyclopedia of Unified Science* in which the SSR appeared. In those letters Carnap did not at all perceive Kuhn's book as undermining his views about science. On the contrary, he especially liked Kuhn's description of scientific revolutions as conceptual framework displacements (ibid.: 266). He also agreed with Kuhn that "the development of theories is not directed toward the perfect true theory, but rather is a process of improvement of an instrument" (ibid.: 267). Indeed, revolutionary changes and the discontinuities they engender find a natural home in Carnap's conception of linguistic frameworks.

According to Carnap, scientific revolutions can occur in two ways: through a change in the logico-mathematical rules of the linguistic framework in which a theory is represented, and through a change in the physical/theoretical postulates (T-postulates) of the theory that express fundamental laws about a certain domain of the phenomena (see Irzik and Grünberg 1995):

> I should make a distinction between two kinds of readjustment in the case of a conflict with experience, namely, between a change in the language, and a mere change in or addition of, a truth-value ascribed to an indeterminate statement (i.e., a statement whose truth-value is not fixed by the rules of language, say by the postulates of logic, mathematics, and physics). *A change of the first kind constitutes a radical alteration, sometimes a revolution and it occurs only at certain historically decisive points in the development of science.* On the other hand, changes of the second kind occur every minute.

> *A change in the first kind constitutes, strictly speaking, a transition from a language L_n to a new language L_{n+1}.*
>
> (Carnap 1963: 921; emphasis added)

Elsewhere, he writes: "This class [the class of the theoretical terms] will generally be changed only when a radical revolution in the system of science is made, especially by the introduction of a new primitive theoretical term and the addition of postulates for that term" (Carnap 1956: 51).

No wonder, then, that Carnap liked Kuhn's idea of scientific revolutions as conceptual framework changes. We can press further. According to Carnap, certain synthetic sentences such as T-postulates behave like analytic sentences because they have a double—a factual and a semantic—function: they not only express fundamental laws but also contribute to the meaning of theoretical terms occurring in them. That is why they are recalcitrant to refutation by experience, like the analytic meaning postulates. Indeed, the previous passage quoted from Carnap (1963) ends by noting just that:

> To be sure, this status [analyticity in a language] has certain consequences in case of changes of the second kind, namely, that analytic sentences cannot change their truth value. But this characteristic is not restricted to the analytic sentences; it holds also for certain synthetic sentences, e.g., physical postulates and their consequences.
>
> (Carnap 1963: 921)

This is exactly what Kuhn says about symbolic generalizations: "They function in part as laws but also in part as definitions of the symbols they deploy . . . I currently suspect that all revolutions involve, among other things, the abandonment of generalizations the force of which had previously been in some part that of tautologies" (1962/1970: 183–4).

It is clear then that, according to Carnap, no linguistic framework principles are immune to change. Moreover, they are all subject to Carnap's principle of tolerance: One can choose the rules of language and thereby logic freely. There is then a plurality of conventionally and pragmatically chosen linguistic frameworks and therefore of logics, none of which is the valid or correct one since the very standards of validity and correctness are relative to the framework chosen. Hence, because framework principles constitute the meanings of "validity" and "correctness" and yet can change, they can be seen as a generalization of Hans Reichenbach's relativized constitutively a priori principles that he employs in his reconstruction of relativity theory (Friedman 2001: 30–32, 2003). As we shall see, in his later writings Kuhn came to view the language of scientific theories in similar terms.

A final point is worth noting. Carnap (1950a) sharply distinguishes between internal and external questions of existence relative to them. For example, the question "Are there electrons?" understood as an internal question relative to a linguistic framework *L* can be answered by carrying out the appropriate experiments, provided that *L* has methodological rules for testing and confirming the claim, "There are electrons." External questions raise the question of the existence of electrons prior to the endorsement of any *L* and are devoid of cognitive meaning. This relativization of existence claims to an *L* is essentially no different from Kuhn's remark that there is no paradigm-independent way to understand phrases like "really there" (1962/1970: 206).

To summarize, when taken collectively, Carnap's views about scientific development start to look very much like Kuhn's even though the two use very different terminology. Both distinguish between two kinds of scientific activity: ordinary and revolutionary. For Kuhn, the former is the normal scientific activity of puzzle-solving during which the paradigm is never tested and abandoned. For Carnap, it is the activity characterized by assigning a truth-value to an indeterminate statement or changing its truth-value, where fundamental laws and framework principles are firmly in place even when there is a conflict with experience. In Kuhn's philosophy, standards of validity and correctness of problem solutions are relative to the paradigm; in Carnap's philosophy, they are relative to the linguistic framework in which the theory is represented. For Kuhn, a scientific revolution occurs through a paradigm shift; for Carnap, it occurs by the replacement of one set of T-postulates or logico-mathematical framework rules by another. Just as for Kuhn, for Carnap too, scientific development is non-cumulative because truth-values of some synthetic statements might change as a result of a change in those postulates or rules.

Meaning incommensurability and meaning holism

Carnap hit upon a rudimentary form of meaning incommensurability (without using that term) as early as 1936 (Earman 1993). Comparing classical and quantum physics, he wrote: "In translating one language into another the factual content of an empirical statement cannot always be preserved unchanged. Such changes are inevitable if the structures of the two languages differ in essential respects" (1936/1949: 126; cf. 1963: 929–30). This follows from the semantic function that Carnap, like Kuhn, attributes to fundamental laws. For if the meanings of certain theoretical terms are determined in part by the sum total of T-postulates in which they occur (local meaning holism), changing those postulates necessarily changes those terms' meanings, so it becomes impossible to translate sentences containing them from one theory to another without residue (local meaning incommensurability). Both Carnap and Kuhn are moderate meaning holists, and meaning holism paves the way for meaning incommensurability (Irzik and Grünberg 1995).

In fact, Carnap's philosophy contains not just one, but two kinds of incommensurability, what I have dubbed "confirmational incommensurability," in addition to meaning incommensurability (Irzik 2003). Carnap explicitly states that his internal-external distinction applies to inductive logic as well (1963: 982). This implies that the degree of confirmation is framework relative. Further support for confirmational incommensurability can also be found in Carnap's new studies of inductive logic (see Hilpinen 1973).

Even though Kuhn rejects the notion of confirmation entirely, thus marking a significant divergence from Carnap, it is nevertheless striking that just as there is no neutral paradigm comparison on the basis of shared values for Kuhn, there is no neutral theory evaluation on the basis of degree of confirmation for Carnap either. This brings us to the topic of rationality and relativism.

Theory choice, relativism, and inductive logic

Kuhn and post-positivists attribute to Carnap the view that theory choice or acceptance is a matter of "experience and logic" alone, based on the degree of confirmation, which with sufficient effort could be reduced to an algorithm. According to them, rationality for Carnap is merely a matter of choosing or accepting that theory which has a greater degree of confirmation.

Carnap's confirmational incommensurability should already cast doubt on such allegations. But there is more. Like Kuhn, Carnap is a relativist about methodology, ontology, and even (perhaps unlike Kuhn) logic. This follows from his principle of tolerance and his internal-external distinction. Nevertheless, he does not hold an "anything goes" position, because he insists that the decision to choose a linguistic framework, while itself not theoretical, can be influenced by theoretical factors such as efficiency, fruitfulness, and simplicity. There is a fact of the matter as to whether framework A or B is simpler, more efficient, and more fruitful, a fact that can be discovered empirically. Therefore, the choice among alternative frameworks can be made rationally on the basis of such criteria, mitigating Carnap's relativism (see Irzik 2003).

As for theory choice, like Kuhn, Carnap too never believed that it can be reduced to an algorithm and explicitly acknowledged the role of many factors, including subjective ones. Carnap's critics fail to understand the task of "inductive logic proper," which is merely to assign a quantitative degree of confirmation to a hypothesis on the basis of a body of evidence, nothing more: "Inductive logic alone does not and cannot determine the best hypothesis on a given evidence, if the best hypothesis means that which scientists would prefer. This preference is determined by factors of many different kinds, among them logical, methodological, and purely subjective factors" (1950b/1962: 221).

Finally, Carnap's anti-realist stance about scientific truth and progress, as stated explicitly in his letter to Kuhn, must also be noted. Whereas Kuhn relativizes truth and existence claims to paradigms, Carnap relativizes them to linguistic frameworks. They are in full agreement that scientific progress is not towards some final perfect theory that is true in the sense of correspondence with a mind-independent world.

Kuhn's linguistic turn

After Kuhn's linguistic turn, his tendency to express the SSR's crucial theses in terms of scientific lexicons in his post-1980 writings, the similarities discussed previously leap to the eye.

Kuhn replaces the talk of paradigms with that of theories and tells us that every major scientific theory has its own distinctive, structured lexicon. Lexicons are taxonomically ordered networks of kind-terms and necessary for the formulation of scientific problems, their solutions, and the description of nature. Since radically different lexicons permit radically different descriptions and generalizations, scientific development is necessarily discontinuous. In line with this, the revolutionary-normal science distinction becomes the distinction between activities that require significant changes in the lexicons of theories and those that do not. The talk of gestalt switches and conversions disappear altogether. So do all forms of incommensurability except that which pertains to meaning, which now becomes "a sort of untranslatability, localized to one or another area in which two lexical structures differ" (2000: 93). Finally, as different lexicons impose different structures on the world, they "resemble[s] Kant's a priori when the latter is taken in its second, relativized sense [in the sense of 'constitutive of the concept of the object of knowledge']" (ibid.: 245). Thus, Kuhn adopts "a sort of post-Darwinian Kantianism," where the lexicon makes knowledge of nature possible, yet does change historically (ibid.: 104).

To realize how much all of this sounds like Carnap, it suffices to replace "lexicons" with "linguistic frameworks."

Comparison with Reichenbach, Frank, and Hempel

It is not just Carnap's views that overlap with Kuhn's. Take, for instance, Hans Reichenbach's formulation of relativity theory in terms of relativized a priori principles. These principles are both constitutive of the concept of the object of scientific knowledge and change historically. As we saw, this is exactly how Kuhn thinks of scientific lexicons. Moreover, Reichenbach later also emphasizes the gestalt character and conceptual framework dependence of our experiences:

"[T]he objects of our sensations have always a gestalt character. They appear as if pressed into a certain conceptual frame" (1938: 221). This is not to say that there are no significant differences between his and Kuhn's views. There's Kuhn's rejection of the context-distinction synonymous with Reichenbach's name and Kuhn would have no truck with his scientific realism. Nevertheless, it is doubtful that Reichenbach would have found SSR as a threat to LE *tout court*.

As for Hempel, despite the fact that he thought confirmation was an important issue and his DN model of explanation implied theoretical accumulation, he received SSR well. He agreed that "[a] large-scale theory that has been successful in many areas will normally be abandoned only when a more satisfactory alternative theory is available" (1966a: 40). He was also aware that many factors such as confirmation, explanatory power, and simplicity all played a role in theory acceptance, and that how to combine them was a difficult and yet unsolved issue (1966b: 131).

Among the logical empiricists who espoused views closest to Kuhn is Philip Frank. Frank believed that a logico-structural analysis of science must be supplemented by a pragmatic-historical one. Hence, before Kuhn, he often appealed to historical cases such as the Copernican revolution for insight into scientific development. He paid particular attention to the issue of theory acceptance and noted, much as Kuhn did years later, that theory acceptance is not just a matter of match between theory and observational facts, because no theory ever agrees with all facts completely, but only more or less. Therefore, he also underlined the importance of the criterion of simplicity and drew attention to the problem of how to weigh different factors in choosing among rival theories, which is exactly how Kuhn formulated the problem (see Nemeth 2007; Reisch 2005: 229–33).

Back to Carnap: Divergences

Despite all the remarkable similarities and affinities, a number of important differences remain between Kuhn's philosophy of science and LE. I already pointed out some of them, yet I cannot do justice to them all here. I will be content with drawing attention to a few more of them by focusing on Carnap. To begin with, Kuhn's view of a scientific theory is entirely different from Carnap's two-tiered picture. According to Kuhn, a theory is a model-theoretic structure consisting of a set of distinct applications which all share the same basic laws and of constraints binding the applications together (1976). Specifying and learning a theory involves mastering a set of exemplary applications through which knowledge of nature is acquired. Such knowledge is simultaneously a mastery of how the theory's terms attach to nature. In effect, this means that "theoretical terms" may attach to nature directly without the mediation of "observation terms," which is a major reason why the theory-observation distinction is untenable. For Kuhn, the acquired ability obtained by studying exemplars is precisely the mechanism which accounts for this attachment mistakenly attributed to the C-rules (1977: 300–6). These concrete model problems and solutions, which are totally absent in Carnap, provide a much more illuminating way of understanding how a theory applies to nature.

Kuhn also believes that we cannot do justice to the history of science if we see the relationship between theory and experimental results in terms of confirmation and disconfirmation. Because Carnap is preoccupied with the logical reconstruction of science exclusively within the context of justification, his image of it is deemed excessively rule-based and ahistorical. Carnap's discussion of scientific revolutions, meaning incommensurability and their implications, pales in comparison to Kuhn's and is not at all central to his understanding of the main task of philosophy of science. Whereas for Carnap that task is "the logic of science," (i.e., the logical analysis of theories, sentences, and concepts of science), for Kuhn it is the historical (hence empirical) investigation of the dynamics of scientific development, an investigation that obliterates the distinction between the context of discovery and the context of justification.

Some commentators have claimed that these last differences in philosophical style are so deep that all similarities and affinities the revisionists have unearthed appear inconsequential or superficial (Oliveira 2007; Tsou 2015). Carnap's aprioristic and exclusive concern with the logic of science, it is said, leaves no room for a Kuhnian historical-empirical analysis of activities of scientific communities. However, as Thomas Uebel has shown, neither Kuhn nor Carnap rejected the context-distinction in the minimal sense of distinguishing between causal-descriptive and normative inquiries. Moreover, Uebel has also demonstrated that Carnap adheres to the project of "bipartite metatheory," i.e., to a theory of science that includes empirical investigations of science as well as the logic of science. His opting for the logic of science and others', such as Philip Frank's and Otto Neurath's, engagement with a socio-historical theory of science should be seen as a reflection of a division of labor rather than of an exclusionist perspective (see Uebel 2011 and CH. 38).

Apparently, Tsou grants substantive similarities between Carnap and Kuhn in light of Kuhn's post-SRR writings, but he dismisses them as superficial when they are placed into the context of Carnap's and Kuhn's "fundamentally different ways of doing philosophy of science" and when we limit our comparison solely to the SSR. However, even then, there is no compelling reason why the existing differences should render the remaining similarities superficial, especially in view of Carnap's and Kuhn's anti-realism and the ensuing relativism, their similar understanding of the rationality and progress of science, their adherence to local meaning holism, and, last but not least, the near identity of their views about the function of fundamental laws of science and how a change in them ushers scientific revolutions, all of which are ignored by Tsou. As we saw, these similarities become much more conspicuous after Kuhn's linguistic turn. Indeed, they culminate in a self-critical appraisal of the historical approach to science in his 1991 lecture with the telling title "The Trouble with the Historical Philosophy of Science." There, Kuhn argues that "many of most central conclusions we drew from the historical record can be derived instead from the first principles. Approaching them in that way reduces their apparent contingency" (2000: 112). Carnap could not have agreed more.

Concluding remarks

Even though LE was already in decline under the pressure of Quine, Hanson, and Feyerabend, there is no doubt that the destructive impact of SSR turned out to be stronger than any of them. It is equally clear that Kuhn set himself an easy target by portraying a simplistic picture of LE, what he later called "that sort of everyday image of logical positivism" (see Richardson 2007). In retrospect, the success of SRR appears to be due less to its attack against LE than to outlining an alternative account of science based on a historical approach that foregrounded a set of problems ranging from explaining scientific change and its rationality to the relationship between history of science and philosophy of science. Thus, SSR put the coffin's nail in "the logic of science" along with the intractable problems associated with it, such as the criteria for cognitive meaningfulness, inductive logic, and confirmation, by making them look irrelevant to a historical understanding of science. Yet, for logical empiricists, logical analysis was not merely a tool for technical philosophizing. As a component of "the bipartite metatheory," it was also to help reinvigorate the Enlightenment ideals and contribute to a scientific conception of the world by fighting against pseudo-philosophical claims, nationalistic hyperbole, obscurantism, mysticism, and dogma. Under the ideology of the Cold War in the US in the 1950s, this aspect of LE disappeared, and consequently LE became a purely technical, apolitical movement (Reisch 2005). There is reason to believe that SSR was not only a product of this ideological transformation, but also in its service by caricaturing LE and contributing to an apolitical professionalization of philosophy of science more generally.

References

Carnap, R. (1936) "Wahrheit und Bewährung," *Actes du Congrès Internationale de Philosophie Scientifique, Sorbonne, Paris 1935*, Facs. IV, "Induction et Probabilité," Paris: Hermann & Cie, pp. 18–23. Trans. with additional material in Carnap 1949.

——— (1936–7) "Testability and Meaning," *Philosophy of Science* 3: 419–71 and 4: 1–40. Excerpts repr. in H. Feigl and M. Brodbeck, eds., *Readings in the Philosophy of Science*, New York: Appleton-Century-Crofts, 1953, pp. 47–92.

——— (1949) "Truth and Confirmation," in H. Feigl and W. Sellars (eds.), *Readings in Philosophical Analysis*, New York: Appleton-Century-Crofts, pp. 119–27.

——— (1950a) "Empiricism, Semantics, and Ontology," *Revue Internationale de Philosophie* 4: 20–40. Repr. in *Meaning and Necessity*, Chicago: University of Chicago Press, 2nd ed., 1956, pp. 205–21.

——— (1950b) *Logical Foundations of Probability*, Chicago: University of Chicago Press, 2nd ed., 1962.

——— (1956) "The Methodological Character of Theoretical Concepts," in H. Feigl and M. Scriven (eds.), *The Foundations of Science and the Concepts of Psychology and Psycho-Analysis*, Minneapolis: University of Minnesota Press, pp. 38–76.

——— (1963) "Replies and Systematic Expositions," in P. A. Schilpp (ed.), *The Philosophy of Rudolf Carnap*, La Salle: Open Court, pp. 859–1013.

——— (1966) *The Philosophical Foundations of Physics*, New York: Basic Books, 2nd ed. as *Introduction to the Philosophy of Science*, New York: Basic Boks, 1974, repr. New York: Dover, 1995.

Earman, J. (1993) "Carnap, Kuhn, and the Philosophy of Scientific Methodology," in P. Horwich (ed.), *World Changes: Thomas Kuhn and the Nature of Science*, Cambridge, MA: MIT Press, pp. 9–36.

Friedman, M. (2001) *Dynamics of Reason*, Stanford: CSLI Publications.

——— (2003) "Kuhn and Logical Empiricism," in T. Nickles (ed.), *Thomas Kuhn*, Cambridge: Cambridge University Press, pp. 19–44.

Hempel, C. G. (1966a) *Philosophy of Natural Science*, Englewood Cliffs: Prentice Hall.

——— (1966b) "Recent Problems of Induction," in R. Colodny (ed.), *Mind and Cosmos*, Pittsburgh: Pittsburgh University Press. Repr. in Hempel, *Studies in Science, Explanation and Rationality* (ed. by J. Fetzer), Oxford: Oxford University Press, 2001, pp. 29–48.

Hilpinen, R. (1973) "Carnap's New System of Inductive Logic," *Synthese* 25: 307–33.

Irzik, G. (2003) "Changing Conceptions of Rationality: From Logical Empiricism to Post Positivism," in P. Parrini, W. C. Salmon, and M. H. Salmon (eds.), *Logical Empiricism: Historical and Contemporary Perspectives*, Pittsburgh: University of Pittsburgh Press, pp. 325–46.

——— (2012) "Kuhn and Logical Positivism: Gaps, Silences, and Tactics of SSR," in V. Kindi and T. Arabatzis (eds.), *Kuhn's The Structure of Scientific Revolutions Revisited*, London: Routledge, pp. 15–40.

Irzik, G. and Grünberg, T. (1995) "Carnap and Kuhn: Arch Enemies or Close Allies?" *British Journal for the Philosophy of Science* 46: 285–307.

Kuhn, T. (1962) *The Structure of Scientific Revolutions*, Chicago: University of Chicago Press, 2nd ed., 1970.

——— (1976) "Theory-Change as Structure-Change: Comments on the Sneed Formalism," *Erkenntnis* 10: 179–99.

——— (1977) *The Essential Tension*, Chicago: University of Chicago Press.

——— (2000) *The Road Since Structure* (ed. by J. Conant and J. Haugeland), Chicago: University of Chicago Press.

Nemeth, E. (2007) "Logical Empiricism and the History and Sociology of Science," in Richardson and Uebel (2007), pp. 278–302.

Oliveira, de P. (2007) "Carnap, Kuhn and Revisionism: On the Publication of Structure in Encyclopedia," *Journal of General Philosophy of Science* 38: 147–57.

Reichenbach, H. (1938) *Experience and Prediction*, Chicago: University of Chicago Press.

Reisch, G. (1991) "Did Kuhn Kill Logical Empiricism?" *Philosophy of Science* 58: 264–77.

——— (2005) *How the Cold War Transformed Philosophy of Science: To the Icy Slopes of Logic*, Cambridge: Cambridge University Press.

Richardson, A. (2007) "That Sort of Everyday Image of Logical Positivism: Thomas Kuhn and the Decline of Logical Empiricist Philosophy of Science," in Richardson and Uebel (2007), pp. 346–70.

Richardson, A. and Uebel, T. (eds.) (2007) *The Cambridge Companion to Logical Empiricism*, Cambridge: Cambridge University Press.

Tsou, J. (2015) "Reconsidering the Carnap-Kuhn Connection," in W. J. Devlin and A. Bokulich (eds.), *Kuhn's Structure of Scientific Revolutions—50 Years On*, Cham: Springer, pp. 51–69.

Uebel, T. (2007) *Empiricism at the Crossroads: The Vienna Circle's Protocol Sentence Debate*, Chicago: Open Court.

——— (2011) "Carnap and Kuhn: On the Relation Between the Logic of Science and the History of Science," *Journal of General Philosophy of Science* 42:129–40.

38
THE BIPARTITE METATHEORY CONCEPTION OF PHILOSOPHY

Thomas Uebel

The bipartite metatheory thesis attributes to Rudolf Carnap, Philipp Frank, and Otto Neurath a conception of philosophy of science as a strictly second-order inquiry in which the purely formal-logical analyses of the logic of science are complemented by empirical inquiries into the psychology, sociology, and history of science, i.e., the pragmatics of science. This chapter considers the thesis, the evidence for it, and some of the consequences that follow from it.

Competing metaphilosophies in the Vienna Circle

Different conceptions of post-metaphysical philosophy were on offer in the Vienna Circle. There was Moritz Schlick's Wittgenstein-inspired model of philosophy as meaning determination (first reductive, later use-based), there was Carnap's model of logico-linguistic analysis of conceptual frameworks (first syntactic, later semantic), and there was Neurath's naturalistic program for the epistemology of empirical science. One way of distinguishing them is with regard to the role played by philosophy vis-à-vis the sciences. Whereas Schlick retained an autonomous role for philosophy, both Carnap and Neurath tied philosophy more closely to science. Neither of them considered traditional philosophical problems to be solved by science either, of course, but they understood the linguistic turn which philosophy had taken not as characterizing a categorical difference between science and philosophy, but as characterizing a difference within unified science.

Though it may not have looked this way when their programmatic essays were published (Verein Ernst Mach 1929; Schlick 1930), theirs was not a minor disagreement but indicated the way the Circle was about to split into factions still before its leader was murdered and its members dispersed in exile. That there were such divisions and that there existed a "left wing" of the Vienna Circle was first suggested to the philosophical public by Ernest Nagel in his 1936 report on current philosophy in Central Europe and confirmed by Carnap in his autobiography. It appears that this label was used first by Neurath in private discussions in the early 1930s to designate what he also called "the unified science group" comprised of Carnap and himself together with Hans Hahn and Philipp Frank. Carnap himself used "left wing" to denote the group in their capacity of opposing, already in the early 1930s, Schlick and Friedrich Waismann on the issue of retaining Wittgensteinian views like strict verifiability in their search for "a more liberal criterion of significance" (1963: 57). This was but part of a distinctive wider orientation. It was members of

 DOI: 10.4324/9781315650647-43

the left wing who first argued for the rejection of Wittgenstein's strictures against metalinguistic discourse and who soon came to reject all foundationalist ploys in philosophy. Most important for present purposes was their interpretation of the doctrine of unified science. All in the Vienna Circle objected to the separation of the human sciences from the natural sciences. But for the left wing, unified science also had implications for philosophy itself. For them, philosophy formed part of unified science because it was assimilated to science as its metatheory, as a second-order inquiry of itself a scientific nature employing only formal-logical or empirical resources.

Carnap and Neurath went about this in strikingly different ways, however. Carnap, having abandoned the already non-foundationalist epistemological ambition of the *Aufbau* to ground objectivity in structural features of a cognizer's experience, came to foreswear further concern with matters entangled with psychological considerations and focused on the purely logical analysis of means of representation. At first he developed the logical syntax (1934a), later the semantics of suitably regimented, formal languages of science (1942). Neurath, long suspicious of any foundationalist ambitions—his famous metaphor of seafarers having to repair their ship on the open seas without being able to pull into dry-dock ("Neurath's boat") predates World War I (1913/1998: 215–16)—sought whatever grounds there could be for justifying scientific claims in intersubjectively available evidence, that is, in statements about observable states of affairs and the conditions of their rational acceptance. Both theorists focused on language, but whereas Carnap abstracted from what was required by and of speakers and hearers of a language, Neurath focused on just that.

The point of their investigations was still a philosophical one and not merely descriptive of given or conceivable language-systems or practices of use. They were developed with the point of understanding and legitimating knowledge claims clearly in view. Yet whereas Schlick pursued quasi-phenomenological explorations of the import of verifications (1934), Carnap and Neurath viewed the task of understanding scientific knowledge as itself a task for science. Schlick remained relatively traditionalist in focusing on the individual knower and retaining a rarefied but still broadly correspondentist conception of truth which Neurath rejected as "metaphysical," i.e., beyond empirical control. By contrast, once he had accepted semantics, Carnap settled for a disquotational, deflationary conception and unsuccessfully tried to convince Neurath of its wisdom (see their correspondence in Cat and Tuboly [2018] and CH. 15). This disagreement apart, both Carnap and Neurath sought to overcome the traditional spectator view of knowledge but did so by different means from early on: Carnap by exploring the structures of possible representational systems in abstraction and aiming thereby to delineate conditions of propositional justification, Neurath by exploring the conditions of social interaction that make empirical knowledge possible and aiming to elucidate, to begin with, the acceptance conditions of observational testimony. (More on both follows.)

The question arises how Carnap's and Neurath's conceptions were related to each other beyond their shared opposition to Schlick's. It may easily appear as if what Carnap called "the logic of science" (1934a: §§72–73) and what Neurath called the "behavioristics of scholars" (1936a/1983: 160, 168) were informed by entirely different conceptions of what replaced traditional philosophy: the former as a formal study of symbol systems by the aprioricism of logic, the latter by trans-disciplinary studies involving psychology, sociology, and history (and perhaps also economics and anthropology). To see all that was involved, it is helpful to bring in another player and consider some further history.

Frank and Neurath on the need to complement formal studies

Early on in his 1932 monograph on the use and abuse of notions of causality, Frank suggested adding a sociological dimension to the new philosophy of science. To date it had been developed mainly in terms of an analysis of formal aspects of the symbol system employed, abjuring claims to knowledge of unobservables and regarding it instead as a complex "instrument" for the prediction of observable occurrences.

> The events around Galileo make it clear that the passionate conflicts connected with a physical theory have nothing to do with its suitability to represent natural processes but much more with their relationships to the political and social events of the time. Therefore there is no need to amplify the positivist conception of science by a metaphysical concept of truth but only by a more comprehensive study of the connections that exist between the activity of the invention of theories and the other normal human activities.
>
> (1932a/1998: 14)

About this "positivist" philosophy of science, Frank remarked elsewhere:

> Only the following can meaningfully be meant by "philosophy of natural science" or "philosophy of nature": either *the precise analysis of the meaning of scientific symbols*, i.e. of the relations of coordination, something which really constitutes a significant part of every natural scientific theory, or the investigation of certain observable processes concerning the relation between the theories of natural science and other expressions of human activity, something which is to be regarded as a part of scientific sociology.
>
> (1932b: 156, orig. emphasis)

Understanding typical disputes over science required not metaphysics, but the (broadly speaking) sociological supplementation of the studies of the symbol system of science with investigations of science-related human activities. Without it, the "positivist" philosophy of science remained incomplete.

Frank's remarks draw attention to the relation between the two approaches to the study of science pursued by Carnap and Neurath and a development that was already underway then. In fact, Frank returned to a suggestion that was already contained in the Circle's unofficial manifesto of 1929:

> The scientific world-conception is characterized not so much by theses of its own, but rather by its basic attitude, its points of view and direction of research. The goal ahead is unified science. . . . Clarification of the traditional philosophical problems leads us partly to unmask them as pseudo-problems, and partly to transform them into empirical problems and thereby subject them to the judgement of experimental science. The task of philosophical work lies in this clarification of problems and assertions, not in the propounding of special "philosophical" pronouncements. The method of this clarification is that of logical analysis.
>
> (Verein Ernst Mach 1929/2012: 81–82)

Note that two tasks were here assigned to philosophy as "clarification": first, "unmasking" some traditional philosophical problems as mere pseudo-problems by logical analysis; second, transforming others into empirical questions, i.e., revealing them to be empirical in nature. Note also that at this stage, "logical analysis" was characterized still as concerned with reduction to "the given" (ibid.: 84) in a framework familiar from Carnap's *Aufbau*, aiming to recover the epistemic order of priority of phenomenal over physical states of affairs under the heading of methodological solipsism (1928: §59). The transformation of some philosophical questions into empirical questions was not yet further discussed.

In 1931 Neurath recast a similar two-part agenda that made fewer concessions to the Wittgensteinian wing ("clarification") and also reflected his ongoing debate with Carnap over the form, content and status of the basic evidence statements in science (the protocol-sentence debate; see Uebel 2007) by giving expression to his opposition to methodological solipsism.

> Within a consistent physicalism there can be no "theory of knowledge," at least not in the traditional form. It could only consist in defense actions against metaphysics, i.e. unmasking meaningless terms. Some problems of the theory of knowledge will perhaps be transformable into empirical questions so that they can find a place within unified science.
>
> (1932a/1983: 67)

Again, two tasks were assigned to whatever was to be philosophy's successor discipline in unified science: unmasking meaningless terms and asking empirical questions about knowledge production. Both represent different aspects of what unified science contains alongside all of the first-order disciplines: a scientific metatheory which comprised logical inquires as well as empirical ones. This program fits Frank's agenda well.

Notably, Neurath now also provided a striking example of the transformation of traditional philosophical into empirical questions (1932b). Consider observational knowledge claims. Rather than conceive of their justification as afforded by a phenomenalist reduction such that a direct correspondence of their content with the "given" was established, Neurath developed a scheme that sought to determine when the conditions of intersubjective acceptance of observational testimony were satisfied. In place of simple truth conditions of observation statements, a network of interlocking conditions was presented that accounted for their rational acceptance by providing a checklist of causal environmental as well as psychological and social conditions indicative of reliable testimony. Thus, "justification" was naturalized (see Uebel 2009; Bentley 2022).

By discarding methodological solipsism for all practical purposes in late 1932, Carnap brought his position into broad agreement with Neurath's physicalism. Yet Carnap did not yet fully accept Neurath's argument against methodological solipsism (nor Neurath's specific proposal for how protocol statements were to be analyzed). Carnap added: "this is a question not of two mutually inconsistent views, but rather two different methods for structuring the language of science which are both possible and legitimate" (1932/1987: 457). So even though something of a working agreement on the overall strategy of how to approach the analysis of the language of science was now established, Carnap reserved the right in principle to explore counterfactual suppositions (like private protocol languages). As long noted (Creath 1987), this was the first employment of the Principle of Tolerance later announced in *Logical Syntax* (1934a: §17). Here, Carnap used it to claim a certain autonomy for his austere conception of philosophy as logic of science.

> Philosophy deals with science only from the *logical* viewpoint. *Philosophy is the logic of science*, the logical analysis of the concepts, propositions, proofs, theories of science, both of those occurring in existing science as well as of those which are common to the possible methods of constructing concepts, proofs, hypotheses, theories.
>
> (1934b/1967: 54–55, orig. emphasis, trans. amended)

Revolutionary in other respects, the logic of science was continuous with traditional philosophy with its a priori method. Empirical concerns about what kind of languages humans can employ do not play into the exploration of the consistency of logics and of the expressive power of logico-linguistic frameworks. Nor was the logic of science obliged to take account of conceptions of epistemic priority: such concerns now fell under the rubric of psychologistic fallacies. Carnap wanted to "cleanse" epistemology of psychological concerns, not reject it altogether (1936: 36). When investigating evidential relations, the logic of science was therefore concerned no longer with doxastic but with propositional justification, that is, justification not of individual beliefs but of propositions in light of available evidence where that evidence is conceived of independently of its appreciation by a subject. Beyond that, the logic of science only offered different ways of conceptualizing contested scientific notions: it made "proposals" for language reform and explored conceptual possibilities (later called "explications").

Neurath returned to the matter of metatheoretical methodology at the Paris Congress 1935 when he stated that the "misgiving . . . that logical empiricism could decay into empty scholasticism and dogmatism" required counteracting by work on unified science (1936b/1983: 132). Importantly, such work encompassed two theoretical pursuits:

> If we regard the mass of statements as the result of experiments, travels, or certain other behaviour, then we move in the fields of the significant behaviouristics of scholars, history of science, sociology of science. . . . If we state: "On the basis of experiments a scholar has replaced an earlier published statement by a new one," this is not a statement of the logic of science; but the logic of science can compare this statement with other statements as to its logical content; in the same way it can compare the scholar's earlier statement with his later statement.
>
> (Ibid.: 135, trans. amended)

In a contemporaneous debate with Åke Petzäll about the supposed poverty of his and Carnap's "physicalism," Neurath stated programmatically: "The program of unified science . . . presupposes that . . . one must concern oneself with the 'logical syntax of language' (the title of a work by R. Carnap) and at the same time with a behavioristic study of the actions of the men of science" (1936c/1983: 149).

Precisely that, again, had been Frank's point in 1932. He and Neurath must count as prime movers behind the bipartite metatheory conception: alongside Carnap's "logic of science" there was Neurath's "behavioristics of scholars," what Frank later and more aptly called the "pragmatics of science" (1957/2004: 360). (In these later years Frank often worked on the socio-historical conditions of theory acceptance; see his 1951, 1954.)

Carnap's accommodation of the pragmatics of science

But what about Carnap? To some, his claim that "psychology and sociology are empirical sciences; they do not belong to philosophy" (1934b/1967: 54) suggests that he had no part in the project of bipartite metatheory. Yet Carnap also avowed that "the task of the philosophy of science can be pursued only in a close cooperation between logicians and empirical investigators" (ibid.: 62). Already in *Logical Syntax*, Carnap stressed that the logic of science was not an end in itself: the point was to develop reform proposals "useful and productive in practice" for "particular point[s] of the language of science" (1934a/1937: 332). The logic of science was "an instrument of unified science" serving as a "tool for the construction of a unified science"

(1934c/1987: 56, 66). For instance, "a proposed new syntactical formulation . . . can only be useful and productive in practice if it has regard to the available empirical findings of scientific investigations": "all work in the logic of science" required "close cooperation with the special sciences" (1934a/1937: 332).

Already in 1934, then, Carnap saw the logic and the pragmatics of science as complementary, even though he still conceived of philosophy of science—better: its successor discipline—as logic of science alone. One may wonder whether this was a merely terminological difference, distinguishing a narrow from a broad sense of "philosophy of science." What would seem to stand against it is Carnap's resolute determination to view philosophical inquiries as exclusively concerned with language: prima facie at least, the psychology and sociology of science have a different focus. Yet matters changed with Carnap's increasing appreciation, after his semantic turn in 1935, of the pragmatic dimension of meaning (see Uebel 2013). Using terminology learned from Charles Morris, he wrote in *Introduction to Semantics*:

> It has turned out to be very fruitful to look at the problems of theoretical philosophy from the point of view of semiotic, i.e. to try to understand them as problems which have to do with signs and language in one way or another. Among problems of this kind we may first distinguish between those problems—or components of complex problems—which are of a factual, empirical, rather than logical nature. They occur especially in the theory of knowledge and the philosophy of science. If construed as problems of semiotic, they belong to pragmatics. They have to do, for instance, with the activities of perception, observation, comparison, registration, confirmation, etc., as far as these activities lead to or refer to knowledge formulated in language. On the other hand, we have the problems of logical analysis; they occur in what is known as logic and, combined with problems of the first kind, in the theory of knowledge and the philosophy of science. These problems belong either to semantics or to syntax.
>
> (1942: 245)

Accordingly, Carnap reformulated the task of philosophy.

> *The task of philosophy is semiotical analysis*; the problems of philosophy concern—not the ultimate nature of being but—the semiotical structure of the language of science, including the theoretical part of everyday language. We may distinguish between those problems which deal with the activities of gaining and communicating knowledge and the problems of logical analysis. Those of the first kind belong to pragmatics, those of the second kind belong to semantics and syntax.
>
> (Ibid.: 250, orig. emphasis)

Here, Carnap clearly endorsed the bipartite metatheory conception of philosophy: as semiotical analysis it divides into semantics and syntax belonging to the logic of science and pragmatics encompassing the empirical sciences of science. Note not only the very wide range of problems Carnap here assigned to pragmatics, but also his self-correction: "It is a terminological question whether to use the term 'philosophy' in a wider sense, including certain empirical problems. If we do so then these empirical problems will turn out to belong mostly to pragmatics" (ibid.). This confirms that Carnap's earlier terminological distinction was no longer (if it ever was) hard and fast. He too now was an adherent of the bipartite metatheory conception.

While Carnap soon dropped talk of "semiotics," he retained the viewpoint thereby gained. For instance, in "Empiricism, Semantics and Ontology," Carnap discussed how to decide matters of language choice with several examples, among them the choice of the "thing language," and made explicit the pragmatic nature of the decisions involved.

> The decision of accepting the thing language, although itself not of a cognitive nature, will nevertheless usually be influenced by theoretical knowledge, just like any other deliberate decision concerning the acceptance of linguistic or other rules. The purposes for which the language is intended to be used, for instance, the purpose of communicating factual knowledge, will determine which factors are relevant for the decision. The efficiency, fruitfulness, and simplicity of the use of the thing language may be among the decisive factors.
>
> (1950/1956: 208)

Having specified a purpose—"communicating factual knowledge"—Carnap suggested that the proposals arrived at by the logic of science be subjected to an empirical efficiency test. "The acceptance or rejection . . . of linguistic forms in any branch of science will finally be decided by their efficiency as instruments, the ratio of the results achieved to the amount and complexity of the efforts required" (ibid.: 221). Logical explications require pragmatic validation just as pragmatic innovation demands valid reasons.

Importantly, Carnap's own personal preference for purely logical investigations does not represent his considered views, as shown by this late remark:

> many problems concerning conceptual frameworks seem to me to belong to the most important problems of philosophy. I am thinking here both of theoretical investigations *and of practical deliberations and decisions with respect to an acceptance or a change of frameworks, especially of the most general frameworks containing categorial concepts which are fundamental for the representation of all knowledge.*
>
> (1963: 861, emphasis added)

Not only is pragmatics here counted into philosophy but, along with it, the wide notion of pragmatics (the investigation of "problems which deal with the activities of gaining and communicating knowledge") is endorsed once more.

Conclusion

Even though it was never publicly announced as such, the bipartite metatheory conception of philosophy emerged from seeds planted in the Circle's unofficial manifesto, received its first articulation by Frank and Neurath in the early and mid-1930s, and was adopted by Carnap in the early 1940s and retained by him ever since. This is not to say, of course, that Carnap and Neurath agreed on every detail (see Uebel 2001, 2015). According to the broad outline on which their views converged, however, the logic of science investigated scientific theories, typically in axiomatized form, and considered their internal structure and their relation to their evidential base in purely logical terms (deductive, inductive, abductive), while the pragmatics of science investigated scientific practice by means of linguistics and the empirical sciences of science, primarily the psychology and sociology as well as the history of science. So while the logic of science investigated abstract relations of evidential support, the pragmatics of science investigated concrete conditions of data acceptance, and theory choice and change. (It may be added that both also issued conditional normative prescriptions as proposals that traded on the instrumental value that the adoption of the recommendation was supposed to bestow.)

Recognition of Carnap's, Neurath's, and Frank's bipartite metatheory conception has important consequences for the interpretation of the work of the Vienna Circle. To begin with, it clarifies the topography of the Vienna Circle. It allows us to see what unites the approaches to the philosophy of science on the so-called left wing of the Vienna Circle: Carnap's formalism and Neurath's and Frank's naturalisms complement each other (Hahn died in 1934). Individually and jointly, they make sense of counting philosophy of science as part of science, namely its metatheory, and distinguish it from Schlick's Wittgensteinian approach.

But recognition of the bipartite metatheory conception also casts light on the later stages of logical empiricism. It dissolves the puzzle that Thomas Kuhn's *The Structure of Scientific Revolutions* (1962) should have been published first as a volume of the *International Encyclopedia of Unified Science*, the Neurath-founded flagship of the logical empiricist movement, when it is widely regarded to have sounded the death knell of that movement (see CH. 37). Carnap's adherence to the bipartite conception makes intelligible why he should have been as delighted as he was when he read the manuscript of Kuhn's book as one of the remaining series editors and recommended its acceptance (Reisch 1991). If any "irony" attends Kuhn's book being published there, it is that after World War II, given the death of Neurath and the marginalization of Frank, the logic of science had become the sole representative of logical empiricism for most philosophers of science, so that explorations of what had been their pragmatics of science were now considered adversarial.

What also must not be forgotten is the cooperation demanded by work on the scientific world-conception that the manifesto advertised, indeed, the collective nature of that project. Carnap had extolled it already in the Preface to his *Aufbau* and referred to it again late in his career when faced with Frank's criticism of the ineffectiveness of his own campaign against metaphysics on account of its purely logical nature.

> [T]he Vienna Circle, essentially because of Otto Neurath, did recognize the importance of a sociological analysis of the roots of philosophical movements. But unfortunately, a division of labor is necessary, and therefore I am compelled to leave the detailed work in this direction to philosophically interested sociologists and sociologically trained philosophers.
>
> (1963: 868)

His own predilection apart, Carnap's concentration on the logic of science (and abstention from history and sociology) was an expression of the division of labor that large collaborative projects brought with it. Such an understanding of their joint project (clearly expressed by Carnap also in conversations with the Belgian philosopher Leo Apostel: see Dewulf [forthcoming]) naturally follows from the bipartite metatheory conception.

Unqualified reports of the death of logical empiricism may have been premature. With a suitably informed pragmatics of science as complement—neither their sociology nor their history of science needs to follow existing paradigms (see Uebel 2000; Carus 2013, respectively)—the pluralist logic of science appears fit for further exploration. That Carnap's explications pioneered what nowadays is called "conceptual engineering" is no longer news (see Creath 1990), nor should it be that such interventions cannot limit their resources unduly if they are to succeed. Whether one wants to see the project of scientific philosophy being pursued austerely as by the younger Carnap—freeing reflection about science from extraneous disturbances (Richardson 2013)—or in a more expansive Neurathian sense—either as reigniting a new Enlightenment spirit (Carus 2007: ch. 11) or as targeting specific oppressive social conditions (Dutilh Novaes 2020)—a collaborative and collectivist approach, calling on both the logic and the pragmatics of science, is surely what is required, given the wide ramifications of all such remedial enterprises.

References

Bentley, J. (2022) "Conceptions of Protocol Sentences in Neurath and Carnap and the Bipartite Metatheory Conception," in C. Damböck, J. Friedl and U. Höfer (eds.), *Ways of the Scientific World-Conception. Rudolf Carnap and Otto Neurath*, Leiden: Brill.

Carnap, R. (1928) *Der logische Aufbau der Welt*, Berlin: Weltkreis/Bernary. Trans. *The Logical Structure of the World*, Berkeley: University of California Press, 1967, repr. Chicago: Open Court, 2003.

——— (1932) "Über Protokollsätze," *Erkenntnis* 3: 215–28. Trans. "On Protocol Sentences," *Nous* 21 (1987): 457–70.

——— (1934a) *Logische Syntax der Sprache*, Vienna: Springer. Rev. ed. trans. *The Logical Syntax of Language*, London: Kegan Paul, Trench, Trubner & Cie, 1937, repr. Chicago: Open Court, 2002.

——— (1934b) "On the Character of Philosophic Problems," *Philosophy of Science* 1: 5–19. Repr. in R. Rorty (ed.), *The Linguistic Turn: Essays in Philosophical Method*, Chicago: University of Chicago Press, 1967, pp. 54–62. Orig. "Über den Charakter der philosophischen Probleme," in Carnap, *Scheinprobleme der Philosophie und andere metaphysikkritische Schriften* (ed. by T. Mormann), Hamburg: Meiner, 2004, pp. 111–28.

——— (1934c) *Die Aufgabe der Wissenschaftslogik*, Vienna: Gerold. Trans. "The Task of the Logic of Science," in B. McGuiness (ed.), *Unified Science*, Dordrecht: Reidel, 1987, pp. 46–66.

——— (1936) "Von der Erkenntnistheorie zur Wissenschaftslogik," *Actes du Congrès International de Philosophie Scientifique, Sorbonne, Paris 1935*, Facs. I "Philosophie Scientifique et Empirisme Logique," Paris: Herman & Cie, pp. 36–41.

——— (1942) *Introduction to Semantics*, Cambridge, MA: Harvard University Press.

——— (1950) "Empiricism, Semantics and Ontology," *Revue Internationale de Philosophie* 4: 20–40. Repr. in Carnap, *Meaning and Necessity*, Chicago: University of Chicago Press, 2nd ed., pp. 205–11.

——— (1963) "Intellectual Autobiography" and "Comments and Systematic Expositions," in P. A. Schilpp (ed.), *The Philosophy of Rudolf Carnap*, LaSalle: Open Court, 1963, pp. 3–84, 859–1016.

Carus, A. (2007) *Carnap and Twentieth-Century Thought: Explication as Enlightenment*, Cambridge: Cambridge University Press.

——— (2013) "History and the Future of Logical Empiricism," in E. Reck (ed.), *The Historical Turn in Analytic Philosophy*, Basingstoke: Palgrave Macmillan, pp. 261–93.

Cat, J. and Tuboly, A. (eds.) (2018) "The 1940–1945 Neurath-Carnap Correspondence," in J. Cat and A. Tuboly (eds.), *Neurath Reconsidered: New Sources and Perspectives*, Cham: Springer.

Creath, R. (1987) "Some Remarks on 'Protocol sentences'," *Nous* 21: 471–5.

——— (1990) "Introduction," in R. Creath (ed.), *Dear Carnap, Dear Van: The Quine-Carnap Correspondence and Related Work*, Berkeley: University of California Press.

Dewulf, F. (Forthcoming) "Leo Apostel and Rudolf Carnap: The Development of Logical Empirical Ethics in Post-War Europe," in C. Damböck and A. Tuboly (eds.), *The Socio-Ethical Dimension of Knowledge: The Mission of Logical Empiricism*, Cham: Springer.

Dutilh Novaes, C. (2020) "Carnap's Explication and Ameliorative Analysis: A Systematic Comparison," *Synthese* 197: 1011–34.

Frank, P. (1932a) *Das Kausalgesetz und seine Grenzen*, Vienna: Springer. Trans. *The Causal Law and Its Limits*, Dordrecht: Kluwer, 1998.

——— (1932b) "Naturwissenschaft," in R. Dittler et al. (eds.), *Handwörterbuch der Naturwissenschaften*, Jena: Fischer, 2nd ed., vol. 7, pp. 149–68.

——— (1951) "The Logical and Sociological Aspects of Science," *Proceedings of the American Academy of Arts and Sciences* 80: 16–30.

——— (1954) "The Variety of Reasons for the Acceptance of Scientific Theories," *Scientific Monthly* 79. Repr. in P. Frank (ed.), *The Validation of Scientific Theories*, Boston: Beacon Press, 1956, pp. 3–17.

——— (1957) *Philosophy of Science: The Link Between Science and Philosophy*, Englewood Cliffs, NJ: Prentice-Hall. Repr. Minola: Dover, 2004.

Kuhn, T. S. (1962) *The Structure of Scientific Revolutions*, Chicago: University of Chicago Press, 2nd ed., 1970.

Nagel, E. (1936) "Impressions and Appraisals of Analytic Philosophy in Europe," *Journal of Philosophy* 33: 5–24, 29–53.

Neurath, O. (1913) "Probleme der Kriegwirtschaftslehre," *Zeitschrift für die gesamte Staatswissenschaft* 69: 438–501. Repr. in Neurath, *Gesammelte ökonomische, soziologische und sozialpolitische Schriften, Teil 2* (ed. by R. Haller and U. Höfer), Vienna: Hölder-Pichler-Tempsky, 1998, pp. 201–49.

——— (1932a) "Soziologie im Physikalismus," *Erkenntnis* 2: 393–431. Trans. "Sociology in the Framework of Physicalism," in Neurath (1983), pp. 58–90.

——— (1932b) "Protokollsätze," *Erkenntnis* 3: 204–14. Trans. "Protocol Statements," in Neurath (1983), pp. 91–99.

——— (1936a) "Physikalismus und Erkenntnisforschung," *Theoria* 2: 97–105, 234–7. Trans. "Physicalism and the Investigation of Knowledge," in Neurath (1983), pp. 159–66, 168–71.

——— (1936b) "Einzelwissenschaften, Einheitswissenschaft, Pseudorationalismus," *Actes du Congrès International de Philosophie Scientifique, Sorbonne, Paris 1935*, Facs. I "Philosophie Scientifique et Empirisme Logique," Paris: Herman & Cie, pp. 57–64. Trans. "Individual Sciences, Unified Science, Pseudo-Rationalism," in Neurath (1983), pp. 132–8.

——— (1936c) "L'encyclopédie comme 'modèle'," *Revue de Synthèse* 12: 187–201. Trans. "Encyclopedia as Model," in Neurath (1983), pp. 145–58.

——— (1983) *Philosophical Papers 1913–1946* (ed. by R. S. Cohen and M. Neurath), Dordrecht: Reidel.

Reisch, G. (1991) "Did Kuhn Kill Logical Empiricism?" *Philosophy of Science* 58: 264–77.

Richardson, A. (2013) "Taking the Measure of Carnap's Philosophical Engineering," in E. Reck (ed.), *The Historical Turn in Analytic Philosophy*, Basingstoke: Palgrave Macmillan, pp. 60–77.

Schlick, M. (1930) "Die Wende in der Philosophie," *Erkenntnis* 1: 4–11. Trans. "The Turning Point in Philosophy," in Schlick (1979), pp. 154–60.

——— (1934) "Über das Fundament der Erkenntnis," *Erkenntnis* 4: 79–99. Trans. "On the Foundation of Knowledge," in Schlick 1979, pp. 370–87.

——— (1979) *Philosophical Papers, vol. 2 (1925–1936)* (ed. by H. L. Mulder and B. van de Velde-Schlick), Dordrecht: Reidel.

Uebel, T. (2000) "Logical Empiricism and the Sociology of Knowledge: The Case of Neurath and Frank," *Philosophy of Science* 67: S138–S150.

——— (2001) "Carnap and Neurath in Exile: Can Their Disputes Be Resolved?" *International Journal of the Philosophy of Science* 15: 211–20.

——— (2007) *Empiricism at the Crossroads: The Vienna Circle's Protocol-Sentence Debate*, Chicago: Open Court.

——— (2009) "Neurath's Protocol Statements Revisited: Sketch of a Theory of Scientific Testimony," *Studies in History and Philosophy of Science* 40: 4–13.

——— (2013) "Pragmatics in Carnap and Morris and the Bipartite Metatheory Conception," *Erkenntnis* 78: 523–46.

——— (2015) "Three Challenges to the Complementarity of the Logic and the Pragmatics of Science," *Studies in History and Philosophy of Science* 53: 23–32.

Verein Ernst Mach (1929) *Wissenschaftliche Weltauffassung. Der Wiener Kreis*, Vienna: Wolf. Trans. "The Scientific Conception of the World. The Vienna Circle," in O. Neurath, *Empiricism and Sociology* (ed. by R. S. Cohen and M. Neurath), Dordrecht: Reidel, 1973, pp. 299–318; rev. trans. (with orig. annotated bibliography) "The Scientific World-Conception. The Vienna Circle," in F. Stadler and T. Uebel (eds.), *Wissenschaftliche Weltauffassung. Der Wiener Kreis. Hrsg. vom Verein Ernst Mach (1929)*, Vienna: Springer, 2012, pp. 75–116.

39

LOGICAL EMPIRICISM AND FORMAL EPISTEMOLOGY

Sahotra Sarkar

The term "formal epistemology" has become fashionable since the turn of the century, but the area of work to which it refers is of much older vintage. Formal epistemology consists of the deployment of formal techniques (including those from, but not limited to, logic, mathematics, computer science, and operations research) to the analysis of epistemological problems. Formal techniques are typically quantitative (or "cardinal"), that is, they involve the manipulation of numerical values, for instance, probabilities intended to estimate degrees of confirmation. However, formal techniques can also be used to model qualitative reasoning including comparative (or "ordinal") reasoning, for instance, through the use of comparative probabilities; or classificatory structures (e.g., "yes-no" assessments such as those found in Hempel's models of confirmation—see later in this chapter). A direct lineage for formal epistemology can be traced back, among many other sources, to inductive logic in the nineteenth century (e.g., in some of the work of Venn) as well as to probability theory in the seventeenth century (e.g., in Arnaud's *Port Royal Logic* from 1662). In the twentieth century, a focus on formal epistemology was part of the core of the "mature" program of logical empiricism (broadly construed)—in the work, for instance, of Carnap, Reichenbach, and Hempel, and their associates and students including Nagel, Jeffrey, and Kemeny (among several others).

However, not all logical empiricists fully embraced what we now call formal epistemology. Though members of the Vienna Circle and their associates generally accepted the centrality of modern post-Frege formalized logic for the analysis of philosophical problems, the extent to which they deployed formal tools varied. Carnap used formalized logical languages wholeheartedly starting in the 1930s and for the rest of his career. Reichenbach used formal methods extensively but did not restrict himself to logic. Schlick barely touched formal methods in practice and Neurath employed them only early on (see Cat 2019). Hempel and Nagel used only the most rudimentary formal methods—almost never anything beyond first-order logic—but their non-formal discussions of the relevance and value of their formalism remain seminal contributions to the philosophy of science. Jeffrey, Kemeny, and others who developed Carnap's program in inductive logic similarly combined formal and non-formal analyses though their formalism was more sophisticated and also not restricted to logic.

This chapter is intended as a critical but selective overview of the reciprocal influence of logical empiricism and recent formal epistemology on each other, though the emphasis will be on work on formal epistemology within the logical empiricist tradition. In one direction, using

 DOI: 10.4324/9781315650647-44

the figures mentioned in the last paragraph as exemplars, it shows how formal epistemology was central to logical empiricist projects. This assessment should not be controversial. In the other direction, it asks to what extent logical empiricism has contributed to contemporary formal epistemology today. The verdict here must be mixed, as what follows will show. The emphasis throughout this chapter will be on common themes found in the logical empiricist corpus rather than on how individual contributions differ in emphasis.

Evidence, cognitive significance, and confirmation

No discussion of formal epistemology in the context of logical empiricism would be complete without mention of Carnap's *Aufbau* (1928). Carnap followed Russell's dictum of scientific philosophizing ("Wherever possible logical constructions are to be substituted for inferred entities") in an attempt to construct the physical world from an "autopsychological" or phenomenalist basis, more specifically, from elementary experiences and one primitive relation of recollection of similarity. His method of "quasi-analysis" used the logic of the *Principia*.

Yet, no matter how the epistemology of the *Aufbau* should be interpreted, empiricism requires that theoretical claims be based on empirical data. In mature logical empiricism, these data were supposed to be recorded in observation sentences formulated in what Carnap and Neurath called a physical language, that is, as claims about everyday objects, for instance readings on dials of measuring instruments. (This "thing language" is not the language of theoretical physics that Carnap still preferred as basic in 1932; rather it is what we would naïvely take to be our most usual observation language.) Because of this commitment to a physical language also in formal epistemology after the *Aufbau*, no special veridical status was attributed to traditional phenomenalism.

For the logical empiricists, not only was the existence of an appropriate empirical basis central to the acceptability of scientific hypotheses, but it was also central to the meaningfulness of concepts. A criterion of meaningfulness, eventually called cognitive or empirical significance, was supposed to demarcate legitimate science from spurious metaphysics. The problem of formulating a precise criterion turned out to be unexpectedly recalcitrant (see CH. 16). This chapter turns to the problem of theory confirmation which became separated from the problem of significance in the 1930s (see CH. 25).

Formal models of confirmation can be qualitative or quantitative. Both were explored within logical empiricism, the former most actively by Hempel who developed an intuitively plausible view based on the deductive consequences of a theory (see Hempel and Oppenheim 1948). In Hempel's deductive-nomological formal model, if one of the deductive consequences of a theory consisted of a statement that was empirically correct, that fact confirms the theory. If the statement is empirically incorrect, that fact disconfirms the theory. Though this approach is intuitively plausible (besides being deceptively simple), it leads to many well-known paradoxes, for instance, that the hypothesis "All ravens are black" is not only confirmed by a black raven but also by any non-white non-raven. (This follows from the trivial logical fact that a theory $P \rightarrow Q$ is logically equivalent to $\neg Q \rightarrow \neg P$ and, therefore, has the same deductive consequences.) Hempel added less intuitive conditions to his original criterion (which he called Nicod's criterion) in an attempt to fend off such problems, but the model still fell afoul of examples such as Goodman's (1955) grue-predicate. Eventually, Hempel took recourse to a quantitative model of confirmation (Rescher 1997). Hosiasson-Lindenbaum (1940) proposed an alternative way out that finds its natural place in a Bayesian framework (Galavotti 2007).

Quantitative models of confirmation emerged as the holy grail of logical empiricism in the 1940s and 1950s. These were also based on hypothetico-deductivism, with explicit attempts going back to Hahn (1933). In particular, von Mises (1928) and Reichenbach (1935, 1938) developed a frequency interpretation of probability; the latter attempted to construct a theory of confirmation within that context starting in the 1930s (Galavotti 2007). For Reichenbach, probability estimates were always *a posteriori*, obtained through "induction by enumeration," a rule that ultimately must have pragmatic justification. A probability assignment was a "posit," that is, "a statement with which we deal as true, although the truth value is unknown" (1935: 373). Recourse to posits allows the attribution of probabilities to single cases even within a frequency interpretation of probability and allows Reichenbach to pursue a semi-formal account of confirmation (see CH. 28).

Turning to Carnap, in the 1930s, during what is typically referred to as his "syntactic phase," he had held that questions about the extent to which evidence supported a theory were pragmatic and, therefore, not fruitfully studied formally. Yet, from 1941 onwards, Carnap also began a systematic attempt to analyze concepts of probability and to formulate an adequate inductive logic intended to provide a logic of confirmation, a project that occupied him for the rest of his life. Carnap viewed this work as an extension of the semantic methods that he had been developing during the previous decade. There was thus an interesting pattern in Carnap's intellectual development. Until the late 1930s Carnap only viewed syntactic categories as formally specifiable; questions of truth and confirmation were viewed as pragmatic (Sarkar 1992). His conversion to semantics saw the recovery of truth from the pragmatic to the semantic realm. Now, confirmation followed truth down the same pathway.

Unlike Reichenbach, in *Logical Foundations of Probability* (1950), Carnap distinguished between two concepts of probability: "statistical probability," which was the relevant concept to be used in empirical contexts and generally estimated from the relative frequencies of events; and "logical probability," which was to be used in contexts such as the confirmation of scientific hypotheses by empirical data (see CH. 23). Though the latter concept went back to Keynes (among others), Carnap provided its first systematic explication. Logical probability was explicated from three different perspectives (1950):

- As a conditional probability $c(h; e)$ which measured the degree of confirmation of a hypothesis h on the basis of evidence e (if $c(h; e) = r$, then r was determined by logical relations between h and e).
- As a rational degree of belief or fair betting quotient (if $c(h; e) = r$, then r gave a fair bet on h if e correctly described the total knowledge available to a bettor).
- As the limit of relative frequencies in some cases.

According to Carnap, the first of these, which specified a confirmation function ("c-function"), was the concept that was most relevant to the problem of induction. In the formal development of the theory, probabilities were associated with sentences of a formalized language. In *Foundations*, Carnap believed that a unique measure $c(h; e)$ of the degree of confirmation would be found, and he even proposed one (Laplace's rule of succession) though he could not prove its uniqueness satisfactorily. His general strategy was to augment the standard axioms of the probability calculus by a set of "conventions on adequacy" which turned out to be equivalent to assumptions about the rationality of degrees of belief that had independently been proposed by both Ramsey and de Finetti (Shimony 1992).

Subsequently, in *The Continuum of Inductive Methods* (1952), using the conventions on adequacy and some plausible symmetry principles, Carnap managed to show that all acceptable c-functions could be parameterized by a single parameter, a real number, $\lambda \in [0, \infty)$. The trouble remained that there is no plausible *a priori* strategy to restrict λ to some preferably very small subset of $[0, \infty)$. Carnap even speculated that it would have to be fixed empirically. Unfortunately, some higher-order induction would then be required to justify the procedure for its estimation and, potentially, this would lead to infinite regress.

In the 1950s and 1960s Carnap made significant modifications to his scheme for inductive logic. Impressed by the earlier work of Ramsey and de Finetti, Carnap (1971) returned to the second of his three 1950 explications of logical probability and emphasized the use of inductive logic in decision problems. Most importantly for the subsequent development of formal epistemology, in "A Basic System of Inductive Logic" Carnap (1971, 1980) finally recognized that attributing probabilities to sentences was too restrictive. If a conceptual system used real numbers and real-valued functions, no language could express all possible cases using only sentences or classes of sentences. Because of this, he now began to attribute probabilities to events or, equivalently, propositions.

This finally brought some concordance between his formal methods and those of mathematical statisticians interested in epistemological questions. Propositions were identified with sets of models; however, the fields of the sets were defined using the atomic propositions of a formalized language. Thus, though probabilities were defined as measures of sets, they still remain relativized to a particular formalized language. Because of this, and because the languages considered remain relatively simple (mostly monadic predicate languages), much of this work remained similar to the earlier attempts (for more detail, see Hilpinen 1973).

Though Reichenbach and Carnap differed in their interpretations of probability, they agreed on the use of Bayesian methods for updating probabilities related to the relevant measures of confirmation. This brought the methodology of the logical empiricists, particularly inductive logic as construed by Carnap, into greater concordance with standard statistical practice of the 1940s and 1950s. Arguably, most subsequent work on Bayesian epistemology—which is part of the core of formal epistemology today—has these logical empiricist roots, however limited they are (see later in this chapter). In passing, it should be noted that the logical empiricists' attempts at developing accounts of confirmation throughout this period were vehemently rejected by Popper, who endorsed a hypothetico-deductive methodology for theory tolerance based on a (not very successfully) formalized notion of corroboration—but that is beyond the scope of this chapter (for more detail, see Thornton 2006).

Scientific theories, explanation, and inter-theoretic relations

From our contemporary perspective, the interpretation of scientific theories, their structure, their use in explanation, prediction, and so on, and the relations between theories raise a constellation of epistemological and metaphysical considerations that cannot often be easily (or usefully) kept distinct. Given their anti-metaphysics rhetoric, the logical empiricists predictably viewed all cognitively cogent questions about theories as epistemological, questions that could be clarified and perhaps even resolved through logical analysis. For the purpose of this chapter, we will accept this logical empiricist assumption. From this perspective, whenever formal techniques are deployed to address questions about the nature of scientific theories, we will be within the realm of formal epistemology. However, as the following discussion will indicate, formal innovation was largely absent in the logical empiricist work on theories, and it is unclear, though elucidation of formal criteria were omnipresent, whether formal epistemology made any telling contribution.

With regard to the nature of scientific theories, supporters and critics of logical empiricists uniformly agree on one claim: that, by and large, logical empiricism viewed scientific theories as axiom systems (Mormann 2007). The most notable exception from this view was Neurath, but for our purposes, what matters is that even he acknowledged some role for logic and axiomatics in the analysis of scientific theories. The axioms of a scientific theory were supposed to provide representations of the laws of nature formalized as universally quantified sentences (after suitable interpretation). In a good axiomatization, each axiom was also supposed to be logically independent of the other ones. The theory, as a whole, could then be viewed as the deductive closure of these axioms. These assumptions were implicit in the work of Carnap, Reichenbach, and Schlick in the 1920s and 1930s and were made explicit especially by Hempel in the 1940s. In most of later logical empiricist work, these assumptions about the axiomatic structure of a theory were so pervasive that it becomes difficult to attribute them to a single source, though, ultimately, Hilbert's influence from the turn of the twentieth century was clearly important. (The axiomatization of physics was included as part of the sixth of the 23 open problems with which Hilbert tried to set the agenda for twentieth-century mathematics in 1900.) Reichenbach (e.g., 1924) explicitly attempted an axiomatization of the theory of relativity in the early 1920s. However, in later work, though they continued to pay lip service to the value of axiomatization, actually axiomatizing parts of the empirical sciences was no longer a major part to the practice of logical empiricism.

For instance, though piecemeal axiomatization of theories continued to be commonplace enough in mathematical physics in the mid- and late twentieth century, probably because mathematical physics was still influenced by Hilbert, this work was not part of the logical empiricist corpus. In other fields, attempts at axiomatization of parts of biology (in the language of the *Principia*) was largely confined to the efforts of Woodger (e.g., 1937) who did not identify himself as being associated with logical empiricism (despite contributing to the *International Encyclopedia of Unified Science*). In his English textbook of logic, Carnap (1959) adopted Woodger's axioms as an example of axiomatization; however, no further role for axiom systems of empirical science is present in Carnap's work or in that of any other logical empiricist. Instead, with the work of Hempel and Nagel, discussion shifts to non-formal conditions that axiom systems should satisfy to be veridical representations of the laws of nature, for instance, whether a proposed axiom system is of the smallest size among those that have empirical support and the same deductive closure. In other words, the importance of formal considerations pales in comparison to these non-formal ones in the logical empiricist account of scientific theories.

Nevertheless, the assumption that theories consist of axiom systems remains central to logical empiricists' accounts of explanation and explication of inter-theoretic relations. By and large, the logical empiricists endorsed a deductive-nomological (DN) model of explanation (see CH. 19). This model goes back at least to Carnap (1939), though it was first systematically developed by Hempel and Oppenheim (1948). What is to be explained, the *explanandum*, is an individual fact, for instance why Halley's comet was visible in 1986. What does the explaining, the *explanans*, consists of axioms representing laws of nature and initial or boundary conditions, for instance, Newton's Laws of Motion (including Universal Gravitation) and the various masses and positions and momenta of the sun and other planetary bodies at a specified time. The laws of nature are assumed to be deterministic and represented by universal generalizations; however, the initial or boundary conditions would typically be represented as individual sentences. Deterministic explanation occurs when it is shown that that the explanandum is a logical consequence of the explanans.

Thus, the formal structure of a deterministic explanation is quite straightforward, though these formal characterizations must be complemented by a host of non-formal conditions for explanations to be significant (see Salmon 1989). However, the DN model does not capture the structure of statistical explanations in science (including those provided by quantum mechanics). Hempel and Oppenheim (1948) produced a structurally similar inductive-statistical (I-S) model in which a statistical law replaces the deductive one in the explanans. Their account of how the *explanans* entails the *explanandum* remained verbal. Hempel was never fully satisfied with this model and, over the years, it has found few defenders. Even within logical empiricism, even if we include those who were only partly sympathetic to it, Hempel and Oppenheim's account was challenged by models such as those of Salmon (1971).

Theory reduction was the most important inter-theoretic relation that drew the attention of the logical empiricists. Three different models were offered. The most influential of these was that proposed by Nagel, initially in the 1940s, and systematized in the 1960s (Nagel 1961) though Woodger (1952) independently made a similar proposal. This model is best viewed as a modification of the DN model of explanation in which the *explananda* consists of the laws of the theory to be reduced. The *explanans* include the laws of the reducing theory. The most important type of reduction is one in which the reduced theory is one about composite wholes and the reducing theory one about their constituent parts. For reduction to be carried out, following the DN model, the reduced theory must be derived from the reducing one. Nagel called this the condition of derivability. However, to attempt a derivation, the terms of the reducing and reduced theories must be systematically connected to each other through what were called bridge laws. Nagel called this the condition of connectability. Nagel supplemented these two formal requirements on reduction with a penetrating discussion of non-formal conditions that reductions must satisfy to be of scientific value. In particular, he stressed that reduction was not to be viewed as a replacement of the reduced theory by the reducing one; rather, reductions showed why, and to what extent, reduced theories were correct and should continue to be used in practice. From the Nagelian perspective, there is no motivation to discard the ontology of the reduced theory. Rather, reduction provides epistemological warrant for using that ontology with greater confidence in its appropriate domain (Sarkar 2015).

An alternative to the Nagel model of reduction was proposed by Kemeny and Oppenheim (1956), who eschewed any direct relationship between the reduced and reducing theories, though both would have to be connected to the same observational domain. Rather, they proposed that a reducing theory must be better "systematized" than the theory it reduces; by this they meant that the former must either be simpler than the latter or have greater explanatory strength. Simplicity and explanatory strength were not formalized. Using this model of reduction, Oppenheim and Putnam (1958) defended the unity of science through reduction. They recognized six levels of organization: social groups, (multicellular) living things, cells, molecules, atoms, and elementary particles. At each level, the theories at that level were better systematized than those immediately above it. Thus, reduction was supposed to be possible and lead to ontological parsimony. Surprisingly, it turns out that Kemeny and Oppenheim's model imposes weaker requirements on reduction than Nagel's model; moreover, as Sarkar (1998) and others have pointed out, the Kemeny–Oppenheim model is a recipe for theory replacement rather than reduction. It appears to be the only formal model of theory replacement offered within the logical empiricist canon. Pointedly, Oppenheim and Putnam did not explicitly call for immediate elimination of upper levels to lower ones.

In contrast to both these models, Suppes (1957) proposed a third alternative: reduction requires the establishment of a suitable isomorphism between the mathematical structures of two theories (characterized set-theoretically). Like Kemeny and Oppenheim's model, this one also requires that the intended structures have the same observational domain. This intriguing view was further developed by Balzer and Dawe (1986), among others, in the 1980s (Sarkar 1998).

Legacy

In the 1950s, Suppes, himself a student of Nagel's, is supposed to have coined and popularized the slogan: "the correct tool for philosophy of science is mathematics, not metamathematics" (van Fraassen 1989: 221). Suppes's dictum can be interpreted in different ways. Van Fraassen takes Suppes to be rejecting the view of theories outlined in the last section, which was derogatorily labeled as "syntactic." The contrast is supposed to be a "semantic" view of theories in which a theory is not a syntactic structure but a set of models. Whether the semantic view of theories is coherent or, in any sense, an improvement over the logical empiricist view remains a matter of dispute. But the issues germane to that dispute are beyond those that directly concerned the logical empiricists (or this chapter). Suffice it here to note that, to the limited extent that the semantic view of theories has received worthwhile formal analysis, the techniques have been identical to those espoused by logical empiricists. But it remains doubtful whether formal analysis has had much to contribute in this context.

However, the most serious objection to the logical empiricist interpretation of theories and DN explanation is that derivations in science rarely resemble logical deductions. Rather, the formal techniques used involve taking limits, making approximations, etc. The epistemological status of these practices, that is, their reliability and warrant, remains murky and in need of rigorous justification (as Hilbert had realized when he also made this issue part of the problem of axiomatizing physics in 1900). In his discussion of non-formal conditions relevant to theoretical reasoning, Nagel (1961) also acknowledged this issue and underscored its importance. Formal epistemology has only recently begun the process of reconstruction of these processes, and what has been achieved remains rudimentary (Sarkar 1998, 2015). However, these new formal analyses embrace all of mathematics and are not limited to techniques of logic; nor is there any stake in translation of the relevant formal structures into logical languages, and this is another, perhaps more fruitful, way of interpreting Suppes's dictum. The influence of logical empiricism in these developments is indirect and somewhat ironic: while the issues came to be framed in this way because of the DN model of explanation, the new formal analyses have become interesting because of a perceived need to go beyond that model (and reject it to some extent).

When we turn to recent formal analyses of reduction, we begin to see a much more lasting positive contribution of logical empiricism. Following the influential critique of Wimsatt (1976), who offered a different model of reduction, Nagel's model was viewed to be irrelevant and treated with disdain for a generation during a period when the philosophy of science was dominated by a vocal rejection of logical empiricism. However, for the last two decades, a new generation of philosophers interested in reduction, many of whom are explicitly associated with formal epistemology, have defended and revived Nagel's model. Today it is considered by most commentators as the most appropriate formal model of reduction available, even though it has needed correction, particularly with respect to the distinction between scientific derivation and logical deduction noted earlier (see Sarkar 2015). Much of this work has been accompanied by a reassessment of the achievements of logical empiricism and a growing recognition by a new generation of philosophers of science that the post-positivist philosophy of science of the 1970s and 1980s contributed little of value compared to logical empiricism itself (Mormann 2007).

However, the degree to which logical empiricism has influenced contemporary formal epistemology must primarily be judged by its role in discussions of scientific inference, because those are central to the latter discipline today. It is probably uncontroversial that Bayesian analyses are central to contemporary discussions of scientific inference and other topics within formal epistemology (Sprenger and Hartmann 2019). Given that central figures in formal epistemology within logical empiricism, most notably Carnap and Reichenbach, endorsed Bayesian analysis, it is tempting to attribute its centrality in contemporary formal epistemology to the logical empiricists. While at least some such influence is probably present (as noted earlier), there are ample grounds for caution about making facile claims of omnipresent logical empiricist influence.

Work in Bayesian epistemology has a continuous history going back at least to the 1960s, for instance in Jeffrey's work on Bayesian decision theory. Jeffrey was a student of Carnap and Hempel, but much of the work on Bayesian epistemology was also critically influenced by scientists and statisticians such as Harold Jeffreys and I. J. Good, who are not associated with logical empiricism. Since the 1990s, there has been an explosion of Bayesian statistics within the sciences, particularly in biology. However, philosophical considerations have played no visible role in this veritable Bayesian revolution. Rather, it has been driven by computational innovations, especially the use of Monte Carlo sampling techniques to estimate likelihood functions. What this work has arguably shown is the relative irrelevance of philosophical problems of Bayesian inference in practical scientific contexts. One example will suffice: philosophers have long debated the proper way to select a prior probability distribution during a first Bayesian analysis of a problem. What the new computational Bayesianism has shown is that the choice of prior is typically of marginal relevance in almost all experimental contexts. As a consequence of such developments, there has emerged ample scope for Bayesian statistics and Bayesian formal epistemology to cross-fertilize each other. But logical empiricism *per se* will have had little to do with these developments.

References

Balzer, W. and Dawe, C. M. (1986) "Structure and Comparison of Genetic Theories. 2: The Reduction of Character-Factor Genetics to Molecular Genetics," *British Journal for the Philosophy of Science* 37: 177–91.

Carnap, R. (1928) *Der Logische Aufbau der Welt*, Berlin: Weltkreis-Verlag. Trans. *The Logical Structure of the World*, Berkeley: University of California Press, 1967. Repr. Chicago: Open Court, 2003.

——— (1939) *Foundations of Logic and Mathematics*, Chicago: University of Chicago Press.

——— (1950) *Logical Foundations of Probability*, Chicago: University of Chicago Press, 2nd ed., 1962.

——— (1952) *The Continuum of Inductive Methods*, Chicago: University of Chicago Press.

——— (1959) *Introduction to Symbolic Logic and Its Applications*, New York: Dover.

——— (1971) "A Basic System of Inductive Logic. Part I," in R. Carnap and R. C. Jeffrey (eds.), *Studies in Inductive Logic and Probability*, Berkeley: University of California Press, vol. 1, pp. 33–165.

——— (1980) "A Basic System of Inductive Logic, Part II," in R. Carnap and R. C. Jeffrey (eds.), *Studies in Inductive Logic and Probability*, Berkeley: University of California Press, vol. 2, pp. 5–31.

Cat, J. (2019) "Neurath and the Legacy of Algebraic Logic," in J. Cat and A. T. Tuboly (eds.), *Neurath Reconsidered: New Sources and Perspectives*, Cham, Springer, pp. 241–338.

Galavotti, M. C. (2007) "Confirmation, Probability, and Logical Empiricism," in Richardson and Uebel (2007), pp. 117–35.

Goodman, N. (1955) *Fact, Fiction, and Forecast*, Cambridge, MA: Harvard University Press.

Hahn, H. (1933) *Logik, Mathematik und Naturerkennen*, Vienna: Gerold & Co. Trans. "Logic, Mathematics, and Knowledge of Nature," in B. McGuinness (ed.), *Unified Science*, Dordrecht: Reidel, 1987, pp. 24–45.

Hempel, C. G. and Oppenheim, P. (1948) "Studies in the Logic of Explanation," *Philosophy of Science* 15: 135–75. Repr. with a postscript in Hempel, *Aspects of Scientific Explanation and Other Essays*, New York: Free Press, 1965, pp. 245–96.

Hilpinen, R. (1973) "Carnap's New System of Inductive Logic," *Synthese* 25: 307–33.

Hosiasson-Lindenbaum, J. (1940) "On Confirmation," *Journal of Symbolic Logic* 6: 133–48.

Kemeny, J. G. and Oppenheim, P. (1956) "On Reduction," *Philosophical Studies* 7: 6–19.

Mormann, T. (2007) "The Structure of Scientific Theories in Logical Empiricism," in Richardson and Uebel (2007), pp. 136–62.

Nagel, N. (1961) *The Structure of Science: Problems in the Logic of Scientific Explanation*, New York: Harcourt, Brace and World.
Oppenheim, P. and Putnam, H. (1958) "The Unity of Science as a Working Hypothesis," in H. Feigl, M. Scriven and G. Maxwell (eds.), *Concepts, Theories, and the Mind-Body Problem*, Minneapolis: University of Minnesota Press, pp. 3–28.
Reichenbach, H. (1924) *Axiomatik der relativistischen Raum-Zeit-Lehre*, Braunschweig: Fried. Trans. *Axiomatization of the Theory of Relativity*, Berkeley: University of California Press, 1969.
——— (1935) *Wahrscheinlichkeitslehre*, Leyden: Sijtho.
——— (1938) *Experience and Prediction*, Chicago: University of Chicago Press.
Rescher, N. (1997) "H2O: Hempel-Helmer-Oppenheim, an Episode in the History of Scientific Philosophy in the 20th Century," *Philosophy of Science* 64: 334–60.
Richardson, A. and Uebel, T. (eds.) (2007) *The Cambridge Companion to Logical Empiricism*. Cambridge: Cambridge University Press.
Salmon, W. C. (1971) *Statistical Explanation and Statistical Relevance*, Pittsburgh, PA: University of Pittsburgh Press.
——— (1989) *Four Decades of Scientific Explanation*, Minneapolis: University of Minnesota Press.
Sarkar, S. (1992) "'The Boundless Ocean of Infinite Possibilities': Logic in Carnap's *Logical Syntax of Language*," *Synthese* 93: 191–237.
——— (1998) *Genetics and Reductionism*, New York: Cambridge University Press.
——— (2015) "Nagel on Reduction," *Studies in History and Philosophy of Science* 53: 43–56.
Shimony, A. (1992) "On Carnap: Reflections of a Metaphysical Student," *Synthese* 93: 261–74.
Sprenger, J. and Hartmann, S. (2019) *Bayesian Philosophy of Science*, Oxford: Oxford University Press.
Suppes, P. (1957) *Introduction to Logic*, New York: Van Nostrand.
Thornton, S. P. (2006) "Karl Raimund Popper," in S. Sarkar and J. Pfeifer (eds.), *The Philosophy of Science: An Encyclopedia*, New York: Routledge, pp. 571–8.
van Fraassen, B. (1989) *Laws and Symmetry*, Oxford: Clarendon Press.
von Mises, R. (1928) *Wahrscheinlichkeit, Statistik und Wahrheit*, Berlin: Springer, 3rd ed., 1951. Trans. *Probability, Statistics, and Truth*, London: George Allan & Unwin, 1957, repr. New York: Dover, 2003.
Wimsatt, W. C. (1976) "Reductive Explanation: A Functional Account," in R. S. Cohen, C. A. Hooker and A. Michalos (eds.), *PSA 1974*, Dordrecht: Reidel, pp. 671–710.
Woodger, J. H. (1937) *The Axiomatic Method in Biology*. Cambridge: Cambridge University Press.
——— (1952) *Biology and Language*, Cambridge: Cambridge University Press.

40
CARNAP'S CONCEPTION OF REASON

A. W. Carus

Richard McKeon, chair of the philosophy department at the University of Chicago, had a rather low opinion of his colleague Rudolf Carnap. In McKeon's classification system, one criterion ranked philosophers from the "holoscopic" ("whole-seeing") to the "meroscopic" ("part-seeing")—and in McKeon's view, Carnap was at the "meroscopic" extreme. He was not a big-picture mind, in other words, but worked only on small, isolated technical problems.

The recent revival of interest in Carnap and logical empiricism has shown McKeon's judgment to be very wide of the mark. In this chapter, we will see that Carnap devised a new approach to one of the oldest and most central problems of the western intellectual tradition, the problem of how to make the idea and the practice of "reason" or "reasoning" precise enough to bring it to bear on the urgent problems faced by our species. This had been Plato's task, of course, and he had tried to address it by applying the newly discovered art of geometrical reasoning from a few obvious axioms to highly counterintuitive—but true—conclusions. This same kind of reasoning, Plato thought, could be applied to all the other big problems of life.

This Platonic approach appeared to be triumphantly vindicated when Galileo, two millennia later, applied it to falling bodies and the trajectories of projectiles, convincing himself and many readers that the book of nature was written in mathematics. Within a century, Newton had taken this idea so far that he was able to establish, for the first time ever, something like secure knowledge of the place of our world in the universe. The response to this and parallel developments in other sciences was a movement—the Enlightenment—that stood for replacing traditional forms of knowledge by the new, more scientific and empirical knowledge, and using it to underpin personal autonomy by overcoming passive acceptance of faith, tradition, authority, and popular prejudice through the use of reason.

But there was a major catch, which was that the leading lights of the Enlightenment couldn't agree on what they *meant* by "reason." Since at latest the rival conceptions of Hume and Kant, the very concept of reason itself—the Enlightenment's guiding inspiration—has been up for grabs. The model for both Hume and Kant was Newtonian science, but what they extrapolated from that model couldn't have been more different. Hume saw reason as the skeptical habit of sticking to the facts and not letting our speculations get too far beyond that secure bedrock. Kant, on the other hand, saw those facts themselves as largely shaped by our interpretations of them, so he gave more room to the overall conception of reason (*Vernunft*) of which scientific

DOI: 10.4324/9781315650647-45

reason—sticking to the facts and reasoning from them mathematically—was a mere subsidiary part, which he called the "understanding" (*Verstand*).

Nor has there been any convergence since then. The Lockean or Humean suspicion of any reasoning that isn't strictly mathematical led to nineteenth-century positivism, while Kant's more constructivist approach to reason led to German idealism. Scientists mostly stayed out of the struggle between these two increasingly entrenched positions. Most were instinctively on the side of the skeptics and empiricists, but some notable figures, especially Hermann von Helmholtz, tried to steer a middle way between the extremes.

Today "reason" remains disputed territory (e.g., Nozick 1993), but the historical context has faded into the background. Attention is focused on "rationality" and on such questions as whether economic rationality, in any of its more common axiomatic explications, is itself "reasonable" or "rational" in some larger sense, or whether any of the axiomatic explications of rationality underlying economic theory or decision theory is realized in actual human decision making (Bermúdez 2009). What tends to get lost, though, is the task (so central to the Enlightenment and even to nineteenth-century positivism) of finding a place for values—normative standards and arguments—within the normatively indifferent world described to us by science. (Huw Price [2013] calls these "placement problems.")

And logical empiricism? If it is seen as having a conception of reason at all, logical empiricism is put at the extreme positivist end, allowing *only* scientific reason and excluding all else as frivolous. In Kantian terms, logical empiricism supposedly claimed that there is no more to *Vernunft* than just *Verstand*: scientific reason is all the reason there is. Even Comte, Mill, and Ostwald had allowed a role for "non-cognitive" discourse, meaningful language in which values, art, difficult choices, literary treatments of moral quandaries, and so on could be coherently discussed. The logical empiricists seemed to deny this.

In Carnap's case, though, we will see that this first impression is highly misleading, and that his conception of reason was more of a middle way along the lines of Helmholtz—that he tried, in a rather ingenious way, to find a synthesis between the Kantian and the Humean tendencies that had dominated the previous two centuries.

Logic and Kantian noncognitivism

Carnap is best known for bringing logic to bear on philosophical questions, notably the question of what should even count as a question. This idea goes back to Aristotle, whose logic, however, suffered from a major defect: it could not spell out the reasoning used even in ancient geometrical proofs, never mind later mathematics. This did not become a big issue until Kant, who concluded that since logic could not account for mathematical reasoning, the foundations of mathematics must be "synthetic a priori"—they express basic features of the structure of the world around us that are not inherent in the things themselves but are projected onto the things (the world) by us. With this "Copernican revolution," Kant set the agenda for the next century or so of philosophy.

Carnap, too, grew up within a philosophical discourse still dominated by Kantian questions. But he benefited greatly from the nineteenth-century progress in clarifying these. Among other advances, logic had finally caught up with mathematics, and Frege had been able to show that arithmetic (Kant's prime example of the synthetic a priori) was not synthetic at all but analytic; you could after all prove by logic alone (or close enough) that 7 + 5 = 12, which Kant had denied. There were complications, of course, but by the time Carnap studied with Frege in Jena in 1911–13, the new logic had become a powerful tool for the clarification of philosophical problems—as Russell would programmatically argue in his book *Our Knowledge of the External*

World as a Field for Scientific Method in Philosophy (1914), which put Carnap on the path that would lead him to his own first major book, the *Aufbau*.

The *Aufbau* became the paradigm of the Vienna Circle's program of "rational reconstruction," applying "reason" (in the form of logic) to the reconstruction of relatively vague, informal concepts (in the *Aufbau* case, the empirical basis of everyday and scientific knowledge). This placed the Vienna Circle squarely in the reconstructive ethos of the *Encyclopédie* and then the *École polytechnique* of the early nineteenth century, which radicalized the aspirations of the encyclopedists to reform and replace colloquial language and its impoverished, backward-looking repertoire of concepts by a better, more scientific, and more rational one. This program had been most explicitly initiated by d'Alembert's student Condorcet, who in turn transmitted it to Saint-Simon and thence to Comte. As a conception of reason, it is self-consciously revolutionary. The idea is that since humanity has embarked on a new phase, it requires a new conceptual medium in which to organize and conduct social, political, and educational life. And rather than relying on the messy categories of the languages and concepts handed down to us, we must self-consciously *construct* our new medium of cognitive, moral, and political interaction.

The Carnapian program of rational reconstruction was even more revolutionary, as it proposed not only to reconstruct the human conceptual apparatus, but to do so within the framework of a single, overarching deductive system. It envisaged a piecemeal replacement of vague inherited concepts, one by one, by concepts that had a place in the unified science expressed in a deductive system, the "language of science." Participation in this language offered not only an escape from the concepts and categories of traditional society and the weight of the authoritarian past, but an escape from the prison of subjectivity, since instead of, e.g., merely feeling hot or cold, the rational reconstruction of such subjective concepts in objective (intersubjective) science offered quantitative tools for measurement (temperature, in this case), allowing for mutual agreement on publicly available standards.

While this new conception of reason took the Enlightenment and Comtean tradition a step further, then, it also exacerbated some of the weaknesses of that tradition. This provoked a reaction against its one-sided emphasis on (its own very clearly defined conception of) reason, for this emphasis appeared to come at the expense of *applying* reason to practice; it seemed to exclude reasonable discourse in which reason could be applied to argument about human conduct, e.g., legal or political argument, or reasoning about the advantages of different ideals, or their articulation in literature and art. The reaction provoked by this apparent one-sidedness is epitomized by Horkheimer and Adorno in their wartime book *The Dialectic of Enlightenment* (1947), an explicit attack on the Vienna Circle. They point out that while the Enlightenment with its scientifically oriented conception of reason began as an anti-establishment, anti-traditional opposition movement, it then became the ideology of the establishment, and was assimilated to *what is*—rather than leaving room for better alternatives that *might be* or *should be*. How can this establishment ideology be the basis for reasoned argument for a better world? Horkheimer and Adorno (among many others) argued for a broader and less tainted conception of reason, reminiscent of Kant's *Vernunft*, in post-Kantian versions such as those of Hegel or Marx.

Carnap, however, retained a characteristic attitude—barely visible at times through the fog of Vienna Circle rhetoric, but nonetheless unmistakable—that could be called a kind of "functionalism" (Creath 1994; Price 1997), in which the Vienna Circle pronouncements about the "meaninglessness" of ethical and value statements—thrown into one pot with metaphysics, it seemed—are not to be taken literally. These remarks, as Carnap later admitted (e.g., 1963: 45), should have been put differently: while normative statements lack *cognitive* meaning, they do have a *different sort* of meaning—they have a different *function* from the cognitive one of conveying true (or false) statements. In effect, he was already pursuing this strategy in the few passages about values we have from the Vienna Circle period (e.g., Carnap 1934b; see Richardson 2007). This same view was widely discussed in Britain after G. E. Moore in 1905 identified and criticized the "naturalistic fallacy" (the assumption that cognitive meaning is the only kind of meaning there could be) in his *Principia Ethica*. It was put on a more secure basis by R. M. Hare and others in the 1950s. In his own later exposition of it, Carnap (1963) would call such a view "noncognitivist," a name that stuck.

There is an obvious parallel, then, between this noncognitivism about normative statements and Carnap's better known "noncognitivism" about logic and mathematics, derived from Wittgenstein. Here, too, the tautologies of logic (which the Vienna Circle, by dint of their logicism, extended to the whole of mathematics) were meaningless, in the sense that they did not in themselves refer to anything, but merely had the *function* of reorganizing the cognitively meaningful sentences of the empirical sciences.

This functionalist conception of logic and mathematics, underlying Carnap's famous distinction between analytic and synthetic statements, survived into Carnap's later conception of reason and was notoriously misunderstood by Quine (Creath 1994). But even the earlier, rational-reconstruction functionalism rejected any form of metaphysical realism as firmly as it rejected idealism or any other notion of an ultimate reality; in this sense, Carnap continued to embrace a "Kantian humility." No metaphysical conception of ultimate reality is compatible with a genuine functionalism (Price 1997), since the claim that one function identifies or connects with "reality" privileges it over other functions, implying that they are reducible to it. Functionalism dispenses with this; the "placement problems" disappear. This radically anti-ontological aspect of Carnap's conception of reason may also have been what so disturbed Horkheimer and Adorno, who shared the widespread prejudice that a conception of ultimate reality is indispensable to any idea of reason. In any case, Carnap arrived at an entirely new approach in 1931–2 that put him in a position to overcome the one-sidedness of rational reconstruction, and led to yet another new conception of reason.

Functional pluralism and explication

In late 1932 Carnap abandoned his insistence on a single canonical language of science and instead embraced his "principle of tolerance." Up to then, though conventional stipulation had played a major role in the construction of scientific knowledge, the underlying logic had remained fixed. Like Wittgenstein and all members of the Vienna Circle, Carnap had assumed that the Russellian theory of relations, situated within some form of type theory, just *was* logic. When they became aware of Brouwer in the later 1920s, and Hilbert attempted to incorporate elements of Brouwer's intuitionism into his finitist metamathematics, different forms and articulations of logic began to proliferate. The question became, which of these was *right*? Carnap began by participating in this debate (Carnap 1931), just as he began by participating in the Vienna Circle's "protocol-sentence debate" about the *right* way to represent observation statements (Uebel 2007). But he soon saw that in both these debates, the question wasn't one of "correctness" (or adequacy to some "reality" out there), but rather a question of which *language* should be preferred. In the mathematics case, should it be a formalist, an intuitionist, or a logicist one? In the protocol-sentence case, should it be a phenomenalist or a physicalist one? But, Carnap realized, these were not cognitive (descriptive) questions; they were *normative* questions—and "in logic there are no morals," as he would soon put it in his best-known formulation of the principle of tolerance (1934a/1937: 45).

This principle would eventually provide a much more suitable basis for Carnap's functionalist conception of language and meta-discourse than rational reconstruction. In the short term, though, the functional nature of the overall conception of reason was obscured at least partly by Carnap's own rhetoric, as he persisted in the misleading rhetoric he would later come to regret, calling normative and value statements "meaningless" without qualification.

It took time even for Carnap himself to assimilate and accommodate the far-reaching consequences of his principle of tolerance; in fact, he never really arrived at a clear formulation of the functional pluralism implicit in his later writings. The functionalist conception of logic and mathematics was expressed clearly enough in his famous paper "Empiricism, Semantics, and Ontology" (1950a), and his (functionalist) conception of noncognitivism about value statements was outlined in his reply (1963) to Kaplan. In the opening chapter of *Logical Foundations of Probability* (1950b), rational reconstruction is rearticulated within the new setting of logical and linguistic tolerance, and renamed *explication*. Superficially, there is a great deal of continuity between rational reconstruction and explication, but there are also fundamental differences. Carnap now gives much more attention, for instance, to the *clarification* of the explicandum that must precede its actual replacement by a more precise explicatum, and also makes clear (as he hadn't previously) that there can be no criteria of "correctness" of an explicatum; whether an explicatum is appropriate or suitable or adequate to serve the purposes previously served by the explicandum is a *practical* question, i.e., a normative one, not a cognitive one.

As Howard Stein, a student of Carnap's, later pointed out, this means that explicandum and explicatum are framed in different languages; so in the terms of "Empiricism, Semantics, and Ontology," the question of the suitability of an explicatum for an explicandum is an *external* question. Furthermore, as Stein points out, this gives rise to an implicit *dialectic* between the cognitive and practical realms; on the one hand, our knowledge informs us of the consequences of our choices and enables us to make informed choices, but on the other hand, we use pragmatic or normative criteria to choose explications of cognitive concepts (Stein 1992; Carus 2007). In short, we use the cognitive to inform the practical and the practical to articulate the cognitive. There is a continuous mutual feedback between the two.

This conception of reason goes a considerable way toward allaying the worries expressed about the "dialectic of enlightenment" inherent in rational reconstruction by critics such as Horkheimer and Adorno. Science is certainly still regarded as the model of cognitive rationality, in Carnap's conception, but cognitive rationality is no longer (as it seemed to be under rational reconstruction) the whole of reason; it becomes a subordinate part of the overall dialectical give and take. A minimal framework of the Kantian relation between *Verstand* (scientific rationality) and *Vernunft* (reason) is thereby re-established, but without any metaphysical *Anfangsgründe* (foundations) of science on the one side, and without any groundwork of the metaphysics of morals on the other—i.e., without any aspiration to arrive at a unique highest principle of morality (such as the categorical imperative) by reason alone. We have a "relativized a priori" (Friedman 2001) on *both* sides, which are in turn relativized to each other. The Kantian *framework* has been freed of the residual metaphysical lumber of the Kantian *doctrine*.

Clearly specified linguistic frameworks are constitutive of rationality in the later Carnap. Our concepts have meaning only relatively to a particular framework, and we can only ask about the existence of this or that item, or the truth of this or that statement, within the terms of such a framework (Carnap 1950a). Many have thought (e.g., George 2012; Steinberger 2016) that in a Carnapian setting there can therefore be no rational way of choosing *among frameworks* (or, on a more local scale, *among* candidate explications of particular informal concepts), or any rational basis of values. In the absence of any guidance from a categorical imperative or some substitute, how are we to choose? And doesn't this also make the principle of tolerance then actually rather empty, since the choice of explication or framework can be arrived at only from irrational momentary whim rather than reasoned deliberation? Doesn't Carnap's noncognitivism preclude rational argument about choices of explications or frameworks, let alone about values?

Such questions overlook two things. First, they overlook Carnap's subordination of scientific rationality to a larger realm of reason, as just described; and second, they miss the dialectic implicit in Carnap's conception between the cognitive (theoretical) and the normative (practical), pointed out by Stein. Regarding the subordination of *Vernunft* to the interplay between *Verstand* and *Vernunft*, it is true that Carnap rarely alludes to it, even implicitly, and it is not easy to discern in his later writings. But there is now explicit textual evidence for it. In a draft continuation of his reply to Kaplan recently found in his notes, after defining value functions for specific, narrowly focused values, and then defining "comprehensive value functions" on that basis, Carnap asks "Are there also *standards of rationality for value functions?* The . . . standards of inductive logic are not applicable here. The acceptance of a value function is completely independent of factual questions" (2017: 192, orig. emphasis). And then he answers his question in the affirmative: "Although all logic, including inductive logic, and factual knowledge are irrelevant, it nonetheless seems to me that there are other, *purely valuational criteria* by which to judge a value function as more or less rational than another" (ibid.: 193, emphasis added). There is thus an asymmetry; comprehensive value functions guide the selection of explications and language frameworks, but the knowledge framed in these can only *inform* the choice of framework; it cannot make the choice, which is guided rather by "purely valuational criteria"—*Verstand* is dialectically subordinate to (*Verstand*-informed) *Vernunft*.

The second overlooked aspect of Carnap's approach (the dialectic between the cognitive and the normative) seems to have remained implicit in Carnap's later thought without ever being fully articulated. Certainly it is not in contradiction with anything he actually said, and a number of persistent themes in his later writings are hard to reconcile any other way. But it should perhaps be regarded as a sympathetic extension or continuation of Carnap's thought by Stein and others rather than attributed to Carnap himself.

Carnap certainly conceived of practical choices, including those among frameworks or among explications, as subject to rational guidance and reasoned deliberation; he made this quite explicit (1963: 982; see also Carus 2017: 172) and also made clear that our provisionally most advanced scientific knowledge should be brought to bear on such choices. In Carnap's conception of meta-discourse, there is a category for this form of deliberation; he called it "pragmatics," a category of meta-discourse distinct from syntax and semantics but of equal importance—in principle at least. In Carnap's actual writings, it is hardly obvious on the surface how central pragmatics had in fact become, and that it now included not only epistemology, methodology, practical philosophy, and much of the philosophy of language (indeed, the philosophy of philosophy), but all *meta*-semantic deliberation about choices of frameworks or explications. When Stein, as a graduate student at the University of Chicago in the 1940s, pointed this out to him in a student paper, Carnap agreed enthusiastically (Carus 2010: 619–20). As in syntax and semantics, Carnap distinguished pure from descriptive pragmatics. The empirical knowledge brought to bear on the deliberation process of clarifying the explicandum and choosing among candidate explications belong to descriptive pragmatics—including, for instance, the entire history of science, the field of descriptive pragmatics that Stein himself would most intensely focus on (Carus 2010, 2013). It includes all the empirical knowledge we have, including all our knowledge about knowledge.

Naturalism without metaphysics

This Carnapian conception of reason has been criticized from various perspectives. The best-known critique was Quine's; in the present context, Quine's salient claim was that because Carnap allowed for an external perspective on science, he left room for a metaphysical "first philosophy" *outside* of our actual knowledge. In Quine's view, Carnap did not rest content with a standpoint internal to science, *in mediis rebus* (Dreben 1994), but insisted on stepping outside the "conceptual scheme" of working science and appraising it from an external perspective.

In fact, Carnap considered our cognition to be just as entangled, just as immersed *in mediis rebus*, as Quine did. For Carnap, though, language users are also *participants* in the world that their language is about. They are not just representers, and consumers of representations; they are also *agents* in the world so represented. And unlike Quine, Carnap thought that we *can* step back from our entanglement in this messy predicament of having to act and facing choices. We can abstract from it and isolate certain of its features from our immediate, deeply entangled experience. We can separate out a purely representational (or semantic) component of language from its pragmatic embeddedness in a world of agent-users. (Had we been incapable of this, our species would have been trapped from the outset in Parmenidean paralysis; cognitive life could never have got off the ground.) For everyday scientific and technological purposes, this semantic abstraction works smoothly and requires no help from pragmatics. Indeed, the cognitive progress of the species depends, to a large degree, on our institutionalization of this and similar feats of abstraction. But when the need for explication or language change arises, semantics is no longer self-sufficient; it needs to resort to the context of clarification afforded by descriptive pragmatics.

Quine, who excluded *all* external discourse, thought our only recourse at such points is a retreat to the "mother tongue"—our colloquial language, suitably trimmed. Carnap *also* excluded external discourse, but for him this exclusion extended only to semantics, where the user-as-agent is abstracted away. Faced with a problem that points beyond semantics, a problem that puts the concepts themselves or the constitutive rules of the relevant language into question, we retreat—not to a "mother tongue," but to *pragmatics*, where the user is once again an actor in the world represented, and there are choices to be made. Friedman's (2001: 47–68, 105–15) argument to the effect that object-level Kuhnian incommensurability, e.g., during scientific revolutions, can coexist with meta-level rationality, where the meta-level is that of philosophical or meta-scientific debate—e.g., the debate about geometry between Poincaré and Helmholtz as a backdrop to Einstein's introduction of relativity—makes essentially this same point; Carnap would have classified such meta-scientific debate as belonging to descriptive pragmatics.

The inescapability of choice does not, in Carnap's view, force us to retreat to the arbitrariness and inconsistency of ordinary language. Pragmatics, unlike the mother tongue, affords us tools to override the intuitions anchored in ordinary language when we need to. (Just as language itself, especially written language, gives us tools to transcend the limitations of immediate subjective consciousness.) All the tools of scientific inquiry (including those of syntax and semantics), and all the tools of practical deliberation are at our disposal in the realm of pragmatics. Descriptive pragmatics draws on our entire fund of knowledge, including not only empirical and theoretical scientific knowledge, but all our tools of logical and mathematical reasoning, as well as our knowledge about knowledge (how we got it, how secure it is, how its parts relate to each other, etc.). Pure pragmatics—explication and formalization of practical concepts and normative languages of practical decision—enables us, by whatever standards we set ourselves, to understand and act in conformity with the consequences of our provisionally chosen values (and to revise them if we cannot accept those consequences).

In this limited pragmatic sense, though, Carnap *did*, in contrast to Quine, insist on the availability of an external perspective, from which criticism of our knowledge and our language is possible (e.g., George 2012). An example is his posthumously published discussion of entropy in thermodynamics (1977), where he criticizes some then-recent developments in physics from both a logical and a history-of-science perspective, i.e., using descriptive pragmatics. Isn't such a use of logic and factual knowledge excluded, though, by Carnap's statement (in the "Value Concepts" fragment mentioned earlier) that "all logic, including inductive logic, and factual knowledge are irrelevant" to any constraints on the rationality of comprehensive value functions? But this can hardly be taken literally, as no such constraint can be expressed in "pure optatives" alone, i.e., without some factual component, as Carnap recognized in labeling that component "descriptive pragmatics."

Nor is it circular to use logic or facts as parts of descriptive pragmatics. What we use at the pragmatic level, for purposes of choosing among frameworks or explications, as our working conception of reason, is framed in a provisionally fixed language. We have at our disposal all the knowledge available via (i.e., framed in the terms of) that provisionally fixed meta-framework, to inform us, as well as is currently possible, about the consequences of selecting any given framework or any given explication. This does not mean that either the meta-framework or the knowledge framed in it is inviolate or frozen in place; it is a provisional starting point. (As Stein puts it, quoting an apothegm of William Blake, "Reason or the ratio of all we have already known, is not the same that it shall be when we know more.") Nor can the meta-framework or the knowledge expressed in it be applied inductively to select a framework that could, in turn, define what is to be counted as knowledge. The role of descriptive pragmatics can never be more than supportive and "heuristic." Neither the logical and empirical knowledge employed as descriptive pragmatics nor the reasoning used in pure pragmatics can decide the issue; it can only inform people of the consequences of their choices and reveal possible courses of action that unaided intuition might overlook.

To some philosophers in the Kantian tradition such as Rawls or Habermas, in fact, such a Carnapian external perspective is far too weak and unambitious. It fails to give practical reason enough scope to yield substantive principles of morality, even of the extremely formal kind that Kant found in the categorical imperative. Complaints about Carnapian rationality from this viewpoint are just the opposite of Quine's. And while it is true that Carnap, in contrast to Quine, left room for an external perspective, even a critical and valuational perspective, on scientific knowledge, it is also true that in contrast to Rawls or Habermas, Carnap did not think such a perspective could yield anything *more* than practical critique. He did not think we could actually *find out* anything, either about facts or about values, from such an external perspective. He did not think that *Vernunft* could give us a categorical imperative any more than he thought it could give us synthetic a priori knowledge. The most it could do—though Carnap thought this of overriding importance—was to guide *Verstand*, the search for knowledge, by the wider view it offered. And this wider view afforded by the perspective of *Vernunft* could in turn be informed and disciplined by the progressively better and more comprehensive picture of the world given us by (*Vernunft*-guided) *Verstand*. The modesty of this conception of reason sets it apart from its predecessors in the philosophical tradition. It is an ideal of reason that vindicates the Enlightenment ambition to reconstruct our cognition and our values to bring about the further progress of the species, but without the cognitive authoritarianism that had discredited earlier Enlightenment and positivistic attempts to formulate such an ideal.

References

Adorno, T. and Horkheimer, M. (1947) *Dialektik der Aufklärung*, Amsterdam: Querido. Trans. *Dialectic of Enlightenment*, New York: Seabury Press, 1972.

Bermúdez, J. L. (2009) *Decision Theory and Rationality*, Oxford: Oxford University Press.

Carnap, R. (1931) "Die logizistische Grundlegung der Mathematik," *Erkenntnis*, 2: 91–105. Trans. "The Logicist Foundations of Mathematics," in P. Benacerraf and H. Putnam (eds.), *The Philosophy of Mathematics*, Englewood Cliffs: Prentice-Hall, 1964, pp. 31–41.

——— (1934a) *Logische Syntax der Sprache*, Vienna: Springer. Expanded ed. trans. *The Logical Syntax of Language*, London: Kegan Paul, Trench, Trubner & Cie, 1937, repr. Chicago: Open Court, 2002.

——— (1934b) "Theoretische Fragen und praktische Entscheidungen," *Natur und Geist*: 257–60. Repr. in C. Damböck (ed.), *Der Wiener Kreis. Ausgewählte Texte*, Stuttgart: Reclam, 2009, pp. 173–6.

——— (1950a) "Empiricism, Semantics, and Ontology," *Revue Internationale de Philosophie* 4: 20–40. Repr. in Carnap, *Meaning and Necessity*, Chicago: University of Chicago Press, 2nd ed., 1956, pp. 205–21.

——— (1950b) *Logical Foundations of Probability*, Chicago: University of Chicago Press, 1962, 2nd ed.

——— (1963) "Abraham Kaplan on Value Judgments," in P. A. Schilpp (ed.), *The Philosophy of Rudolf Carnap*, LaSalle, IL: Open Court, pp. 999–1013.

——— (1977) *Two Essays on Entropy* (ed. by A. Shimony), Berkeley: University of California Press.

——— (2017) "Value Concepts (1958)," *Synthese* 194: 185–94.

Carus, A. W. (2007) *Carnap and Twentieth-Century Thought: Explication as Enlightenment*, Cambridge: Cambridge University Press.

——— (2010) "The Pragmatics of Scientific Knowledge: Howard Stein's Reshaping of Logical Empiricism," *Monist* 93: 618–39.

——— (2013) "History and the Future of Logical Empiricism," in E. Reck (ed.), *The Historical Turn in Analytic Philosophy*, Basingstoke: Palgrave Macmillan, pp. 261–93.

——— (2017) "Carnapian Rationality," *Synthese* 194: 163–84.

Creath, R. (1994) "Functionalist Theories of Meaning and the Defense of Analyticity," in W. Salmon and G. Wolters (eds.), *Logic, Language, and the Structure of Scientific Theories*, Pittsburgh: University of Pittsburgh Press, pp. 287–304.

Dreben, B. (1994) "In Mediis Rebus," *Inquiry* 37: 441–7.

Friedman, M. (2001) *Dynamics of Reason*, Stanford: CSLI Publications.

George, A. (2012) "Opening the Door to Cloud-Cookoo-Land: Hempel and Kuhn on Rationality," *Journal for the History of Analytical Philosophy* 1 (4): 1–17.

Nozick, R. (1993) *The Nature of Rationality*, Princeton: Princeton University Press.

Price, H. (1997) "Naturalism and the Fate of the M-Worlds," *Proceedings of the Aristotelian Society* 71: 247–67, repr. in Price, *Naturalism without Mirrors*, Oxford: Oxford University Press, 2011, pp. 132–48.

——— (2013) *Expressivism, Pragmatism, and Representationalism*, Cambridge: Cambridge University Press.

Richardson, A. (2007) "Carnapian Pragmatism," in M. Friedman and R. Creath (eds.), *The Cambridge Companion to Carnap*, Cambridge: Cambridge University Press, pp. 295–315.

Russell, B. (1914) *Our Knowledge of the External World*, Chicago: Open Court.

Stein, H. (1992) "Was Carnap Entirely Wrong, After All?" *Synthese* 93: 275–95.

Steinberger, F. (2016) "How Tolerant Can You Be? Carnap on Rationality," *Philosophy and Phenomenological Research* 92: 645–68.

Uebel, T. (2007) *Empiricism at the Crossroads: The Vienna Circle's Protocol-Sentence Debate*, Chicago: Open Court.

41
RETHINKING THE LEGACY OF LOGICAL EMPIRICISM IN NORTH AMERICA

Alan Richardson

In 1958, in his short book *English Philosophy since 1900*, G. J. Warnock offered this assessment of the fortunes of logical positivism: "The gradual accumulation of difficulties in maintaining the simple purity of philosophical doctrine, combined with the physical dissolution of the group as a result of the war, brought about the virtual extinction of the Logical Positivism as a coherent movement" (1958: 51). Warnock's remarks came in a chapter that concluded the section of his book on "the philosophy of yesterday" (ibid.: 52). As the title of Warnock's book indicates, his chief concern is English philosophy, and thus his account of logical positivism relates it mainly to other English philosophies, chiefly those of Moore and Russell, and is inflected by logical positivism's leading English recruit, A.J. Ayer. My concern is different in topic: I wish to consider the legacy of logical positivism or logical empiricism, as it came to be more capaciously known, in North America. But beyond this topical difference, I wish to offer a different way of approaching the historical question of the decline of logical empiricism from the one offered by Warnock. I offer two main considerations contra Warnock's framing of the issue. First, there never was a purity of philosophical doctrine in logical empiricism; the movement was never organized according to doctrinal purity, and thus no difficulties on that score can lead to extinction. Second, whatever its trajectory as a living movement, logical empiricism's influence went far beyond what Warnock suggests: because of the activities undertaken under its auspices, it has had lingering effects even after its "extinction."

Before continuing, it is worth noting that in Warnock's estimation, logical positivism was dead in 1958, the year *before* Ayer's well-known anthology with that title appeared (Ayer 1959). It seems as little possible in the philosophical world as in the biological one to be both ascendant and extinct. There seems to be very little clarity about just when logical empiricism was a thriving movement and when it was dead. In fact, it is only rarely the case that the most general framing terms in which the history of logical empiricism has been written—life and death—have been interrogated. I take it to be a virtue of my approach that it doesn't confuse legacy and influence with any form of active philosophical commitment to doctrine or continuity of personnel.

DOI: 10.4324/9781315650647-46

Reframing logical empiricism in North America: organization and goals

In 1939, writing from his new academic home at UCLA, Hans Reichenbach ended his essay in the Schilpp volume, *The Philosophy of John Dewey*, with a rousing call to alter the organization of US philosophy departments:

> The early period of empiricism in which an all-around philosopher could dominate at the same time the fields of scientific method, of history of philosophy, of education and social philosophy, has passed. We enter into the second phase in which highly technical investigations form the indispensable instrument of research, splitting the philosophical campus into specialists of its various branches. We should not regret this unavoidable specialization which repeats on philosophical grounds a phenomenon well known from all other fields of scientific inquiry.
>
> (1939: 192)

There might be any number of reasons to issue a call for specialization in philosophy, but Reichenbach's was both a long-held commitment and backed by a highly specific project. Reichenbach had already long argued for specialization as an aspect of how the philosophical community could begin to organize itself on the model of the scientific community (see his 1928, 1929). This is necessary because philosophy, if it is to participate in the production of well-founded knowledge, needed to become a science. For Reichenbach, logical empiricism was, pre-eminently, an effort to bring the standards of evidence, the forms of reasoning, and the methods and goals of science into philosophy. (By contrast, the ordinary language philosopher Gilbert Ryle [1963] argued that specialization was a result of professionalization and posited a form of specialization akin to the various branches of law organized by precedent and argument rather than to a splintering philosophical project induced by adoption of detailed technical scientific methods.)

In hewing to scientific precedent, Reichenbach was not alone. Rudolf Carnap also viewed introducing scientific methods and standards into philosophy as the most significant aspect of the logical empiricist project. Carnap's *Aufbau* was meant to be both an example of the good that could accrue to philosophy if it modeled itself on the exact science and a call to further such activity (1928: Preface). Again, this was not a view that Carnap gave up or modified—right through his career, his model of the way to engage in philosophical work was work in the exact sciences. For example, one of the great benefits in his view to the metalogical turn that he took in earnest in *Logical Syntax* (1934) was that it allowed logical empiricism more precisely to follow what was right in Wittgenstein's *Tractatus* (its understanding of the uniqueness and significance of logico-mathematical structure for knowledge) and to drop off what was most troubling (his strange mysticism that disallowed philosophical propositions to be stated at all). Statements about logical form are not impossible attempts to say the inexpressible; they are simply sentences (in the metalanguage) about sentences (in the object language); they are the sentences of logical syntax. Thus, we can safely say that logical empiricism was, for both Reichenbach and Carnap, a specific form of scientific philosophy.

In the development of logical empiricism in North America, as it grew into the project that became (in)famous, it was Carnap's and Reichenbach's visions of logical empiricism that were most influential. Carnap's general account of the methods of logical empiricism were from the 1930s onward always the most commonly cited, positively or negatively; this focus on Carnap's methodological pronouncements continues right up to the present day. Meanwhile, of the founding generation of logical empiricists, it was Carnap and Reichenbach who had the most direct influence upon generations of graduate students and the general philosophical community in the USA. The only two philosophers who arguably had as much influence on the development of the project both doctrinally and institutionally were the younger émigré scholars of the next generation, Carl Hempel and Herbert Feigl. We can draw the inference that the logical empiricism that Carnap and Reichenbach sought to institutionalize in the USA was a form of scientific philosophy. For example, Reichenbach was serious in his suggestions at the end of the Dewey essay and, moreover, he worked to alter the structure of the philosophical community in a way consistent with this vision.

This being the case, we may inquire into the success or failure of logical empiricism in North America on precisely this point. So, the questions of this essay are two. First, to what extent were American philosophers cognizant of this ambition of logical empiricism, and, if they were cognizant, how convinced were they that philosophy should become a science and, indeed, the very science logical empiricism claimed to be? Second, to what extent did the activities of the logical empiricists to bring about their mission alter the terrain of American philosophy—in particular, advance the cause of specialization in philosophy—even if the vision of scientific philosophy informing their version of specialization was never universally shared nor ultimately carried through?

The questions I have proposed are different from those that typically frame logical empiricism's trajectory in American philosophy. As we have seen in Warnock earlier, logical empiricism is typically taken to be a movement in philosophy that, like other movements in philosophy, is characterized by paradigmatic philosophical doctrines and arguments for said doctrines. Among the doctrines scouted in a typical history of logical empiricism are the verifiability criterion of meaning, the claim that there is a principled and philosophically important (because it is able to distinguish logico-mathematical and empirical claims) analytic/synthetic distinction, unity of science, and a few others (perhaps sentence-by-sentence reductionism into a language of pure sensation). The significance of logical empiricism, then, is as a chapter of analytic philosophy that promoted those doctrines, helped forge analytic philosophy, and then was superseded as those doctrines came to seem both false and unnecessary for the analytic project. This was Warnock's vision, and it is repeated throughout 60 years of histories of analytic philosophy. (It may be noted that the lack of doctrinal unity in logical empiricism was already implicit in the Vienna Circle's *Aufruf, Scientific World-Conception* [Verein Ernst Mach 1929] and was explicitly and repeatedly thematized in Otto Neurath's encyclopedism in the 1930s [e.g., Neurath 1938]. Large historical works that have thematized differences among the logical empiricists on core doctrines include Stadler [1997], Reisch [2005], and Uebel [2007]. For the persistence of histories of logical empiricism organized in ways akin to Warnock's, see Stroll [2000] and Soames [2017].)

The problem with such a framing is not that it is wholly misleading, but that it is insufficiently curious. For example, within a project of historical understanding so framed, the importance of logical empiricism in forging new philosophical fields of study—philosophy of science, philosophy of special sciences like physics, mathematics, and psychology, semantics, formal epistemology, to name a few—barely gets a look in. This is because new fields of study like semantics are not forged due to a commitment to a specific philosophical doctrine; semantics as a field of study arose in the work of Carnap, Tarski, other Polish logicians, and some others not due to some shared philosophical claim they agreed upon, but because it was a technical approach to philosophical issues they each saw promise in. Similarly, a search for a set of doctrines that logical empiricists all agreed about—so that commitment to those doctrines constitute what logical empiricism was (and, presumably, is)—has come up empty. Doctrines such as the verifiability criterion were not universally shared among those who called themselves (and are understood by other logical empiricists to be) logical empiricists. For some, like Schlick, verifiability was a core doctrine; for Reichenbach, it was a mistaken approach to semantics; for Carnap after the *Aufbau*, it quickly became first not a doctrine but a proposal, and then a failed proposal that was superseded by more nuanced ways to structure formal languages for science.

Another problem with understanding philosophical projects to be organized as doctrinal-cum-argumentative unities is that other activities of philosophers over and above arguing for doctrines drop out as irrelevant. But those activities are both a very large part of what philosophers do and a very large part of what the legacy of a philosophical movement might be. The articulation of curricula for undergraduate and graduate students, for example, is certainly guided by philosophical commitments but can scarcely be said to express a unanimous agreement upon philosophical doctrines, nor is teaching philosophy exhausted by providing arguments for positions. To take a specific issue: there was in American philosophy never a department in which logical empiricists formed a majority, much less the totality, of the department membership; so however much new curricula after the Second World War aided the consolidation of logical empiricism as a movement in American philosophy, not a single departmental curriculum could be said to express a commitment to logical empiricism or serve as a training ground for it. Similarly, there was never a single journal in American philosophy that was edited by a team that included only logical empiricists, published only logical empiricist work, or served as the logical empiricist house organ. Nonetheless, various American journals were influential in giving the movement venues for publication and helped to advance and consolidate the project; these journals include *Philosophy of Science*, *The Journal of Philosophy*, and *Philosophical Studies*.

Since many stories have been told of the place of logical empiricism, considered as a doctrine-cum-argument project within a history of analytic philosophy, my project here is to augment those stories with some gestures in the direction of answering the previously enunciated questions. They are only gestures, because these questions are at once empirical questions and large, complicated questions not amenable to the sorts of historical work that goes into histories of philosophy organized by doctrine and argument. The gestures are meant to provide new resources for and new claims regarding the legacy of logical empiricism in North American (and ultimately Anglophone) philosophy.

Scientific philosophy as a topic in the reception of logical empiricism

Our first question is the extent to which logical empiricism was understood among American philosophers to be offering (by at least Carnap and Reichenbach) a scientific philosophy during the period in which the logical empiricists were leaving Germany and Austria and arriving in the Anglophone world. The answer is that there were some then prominent philosophers who did understand this aspect of the project and who framed their reception of logical empiricism precisely on this point. I will briefly discuss four such philosophers—Charles Morris, Curt Ducasse, Errol Harris, and Max Black. These philosophers are of interest on this topic since they, variously, used the project of scientific philosophy either to embrace logical empiricism while also broadening and correcting it (Morris), or to offer a scientific alternative to it (Ducasse), or to argue for a philosophical incoherence in the project, albeit from very different vantage points (Harris, Black).

The American figure who most securely fastened up the scientific ambitions of logical empiricist scientific philosophy and who most clearly associated himself with those ambitions was Charles Morris. His synthetic 1935 essay in *Philosophy of Science* that aimed to show connections between logical empiricism and various forms of Anglophone philosophy, especially pragmatism, had the telling title "Philosophy of Science and Science of Philosophy." In this essay he discussed four versions of scientific philosophy: philosophy as the logic of science (which he rightly associate with Carnap), philosophy as the clarification of meaning (the Schlick–Waismann Wittgenstein-inflected version of logical empiricism as an activity rather than a body of doctrine—but an activity that occurs also in the sciences; Peirce's general semiotics), philosophy as formal axiology (under which he placed Schlick's philosophy of life and Dewey's vision of science as "institutionalized intelligence" [Morris 1935: 279]), and philosophy as empirical cosmology (Whitehead's philosophy of nature; Peirce's metaphysics). Morris's main points in this essay and related work are two. First, he wishes to find a commonality of scientific aim in a wider variety of philosophy. As he put it: "*Formalism, empiricism, and pragmatism are complementary phases of the scientific temper.* I propose to recognize this fact by designating the implied philosophical attitude as *scientific empiricism*" (1935: 285, orig. emphasis). Morris's second and most well-known intervention in the history of logical empiricist philosophy was in line with his synthetic tendencies: he sought to bring some lessons from Peircean semiotics into the study of scientific language by insisting on the scientific study of the pragmatics of language, that is, of the use of language by communities of speakers.

Curt Ducasse was, in the 1930s and 1940s, a well-known American philosopher working at Brown University; he was President of the Eastern Division of the American Philosophy Association in 1939–40. In 1941 he published a book, *Philosophy as a Science* (Ducasse 1941), that sought to present a coherent vision of precisely that: philosophy as a science. His entering wedge for that endeavor was the scientific ambition for philosophy that he saw in both the work of Dewey and the work of Carnap. He wished to disagree with them both on what the science of philosophy was while agreeing with them that philosophy ought to organize itself as a science. Ducasse comes down on the side of philosophy as the science of appraisals, but neither the details of that idea nor his objections to Carnap need detain us here. What is at stake is simply that the idea that Carnap sought to make philosophy into a science (and indeed what science he thought philosophy was) was not unknown in the early reception in the USA of his work. Sometimes, at least, this scientific ambition was shared even if logical empiricist scientific philosophy was not embraced.

The idea of "scientific philosophy" was, however, at least as often roundly rejected, often from very different points of view. It is perhaps not surprising that the Hegel scholar and metaphysician of science, Errol Harris, found the vision of "scientific philosophy" in Reichenbach's 1951 book *The Rise of Scientific Philosophy* to be thin, scientistic, and contrary to the spirit of the best of the philosophical enterprise. To Harris, to propose making philosophy into a science is to drain it of its sustaining ethos and its importance: "The desert sands in which the river of philosophical thought is choked are the arid wastes of self-styled 'scientific philosophy'" (1952: 165). Perhaps more unexpectedly, some of the British analysts who would help forge the consensus on analytic philosophy in mid-twentieth-century America argued that scientific philosophy was not only an odd ambition but, indeed, ultimately an incoherent one. Thus, writing in the introduction to his collection of essays that were meant to demonstrate the philosophical virtues of analysis, Max Black wrote:

> By adopting the scientific method, philosophers are to learn from scientists and mathematicians how to agree; and steady calculation, guaranteed to produce an acceptable answer, is to replace philosophical disputation. If some such hope as this inspired Russell (as it certainly did the Logical Positivists, who learned so much from him) his program was a failure. The merits of his views on philosophical analysis have to be

> argued on *philosophical* grounds; and to baptize them as "scientific" can only generate confusion.
>
> (1950: 6, orig. emphasis)

This claim by Black is about as precisely contrary to Carnap's philosophical vision as it could be: it simply presumes a prior and informal philosophy in which claims are adjudicated before they are scientifically formulated.

Be that as it may, my own claims here are two. First, in the reception of logical empiricism in North America on at least some fairly prominent occasions, the scientific ambitions of the Carnap–Reichenbach variant of logical empiricism were acknowledged and discussed. Second, sympathy for this ambition came from Americans associated with philosophy of science (Ducasse was president of the Philosophy of Science Association in the late 1950s) and pragmatism, while both practitioners of philosophical projects clearly attacked by logical empiricism (such as Harris's Hegelianism) and some practitioners and promoters of the philosophical project into which logical empiricism was eventually assimilated (Black's analytic philosophy) attacked it.

Implementing scientific philosophy: the legacy of logical empiricism in North America

An empirical question—a very hard-to-answer empirical question—is how far the sort of specialization of the field of philosophy that Reichenbach was arguing for in 1939 was responsible for the fracturing of philosophy into the many different areas of specialization that came to characterize American analytic philosophy by the 1960s. Because I do not believe I have anywhere near enough information to answer this question here, I will instead only exhibit the historical fact I am asking for causal information about and make some preliminary gestures at the sort of information we would need to explain that fact.

Philosophy has never been one monolithic form of study—it has always had subdisciplines or fields of study within it. But it has not always had the specific subdisciplines characteristic of mature analytic philosophy within it. Consider the following examples. In 1947, just before the time when analytic philosophy became a widely endorsed project among American philosophers, Dagobert Runes edited a volume titled *Twentieth Century Philosophy: Living Schools of Thought* (Runes 1947). This work clearly did not represent the Reichenbachian vision of the division of philosophical labor, as the term "schools of thought"—a term repugnant to Reichenbach—indicates. It is interesting to look at the titles of the chapters in the Runes book. A first, systematic section contains chapters on the following fields in philosophy: ethics, aesthetics, axiology, philosophy of law, philosophy of history, philosophy of science, philosophy of life, metaphysics, and theology and metaphysics. The living schools were presented in the second section, which, after a general essay by Bertrand Russell on twentieth-century philosophy, included Kantianism, Hegelianism, Thomism, transcendental absolutism (Santayana), personalism, phenomenology, logical empiricism (by Herbert Feigl), American realism, American pragmatism, dialectical materialism, naturalism, and philosophies of China. About twenty years later, after the consolidation of analytic philosophy, Carl Hempel's textbook, *Philosophy of Natural Science* (Hempel 1966), appeared in the Prentice-Hall Foundations of Philosophy series. This series included other short textbooks on the following topics: philosophy of art, philosophy of language, philosophy of social science, theory of knowledge, philosophy of history, social philosophy, philosophy of technology, ethics, philosophy of law, philosophy of religion, political philosophy, philosophy of perception, logic, philosophy of mind, and metaphysics. This series contains most of the major divisions of analytic philosophy up to the present time. This series also does not present Reichenbach's vision—neither the departments nor the methods of scientific philosophy are quite those of analytic philosophy. But the series does show many of the many historical developments in postwar American philosophy—no more "schools of thought," but only topics and fields of philosophy; certainly there was no reason to suggest that Kantianism, Hegelianism, Marxism, or Thomism were anything like live options in philosophy, now conceived as a much more thinly sliced field without any real room for axiology, philosophy of life, theology, or any philosophical perspective that seemed to encode a political point of view.

So, our question is not "How did Reichenbach succeed in implementing his vision of philosophy in American philosophy?" because, evidently, he did not. Nonetheless, his vision of a philosophy of narrow specialisms with expert practitioners schooled in each division's technicalities is far closer to the analytic philosophical view than is Runes's view. Moreover, the Prentice-Hall series did not simply represent a desired state of affairs but rather pretty much an accomplished one: the specialization had proceeded to the point that each of these subfields could be presented in textbooks to undergraduates. Understanding these developments requires not merely a better sense than we have about how persuasive Reichenbach's vision was, but also a better sense of the activities he and his fellow-travelers engaged in to actualize that vision. Interestingly, even books such as John McCumber's recent *The Philosophy Scare* (McCumber 2016: esp. ch. 4) that takes the UCLA Philosophy Department during Reichenbach's and then Carnap's tenures there as its topic concentrate far more on the vision than the actual activities. But it is at UCLA, if anywhere, that the Carnap/Reichenbach vision might have been realized, not least since the UCLA university archives themselves suggest that after his arrival in 1938 Reichenbach took a firm lead in curricular matters, especially at the graduate level:

> The Department of Philosophy at UCLA began operating as a separate entity in 1924. In 1929 the campus moved to Westwood and an undergraduate Philosophy curriculum was created; shortly after, a graduate program was developed in 1933. Professor Hans Reichenbach greatly accelerated the program by introducing a strong curricular tradition rooted in studies in logic and the philosophy of science.
>
> (UCLA 1930–87)

To see one dimension upon which Reichenbach's vision and his attested curricular activity differed from the sort of division of the philosophical field enshrined in analytic philosophy, let us look at greater detail of his 1939 recommendations. In his account of the necessary education of a modern scientific philosopher, Reichenbach emphasized the need for a robust education in mathematics and logic: "In as much as modern physics in particular is intrinsically mathematical, philosophic analysis of modern science cannot be achieved without a profound study of mathematical methods. This is why in our time a qualified philosopher has to be a good mathematician—a maxim which our students of philosophy should note" (1939: 191).

While it is not easy to know exactly what curriculum changes Reichenbach was responsible for, we do have testimony that Reichenbach took his own recommendation seriously. For example, a student of his during his time in Istanbul, Matild Kamber, claimed that Reichenbach implemented this form of training in Istanbul:

> Like many European universities, the University of Istanbul was divided into two faculties: The Faculty of Letters and the Faculty of Science. Philosophy belonged to the Faculty of Letters, yet Professor Reichenbach put philosophy into a unique position. Like regular science students, philosophy students had to attend classes in the Faculty of Science. To be more specific, they had to study two theoretical sciences and one

> experimental one each semester for four years. So they studied mathematics, biology, physics, medicine, chemistry, physiology, and genetics in addition to the different humanity courses such as literature, sociology, psychology, history, etc.
>
> (Reichenbach and Cohen 1978: 37)

While the various reminiscences of Reichenbach's graduate students at UCLA are not as clear as this and do not suggest that Reichenbach fundamentally shifted the curriculum at UCLA, they do make clear that Reichenbach continued to demand a rather extraordinary level of mathematical and scientific engagement in his courses (see ibid.: 45–64). Moreover, in his remarks, Donald Kalish, first a graduate student and then a faculty member at UCLA, does say that Reichenbach "undertook to make [UCLA Philosophy] a department of scientific philosophy" (ibid.: 47). It is to such undertakings that our philosophical curiosity and historical effort should be applied.

We can set aside the idea that Carnap and Reichenbach argued for a scientific community in philosophy but did not strive to create one. We are left, then, with a few options, which only more detailed scholarship of things like graduate curricula, reading lists, and teaching and research practices could help us sort through the evidence for. First, the division of labor and scientific skills of the logical empiricist scientific philosopher were absorbed into both an ethos of specialization and into specific specializations (logic, technical philosophy of science and philosophy of special sciences, formal epistemology) in the analytic division of the philosophical field. Surely at least, this did in fact happen. Second, because both the non-formal-logical, non-mathematical specialist techniques within certain fields of analytic philosophy and the very subject matters of those fields presume the falsity of logical empiricism, specialization as it actually happened helped cause the decline of logical empiricist philosophy. Third, within this more catholic set of specializations and philosophical techniques, Quinean naturalism became a more persuasive form of "scientific philosophy" not because of the success of Quine's arguments against Carnap, but precisely because naturalism's less clear demands of subject matter and method made it a more flexible and expansive framework to deploy among those who still wished to consider analytic philosophy as following a scientific ideal. It also could appease those like Black who found such ambitions incoherent: philosophy is continuous with science or somehow forms a whole with science but need not cede its own ground; indeed, Quine's objections to Carnap are similar to Black's in simply presuming the antecedent meaningfulness of informal philosophical terms. Whatever may be the case, the fragmentation of the philosophical field into a broad range of specialties is an accomplished fact of analytic philosophy. Moreover, it is not obvious that the set of specialties in analytic philosophy is consistent with any clear philosophical doctrine; it seems more a system wherein philosophical differences can find a place within the specialties and not one in which any philosophical doctrine grounds or justifies the specialties.

My topic in this section of the chapter was to raise the question of how the Carnap–Reichenbach version of specialization played into the actual botanization of analytic philosophy (and to demur from answering that question). That is one legacy of logical empiricism that still needs investigation. I would be remiss to fail to note that it is also in the spirit of a scientific approach to philosophy to demand that histories of analytic philosophy stop being written as "great man" histories but rather should attend precisely the social dynamics of large-scale changes in philosophy such as those that give rise to specialization itself. This sort of history, if undertaken with diligence and robust attention to empirical facts, could, too, become a legacy of scientific philosophy in both its pragmatist and logical empiricist forms.

References

Ayer, A. J. (ed.) (1959) *Logical Positivism*, New York: The Free Press.

Black, M. (1950) "Introduction," in M. Black (ed.), *Philosophical Analysis*, Englewood Cliffs, NJ: Prentice-Hall, pp. 1–13.

Carnap, R. (1928) *Der Logische Aufbau der Welt*, Berlin: Weltkreis-Verlag. Trans. *The Logical Structure of the World*, Berkeley: University of California Press, 1967. Repr. Chicago: Open Court, 2003.

——— (1934) *Logische Syntax der Sprache*, Vienna: Springer. Rev. ed. trans. *The Logical Syntax of Language*, London: Kegan Paul, Trench, Trubner, & Co., 1937. Repr. Chicago: Open Court, 2002.

Ducasse, C. J. (1941) *Philosophy as a Science*, New York: Oskar Piest.

Harris, E. (1952) "Scientific Philosophy," *Philosophical Quarterly* 2: 153–65.

Hempel, C. G. (1966) *Philosophy of Natural Science*, Inglewood Cliffs, NJ: Prentice-Hall.

McCumber, J. (2016) *The Philosophy Scare: The Politics of Reason in the Early Cold War*, Chicago: University of Chicago Press.

Morris, C. (1935) "Philosophy of Science and Science of Philosophy," *Philosophy of Science* 2: 271–86.

Neurath, O. (1938) "Unified Science as Encyclopedic Integration," in O. Neurath et al. (eds.), *Encyclopedia and Unified Science*, Chicago: University of Chicago Press, 1–27.

Reichenbach, H. (1928) *Philosophie der Raum-Zeit-Lehre*, Berlin: De Gruyter. Trans. *The Philosophy of Space and Time*, New York: Dover, 1958.

——— (1929) "Neue Wege der Wissenschaft: Philosophische Forschung," *Vossische Zeitung*, 16 June 1929. Trans. "New Approaches in Science: Philosophical Research," in Reichenbach 1978, pp. 249–53.

——— (1939) "Dewey's Theory of Science," in P. A. Schilpp (ed.), *The Philosophy of John Dewey*, LaSalle, IL: The Open Court, pp. 159–92.

——— (1978) *Selected Writings, 1909–1953* (ed. by M. Reichenbach and R. S. Cohen), Dordrecht: Reidel, vol. 1.

Reichenbach, M. and Cohen, R. S. (1978) "Memories of Hans Reichenbach," in Reichenbach (1978), pp. 1–86.

Reisch, G. (2005) *How the Cold War Transformed Philosophy of Science*, Cambridge: Cambridge University Press.

Runes, D. (1947) *Twentieth Century Philosophy*, New York: Philosophical Library.

Ryle, G. (1963) "Introduction," in A. J. Ayer et al. (eds.), *The Revolution in Philosophy*, London: Macmillan, pp. 1–11.

Soames, S. (2017) *The Analytic Tradition in Philosophy, Volume 2: A New Vision*, Princeton: Princeton University Press.

Stadler, F. (1997) *Studien zum Wiener Kreis. Ursprung, Entwicklung und Wirkung des Logischen Empirismus im Kontext*, Frankfurt a. M.: Suhrkamp, 2nd abridged ed., Cham: Springer, 2015. Trans. *The Vienna Circle: Studies in the Origins, Development, and Influence of Logical Empiricism*, Cham: Springer, 2015.

Stroll, A. (2000) *Twentieth-Century Analytic Philosophy*, New York: Columbia University Press.

UCLA (1930–1987) *UCLA Philosophy Department Chair Correspondence Files, 1930–1987*, Los Angeles: University of California, www.oac.cdlib.org/findaid/ark:/13030/c8862hzk/; accessed 18 June 2018.

Uebel, T. (2007) *Empiricism at the Crossroads*, Chicago: The Open Court.

Verein Ernst Mach (1929) *Wissenschaftliche Weltauffassung. Der Wiener Kreis*, Vienna: Wolf. Trans. "The Scientific Conception of the World. The Vienna Circle," in O. Neurath, *Empiricism and Sociology* (ed. by R. S. Cohen and M. Neurath), Dordrecht: Reidel, 1973, pp. 299–318; rev. trans. (with orig. annotated bibliography) "The Scientific World-Conception. The Vienna Circle," in F. Stadler and T. Uebel (eds.), *Wissenschaftliche Weltauffassung. Der Wiener Kreis. Hrsg. vom Verein Ernst Mach (1929)*, Vienna: Springer, 2012, pp. 75–116.

Warnock, G. J. (1958) *English Philosophy Since 1900*, Oxford: Oxford University Press.

INDEX

Note: Page numbers in **bold** indicate a table on the corresponding page

Taylor & Francis Group
an informa business
Taylor & Francis eBooks
www.taylorfrancis.com
A single destination for eBooks from Taylor & Francis with increased functionality and an improved user experience to meet the needs of our customers.
90,000+ eBooks of award-winning academic content in Humanities, Social Science, Science, Technology, Engineering, and Medical written by a global network of editors and authors.
TAYLOR & FRANCIS EBOOKS OFFERS:
A streamlined experience for our library customers
A single point of discovery for all of our eBook content
Improved search and discovery of content at both book and chapter level
REQUEST A FREE TRIAL
support@taylorfrancis.com
Routledge
Taylor & Francis Group
CRC Press
Taylor & Francis Group

Made in the USA
Las Vegas, NV
10 June 2026

48463606R00234